Theoretical Systems in Biology
Hierarchical and Functional Integration

Volume I   **Molecules and Cells**

Some related Pergamon titles

Pergamon Studies in Neuroscience Series

No. 1. **Function and dysfunction in the basal ganglia** ed. A. J. Franks, J. W. Ironside, R. H. S. Mindham, R. F. Smith, E. G. S. Spokes and W. Winlow

No. 2. **Comparative aspects of neuropeptide function** ed. Ernst Florey and George B. Stefano

No. 3. **Neuromuscular transmission: basic and applied aspects** ed. Angela Vincent and Dennis Wray

No. 4. **Neurobiology of motor programme selection** ed. Jenny Kien, Catherine R. McCrohan and William Winlow

No. 5. **Interleukin-1 in the brain** ed. Nancy Rothwell and Robert Dantzer

No. 6. **Neurophysiology of ingestion** ed. D. A. Booth

No. 7. **A theory of the striatum** ed. J. Wickens

No. 8. **Neuroregulatory mechanisms in ageing** ed. Maynard H. Makman and George B. Stefano

No. 9. **Thalamic networks for relay and modulation** ed. Diego Minciacchi, Marco Molinari, Giorgio Macchi and Edward G. Jones

No. 10. **Glycobiology and the brain** ed. M. Nicolini and P. F. Zatta

No. 11. **Neural modelling and neural networks** ed. F. Ventriglia

No. 12. **Information processing underlying gaze control** ed. J. M. Delgado-García, E. Godaux and P.-P. Vidal

Wenner-Gren International Series

Vol. 60. **Neuro-immunology of Fever** ed. T. Bartfai and D. Ottoson

Vol. 61. **Functional Organisation of the Human Visual Cortex** ed. B. Gulyás, D. Ottoson and P. E. Roland

Vol. 62. **Trophic Regulation of the Basal Ganglia** ed. K. Fuxe, L. F. Agnati, B. Bjelke and D. Ottoson

Vol. 63. **Light and Biological Rhythms in Man** ed. L. Wetterberg

Vol. 64. **Eye Movements in Reading** ed. J. Ygge and G. Lennerstrand

Vol. 65. **Active Hearing** ed. Å. Flock, D. Ottoson and M. Ulfendahl

Vol. 66. **Challenges and Perspectives in Neuroscience** ed. D. Ottoson, T. Bartfai, T. Hökfelt and K. Fuxe

Vol. 67. **Life and Death in the Nervous System** ed. C. F. Ibáñez, T. Hökfelt, L. Olson, K. Fuxe, H. Jörnvall and D. Ottoson

IBRO Series

Muscle afferents and spinal control of movement ed. L. Jami, E. Pierrot-Deseilligny and D. Zytnicki

Theoretical Systems in Biology
Hierarchical and Functional Integration

Volume I   Molecules and Cells

G. A. Chauvet

Institute of Theoretical Biology,
Faculty of Medicine, University of Angers, France

and

Department of Biomedical Engineering,
University of Southern California,
Los Angeles, California, U.S.A.

translated by
K. Malkani
Department of Histology, Embryology and Cytology,
Faculty of Medicine, University of Angers, France

PERGAMON

UK	Elsevier Science Ltd, The Boulevard, Langford Lane, Kidlington, Oxford OX5 1GB, UK
USA	Elsevier Science Inc., 660 White Plains Road, Tarrytown, New York 10591-5153, USA
JAPAN	Elsevier Science Japan, Tsunashima Building Annex, 3-20-12 Yushima, Bunkyo-ku, Tokyo 113, Japan

---

First edition published in French by Masson Editeur, Paris 1986
    (French edition titled *Traité de Physiologie Théorique*)
Revised, updated and translated into English for this Elsevier Science Edition 1996

Library of Congress Cataloging in Publication Data
Chauvet, G. A. (Gilbert)
    Theoretical systems in biology: hierarchical and functional
integration / G. A. Chauvet; translated by K. Malkani
    Includes bibliographical references and index.
    Contents: v. 1. Molecules and cells – v. 2. Tissues and organs –
v. 3. Organisation and regulation
    1. Physiology–Mathematical models.   2. Molecular biology–
Mathematical models.   I. Title.   II. Series.
QP33.6.M36C473   1995   95-30324
574'.01'1–dc20

British Library Cataloguing in Publication Data
A catalogue record for this book is available from the British Library.

ISBN 0 08 041992 5 (Volume I)
ISBN 0 08 041995 X (3 volume set)

Printed in Great Britain by Alden Press, Oxford

Contents

Preface   xiii

Foreword to Volume I   xxiii

PART I: MATERIALS AND METHODS IN BIOLOGICAL
DYNAMICS   1

Introduction to Part I   3

1   Macromolecular Components and Interactions in Living Organisms   5
    I.   Structure of nucleic acids   6
    II.   Structure of proteins   9
           1.   Description of protein structure   9
           2.   Theory of the conformation of biological macromolecules   17
                *a.   Physical description of a macromolecular system in
                     solution*   17
                b.   Free energy variation during a conformational change 19
           3.   Molecular conformation and biological activity   20
    III.   Molecular interactions in proteins: protein–ligand interactions 21
           1.   Thermodynamic theory of molecular interactions   22
                a.   Protein–ligand associations (homotropic effects)   22
                b.   Consequences   25
                c.   Allosteric models generally used   27
           2.   The cooperative effect   27
                a.   The concept of cooperativity   27
                b.   Cooperativity and the binding polynomial   30
                c.   An illustration of the cooperative effect   34

2   The Internal Chemistry of Cells   37
    I.   Catalytic function of the enzymic reaction   37
           1.   Henri–Michaelis–Menten equations: equilibrium conditions   39
           2.   Briggs–Haldane equations: quasi-stationary state conditions 41
           3.   The mathematical viewpoint: the pseudo-steady state
                hypothesis   43
    II.   Regulatory function of the enzymic reaction   44
    III.   Molecular interactions and the active site concept   52
    IV.   Coupled chemical reactions: reaction–diffusion equations   58
           1.   General equation for a chemical transformation   58
           2.   The physics of diffusion: Fick's equation   60

3. The mathematics of a reactional–diffusional system   62
4. An autocatalytic reaction   65

3   Methods in Biological Dynamics   71
    I.   Biology and complexity   71
    II.   Relational theories   74
          1. Rosen's theory of abstract biological systems   74
          2. Delattre's theory of transformation systems   78
              a. *Axioms*   79
              b. *Graphic representation*   81
              c. *An application showing the impossibility of spontaneous,*
                  undamped oscillations in physically linear systems and
                  the existence of oscillations in non-catalytic systems   84
    III. Thermodynamic theory   87
          1. Principles of equilibrium thermodynamics   88
          2. The entropy production term. Consequence in the linear
              field: non-equilibrium stationary states in the neighbourhood
              of equilibrium   89
          3. Non-linear thermodynamics of chemical reactions:
              non-equilibrium stationary states far from equilibrium   92
              a. *The universal criterion of evolution*   93
              b. *The Glansdorff–Prigogine functional*   93
              c. *Physical interpretation*   94
              d. *Network thermodynamics*   96
                  α. *Characteristics of an element in a chemical network*   98
                  β. *Bond graphs*   101
    IV.   Thom's theory of elementary catastrophes   104

Summary of Part I   115

PART II: THE MOLECULAR ORGANISATION OF LIVING
MATTER   117

Introduction to Part II   119

4   Organisation of Biological Systems   121
    I.   A formal definition of self-organisation   122
          1. Organisation   122
          2. Self-organisation   123
    II.   Biological organisation and information theory   126
          1. Von Foerster's self-organising system   126
          2. The Yockey–Atlan theory of self-organisation   129
    III. A theory of the functional organisation of formal biological
          systems: some concepts and definitions   135
          1. Introduction to the functional organisation of biological
              systems   135

2. The problem of representation in biology   137
    a. Notions of system, structure, function and evolution   138
        α. *System: general considerations*   138
        β. *Structure and system*   138
        γ. *Function*   139
        δ. *Evolution*   139
    b. A representation of physiological systems: the hypothesis of associative functional self-organisation   140
        α. *Biological structure and systems*   140
        β. *The hypothesis of associative self-organisation*   140
        γ. *The physiological system as a hierarchical system. Functional interactions, levels of organisation*   141
3. Mathematical representation of the functional organisation   143
4. Dual representations: (N, a) and (ψ, ρ) 147
IV. Spatial organisation in the cell: concept of structural discontinuity   149
  1. On the existence of non-local interactions: the concept of the active site re-examined   149
  2. Enzyme organisation: microcompartmentation and the example of 'channelling'   152
    a. Introduction to the problem of enzyme organisation   152
    b. Relation with physiology and physiological consequences   153
    c. Definitions of channelling on biophysical bases   154
        α. *Metabolic organisation*   154
        β. *Application of Curie's principle of symmetry*   155
        γ. *The respective roles of diffusion and reaction*   156
        δ. *How should the diffusion process be characterised in order to account for the existence of molecular channels in the absence of a physical limit such as a membrane?*   157
  3. On the functional organisation in a biological structure: the example of enzyme organisation   157
    a. Definition of a metabolic pathway as a structural unit   158
        α. *Michaelis enzymatic reactions: metabolic flux and transport between the local medium and the bulk phase*   158
        β. *Stability of the dynamics of a step in the metabolic chain*   160
        γ. *Stability of a metabolic pathway with allosteric control of production*   162
  4. A paradigm for the creation of functional interactions: the self-association hypothesis   165
    a. The structural unit and the physiological function   165
    b. The self-association hypothesis   166
    c. On the nature of a break in the self-association   167

    α. *A break in the functional interaction of the*
      metabolic pathway 167
    β. *Basic mechanisms of the association* 167
   5. Functional association between two metabolic pathways
     defined as structural units 168
    a. *A general and generative schema of the association* 168
    b. *Mathematical study of the dynamics in a u_2-unit:*
      a specific system 170
    c. *Numerical study of the dynamics in a u_2-unit: a general*
      dynamical system 172
   6. The paradigm of self-association applied to the enzyme
     organisation: role of local and bulk phase 173

5 The Replication–Translation Apparatus 177
 I. The 'central dogma' of molecular biology 177
   1. Genes and chromosomes 178
   2. Replication, transcription and translation 181
   3. Computational methods of determining the sequencing
     properties of DNA 189
 II. Information theory and the genetic code 194
   1. Measurement of the quantity of information in the genetic
     code 194
   2. The genetic code 194
 III. Chemical dynamics of heredity 198
   1. Generalities 199
   2. DNA replication 202
   3. Protein synthesis 206
 IV. Topological, structural and functional implications of nucleic
   acid chains 209
   1. Topological concepts in DNA replication and structural
     consequences 209
    a. *Topological findings* 210
    b. *Experimental data* 214
   2. Topology and protein biosynthesis 217
    a. *The concept of contractibility and its consequences* 218
    b. *The concept of restorability and its consequences* 219
 V. The hierarchical organisation of the replication–translation
   apparatus 221

6 Molecular Evolution and Organisation 225
 I. Evolution of self-instructing information carriers 226
   1. Phenomenological description of the evolution of chemical
     species 226
   2. Solution of the system of equations 231
   3. Explicit solutions 232
   4. Consequence: selection in molecular systems 234

II.   Evolution with complementary instruction. The case of DNA or
      RNA replication   236
III. Protein biosynthesis: self-organising enzymic cycles   240
      1.   Catalytic protein cycles   240
      2.   Self-reproducing hypercycle   243

7   Evolution and Physiology   247
    I.   Evolution and self-organisation of molecular systems   247
         1.   Introduction   247
         2.   The three phases of evolution   248
         3.   Darwinian systems   252
    II. A coherent interpretation of evolution   253
         1.   Creation of thermodynamically stable spatiotemporal
              structures   253
         2.   Self-organisation and the evolution of molecular biosystems 254
         3.   Evolution of the species in terms of information theory   257
         4.   Recapitulation: the scenario of evolution   259
    III. Functional biology and evolutionary biology   262
         1.   Darwinism and physiology: the principle of vital coherence   262
         2.   An elementary model of evolution   265
              a.   Formalisation of the principle of vital coherence in
                   terms of the levels of organisation   267
              b.   Description of an elementary model of evolution   269
              c.   Discussion and results: can this model be generalised? 273

Summary of Part II   279

PART III: CELLULAR ORGANISATION OF LIVING MATTER   281

Introduction to Part III   283

8   Cellular Organisation   285
    I.   Cell description   285
    II. Cellular organisation and regulation   287
         1.   Formation of structures at the cellular level   287
         2.   Regulation and metabolic pathways   289
              a.   The glucose-6-phosphate pathway   289
              b.   The Krebs cycle (the citric acid cycle)   290
              c.   Interpretation of regulatory phenomena   291
              d.   The phenomenon of inverse regulation   296
    III. Cell growth: an introduction   297
         1.   Growth   298
         2.   Development   298
         3.   Differentiation   298
         4.   Morphogenesis   299

9   Regulation of Cell Function through Enzyme Activity   301
     I.   Introduction to the regulation of enzyme synthesis   301
     II.   Theoretical model of regulation   305
     III. Regulation of protein biosynthesis in higher organisms   309

10   Cell Growth and Morphogenesis   315
     I.   General aspects   315
     II.   Unicellular organisms   317
           1.   Differentiation in the Acrasiales   317
           2.   Human red blood cells   320
           3.   Some examples of cell differentiation at the molecular level   321
     III. Higher organisms   322
           1.   Embryogenesis   322
                a.   *Principal steps of embryogenesis*   322
                b.   *Concepts of cell differentiation*   327
           2.   Positional information and cell differentiation   328
           3.   Control and cell differentiation   334
                a.   *Transcriptional control*   334
                b.   *Post-transcriptional control*   335
                c.   *Post-translational control*   335
                d.   *Mathematical models of control in cell differentiation*   336
     IV.   Morphogenesis: Turing's theory   338
           1.   General aspects   339
           2.   Theoretical models of morphogenesis   340
                a.   *Creation of the gradient*   340
                b.   *Interpretation of the gradient of positional information*   343
           3.   Morphogenesis: local and global theories   348
                a.   *Turing's theory: a synthesis of theoretical models*   348
                b.   *Thom's theory: towards a general theory of development*   353
     V.   A description of growth for functional organisation   357
           1.   A break in the functional interaction: consequences on the stability of biological systems   357
                a.   *A break in the functional interaction: the choice between Life and Death*   357
                b.   *Functional hierarchical organisation: the consequence of the choice*   358
           2.   *Evidence for the existence of self-association: an increase in stability*   361

11   Cell Division   363
     I.   The cell cycle   364
           1.   Description   364
           2.   Models of the cell cycle   368
           3.   The limit cycle model of biochemical oscillations   369

II.   Development of a cell population   372
   1. Kinetics with variables (t, a) 373
   2. Kinetics with variables (t, μ) 375
III. Analysis of the cell cycle: population theory   378
   1. Leslie matrices applied to population studies   379
   2. Interpretation of the FLM curve by population theory   383

12 Cell Growth, Division and Differentiation   387
I.   Asymmetrical cell division   387
II.   Cell growth   390
   1. Analytical description: mass and volume   390
   2. Global description of the behaviour of a cell   396
III. Mechano-chemical approach to morphogenesis: Murray's
mechanical model for mesenchymal morphogenesis   400
IV. Topological description of developmental dynamics: potential
of functional organisation   407
   1. Introduction: variational principles in biology   407
   2. The potential of functional organisation   409
      a. *The nature of the concept: the combinatorial approach
and non-symmetry*   409
      b. *Definition and formulation*   410
   3. Criterion of maximality for the potential of organisation:
a class of biological systems   411
      a. *State of maximum organisation*   411
      b. *The extremum hypothesis: a class of biological
systems*   413
        α. *The organisational state is an attractor*   413
        β. *Consequence: the extremum hypothesis for the time-
variation of the number of sinks*   413
   4. Criterion of evolution for the functional organisation:
orgatropy   414
      a. *The concept of 'orgatropy'*   414
      b. *Does orgatropy provide a criterion for the time-variation
of the (O-FBS)?*   417
   5. Criterion of specialisation and reorganisation of the (O-FBS)
during development   417
      a. *Criterion of specialisation*   417
        α. *The concept of specialisation*   417
        β. *The relation between specialisation and
hierarchisation*   419
      b. *Consequence: mathematical expressions of
specialisation and emergence of a level of
organisation*   420
      c. *Functional order*   421
      d. *Time-variation of an (O-FBS) during development*   424

V. A comparison between biological and physical systems   425
    1. Structural entropy and functional orgatropy   425
    2. The consequence of the optimum principle   426
    3. On the meaning of the optimum principle   427

Summary of Part III   429

Conclusion to Volume I: Unity at the Gene Level   431

Mathematical Appendices   435
Appendix A: Vector analysis   435
    1. Gradient of a function $U(x, y, z)$ 435
    2. Divergence of a vector   436
    3. Green's theorem   438
    4. The Laplace function: the second-order scalar operator   438
    5. Summary   439
Appendix B: Dynamic systems   440
    1. Notion of a dynamic system   440
    2. The Hamiltonian form. Conservative systems   441
    3. Stability: Lyapunov functions   442
    4. Limit cycles, critical points, Jacobian, Hessian   444
    5. Partial differential equations   447
    6. Some notes on the terminology of ordinary differential
        equations. Compact differential manifolds   448
Appendix C: Notations in matrix algebra   451
Appendix D: Probability and information theory   452
    1. Probability   452
    2. The Shannon function of information   453

Symbols and constants (mainly in CGS units)   457

General reading   461

Bibliography   463

Index   479

Preface

The use of models in our approach to human physiology is aimed at laying down the methodological bases for the interpretation of experimental results, both old and new. True, the title is likely to cause some surprise, so some justification may be required: first, we propose to present formalised biological theories at various levels of description, ranging from the molecular level to that of the whole organism; and, secondly, we shall consider certain aspects of contemporary biology, selected not only for their intrinsic importance but also for their capacity to generate new insights. And all these are, of course, fundamental to theoretical biology, a discipline analogous in nature to theoretical physics in its relationship to experimental physics.

Although this work is not meant to be an exhaustive treatise, an attempt has been made to cover all the subjects of 'classical' biology in a logical manner, going from the most elementary level — the molecular level — up to the control systems of the entire organism. Thus, a succinct description of each of the principal physiological phenomena is followed by a formalised explanation, in so far as this is possible in the present state of knowledge. The choice of subjects may seem to be somewhat arbitrary, but the main criterion used has been the didactic aspect of the topic. For example, certain formalised theories that are now fairly old, such as those of DNA replication and membrane excitability, have been duly treated in detail. But very recent theories, such as that of DNA topology, have also been given careful attention. For how can we be sure which of these approaches will prove the more useful in the future? Similarly, except in a few special cases, we have preferred the use of deterministic methods to stochastic processes. One reason for this is that the stochastic formalism is generally less well known and does not always carry an obvious advantage, at least for the time being. We have tried — but perhaps with limited success — to conserve the necessary mathematical rigour without going into too much detail, and to recall the essentials of biological phenomenology without striving to explore all the finer points. This is, of course, a delicately balanced task and the results may annoy 'pure' mathematicians as well as 'experimental' physiologists. The point of view here is rather that of a physicist attempting to describe natural phenomena through abstract representation expressed in concise language. We hope this interdisciplinary approach will not appear too esoteric to some readers or too lacking in rigour to others. The basic requirement for understanding the text is a sound knowledge of physics and mathematics at the undergraduate level, and of physiology as treated in standard textbooks.

This three-volume work corresponds to the usual levels of structural organisation in biology. Volume I describes molecular and cellular aspects (Chapters 1 to 12). Volume II examines the intercellular relationships within organs (Chapters 1 to 5) as well as the major functional systems of the organism: energy metabolism, respiration, blood circulation, renal activity (Chapters 6 to 9). Chapter 10 introduces the important concepts of non-symmetry, non-locality and structured discontinuity. These concepts are used in Volume III which addresses the delicate problem of shifting from one biological level to another. Volume III contains a discussion of the mechanisms of control and regulation exercised by the nervous and endocrine systems (Chapters 1 and 2). The concluding chapter proposes a method of vertical functional integration in a multiple-level hierarchical system (Chapter 6). The formalisation necessary for certain physiological problems, particularly those involved in the regulation of the organism, calls for new methods and concepts. Thus, the notion of the integron, proposed by Jacob in *La logique du vivant* (1970) has been largely used. The regulatory functions of respiration, blood circulation and renal activity are integrated into two major equilibria of the organism: the hydroelectric equilibrium and the acid–base equilibrium (Chapter 4). Some of the notions of mathematics and physics used are briefly recalled in the appendices of each volume. It is hoped that these, together with the comprehensive index and the list of the principal symbols and units used, will be of some help to the non-mathematical reader.

Let us now try to justify the choice we have made. Why, indeed, bring up the idea of a theoretical physiology? First, because we are more interested by the functional than the descriptive aspect of biology; and, secondly, because we have deliberately sought the mathematical formalisation of physiological phenomena. Here, an obvious difficulty arises since this choice requires the contribution of all the other sciences — mathematics, physics and chemistry — and demands an interdisciplinary interpretation. Several reasons lead us to believe that the evolution of physiology towards greater formalisation is unavoidable: (i) the rapidly increasing number of experimental results for which no interpretation is available because of the multiple factors involved; (ii) the continuing technological advances in instrumentation giving finer results than ever before; (iii) the necessity of integrating the results obtained to counteract the reductionist tendencies of specialised disciplines with divergent objectives. However, these are not the only reasons although they are certainly most evident. Indeed, there are other, deeper reasons of an epistemological order which we shall now discuss.

Of course, it is possible to explain without formalisation, and indeed up to now this has been the principal approach in biology. But what is the actual nature of the 'explanation' in biology? Everybody knows, for example, the theory of evolution and the theory of gene regulation in procaryotes, to mention only the best known theories concerning the living world. Clearly, these two qualitative descriptions cannot be considered to have the same level of intelligibility. The former rests on observations on the scale of geological time and on considerations of a rational order, while the latter stems from rigorous experimentation in a 'molecular' context, the results of which are unanimously accepted. Indeed, the reticence of many scientists with respect to the Darwinian theory of evolution contrasts sharply with the general approval of the model proposed by Jacob and Monod, at least as far as it applies to procaryotes.

These examples are characteristic of non-formalised theories, even though they describe 'reality' — or what can so be considered, as we shall see below — at different levels of 'certitude'. As opposed to theoretical concepts which lead to the induction of theoretical laws capable of generating new empirical laws, non-formalised theories in fact introduce elementary mechanisms which, taken together, are difficult to generalise under the form of a theoretical law.

From this point of view, the problem of biological evolution is exemplary and is considered in detail in Volume I, Part II. Of course, experimental descriptions and experimental verifications are indispensable to science, but it has to be admitted that formalisation is far more useful than rigorous taxonomy. We merely need to think of the known results of physics and the difficult objective of theoretical physics (not necessarily the same as in theoretical biology) which is the search for the great universal laws underlying the reality of the material world. Several epistemologists have examined this problem, in particular the physicist d'Espagnat who explains his philosophical point of view in *A la recherche du réel* (1979). It may be objected by some that physics, the science of inanimate matter, is obviously a great deal 'simpler' than physiology, and therefore, even in the best of cases, the formal description of physics will not be applicable to biology, so that it may be preferable to give a literary description of biological phenomena rather than to introduce some useless, esoteric formalism. In answer to this we would make the following points:

(1) The abundance of experimental results does not in itself lead to a better understanding of the phenomena studied but rather calls for a synthetic interpretation. Indeed, new concepts introduced into a theory enhance the value of the observed results.

(2) A good qualitative or quantitative formalisation permits a synthetic view of phenomena which are unrelated *a priori*, thus generating various new laws. It leads to the rigorous description of the phenomenon observed in terms of the hypotheses used.

(3) The enunciation of sufficiently general theoretical laws allows us to imagine new experiments, and vice versa.

While considering the merits of formalisation in physiology, it would be well worth bearing in mind the epistemological notions concerning the relationships between empirical laws and theoretical laws, between theories and models in the science in which experimentation has always played the foremost role. The reader may profitably consult some of the excellent contributions to scientific epistemology dealing with this subject (Delattre, 1981, Volume I).

To illustrate this, let us go back to the two examples above. We know that a theory of evolution, based on transformism and natural selection, introduces observable dimensions obtained directly from palaeontological or biological observation. However, such a theory is practically powerless in the induction of new empirical laws. But a theory of evolution, formalised in terms of concepts such as those of self-organisation or of selective value, is seen to be quite potent (Volume I, Chapter 7). And the theory of gene regulation in bacteria, established in terms of molecular concepts, reveals a far greater predictive value. Moreover, a quantitative formalisation of this phenomenon leads to empirical laws which actually justify the initial hypothesis (Volume I, Chapter 9).

It should, however, be observed that most of the current biological hypotheses, whether formalised or not, depend on fundamental physico-chemical knowledge. Such hypotheses therefore rely on already existing theories of matter. We believe it should be possible to express a fecund biological theory in terms of non-observables specific to biology, according to theoretical concepts of which the rules of correspondence with objective reality would be unique and not simply borrowed from other sciences. As proof of this, we consider two examples in detail: the morphogenetic field in developmental biology (Chapter 10), and the neural field in the central nervous system (Volume III, Chapter 2). Working on this basis, we have tried to develop a theory of functional organisation in multiple-level hierarchical systems (Volume III, Chapter 6).

Is biological reality 'veiled'?

The problem of biological 'reality', mentioned above, remains to be solved. But what reality are we actually referring to? We know, of course, what a controversial subject this has been for philosophers all through the ages. D'Espagnat (1979) comes to the conclusion that non-physical realism is the only conception that appears to fit all the facts. The philosophy of a 'veiled' reality should inspire considerable modesty. However, this is a physicist's point of view and would therefore need to be qualified in terms of the biological perspective. But, finally, do we not today perceive fundamental incertitudes in the living as well as in the non-living world?

Prigogine (1980, Volume I), working on classical dynamic theory and taking fluctuations into account, has recently added a new indeterminism alongside the already known indeterminism of quantum theory. Transposed to the biological world, may not the variability of living organisms be just one form of this incertitude, or on the contrary could it be our degree of ignorance that leads us to this postulate? The latest theories of matter seem to answer this important question through a statistical view of fundamental concepts. We shall have to take this into account, for example, in considering a formalised theory of the evolution of the species.

Some comment may be made on the imprecise use of the terms: theories and models. Mathematical models, physical models, chemical models, and so on, are being increasingly used in biological work. But when can a model be considered to constitute a theory? Indeed, if we wish to avoid errors of interpretation of facts — not to mention the underlying reality — we should be careful to distinguish between the explicative models, with which we are directly concerned in this work, and other models that are merely circumstantial. For instance, we refer to a statistical model when, on the basis of a large number of experimental results, we seek to verify a hypothetical mathematical relationship between various dimensions. Although often necessary at the beginning of any scientific investigation, this kind of analysis does not usually generate a theoretical law. Theoretical biology is surely not a mere veneer of mathematical methods applied to biological observations.

A most interesting analysis of the distinction between theories and models has been made by Delattre (1981, volume I), who raises the following questions: Is there an ideal form for the explanation of phenomena? If there is, can we propose, within the framework thus defined, a more precise distinction than currently available between the notions of theories and models? With the same hypothesis, can we, for a given discipline, claim to achieve right away the best equilibrium between theoretical endeavour and experimentation, i.e. that capable of leading the most directly to the best form of theoretical explanation? According to Delattre the concept of the theory applies best at the level of the general language of description, the theory then including the inductive synthesis which justifies the choice of the definitions and their internal coherence. The explanation always implies the involvement of the constituent parts and of processes causing interactions between the parts.

Finally, we may add a few words here on the relationship between formalised theoretical physiology and medicine. There now exists a considerable gap between medical care-giving and the increasingly refined and complex knowledge that underlies medical activity. While the general practitioner can hardly be required to master the fundamentals indispensable to a formalised understanding of physiological functions, we believe that biologists and other users of advanced techniques in genetic and medical engineering should acquire a sound working knowledge in this field. Like the experimental physicists, they will soon discover the advantages of a formalised, synthetic approach. Indeed, the second half of the twentieth century is a major turning point for biology, just as there was one for physics some hundreds of years ago. It requires no extraordinary vision to predict that the unfortunate division between the so-called 'exact sciences' and 'natural sciences' will continue to decrease, and that the outcome of predictions in biology, as in physics, will become more and more certain in spite of the multiple levels of description involved. Does this mean, for example, that we shall succeed in controlling the conditions of biological variability? Perhaps not, but, like the fundamental problems concerning reality and interpretation that have appeared in physics, similar questions are likely to arise in biology, connected with the very nature of the self-organisation of living organisms and structure–function relationships. Undoubtedly, the difficulty lies in the multiple levels of biological description and the formalism used, but the formidable immensity of the task is more than compensated by the fascinating beauty of the functioning of living organisms. In this perspective, and in spite of difficulties of another order due to the novelty of the discipline, let us hope that more and more biologists will become interested in these problems since, as a reading will show, our work surely raises far more questions than it provides answers.

I would like to thank all those who have helped in this long work through their advice and encouragement: J. A. Jacquez, Professor of Physiology at the University of Michigan, P. Delattre, who pioneered theoretical biology in France, T. W. Berger, Professor of Neuroscience at the University of Southern California, and J. D. Murray, Professor of Mathematical Biology at the University of Seattle. I am particularly indebted to Dr. A. Tadei, Professor of Cardiology at the University of Angers, whose dynamism and competence have always been an outstanding example of the ideal medical research worker, teacher and practitioner. My wife, with constant understanding, never failed to provide full moral support. May this work bear witness to our affection.

This edition of *Theoretical Systems in Biology, Hierarchical and Functional Integration* contains all the topics presented in the original three-volume French edition entitled '*Traité de physiologie théorique*', published by Masson & Cie., Paris (1987–1989). The English translation, kindly undertaken by K. Malkani, my friend and colleague at the University of Angers, has provided an opportunity for updating some sections, particularly in the chapters on the organisation of biological systems at the molecular, cellular and organismal levels. Although Volume I may be read as an independent text, it should be observed that the mathematical models introduced here, as well as in Volume II, were essentially chosen with the idea of constructing a theory of functional organisation. A 'bottom-up' approach was initially used to extract properties common to the models selected so as to draw up the general principles which are finally stated in an abstract, 'top-down' form in the concluding chapter of Volume III. The present edition has allowed us to integrate these properties into the discussion of the different models. We hope this will make for an easier understanding of the whole work. Differing considerably from the existing structurally oriented theories set out in Volume I, Chapter 3 of this volume, the theory of functional organisation presented in Volume III views a biological system as consisting of two subsystems, one describing its topology and the other its dynamics. The stability of the biological system would thus depend on the conditions of stability of the corresponding subsystems. Specifically biological concepts, such as those of *non-symmetry* and *non-locality* of the fundamental interaction or *functional interaction*, or that of *structural discontinuity*, emerge progressively from the treatment of the subject in the first two volumes. The most important consequence is that we are obliged to consider the formalism of graphs and fields in hierarchical spaces in which a parameter such as time defines a particular level of organisation.

Our theory of *functional organisation* (Volume I, Chapter 4, Section 3) may be summed-up simply as follows. From the diversity of processes occurring in biological organisms, we have extracted two concepts: on one hand, the concept of a *functional interaction* with a property of *non-symmetry*, and on the other hand, the concept of a *hierarchical system* with a property of *non-locality*. The functioning of a living organism depends on two types of organisation. The first is the structural organisation corresponding to the ordered spatial distribution of the various structural units of the organism, such as cells, tissues and organs. The second is the functional organisation, resulting from the coordination of a set of interactions between the structural units. A convenient way of studying the relations between the structural and the functional organisations is by means of a graphical representation. The points of the graph represent the structural units, and the arcs represent the elementary physiological functions, i.e. the relations between the structural units. The graphs may be used in at least two ways. The first, which scarcely calls for the mathematical properties of graph theory, depends on a computer programme to organise the physiological functions between the structural units so that the functional hierarchy is automatically displayed. The second, however, fully exploits graph theory to search for specific substructures, such as cyclic subgraphs, the best path in the graph for a given constraint, and so on. Just as there exists a structural or anatomical hierarchy, i.e. a group of more or less similar units at different levels of organisation, there also exists a functional hierarchy. Indeed, it is precisely the existence of interlinked functional hierarchies that complicates the representation of the functional organisation of living organisms. Moreover, in most cases, the functional hierarchy does not coincide with the structural hierarchy.

In the third part of Volume I (Chapter 12) we examine a property of the variation of the functional hierarchy during the development of an organism. This approach is based on a *principle of invariance* of the physiological function and on the consequences that may be observed in a given species. For example, an aerobic organism needs oxygen in order to survive, it has to self-replicate to perpetuate the species, and so on. This invariance can only be expressed if the physiological function can be mathematically defined. The presence of the genetic blueprint in all the self-reproducing elements of an organism has led us to formulate the conception of a *potential of organisation of physiological functions*, with a property of optimality, which may be considered to be a general principle governing all living organisms. These principles of invariance and optimality will be validated inasmuch as their consequences can be experimentally verified. A crucial problem raised by the theory proposed lies in the identification, or rather in the deduction of the mathematical structure, of the 'mechanisms' which are at the origin of the existence of the functional interaction on the one hand, and on the construction of the functional hierarchy on the other.

We know that before reaching adulthood, the organism passes through a developmental phase during which its structural and functional organisations are modified under the control of a genetic programme. But why does this programme actually work as it is observed to do? To answer this question, we have proposed the *hypothesis of self-association*, which may be stated as follows: *The functional interaction is created, in other words it exists so that the domain of stability of the physiological functioning increases, or is at least maintained, in spite of the increase in complexity due to the increase in the number of interactions.*

Having suggested the causal mechanism at the origin of a functional interaction, it remains to be seen how the functional hierarchy varies with time. During the development of an organism, we observe the growth of a particular hierarchical organisation. We may wonder why it is precisely this and not some other hierarchical organisation which develops. A similar problem encountered in physical systems has been resolved by the principle of least action which imposes a pathway, which is in fact the pathway actually observed, among a set of possible pathways. The fundamental reason for this compulsory 'choice' lies in the geometry of the space in which the movement occurs. In the case of biological systems in which the functional organisation is represented by a graph of the interactions, we show that the problem may be stated in terms of the stability of the graph. Why, during the development of the organism, does a certain structural unit become a source, and another a sink? For purposes of reasoning, we may separate two processes which are in fact closely dependent: on the one hand, the modification of the number of receptor units of the products (the sinks of the graph), and on the other, the modification of the total number of units (the summits of the graph). We have analysed the consequences of these two variations with time and since the results are of considerable importance, let us now present them briefly.

When the number of structures receiving the product (the sinks) varies during development, the fact that a particular functional organisation is observed among several possible organisations means that there exists a potential of organisation, i.e. a range of potentialities for the organisation of the system, and that there must also exist a cause leading to the organisation observed. The apparent number of the structures, e.g. organelles, cells and tissues, evidently varies, but so does the *quality* (source or sink) of the units of functional organisation. However, as the quality is less apparent, the variation is far less evident. We have shown that biological systems possessing the property of self-replication, and in which the functional interactions are created according to the hypothesis of self-association, have a maximum potential of organisation. Of course, only the experimental verification of the mathematical consequences of this property would validate our theory.

When the number of structural units varies during the course of development, two cases may arise according to whether or not the units are reorganised. Here again, this reasoning is useful for understanding and demonstrating the mathematical property of the variation of functional hierarchy. Let us suppose that the system does not undergo reorganisation, i.e. that the sources remain sources, and the sinks remain sinks, so that the quality of the units is not modified. In this case, the system is governed by what we have called an *orgatropic* function, the time-variation of which is always positive. In reality, the system undergoes simultaneous reorganisation. We have demonstrated the existence of a function, the *functional order*, which describes the time-variation of the biological system through the emergence of various levels of organisation with time. Finally, the hierarchy of the organisation, i.e. the hierarchy of the graph, develops in such a way that the physiological function remains invariant during the successive transformations of the functional organisation. Since the functional order varies positively with time, it indicates the direction of the evolution of the biological system. In addition, the functional order may be used as a criterion of comparison between biological and physical systems.

As mentioned above, biological processes can be associated with graphs of functional interactions. We therefore have to seek a spatiotemporal representation of these processes, i.e. a dynamic representation of the products exchanged between the source of an interaction and the sink (or sinks). We propose to do this by means of the field theory (Volume III). Mathematically speaking, a field is a quantity that varies at each point in space. Subjected at a given instant and at a given point in space (the source), to the action of an operator which, at the following instant, propagates it towards another point (the sink), this quantity depends on certain transformations occurring in the source. Of course, the property of non-locality of the functional interaction leads to complications since the operator must also be nonlocal, making its determination difficult. We have applied the field theory in three interesting cases: the nervous system, the phenomenon of aging, and the evolution of the species.

The nervous system appears to be the biological system with the maximum potential of organisation since the mathematical consequences of the field theory, for the properties considered, are in good agreement with experimental observations. Moreover, the consideration of nerve impulses being propagated in the form of fields in the nervous system, viewed as a hierarchical system system, allows us to interpret the form of electrical potentials measured in a population of neurons. The phenomenon of ageing appears to be linked not only to the genetic blueprint but also to the fluctuation of a considerable number of structural units involved in the physiological function. We show that the fundamental cause of ageing, beyond an immediate genetic cause, lies in the ineluctable deterioration of the structural units involved in the regulatory physiological mechanisms. In the case of the evolution of the species, the field theory provides an interpretation of the existence of evolutionary jumps as a consequence of reciprocal effects between the dynamics of physiological processes and the variable number of structural units involved taking into account the condition of the invariance of the physiological function.

I would like to express my gratitude to J.-M. Chrétien for the illustrations, to A. Breteau for revising the reference section, and to S. Robert-Lamy and D. Bordereau for re-organising the manuscript. I thank the editorial and production staff of Elsevier Science, particularly Mr T. Merriweather, Ms E. Lawrence and Ms A. Hall, for all their patience and help during the preparation of this book. I fully acknowledge the generous support of the Conseil Général de Maine-et-Loire all through this work.

G. Chauvet
Saint Aubin de Luigné
January 1995

Foreword to Volume I

Volume I is divided into three parts as follows:

Part I lays down the macromolecular bases used all through this work. Chapter 1 describes the relationship between the structure of biological macromolecules and the molecular mechanisms of the functioning of organisms: first, the regulatory role through the allosteric effect, by inhibition and competition, and, secondly, the catalytic role. The chemical systems of reaction–diffusion are presented in Chapter 2, together with some notions of mathematical solutions. Chapter 3 goes to the heart of the subject with a discussion of the theories and formalisms currently used in theoretical biology.

Part II takes up the major problem of the self-organisation of biological systems at the molecular level. Chapter 4 examines the problem from the general standpoint as well as from the point of view of information theory. Chapter 5 describes an example of self-organisation: the cellular DNA replication–translation apparatus. In Chapter 6 the discussion begun at the molecular level is carried further in terms of the evolution of macromolecular systems. Given the importance of the subject, in Chapter 7 we have pursued the study of the evolution of the species within the framework of a neo-Darwinian, global synthesis of current evolutionary theories. The reader may perhaps wonder why this theme is introduced in a work on functional biology. The reason will be found in the introductory section of Chapter 3, entitled 'Biology and complexity'. In fact, it is rather surprising to find that the set of processes of physiological regulation — which we call the *vital coherence* — has been 'overlooked' among the postulates of the theories of evolution. As indicated in the Preface, we believe that a non-formalised theory will be unable to take into account a principle based on vital coherence. The consequences of this are analysed and followed up by a formal statement of our thesis of a permanent evolution towards greater complexity (which remains to be defined) through an increase in the degree of self-organisation compatible with the stability of the living organism. At the molecular level, we discuss a formalised description of the DNA replication–translation apparatus, which unfortunately has been somewhat neglected in recent years. We present the topological, structural and functional implications of this fundamental mechanism.

Part III deals with biological organisation at the cellular level. Chapter 8 introduces the central problem of cell differentiation and its relationships with the physiological functions of the cell. Chapters 9 to 12 emphasise the concepts of cell morphogenesis, growth, division and differentiation.

   While the key problem of biology is that of our own evolution, i.e. how a thinking human being actually managed to emerge from among a multitude of living species, the central question at the level of the autonomous organism is that of its development, which is precisely the field in which the implications of cellular physiological mechanisms are determinant.

Part I: Materials and Methods in Biological Dynamics

Cette application des mathématiques aux phénomènes naturels est le but de toute science, parce que l'expression de la loi des phénomènes doit toujours être mathématique.

This application of mathematics to natural phenomena is the aim of all science, since the expression of the law of the phenomena must always be mathematical.

Claude Bernard, *Introduction à l'étude de la médecine expérimentale* (1865)

Introduction to Part I

The nineteenth century saw the early development of biochemistry, the fundamental science which, on the bases of the cell theory proposed by Schleiden and Schwann in 1839, attempted to explain the processes of synthesis and the transformation of constituents in living organisms. Lavoisier and Liebig did pioneering work on chemical energetics through their studies of *intermediate metabolism*. This was also the epoch of the early work of Claude Bernard on the glycogenic function of the liver. The idea of the cell as an assembly of molecules was already taking hold but it was not until the first half of the twentieth century that the notions of macromolecules, nucleotide polymers and polypeptides began to spread. Thus, biochemistry was destined to become part of *molecular biology*, which is based on the two great discoveries in modern biology: that due to Watson, Crick and Wilkins in 1953 of the double helix structure of deoxyribonucleic acid (DNA), and that due to Monod, Wyman and Changeux in 1963 of the regulation of enzyme reactions at molecular level through the *allosteric effect*. The former has considerably influenced the development of molecular genetics, and the latter has truly inaugurated the reign of molecular biology.

After a description of the molecular structures found in living organisms—polynucleotides and polypeptides—we present Wyman's formalisation of the allosteric effect (Chapter 1) together with the formalised aspects of the chemistry of living organisms, including, in particular, a brief study of the systems of reaction–diffusion and their stability conditions (Chapter 2). The mathematical methods currently used (Murray, 1989) depend on various formalisms which will be encountered throughout this work: relational theories, thermodynamical theory, the theory of information and the theory of catastrophes (Chapter 3). All of these formalisms are based on mathematics involving ordinary differential equations and partial differential equations. This is a rapidly developing field, and although some of the biological problems presented here cannot for the moment be completely solved they will serve to illustrate certain aspects of mathematical research and show up some of the defects encountered with currently available tools. However, as we shall see, considerable progress has been made in theoretical biology since the pioneering work of the great precursors such as Rashevsky (1961).

1

Macromolecular Components and Interactions in Living Organisms

The components of living matter may be classified into three categories, the first category corresponding to more than 90% of the atoms of a living organism (hydrogen, carbon, nitrogen and oxygen), the second corresponding to trace elements found in quantities between 0.02 and 0.1% (sodium, magnesium, phosphorus, sulphur, potassium and calcium), and the third corresponding to oligoelements found in quantities of less than 0.001% (such as boron, silicon, vanadium, manganese, iron, cobalt, copper, zinc, molybdenum, and so on). Although the presence of oligoelements with a high atomic mass may sometimes be difficult to establish, they are of considerable importance and play a decisive role in macromolecular structure and thus in physiological function. Macromolecular structures are created through spheres of coordination centred on heavy metal ions. A classical example is that of the porphyrins, among which the best known is haemoglobin with its iron atoms.

The molecules of living matter are made up from a set of about 20 elements either in the form of ions such as Na^+, Mg^{2+}, PO_4^{3-}, SO_4^{2-}, Cl^-, K^+, Ca^{2+}, CO_3^{2-} and NO_3^-, or in the form of molecules of various sizes and degrees of complexity. Not only are these molecules the main building blocks of cells but they also act as agents of biological specificity.

All the properties and functions of living organisms are adapted to the environment by the action of a variety of molecular structures and complex regulatory mechanisms. It is indeed striking that the carbon atom plays the key role in all organic compounds. In fact, because of the carbon atom, these compounds possess two properties essential to life: *slow reactivity* and *great diversity*. Moreover, the molecules basic to all structures in the animal and vegetable kingdoms are identical.

Such fundamental unity justifies the current hypotheses on the origin of life. These will be discussed in Chapter 7 which deals with the relationship between evolution and physiology. For the present, let us briefly review the structural characteristics of nucleic acids and proteins. These macromolecules are essential cell components with a *functional specificity* closely linked to their structure.

I. Structure of nucleic acids

Two classes of nucleic acids, deoxyribonucleic acid (DNA) and ribonucleic acid (RNA), are produced by the polymerisation of nucleotides:

$$\text{Polymer} = \underbrace{\text{Phosphate–Nucleoside}_1}_{\text{Nucleotide}_1} \ldots \underbrace{\text{Phosphate–Nucleoside}_n}_{\text{Nucleotide}_n}$$

where:

$$\text{Nucleotide} = \text{Phosphate–deoxyribose–(DNA)} \begin{cases} \begin{array}{ll} \text{purine} & \text{A} = \text{adenine} \\ \text{nitrogen bases} & \text{G} = \text{guanine} \\[1ex] \text{pyrimidine} & \text{C} = \text{cytosine} \\ \text{nitrogen bases} & \text{T} = \text{thymine} \end{array} \end{cases}$$

$$\text{Nucleotide} = \text{Phosphate–ribose–(RNA)} \begin{cases} \begin{array}{ll} \text{purine} & \text{A} = \text{adenine} \\ \text{nitrogen bases} & \text{G} = \text{guanine} \\[1ex] \text{pyrimidine} & \text{C} = \text{cytosine} \\ \text{nitrogen bases} & \text{T} = \text{uracil} \end{array} \end{cases}$$

which is schematically represented in Fig. 1.1, where $D_{(i_1,i_2,i_3,i_4)}$ corresponds to any one of the four bases A, G, C and T of DNA, and $R_{(i_1,i_2,i_3,i_4)}$ corresponds to any one of the four bases A, G, C and U of RNA. Table 1.1 shows the structure of these bases. In terms of chemical bonding, a polynucleotide segment is made up of nucleosides linked together by 3′–5′ phosphodiester bonds. This general law is found to be observed in all living organisms.

Nucleotide structure is now well established, and it is accepted that DNA contains almost the entire hereditary information and thus represents the genetic material of all living organisms. The information is transmitted by the coding of a one-dimensional structure, i.e. the order of the nitrogen bases on the DNA molecule. Before going into the details of the physiological phenomena of hereditary transmission, let us recall the polymeric structure of nucleic acids.

Watson and Crick, following the work of Pauling, deduced the bicatenary, helicoidal nature of DNA from the results of X-ray diffraction. The structure, now known as the *double helix*, is characterised by:

- hydrogen bonds between A and T, and between G and C (Fig. 1.2);
- anti-parallel strands (Fig. 1.3);
- bases turned in towards the helical axis.

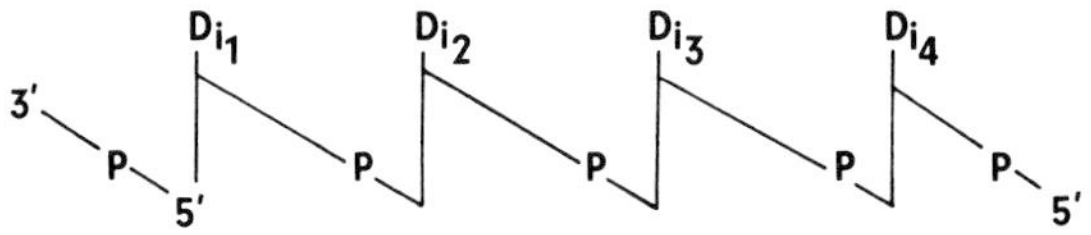

Fig. 1.1. Schematic form of a DNA polynucleotide. D_i represents one of the four bases A, G, C or T.

Table 1.1. Chemical formulae of the fundamental components of nucleic acids (after Loewy and Siekevitz, 1974).

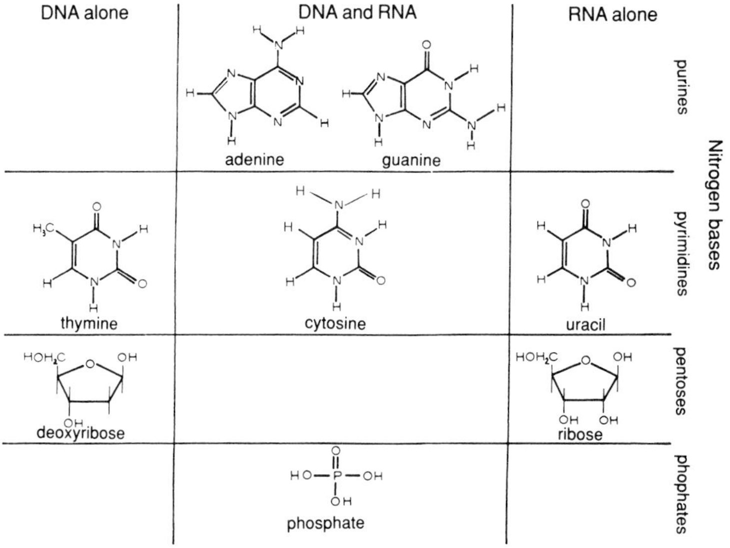

They were then able to explain not only the mechanism of DNA stabilisation but also that of gene reproduction, by the separation and duplication of the strands of the double helix. This is what we usually term the *hereditary transmission of information*. But, as we shall see in Chapter 5, this calls for some qualification, particularly with respect to the definition of the quantity of information 'contained' in a DNA molecule (Gatlin, 1968, 1972).

Whereas there is only one class of DNA with its specific biological function, this is not true of RNA, the physical and biological properties of which have led to the definition of four types:

• *Viral RNAs*, often wrapped in a protein coat, are single-chain macromolecules which may in places be paired to form a double helix. It is believed that the monocatenary structure corresponds to the most elementary form of nucleic acid and that DNA represents a more advanced evolutionary state stabilised by the coupling of the two strands. The current idea is that viral RNA results from a parasitic degeneration of a eucaryotic cell. The reverse transcription necessary would thus be a highly advanced acquisition.

Thymine (T)   Adenine (A)

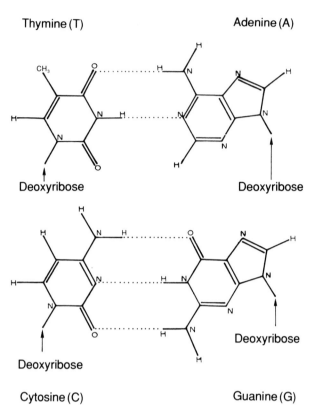

Fig. 1.2. Hydrogen bonds between adenine (A) and thymine (T), and between cytosine (C) and guanine (G).

- *Messenger RNAs* (mRNA), after being synthesised on a DNA template, fix on to a ribosome, a structure essential to the biosynthesis of proteins as we shall see in Chapter 5.
- *Ribosomal RNAs* (rRNA), which are part of the ribosome, are associated with several protein chains. More precisely, the procaryotic ribosome is made up of two parts: a big subunit with two rRNA molecules and a small subunit with one rRNA molecule, each molecule being associated with specific protein chains.
- *Transfer RNAs* (tRNA), or adaptors, are molecules that attach to amino acids. Each tRNA has a structure adapted to one of the 20 amino acids and is made up of about 80 nucleotides assembled in a single chain with covalent bonds. The three-dimensional structure of some of these nucleic acids has been determined by X-ray diffraction analysis. The secondary structure of tRNA is characteristic: the cloverleaf pattern represents the best possibility of obtaining the maximum number of paired bases and producing identically shaped molecules. A precise description of tRNA structure and the physico-chemical experiments that led to its discovery will be found elsewhere (Loewy and Siekevitz, 1974; Watson, 1978).

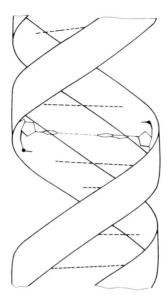

Fig. 1.3. Helical structure of DNA formed by two anti-parallel strands. Broken lines indicate hydrogen bonds between base pairs.

II.   Structure of proteins

Proteins, synthesised according to the nucleotide sequences of the genetic message, have a structure and behaviour specific to a given *physiological function*. We shall give the term 'physiological function' a more precise sense than is usually the case when it is merely used to describe a set of properties observed in living organisms. This new concept will be discussed in Chapter 4. In view of this, considerable importance will be attached to the study of protein *conformation*.

1.   *Description of protein structure*

Proteins, the essential components of a cell, play a *catalytic* and *regulatory* role by means of their enzymic specificity. This, of course, is not the only role of proteins; indeed, several other functions are well known, such as tissue and cell support, contractility, molecular transport, and so on. However, it may be considered that, with the discovery of the regulatory function, molecular biology has touched on one of the secrets of life. We shall therefore emphasise this level of cell physiology.

   Proteins differ from other biological macromolecules in their composition. They are made up of *amino acids* (of which there are 20 varieties) or, as in certain organisms, of derivatives of the fundamental amino acids. What is the chemical structure of an amino acid? An amino acid is *amphoteric*, i.e. it carries a positively charged ammonium group and a negatively charged carboxyl group. Table 1.2 shows

Table 1.2.   Chemical formulae of amino acids (after Loewy and Siekevitz, 1974).

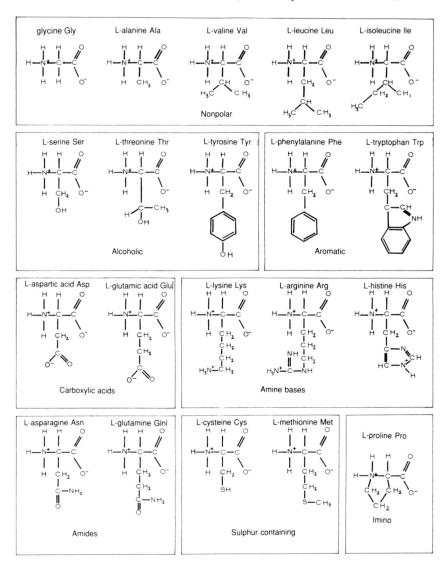

the chemical formulae of the amino acids. Symbolically, R is a characteristic radical of the type R_i, where $i = 1$ to 20, with:

R_1 = GLY	R_2 = ALA	R_3 = VAL	R_4 = LEU	R_5 = ILE
R_6 = SER	R_7 = TRP	R_8 = TYR	R_9 = PHE	R_{10} = TYR
R_{11} = ASP	R_{12} = GLU	R_{13} = LYS	R_{14} = ARG	R_{15} = HIS
R_{16} = ASN	R_{17} = GLN	R_{18} = CYS	R_{19} = MET	R_{20} = PRO

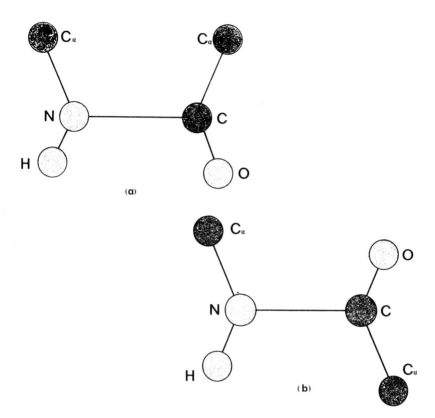

Fig. 1.4. Molecular configurations of the peptide bond: (a) the *cis* isomer form, and (b) the *trans* isomer form (after Yon, 1969).

These 20 amino acids constitute an alphabet, and each word made up of letters from this alphabet corresponds to a protein. In short, protein synthesis is the transfer of information, through enzymic activity, from DNA to cell products, i.e. to a physiological function.

Thus, whereas nucleic acids are composed from an alphabet of four elements, A, T, G and C, proteins derive from an alphabet containing 20 elements $A–R_i$ where A is the portion of the amino acid without the radical R_i. Proteins are constituted by a sequence of amino acids linked by a covalent peptide bond:

$$\begin{matrix} & O & & \\ & \| & & \\ -\!\!&C\!\!&-N- & \\ & & | & \\ & & H & \end{matrix}$$

The C, O, N and H atoms all lie in the same plane, O and H being on opposite sides of the N–C bond (the *trans* position). The C_α carbon atoms, linked to amino acid residues, lie in this plane (Fig. 1.4). Interatomic distances and angles between the valence bonds are now well known (Fig. 1.5). Rotation is possible about the

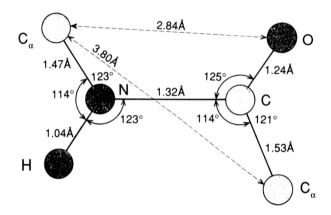

Fig. 1.5. Interatomic angles (± 4°) and distances (± 0.03 Å) in a peptide bond (after Watson *et al.*, 1987).

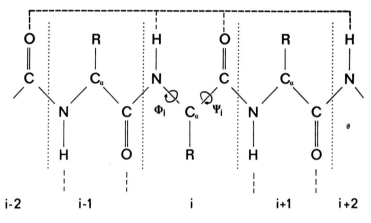

Fig. 1.6. Polypeptide chain developed to show hydrogen bonds (broken lines) and rotations about the N–C_α and C–C_α bonds. R = amino acid residue (after Yon, 1969).

N–C_α and the C–C_α bonds (Fig. 1.6). In the system of internal coordinates, formed by the interatomic distances and the angles between valence bonds, the angles of rotation about N–C_α and C–C_α are usually called ϕ and ψ.

The polypeptide sequence is called the *primary structure* of the protein. The primary structure is of importance because it is exclusively determined by the nucleotide sequence of DNA and because of the consequent existence of a fundamental relationship between the hereditary genetic material and the physiological function. Thus, a genetic mutation leads to a modification of the amino acid sequence and thence to a disturbance of physiological function. Several 'molecular diseases' due to such causes have now been identified.

Another type of protein structure is the *secondary structure*. This is defined as the structure resulting from the formation of hydrogen bonds between the oxygen of the

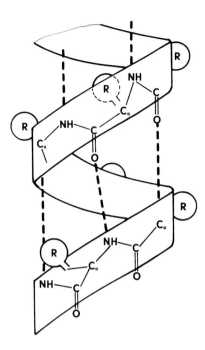

Fig. 1.7. Secondary structure of a protein produced by hydrogen bonds O–H, shown by broken lines (see also Fig. 1.6).

carboxyl groups and the hydrogen of the amino groups of the peptide chain (Fig. 1.7). Pauling showed that the structure is most stable when it takes the form of either a folded sheet or an α-helix, in which the stability is ensured by hydrogen bonds. The α-helix has a pitch of 5.4 Å and contains 3.6 residues per helical turn (Fig. 1.8). The stability of the folded or β-sheet structure of certain fibrous proteins, such as β-keratin, is due to hydrogen bonding between CO and NH (Fig. 1.9).

Physical studies, such as the measurement of viscosity, the coefficient of friction, optical diffusion and so on, have shown that many of the proteins are compact and rigid, which means that the helical structure is further modified by the peptide chain folding on itself. This is known as the *tertiary structure*, and results from a compromise between two tendencies: the tendency of the skeleton to form a stable regular helix, and the tendency of distant lateral groups to bind. An example of this is the myoglobin molecule (Fig. 1.10). Various types of bond lead to the formation of the tertiary structure, such as covalent links formed by disulphide bonds, ionic links due to electrostatic attraction between polar groups (NH_3^+–COO^-), bonding between lateral chains, and Van der Waals' forces between hydrophobic lateral chains oriented towards the interior of globular structures. Thus, apart from the covalent links, all the others are weak bonds. This explains the large variety of configurations — or conformational states — possible for protein chains.

In short, a biological macromolecule may be made up of one or more polypeptide chains, each with its own primary, secondary or tertiary structure. Such chains may

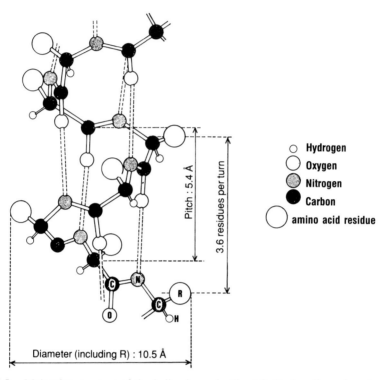

Fig. 1.8.   Molecular structure of the helix shown in Fig. 1.7. Broken lines show hydrogen bonds.

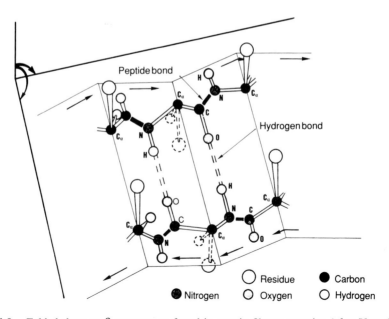

Fig. 1.9.   Folded sheet or β-structure as found in certain fibrous proteins (after Yon, 1969).

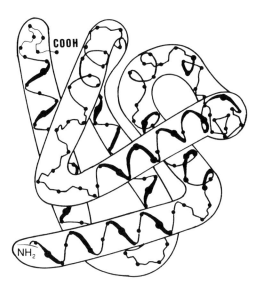

Fig. 1.10.   Tertiary structure of a protein (myoglobin) caused by folding of the polypeptide chain.

further be associated to form larger complexes, the degree of complexity increasing with the number of smaller elements involved. This new structure, called the *quaternary* structure, thus consists of several polypeptide chains which remain independent in terms of non-covalent bonding, i.e. the structure contains *subunits* associated in dimers, trimers, tetramers, and so on. The enzymic activity of such structures is then specific to the individual subunits. Since the biological specificity of molecules is determined by their subunits, let us give precise definitions of the structures involved. The terminology generally used derives from Monod *et al.* (1965).

A *chain* is a polypeptide. A protein liable to dissociate is called an *oligomer*, and the products of the complete dissociation are called *monomers*. The term *protomer* or *subunit* is attributed to an association of monomers defined by their functional role, i.e. polypeptide chains that do not have all the catalytic and regulatory functions of the entire molecule. It follows that a molecule or *unit* is defined at the functional level by its complete catalytic and regulatory activity. Finally, a *complex* is made up of several units, each having a specific activity. A molecule composed of more than one subunit and capable of binding to a protein is termed a *ligand*. If, in addition, the ligand influences the behaviour of the protein, it is called an *effector*, and may be either an *activator* or an *inhibitor*. For instance, ribonuclease consists of a single polypeptide chain. In contrast, haemoglobin is made up of four chains: two chains, α_1 and α_2, each containing 141 residues, and two chains, β_1 and β_2, each containing 146 residues. Each chain is fixed to a corner of a regular tetrahedron with a binary axis of symmetry (Fig. 1.11). α_1, α_2, β_1 and β_2 are the protomers or subunits of the oligomer haemoglobin which, in this terminology, is of course a unit.

The quaternary structure is thus at the heart of biological activity. As we shall see,

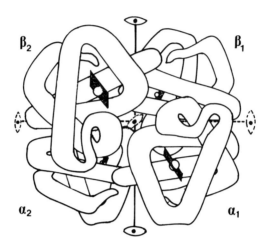

Fig. 1.11.   Quaternary structure of a molecule (haemoglobin) made up of four 'independent' polypeptide chains or subunits.

the interaction of the subunits controls, through the phenomenon of *cooperativity*, the degree of affinity between protein and ligand. *In the present state of knowledge, this is one of the fundamental regulatory mechanisms at the cellular level.* But how can we determine the conformation of a protein? And, considering the delicate nature of the task on account of the structural complexity of proteins, would such an attempt be really worthwhile? A highly theoretical approach, based on a model of protomer interactions, has led Monod and his co-workers to fundamental results concerning structure–function relationships (see, for example, Debru, 1983). These results are globally covered by the term 'allosteric effects'. The initial step of the model is thermo-dynamic as it uses the law of mass action. But although the fundamental principle of enzymic regulation is now established, we still do not know the transformations that occur between the subunits.

Work in this field involves highly complicated calculations to determine the theoretically stable conformations minimising the potential energy of the protein, expressed in terms of its internal coordinates. The large number of variables and interactions involved makes this a most formidable task in spite of the powerful computers now available. Nevertheless, in view of future applications, it may be important to show how the methods of statistical mechanics can help not only to determine the possible conformations of a protein but also to calculate the equilibrium constant of transition between two conformational states. The difficulty, as we shall see in the next section, lies in the fact that the molecular models used presuppose knowledge of the structure of the molecule, of the energy of its different allosteric forms as well as of the density of states.

2. *Theory of the conformation of biological macromolecules*

The three-dimensional structure of proteins is characterised either by the position of its atoms in ordinary space (Cartesian coordinates) or by the interatomic distances and angles (internal coordinates).

A conformational state refers to one of the possible structures defined within the interval of variation of the internal coordinates. The continuity of these structures makes it necessary to postulate a finite but arbitrary number of *conformational classes* in the usual statistical sense, i.e. to suppose the density of states to be known.

But how can a particular macromolecular conformation be described? Because of its importance, we shall study mainly the standard, 'native' conformation, i.e. the highly organised, 'natural' structure responsible for biological activity. For example, the DNA double helix is formed of two strands of nucleotides, but these macro-molecular subunits are capable of dissociation. This is true of all macromolecules, and in general the term *species* is used to describe a molecular combination of sub-units, including the native conformation.

We then face the following problem: if a protein exists in different conformations, can we measure the quantity of a particular species in a given conformation? In particular, can we describe an equilibrium of conformational states in the given environmental conditions? The results of statistical mechanics clearly show that the most stable conformation will be the one that predominates in equilibrium condi-tions. The importance of the study of conformational states has justified the efforts made in this field (Yon, 1969).

a. *Physical description of a macromolecular system in solution*

Let us consider a system containing M identical macromolecules in their native conformation and any number i of molecular species in a solvent S. The system, initially not in thermodynamic equilibrium, tends to equilibrium by passing through a certain number of successive states resulting from two different types of process: (i) dissociation reactions giving rise to new molecular species, and (ii) the shifting of certain species from one conformational state to another.

Let $[i_\alpha]$ be the equilibrium concentration of the species i in the conformational state α, $[i_\beta]$ that in state β, and $[i]$ the total concentration of species i. Classically, for the reaction corresponding to case (ii):

$$i_\alpha \underset{k_-}{\overset{k_+}{\rightleftharpoons}} i_\beta$$

we have the equilibrium constant:

$$K_{i,\alpha\beta} = \frac{[i_\beta]}{[i_\alpha]}$$

since by definition:

$$k_+[i_\alpha] = k_-[i_\beta] \qquad K_{i,\alpha\beta} = \frac{k_+}{k_-}.$$

In general, for r different chemical species of type i at equilibrium, we will have:

$$K = \prod_i [i]^{v_i} \tag{1.1}$$

where v_i is the *stoichiometric coefficient* of the chemical relation for species i in a given conformational state. v_i is, by convention, positive to the right of the reaction and negative to the left. One way of calculating the equilibrium constants is to use the methods of statistical mechanics which allow determination of the potential energy of a macromolecule–solvent system. We may then deduce the *configuration integrals*, which have often been applied in molecular biology.

It is known that the potential energy of a molecular system is additive and that it is the sum of the energy of islolated molecules (which is the energy of electrons in the fundamental state for given atomic positions, i.e. the smallest eigenvalue of Schrödinger's equation describing electron configuration), and the terms of intermolecular energy (due, for example, to Van der Waals' forces).

Now let V_i, the potential energy of a macromolecule–solvent system (for species i), be expressed in the form of a function:

$$V_i\left(x_a^{(M)}, y_a^{(M)}, z_a^{(M)}; x_a^{(S)}, y_a^{(S)}, z_a^{(S)}\right)$$

for all atoms (a) of the macromolecule (M) and the solvent (S). This evaluation obviously supposes it possible to separate small subsystems (Σ) for which the calculation remains possible within the larger system under investigation. This potential energy being known, it is now possible to define:

- the configuration integral $Z_{i,\alpha}$ of the system (Σ) for species i in the conformational state α:

$$Z_{i,\alpha} = \frac{1}{\prod_j N_j!} \int \cdots \int_{\text{Volumes }(M,S)} \exp\left(-\frac{V_i}{k_B T}\right) dV^{(M)} dV^{(S)}$$

where k_B is the Boltzmann constant, T is the absolute temperature and

$$dV^{(M)} = \prod_a dx_a^{(M)} dy_a^{(M)} dz_a^{(M)}$$

$$dV^{(S)} = \prod_a dx_a^{(S)} dy_a^{(S)} dz_a^{(S)}$$

are the elements of volume for the macromolecule (M) and the solvent (S). The order of the multiple integral is three times that of the number of molecules of M and S. The factor $1/N_j!$ for each species j is due to the hypothesis that the molecules in the subsystem (Σ) are indiscernible. Thus the integral remains unaltered when the molecules of the solvent are exchanged for species j. Similarly:

$$Z_i = \sum_\alpha Z_{i,\alpha}$$

is the configuration integral for species i, whatever its conformational state.
- The equilibrium constants are then defined by:

$$K_{i,\alpha\beta} = \frac{Z_{i,\beta}}{Z_{i,\alpha}}$$

- The thermodynamic energy function $W_{i,\alpha}$ is related to the configuration integral $Z_{i,\alpha}$ by:

$$Z_{i,\alpha} = \int \cdots \int_{\text{Volume}(M)} \exp\left(-\frac{W_{i,\alpha}}{k_B T}\right) dV^{(M)} \qquad (1.2)$$

where the function $W_{i,\alpha}$ is such that:

$$\exp\left(-\frac{W_{i,\alpha}}{k_B T}\right) = \frac{1}{\prod_j N_j!} \int \cdots \int_{\text{Volume}(S)} \exp\left(-\frac{V_i}{k_B T}\right) dV^{(S)}.$$

This procedure allows us to separate the calculation of the volume integral, first with respect to the coordinates of the solvent (S), and then with respect to the coordinates of the macromolecules (M). It can be shown that $W_{i,\alpha}$ is the statistical expression of the Helmholtz free energy of the macromolecule–solvent system in a given macromolecular configuration α.

It is often useful to change from Cartesian coordinates to internal coordinates which simplify the interpretation of the terms of the interatomic interaction. In particular, the intramolecular vibration frequencies can be calculated from the frequencies measured by infrared or Raman spectroscopy. This method allows us to assign the frequency observed to a particular type of bond (or the corresponding *representation* of the group of symmetry of the molecule), and to simultaneously deduce the constant of the interatomic force in a molecular grouping (Chauvet, 1974; Wilson *et al.*, 1980).

b. Free energy variation during a conformational change

A change in the conformational state of a molecule evidently produces a variation of the standard free energy $\Delta G^0 = -RT\ln K_e$, where K_e is the equilibrium constant of the chemical reaction that takes species i from conformation α to conformation β. Then:

$$\Delta G^{0}_{i,\alpha\beta} = -RT\ln\frac{Z_{i,\beta}}{Z_{i,\alpha}} \qquad (1.3)$$

where RT is the product of the perfect gas constant R and the temperature T.

It is important to be able to determine the free energy variation of biological macromolecules during the change of state. This determination will be made easier if the variation can be broken down into independent, additive contributions. Thus for the contributions (c):

$$\Delta G^{0}_{i,\alpha\beta} = \sum_{c} \Delta G^{(c)0}_{i,\alpha\beta} = \sum_{c} - RT\ln\frac{Z^{(c)}_{i,\beta}}{Z^{(c)}_{i,\alpha}} \qquad (1.4)$$

where each term is calculated, for example, for a variation between two rigid conformations, or between a rigid conformation and a disordered one.

Before going on to review the biological applications of the theory, let us stress the distinction between the phenomenon of *change in the conformational state* and the phenomenon of *chemical reaction*. In the former the variation of internal coordinates is compatible with the three-dimensional structure, whereas in the latter the variation of the atomic composition of the macromolecule leads to a reorganisation of the coordinates of the product and of the reactant. The rigorous description of such complex phenomena is extremely difficult, if not impossible, and we therefore have to fall back on approximations such as the decomposition of the free energy function into independent components. This approach enables us to relate the rearrangement of coordinates due to a chemical reaction — a broken bond, for example — to $G^{(r)}$, a contribution to the free energy variation $\Delta G_{\alpha,\beta}$. In any case, although the precise structure of macromolecules remains unknown, a good understanding of the regulatory mechanism of enzymic activity has been reached through a phenomenological treatment based on thermodynamics.

3.   *Molecular conformation and biological activity*

Protein activity is closely linked to molecular conformation, and biological specificity depends on the existence of specific interaction *sites* on the protein surface. The catalytic and regulatory properties of an enzyme are modified by a change of conformation. This phenomenon is called the *allosteric effect* while the term *allosteric transition* is used to qualify conformational change. Allosterism is the indirect regulation of an enzymic reaction by an effector (activator or inhibitor), called the *allosteric effector*. The regulatory action of the effector involves neither binding to the specific enzymic site nor any direct substrate–effector interaction. Thus, according to Monod *et al.* (1965), the effect is mediated entirely by the protein, presumably through a conformational change resulting from the binding of the inhibitor, i.e. through an allosteric transition.

In fact, the essential concept appears to be that of the subunit as described by Nemethy (1975). Allosteric regulation appears mainly in oligomers, where we find not only *heterotropic* effects (interaction between different ligands) but also *homotropic* effects (interactions between identical ligands). For a given subunit there is generally one binding site for each ligand, thus the protein must have a quaternary structure since it necessarily has several subunits.

Let us adopt the definition of allosterism given by Nemethy (1975), according to whom allosteric effects are regulatory effects exerted by different or identical ligands bound at distinct sites on an oligomeric protein and accompanied by comformational changes of the protein. However, allosterism, which etymologically refers to the action of an effector 'elsewhere', i.e. other than on the active site at which the substrate reacts, may appear in monomer molecules. Such has been found to be the case for chymotrypsin. But, of course, no cooperative effects will be observed in monomers (see Section III. 2a). Cooperativity is thus essentially linked to the subunit concept.

Finally, one of the main problems in macromolecular physics is the search for the conditions of conformational stability in terms of molecular interactions. This problem remains difficult because the precise structure of the active sites of proteins is unknown, although we can determine the approximate theoretically stable conformation from the potential energy of the molecule, expressed as the sum of a certain number of contributions. We may wonder what modification of behaviour might be produced in a protein by a minimal change in its conformation. Experimentally, an atomic displacement of 0.5 to 1 Å has been found sufficient to maintain an enzymic reaction. So a displacement of this order would be necessary for a qualitative description of the process. However, smaller variations (about 0.001 Å for the Fe ion of myoglobin, for example), which would remain undetected even by X-rays, may lead to considerable effects.

Following Lumry and Biltonen (1969), we may say that:

(i) the conformational changes need not be large to trigger off important biological processes;
(ii) there is no direct relationship between the extent of atomic displacement and the intensity of the allosteric effect; and
(iii) the cooperative phenomena induced by conformational changes are at the heart of the biological reactions essential to life.

This last point will be discussed in the following sections.

III. Molecular interactions in proteins: protein–ligand interactions

We have just seen that a protein may adopt a very great number of conformational states. So the problem is to find the conditions that will restrict this number of states. As the precise structure of proteins is unknown, the description of interactions between proteins, or between proteins and ligands, can only be thermodynamic. We shall therefore use the results of equilibrium thermodynamics for the conformational state.

1. *Thermodynamic theory of molecular interactions*

Among the variety of states in which proteins exist, we may distinguish:

- Proteins in various forms E_j, $j = 0$ to m;
- Ligands S bound to E in i different sites, called S_i, $i = 1$ to n;
- Protein–ligand complexes ES at the i sites, $i = 1$ to n; and
- Protein–ligand–ligand complexes at l sites and k sites of E, called ES_lS_k, $l = 1$ to n and $k = 1$ to n.

a. *Protein–ligand associations (homotropic effects)*

Going from the simpler to the more complex, we may describe three types of reaction:

Reaction R1:

$$E + S \rightleftharpoons ES$$

This is the simplest case of a bond between a protein and a ligand. The association–dissociation *equilibrium constant* is:

$$K_S = \frac{[ES]}{[E][S]}.$$

Reaction R2:

$$E + iS \rightleftharpoons ES_i$$

This is the case of a macromolecular conformation with n sites of E for the ligand S, when i molecules S are bound to a molecule E. Here, the *apparent macroscopic equilibrium constant* is:

$$K_i = \frac{[ES_i]}{[E]s^i}$$

where $s = [S]$ and $K_0 = 1$ by convention. This expression is due to the fact that the reaction R2 is of a global nature, representing a series of reactions of the type:

$$ES_r \rightleftharpoons ES_{r-1} + S$$

where r varies from 1 to i. Let θ_r be the corresponding equilibrium constant:

$$\theta_r = \frac{[ES_r]}{[ES_{r-1}][S]}$$

then:

$$\theta_1 = \frac{[ES_1]}{[E]s}, \qquad \theta_1\theta_2 = \frac{[ES_2]}{[E]s^2}, \ldots, \theta_1\theta_2 \ldots \theta_i = K_i.$$

Between the apparent macroscopic equilibrium constant K_i and the equilibrium constants θ_r ($1 \leq r \leq i$), which may be termed *microscopic*, we thus have the relationship:

$$K_i = \theta_1 \theta_2 \ldots \theta_i$$

The behaviour of the macromolecule, with respect to possible bonds with the ligand, may be described by the *binding polynomial P(s)* defined for the total quantity of protein:

$$\rho_E = \sum_{i=0}^{n} [ES_i] = [E] \sum_{i=0}^{n} K_i s^i = [E] P(s) \tag{1.5}$$

with:

$$P(s) = 1 + K_1 s + K_2 s^2 + \ldots + K_n s^n \tag{1.6}$$

This polynomial 'measures' the concentration of the protein, unbound or bound to the ligand, taking into account all possible states of the protein. We have assumed that the equilibrium constant K_i of the reaction R2 varies according to the number of sites occupied ($0, 1, 2, \ldots, n$), in other words the sites are considered to be *distinct*.

Reaction R3:

$$E_0 \rightleftharpoons E_j$$

$$E_j + iS \rightleftharpoons E_j S_i$$

This is the case of a protein possessing $m + 1$ conformations E_j, where j varies from 0 to m ($j = 0$ corresponds to the *native* conformation E_0), and n sites for the ligand S. Then, with the upper indices f (free) and b (bound):

(i) $K_j^f = \dfrac{[E_j]}{[E_0]}$ is the *transition constant* or the *constant* of intrinsic equilibrium;

(ii) $K_{ji}^b = \dfrac{[E_j S_i]}{[E_j] s^i}$ is the *apparent macroscopic equilibrium constant* as defined above

for reaction R2. This is an *association–dissociation constant* (with respect to the ligand) which is also termed the *binding constant*. The index j varies from 0 to m and the index i varies from 0 to n. We have:

$$K_{j0}^b = 1.$$

This is analogous to the convention $K_0 = 1$ in the case R2 above. The macroscopic equilibrium constant of the total reaction, which describes the transformation of a protein from the standard conformational state E_0 to the *ligand-bound* state $E_j S_i$, can thus be written:

$$K_{ji} = K_j^f K_{ji}^b$$

We deduce that the total quantity of protein calculated as above in (1.5) is:

$$\rho_E = \sum_{j=0}^{m} \sum_{i=0}^{n} [E_j S_i] = \sum_{j=0}^{m} \sum_{i=0}^{n} K_{ji}^{b}[E_j]s^i = \sum_{j=0}^{m} [E_j]P_j^*(s)$$

putting:

$$P_j^*(s) = \sum_{i=0}^{n} K_{ji}^{b} s^i.$$

Writing $[E_j] = K_j^{f}[E_0]$ we have:

$$\rho_E = [E_0]P(s)$$

with:

$$P(s) = \sum_{j=0}^{m} K_j^{f} P_j^*(s).$$

Here we again have the binding polynomial of Eq. (1.6), so that:

$$P_j(s) = \sum_{i=0}^{n} K_{ji} s^i. \tag{1.7}$$

The particular case $j = 0$ leads to:

$$[E_0]K_0^{f} = [E_0] \quad \text{whence } K_0^{f} = 1$$

and we may write the following matrix, called the *binding matrix*, of order $(m + 1, n + 1)$, which completely describes the set of reactions in the case R3:

$$
L = \begin{bmatrix}
1 & K_{01}^{b}s & K_{02}^{b}s^2 & \cdots & K_{0n}^{b}s^n \\
K_1^{f} & K_1^{f}K_{11}^{b}s & K_1^{f}K_{12}^{b}s^2 & \cdots & K_1^{f}K_{1n}^{b}s^n \\
\cdot & \cdot & \cdot & & \cdot \\
\cdot & \cdot & \cdot & \cdots & K_i^{f}K_{ij}^{b}s^j & \cdots \\
\cdot & \cdot & \cdot & & \cdot \\
K_m^{f} & K_m^{f}K_{m1}^{b}s & K_m^{f}K_{m2}^{b}s^2 & \cdots & K_m^{f}K_{mn}^{b}s^n
\end{bmatrix}
$$

More simply: $L = (K_{ji}s^i)$ where $j = 0$ to m, $i = 0$ to n, and $K_{ji} = K_j^{f}K_{ji}^{b}$. This matrix L is general and characterises a macromolecule having $m + 1$ conformational states, not necessarily *independent*, and a ligand S able to bind to n *distinct* sites on the macromolecule. Each line j represents the binding polynomial $P_j(s)$ of the protein in the conformational state j.

b. Consequences

Let us now consider the particular case of a macromolecule, composed of n subunits each possessing *only one site*, for which the overall conformational state is determined by the state of the subunits, i.e. on account of the allosteric effect, by the binding between the subunit and the ligand with:

(i) two states for each subunit: A or B; and
(ii) $n + 1$ conformational states of the polymer: $j = 0, 1, 2, \ldots, n$, such as A_n, $A_{n-1}B$, $A_{n-2}B_2, \ldots, A_1B_{n-1}, B_n$.

For state $A_{n-j}B_j$, for example, this notation signifies that $n - j$ subunits are in state A and that the remaining j subunits are in state B. Let us now describe the possible states by means of the binding polynomial. For a unit in state A, first free and then bound to the ligand S, the binding polynomial according to (1.7) is $(1 + K_A^b s)$. For $(n - j)$ subunits in state A, the number of states possible is:

$$P_{n-j}^A(s) = (1 + K_A^b s)^{n-j}$$

where K_A^b is the binding constant for the link between the unit in state A and the ligand. Similarly, for j subunits in state B, the binding polynomial may be written:

$$P_j^B(s) = (1 + K_B^b s)^j$$

where K_B^b is the binding constant for the link between the unit in state B and the ligand. Then the number of possible states in the conformation $A_{n-j}B_j$ is:

$$P_j^{AB} = (1 + K_A^b s)^{n-j}(1 + K_B^b s)^j = \sum_{i=0}^{n}\left[\sum_{l=0}^{i} C_{n-j}^{i-l}(K_A^b)^{i-l} C_j^l (K_B^b)^l\right]s^i$$

where C is the number of combinations (conventionally, $C_n^p = 0$ for all $p > n$), and:

$$P_j^{AB}(s) = \sum_{i=0}^{n} K_{ji}^b s^i$$

after development of the binomials. Finally, the complete binding polynomial for the $n + 1$ conformational states of the molecule is given by:

$$P(s) = \sum_{j=0}^{n} K_j^f P_j^{AB}(s). \tag{1.8}$$

Here we again find Eq. (1.7) and the matrix that follows.

In short, the construction of the binding polynomial P corresponds to certain schemes of reaction to which a binding matrix can be associated. The first scheme is very simple and general, and the second corresponds to the particular case just described.

Scheme 1: The chemical equilibrium between the reactants is represented by terms (line j, column i) of the type:

$$K_{ji}s^i = K_j^f K_{ji}^b s^i$$

These elements form the binding matrix corresponding to the set of reactions where the macromolecule in the conformational state $E_j (0 \leq j \leq m)$ can bind to i molecules of the ligand $(0 \leq i \leq n)$:

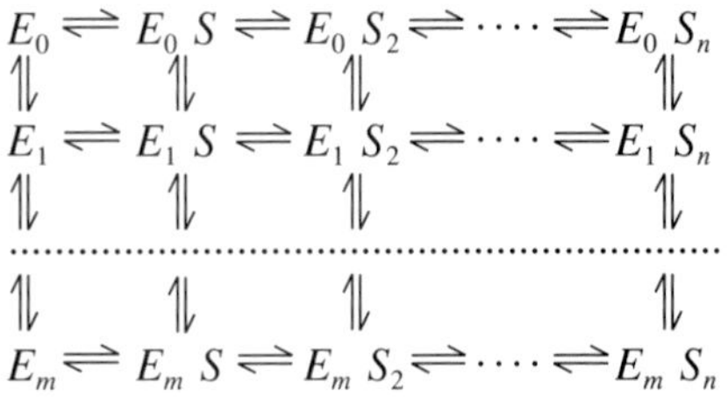

$$
\begin{array}{ccccccc}
E_0 & \rightleftharpoons & E_0\,S & \rightleftharpoons & E_0\,S_2 & \rightleftharpoons \cdots \rightleftharpoons & E_0\,S_n \\
\updownarrow & & \updownarrow & & \updownarrow & & \updownarrow \\
E_1 & \rightleftharpoons & E_1\,S & \rightleftharpoons & E_1\,S_2 & \rightleftharpoons \cdots \rightleftharpoons & E_1\,S_n \\
\updownarrow & & \updownarrow & & \updownarrow & & \updownarrow \\
\cdots & & \cdots & & \cdots & & \cdots \\
\updownarrow & & \updownarrow & & \updownarrow & & \updownarrow \\
E_m & \rightleftharpoons & E_m\,S & \rightleftharpoons & E_m\,S_2 & \rightleftharpoons \cdots \rightleftharpoons & E_m\,S_n
\end{array}
$$

where $E_j S_0 \equiv E_j$ (see above).

So we evidently have (Eq. (1.7)):

$$P(s) = \sum_{j=0}^{m} \sum_{i=0}^{n} K_{ji}s^i = \sum_{j=0}^{m} \sum_{i=0}^{n} K_j^f K_{ji}^b s^i \tag{1.9}$$

Scheme 2: The chemical equilibrium between the reactants is represented by the terms (j, i) where S is bound to subunits either of type A or of type B. The reactions may then be written, with $m = n$:

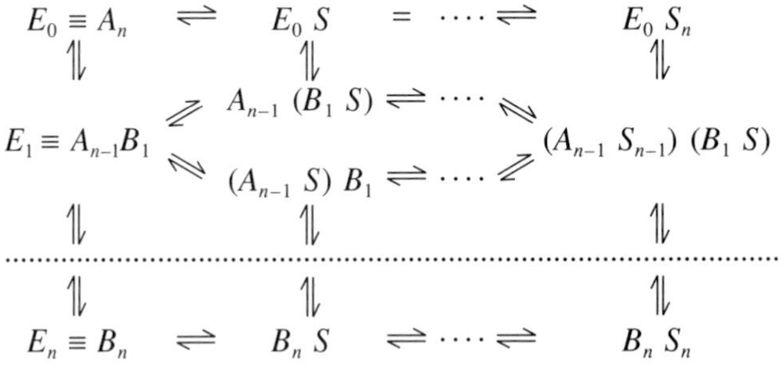

$$
\begin{array}{ccccc}
E_0 \equiv A_n & \rightleftharpoons & E_0\,S & = \cdots \rightleftharpoons & E_0\,S_n \\
\updownarrow & & \updownarrow & & \updownarrow \\
 & \nearrow A_{n-1}\,(B_1\,S) \rightleftharpoons \cdots \searrow & & & \\
E_1 \equiv A_{n-1}B_1 & & & & (A_{n-1}\,S_{n-1})\,(B_1\,S) \\
 & \searrow (A_{n-1}\,S)\,B_1 \rightleftharpoons \cdots \nearrow & & & \\
\updownarrow & & \updownarrow & & \updownarrow \\
\cdots & & \cdots & & \cdots \\
\updownarrow & & \updownarrow & & \updownarrow \\
E_n \equiv B_n & \rightleftharpoons & B_n\,S & \rightleftharpoons \cdots \rightleftharpoons & B_n\,S_n
\end{array}
$$

By arranging the terms s in the order of increasing power i, the element (j, i) can be written:

$$K_{ji}s^i = K_j^f \left[\sum_{l=0}^{i} C_{n-j}^{i-l} C_j^l (K_A^b)^{i-l} (K_B^b)^l \right] s^i$$

and Eq. (1.9) again leads to the binding polynomial of Eq. (1.8). The difference between the two schemes is thus clearly demonstrated. Using similar methods, we may construct binding polynomials, and then determine the quantity ρ_E for more complex schemes, but the calculations are soon likely to become inextricable.

The only practical interest of the schemes lies in their application to the two most widely used allosteric models described below.

c.   Allosteric models generally used

The MWC model (Monod, Wyman and Changeux, 1965). In addition to the hypotheses used in the scheme above, the authors assume a symmetric arrangement of protomers in space. In other words, the subunits are supposed to occupy equivalent positions and the symmetry is maintained for all changes of state. From the physical point of view, all the subunits are in the same state because of strong interaction with neighbouring subunits. To simplify, only two conformations of the polymer are considered possible, *all the subunits being either in state A or in state B* so that the binding matrix contains only the first and the last lines, and the intermediate states are unstable.

Consequently, the binding polynomial (1.8) can be written:

$$P(s) = K_0^f P_0^{AB}(s) + K_n^f P_n^{AB}(s)$$
$$= (1 + K_A^b s)^n + K_n^f (1 + K_B^b s)^n. \tag{1.10}$$

The KNF model (Koshland, Nemethy and Filmer, 1966). Here, the *hypothesis of symmetry* is replaced by the hypothesis that, under certain conditions of partial saturation, mixed conformational states may exist. In other words, subunits may be found in different conformations. An *additional hypothesis* is made: it is supposed that the binding of the ligand S to the subunit leads to a conformational change in the subunit so that increased saturation in terms of the ligand leads to a sequence of conformational changes. This 'sequential model' was first introduced in its simple form, and then used in the generalised form. In the simple form of the model, on the one hand (first hypothesis), the concentration of B is considered negligible in the absence of ligand and the conformation of A is more stable than that of B; on the other hand (second hypothesis), binding is considered to take place exclusively between S and B. In the generalised form of the model, the restriction imposed by the first hypothesis is removed.

The MWC model will be more closely examined below.

2.   The cooperative effect

a.   The concept of cooperativity

There are several types of cooperative phenomenon. Classically, we have:

- The *positive cooperative effect* where binding to a ligand facilitates further binding; and
- The *negative cooperative effect* where binding to a ligand renders further binding more difficult.

The problem is to find out which of these effects may be anticipated with a given molecular model. It can be shown (Goldbeter, 1974) that the MWC model does not predict a negative cooperative effect except in the very special conditions of enzymic catalysis. However, both types of cooperative phenomena may be found with the KNF model, in which the variations of interaction due to conformational changes of subunits cause a bound ligand to modify further binding.

This highly intuitive definition calls for some clarification:

(i)   The total quantity of protein in all the states (bound or unbound) was defined by $\rho_E = [E_0]P(s)$.

(ii) The total quantity of ligand S bound to the macromolecule, in the same conditions, is by definition:

$$\eta(s) = \sum_{i=0}^{n} i[ES_i].$$

(1.11)

Then:

$$\eta(s) = [E_0] \sum_{i=0}^{n} i K_i s^i = [E_0]s\frac{\mathrm{d}P}{\mathrm{d}s}.$$

(iii) The *saturation function* $\overline{Y}_S$ of the macromolecule with n sites for the ligand S is defined by:

$$\boxed{\overline{Y}_S = \frac{1}{n}\frac{\eta(s)}{\rho_E} = \frac{1}{n}\frac{\mathrm{d}\ln P(s)}{\mathrm{d}\ln s}}.$$

(1.12)

(iv) The *state function* $\overline{R}_j$, which is the relative quantity of protein in the conformational state j, is defined by:

$$\boxed{\overline{R}_j = \frac{\rho_{E_j}}{\rho_E} \qquad j = 0, 1, 2, \dots, m}$$

and is such that:

$$\sum_{j=0}^{m} \overline{R}_j = 1.$$

The saturation function has been widely used to determine the nature of the cooperativity. Thus, the shape of the saturation curves shown in Fig. 1.12 is characteristic of zero, positive or negative cooperativity. The interpretation of such curves has been discussed by Bardsley and Waight (1978). Zero cooperativity may be recognised in the Henri–Michaelis–Menten hyperbolic curve for reaction kinetics, corresponding to an equation of the type:

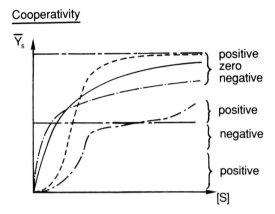

Fig. 1.12. Saturation function $\bar{Y}_S$ for ligand concentration $s = [S]$ (Eq. (1.12)) for zero, positive and negative values of cooperativity.

$$\bar{Y}_S = \frac{K^*s}{1 + K^*s}$$

where K^* is the *effective equilibrium constant* of the form:

$$K^* = \frac{[ES]}{[E_0][S]}$$

(see Eq. (2.8) with $v = v_{max} \bar{Y}_S$ and $K^* = 1/K_m$ where K_m is the *Michaelis constant*).

It is obviously rather risky to draw a conclusion as to the nature of the molecular interaction merely on the basis of the shape of the saturation curve since this may, of course, correspond to very different situations. Thus, the saturation function $\bar{Y}_S$ below is a particular case, for $n = 1$, of:

$$\bar{Y}_S = \frac{K^*s^n}{1 + K^*s^n}$$

which, for $n > 1$, corresponds to the equation $E + nS \rightleftharpoons ES_n$, from which all the intermediate equilibrium constants K_i ($i = 1, 2, \ldots, n - 1$) have disappeared. If $n < 1$, there is of course no physical interpretation. These results may be summarised by putting:

$$n \equiv \frac{\mathrm{d}\ln \dfrac{\bar{Y}_S}{1 - \bar{Y}_S}}{\mathrm{d}\ln s} = \frac{1}{\bar{Y}_S(1 - \bar{Y}_S)} \frac{\mathrm{d}\bar{Y}_S}{\mathrm{d}\ln s}$$

deduced from the expression for $\bar{Y}_S$:

$$\frac{\bar{Y}_S}{1 - \bar{Y}_S} = K^*s^n.$$

Then, in the reference system: $(\ln(\bar{Y}_S/[1 - \bar{Y}_S]), \ln(s))$, we obtain a *Hill graph* for which the slope is $n = 1$ for Henri–Michaelis kinetics. The problem of the type of cooperativity may now be solved by writing:

$$\frac{\mathrm{d}\ln\left(\bar{Y}_S/(1 - \bar{Y}_S)\right)}{\mathrm{d}\ln s} = 1 + f(x) \; . \tag{1.13}$$

The sign of f(x)*, which depends on the sign of the Hessian (see Appendix B) of the binding polynomial* P(s)*, determines the sign of the cooperativity.*

b. Cooperativity and the binding polynomial

The concept of the binding polynomial $P(s)$ (Wyman, 1972) is of great significance, and much credit is due to Bardsley and Waight (1978) for having given a solid foundation to the concept of cooperativity. Polynomial algebra is used to associate the coefficients of P to the nature of the roots, real or complex, and to the shape of the resulting curve.

The binding polynomial is defined in the most general terms by:

$$\rho_E = [E_0]P(s)$$

where ρ_E is the total quantity of protein E, unbound or bound to ligand S, at any number of *distinct* and *dependent* sites. The concept of cooperativity is thus implicit, but notions concerning the physical nature of the binding sites will need further development.

For simplicity, we shall consider only the *homotropic* effects and the case of a macromolecule with n subunits, with one binding site per subunit. Then the binding polynomial is (Eq. (1.8)):

$$P(s) = \sum_{j=0}^{n} K_j^f P_j^{AB}(s) = \sum_{j=0}^{n} K_j^f (1 + K_A^b s)^{n-j} (1 + K_B^b s)^j$$

We may try to determine the form of the polynomial when the protein has n *identical* and *independent* sites on the n subunits, each of which may be in one of two states, A or B.

Let $p = \Pr[A]$ and $q = 1 - p = \Pr[B]$ be the probabilities of states A and B, which can be measured by:

$$p = \frac{[A]}{[E]}$$

and

$$q = \frac{[B]}{[E]}.$$

Then, the 'microscopic' equilibrium constant for the elementary transition $A \rightleftharpoons B$ is:

$$K_{eq} = \frac{[B]}{[A]} = \frac{q}{p}.$$

Let us define the *independence of the subunits* as the independence of states A and B. For example, if A_1 (resp. A_2) is the state of subunit 1 (resp. 2), we have:

$$\mathrm{Pr}[A_1 \text{ and } A_2] = \mathrm{Pr}[A_1] \cdot \mathrm{Pr}[A_2/A_1]$$

which means that the probability of the two subunits being in state A is the product of the probability that the first is in state A and the conditional probability that the second is in state A while the first is also in state A. The independence of the subunits is expressed by:

$$\mathrm{Pr}[A_1 \text{ and } A_2] = \mathrm{Pr}[A^2] = (\mathrm{Pr}[A])^2.$$

More generally, the probability for *one* conformation $A_{n-j}B_j$ is $p^{n-j}q^j$. The number of situations possible being the number of combinations C_n^j, we deduce the binomial probability for the state $((n - j)$ times A and j times $B)$ to be $C_n^j p^{n-j}q^j$.

We can now express the macroscopic transition constant K_j^f in terms of the microscopic constants K_{eq}, functions of p and q. The independence of the n subunits is the experimental expression of the Henri–Michaelis–Menten kinetics of which the saturation function is:

$$\bar{Y}_S = \frac{K_S}{1 + K_S} \qquad \text{with} \qquad K = \frac{[ES]}{[E_0]s}$$

and the total quantity of protein is:

$$\rho_E = [E_0](1 + Ks)^n.$$

We deduce that the *only way* the polynomial $P(s)$, given by Eq. (1.8), can be factorised is by putting:

$$K_j^f = C_n^j p^{n-j}q^j.$$

Then:

$$
\begin{aligned}
P(s) &= \sum_{j=0}^{n} C_n^j p^{n-j}q^j(1 + K_A^b s)^{n-j}(1 + K_B^b s)^j \\
&= \sum_{j=0}^{n} C_n^j(p + pK_A^b s)^{n-j}(q + qK_B^b s)^j \\
&= \left[1 + (pK_A^b + qK_B^b)s\right]^n
\end{aligned}
$$

whence we obtain:

$$P(s) = (1 + Ks)^n$$

by writing:

$$K = pK_A^b + qK_B^b.$$

The polymer thus appears to behave as if it were made up of monomers all in the same state, with the equilibrium constant:

$$K = pK_A^b + qK_B^b = p(K_A^b + K_{eq}K_B^b).$$

The physical significance of this result is evident: the independence of the subunits leads to a statistical distribution of the constant of intrinsic transition, which is a function of the microscopic reaction constants. To carry the interpretation further, we will have to take into account the free energy variations between A and B, between A and A, and between B and B. In general, it can be shown (Laiken and Nemethy, 1970) that there exists a function ϕ_M, corresponding to the Michaelis curve, such that:

$$\phi_j(K_{j,AA}^f, K_{j,BB}^f) = C_n^j \left[\phi_M(K_{j,AB}^f, K_{j,BB}^f) \right]^j$$

where $K_{j,AA}^f$ is the intrinsic constant of the reaction $A \rightleftharpoons A$, the same holding good for AB and BB. This function clearly depends on the nature of the arrangements among the subunits. Thus, the binding polynomial for a protein with n identical and independent sites is:

$$\boxed{P(s) = (1 + Ks)^n} \tag{1.14}$$

where K is the equilibrium constant, identical for each site. This includes the transition constants K_{eq} between subunits as well as the ligand association–dissociation constants, K_A^b and K_B^b.

When the n subunits are independent (i.e. there is equal interaction between the subunits) and distinct, each subunit should be distinguished, so that, for states A_1 and A_2 for example:

$$\Pr[A_1 \text{ and } A_2] = \Pr[A_1] \cdot \Pr[A_2].$$

The number of possible states is then:

$$(1 + K_{A_1}^b s)(1 + K_{A_2}^b s) \dots (1 + K_{A_{n-j}}^b s)(1 + K_{B_1}^b s) \dots (1 + K_{Bj}^b s).$$

But the transition constant no longer follows a statistical law (the binomial distribution), and each state A_j or B_j is defined by a different probability p_j or q_j, whence:

$$K_{eq,j} = \frac{q_j}{p_j}$$

and the factorisation obtained in (1.14) no longer exists. The exact expression of the binding polynomial is thus a delicate matter. However, let us recapitulate the results, giving the form of the binding polynomial when there is one site on each subunit:

n independent, identical sites: $P_1(s) = (1 + K_s)^n = \sum_{i=0}^{n} C_n^i K^i s^i$

n independent, distinct sites: $P_2(s) = (1 + K_1 s) \ldots (1 + K_n s)$

n interacting, distinct sites: $P_3(s) = 1 + K_1 s + \ldots + K_n s^n$

We may now interpret the notion of cooperativity: intuitively, a positive (resp. negative) cooperative phenomenon means that certain sites tend towards increased dependence (resp. independence). This is the same as saying that the polynomial $P_3(s)$ *tends towards factorisation into a smaller (resp. larger) number of real factors.* A rigorous treatment of this notion will be found in the article by Bardsley and Waight (1978), together with mathematical criteria in function of the Hessian (Appendix B).

Finally, this kind of comparison between P_1, P_2 and P_3 can produce interesting results in the very particular cases of the MWC and the KNF models, but we must remember that these results should not be generalised. Let γ_i^1 be the ratio of two successive coefficients of rank $i + 1$ and i in P_1:

$$\gamma_i^1 = \frac{C_n^{i+1}}{C_n^i} K = \frac{n - i}{i + 1} K.$$

Similarly, in P_3:

$$\gamma_i^3 = K_{i+1}/K_i.$$

The *coefficient of cooperativity* between sites $i + 1$ and i is then defined by:

$$\gamma_i = \ln\frac{\gamma_i^3}{\gamma_i^1} = \ln\left(\frac{i + 1}{n - i} \frac{K_{i+1}}{K K_i}\right). \tag{1.15}$$

The coefficient is zero when the ratio γ_i^3 is equal to the 'statistical ratio' γ_i^1. When this holds good for all values of i, the factorisation is complete: there is no cooperativity. The coefficient is negative when:

$$\frac{K_{i+1}}{K_i} < \frac{n - i}{i + 1} K$$

i.e. in the case of increase independence, according to the result above.

The system could be said to be 'anti-cooperative'. The coefficient is positive when:

$$\frac{K_{i+1}}{K_i} > \frac{n - i}{i + 1} K$$

i.e. in the case of cooperativity between sites i and $i + 1$. The cooperativity is said to be *total* when $\gamma_i > 0$ for all values of i from 1 to $n - 1$.

Statistical mechanics has been applied to the binding of ligand molecules to protein molecules (Wang, 1990c). This interesting approach is based on the calculation of the number of 'complexions', i.e. the number of microscopic states, of a given distribution of the assembly. The equilibrium distribution is the one for which the number of complexions is maximum. This maximum is under three constraints: the constancy of the number of protein molecules, that of the number of protein-bound ligands, and that of the total energy. Using multipliers according to Lagrange's method, Wang has determined the equilibrium distribution of the assembly in various cases involving several kinds of ligands, identical and completely independent sites, and identical and interacting sites. These results have been applied to allosteric enzymes (Wang and Kihara, 1990) and to nonspecific inhibitors and activators of enzyme reactions (Wang, 1990b).

c.   An illustration of the cooperative effect

Let us now apply the general notions described above to the MWC model. The expression (1.10) of the binding polynomial allows us to calculate the saturation function (1.12):

$$\bar{Y}_S = \frac{1}{n}\frac{s\,dP}{P\,ds} = \frac{K_A^b s(1 + K_A^b s)^{n-1} + K_n^f K_B^b s(1 + K_B^b s)^{n-1}}{(1 + K_A^b s)^n + K_n^f(1 + K_B^b s)^n} \tag{1.16(1)}$$

which is the ratio of the number of occupied sites to the total number of sites. This is usually simplified by writing $\alpha = K_B^b s$. Then, with $c = \dfrac{K_A^b}{K_B^b}$, we have:

$$\bar{Y}_S = \frac{\alpha c(1 + \alpha c)^{n-1} + K_n^f \alpha(1 + \alpha)^{n-1}}{(1 + \alpha c)^n + K_n^f(1 + \alpha)^n}. \tag{1.16(2)}$$

As K_A^b (resp. K_B^b) is the binding constant for the conformational state A (resp. B), the ratio $\gamma = \ln c$ is given by:

$$\gamma = \ln \frac{K_A^b}{K_B^b}$$

and, with K_n^f, determines the cooperativity. For example, if $\gamma > 0$, the ligand S has a greater affinity for state A than for state B, and will bind more readily to E_0 than to E_1, and so on. The first line of the binding matrix will be 'favoured'. This is the positive cooperative effect.

When $\gamma = 0$, the cooperativity has a zero value and we again find the Michaelis equation:

$$\bar{Y}_S = \frac{K_A^b s}{1 + K_A^b s} \tag{1.17}$$

by putting $K_A^b = K_B^b$ in Eq. (1.16(1)).

A good example of the phenomenon of allosterism is the fixation of oxygen on the haemoglobin molecule, which we shall consider later when we study the respiratory

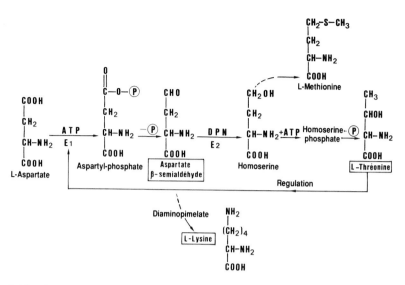

Fig. 1.13. Example of a regulated chain of reactions leading from *L*-aspartate to *L*-threonine in *E. coli* (after Yon, 1969). E_1 = Aspartate kinase I, II and III; E_2 = homoserine dehydrogenase I and II.

system. It was in fact the research on haemoglobin (see for example, Perutz (1948) and Koshland (1958)) that led Monod and co-workers to the definition of the allosteric effect (Debru, 1983).

Haemoglobin is an oligomer with four distinct subunits, each of which has a fixation site. In keeping with the theory, oxygenated haemoglobin has a slightly different conformation compared with that of reduced haemoglobin. Moreover, the binding of the first molecule of oxygen facilitates the binding of the three other molecules: this corresponds to the sigmoidal shape of the saturation curve. It is interesting to compare the example of haemoglobin with that of myoglobin which, having only one subunit and thus only one fixation site, shows no cooperativity.

The regulatory effect of the allosteric phenomenon is particularly effective in the case of enzymic chain reactions of the type:

$$S_1 \xrightarrow{E_1} S_2 \xrightarrow{E_2}, \dots, \xrightarrow{E_{n-1}} S_n$$

where the intermediate reaction $S_i \to S_{i+1}$ involves the allosteric enzyme E_i. This produces an inhibitory retroactive system in which the end product S_n inhibits enzyme E_1. The mechanism is allosteric since the fixation of S_n on E_1 takes E_1 from one conformational state to another, in which the affinity of E_1 for S_1 is modified. Consequently, there is *intrinsic regulation* of the chain of reactions and control over the quantity of product formed. For example, in the set of reactions where $S_1 \equiv$ *L*-aspartate, $E_1 \equiv$ aspartate kinase ..., $S_n \equiv$ *L*-threonine (Fig. 1.13), it is *L*-threonine S_n which retro-inhibits aspartate kinase E_1. Other control mechanisms also intervene at various levels of the chain of reactions. Many examples of this kind have been found. Allosteric interactions are fundamental to epigenetic and metabolic control which we shall consider in greater detail in Part 2.

The mechanism of allosteric interaction is fundamental to the efficient working of living organisms. An example of this is the channel gating process in nerve cells (Chay, 1988). By a change in its conformation, a regulatory protein can respond to another molecule which may be either the substrate of the reaction catalysed or a regulatory signal. Thus, *structural modifications of the protein are induced by its association with the ligand. Several effects follow*:

First, *biological activity is modified* since the physiological function is closely related to molecular structure. For example, molecular activity occurs only within a narrow range of ligand concentration.

Secondly, *biological activity is amplified* by the cooperative interaction of macro-molecular sites. This cooperativity is expressed by a saturation curve (in terms of ligand concentration) which is not hyperbolic but sigmoidal in form. Consequently, for a certain range of ligand concentration, a small variation in ligand concentration leads to a large variation in the concentration of ligand-bound complexes.

The phenomena of interaction, here described at the level of oligomer macro-molecules, are of prime importance in the regulation of biological systems. However, they still require to be integrated into the function of enzymic catalysis, the essential chemical process of life.

2

The Internal Chemistry of Cells

The description of macromolecular interactions in biological activity clearly shows that the dynamic aspects of the complex entities formed by oligomers depend on the laws of chemistry. While notions of the equilibrium constant and the kinetic constant have already been applied above, we still need to understand the physical phenomena at the origin of the intermolecular reactions. The association of a ligand S, the *substrate*, and a protein E at a reaction site takes place according to a process very similar to that suggested by Lindemann (1922) and Hinshelwood (1927) following the initial description given by Henri (1905) and Michaelis and Menten (1913): each chemical reaction in a cell is catalysed by a protein — an enzyme — usually specific to the reaction. Thus in principle there exist at least as many enzymes as there are chemical reactions. The set of reactions is then coordinated either by the *control* of the *quantity* of enzyme synthesised, or by the *regulation* of its *activity*. The former mechanism will be examined in the context of protein biosynthesis in Chapter 6. The latter, as described above, involves direct or allosteric reactions leading to the phenomena of enzymic activation or inhibition. This regulatory function is the second function of the enzyme, the first being catalytic. Incidentally, it may be noted that only a few enzymes (aspartate transcarbamylase, for example) possess distinct catalytic and regulatory subunits.

I. Catalytic function of the enzymic reaction

In living organisms, a highly efficient mechanism — the *enzyme–substrate system* — allows acceleration of the chemical reactions. In the first step, enzymic action reduces

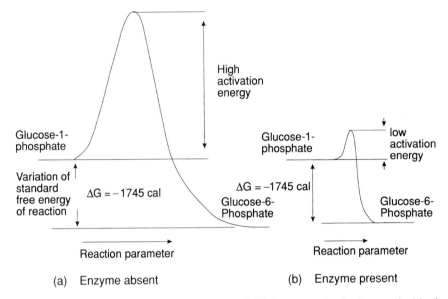

Fig. 2.1.   Energy diagram of enzymic catalysis. (a) High energy of activation required in the absence of enzyme, (b) reduced by enzymic action.

the stability of chemical bonds; this phase activates a molecule of the substrate at a very localised level in space. In the second step, an enzyme–substrate complex is formed in a very similar way (though distinct from a thermodynamical standpoint) to that of the activated complex described by the transitional state theory. In the third step, the enzyme returns to its initial state.

The hypothesis of the activated complex allows us to express the rate constant of the formation of the enzyme–substrate complex ES as a function of the equilibrium constants between ES and the reactants E and S. This is only a very rough hypothesis but its use is justified by the fact that only the rate constants of a reaction can be experimentally determined.

What are the chemical bonds involved? The free energy variation measured ΔG is about 5 to 10 kcal/mol, which shows that E and S are connected by several weak secondary bonds. One way of describing the process is by using the concept of the *active site*, the region in space where a part of the enzyme comes into contact with the substrate, exactly as a key fits a lock (Fischer's hypothesis). At ambient temperature few molecules possess the energy necessary for the formation of an activated complex. In other words, their kinetic energy is too low to overcome the barrier represented by the energy of activation. In fact, it is the catalytic function of the enzyme at the active site that considerably increases the rate of intracellular chemical reactions (Fig. 2.1). In addition to this, the properties of the enzyme allow control and regulation of these reactions, in particular through mechanisms of allosteric interaction.

The proteinic nature of enzymes is demonstrated by the denaturation of enzymes following modification of such factors as acidity and temperature. However, certain enzymes may be associated with non-proteinic groups called prosthetic groups.

The net charge of an enzyme, being due to a polyvalent ion, depends on the acidity. This is why changes in acidity affect enzyme–substrate bonds. Similarly, a rise in temperature increases the number of molecular collisions, modifies the enzyme conformation and directly influences the rate of catalysis.

1. Henri–Michaelis–Menten equations: equilibrium conditions

By applying the law of mass action to the enzyme–substrate complex *ES*, Henri (1905), followed by Michaelis and Menten (1913), found the chemical reaction:

$$(1) \quad E + S \underset{k_{-1}}{\overset{k_1}{\rightleftharpoons}} ES \underset{k_n}{\overset{k_2}{\rightleftharpoons}} E + P \tag{2.1}$$

where E is the enzyme, S the substrate and P the product of the reaction. The experimental conditions are such that, in general, the constant k_n may be neglected, i.e. in the initial conditions of the reaction the concentration $[P]$ is negligible.

The chemical mechanism of this reaction is similar to that described by Lindemann (1922) and Hinshelwood (1927) for a monomolecular reaction:

$$(2) \quad A + A \rightleftharpoons A^* + A$$

$$A^* \rightarrow B + C$$

which may be generalised by taking into account another chemical species X:

$$A + X \rightleftharpoons A^* + X$$

$$A^* \rightarrow B + C$$

and this may be rewritten as:

$$A + X \underset{k_{-1}}{\overset{k_1}{\rightleftharpoons}} A^* \rightarrow B + C.$$

Similarly, the transitional state theory gives the bimolecular reaction:

$$(3) \quad A_1 + A_2 \rightleftharpoons X^* \rightarrow B_1 + B_2. \tag{2.2}$$

Thus the complex *ES* can be compared to the activated complex of modern kinetic theory: the first step, with a rate constant k_1, is interpreted in the reactions (1), (2) and (3) as being respectively a *complexation*, an *activation* or an *association* in each of the three theories above. The second step, with a rate constant k_2, similarly corresponds to a *monomolecular dissociation*, a *deactivation* or a *dissociation*. We see that the monomolecular decomposition corresponds to deactivation of the activated complex. The results obtained may be readily extended to the case of an enzymic reaction. The general differential equations below describe the process:

$$\left. \begin{array}{rl} (1) & \dfrac{d[E]}{dt} = -k_1[E][S] + (k_{-1} + k_2)[ES] \\[2mm] (2) & \dfrac{d[S]}{dt} = -k_1[E][S] + k_{-1}[ES] \\[2mm] (3) & \dfrac{d[ES]}{dt} = k_1[E][S] - (k_{-1} + k_2)[ES] \\[2mm] (4) & \dfrac{d[P]}{dt} = k_2[ES] \end{array} \right\}. \tag{2.3}$$

By adding (1) and (3) we have the law of the conservation of the total quantity of enzyme, bound or unbound, that is:

$$\frac{d[E]}{dt} + \frac{d[ES]}{dt} = 0.$$

By adding (2), (3) and (4), we have the law of the conservation of matter on substrate S:

$$\frac{d[S]}{dt} + \frac{d[ES]}{dt} + \frac{d[P]}{dt} = 0. \tag{2.4}$$

In particular, for $t = 0$ this constant is $[S_0]$, the initial concentration of the substrate. *In this model, the following hypotheses may be admitted:*

(1) Only the initial rates of reaction are considered. Consequently, the quantity of substrate transformed into product P is negligible, so that the concentration of P remains weak ($k_n[P]$ negligible).
(2) There is an excess of substrate S with respect to enzyme E.
(3) The equilibrium of the reaction $E + S \rightleftharpoons ES$ is quickly reached and is maintained during the entire reaction; in other words k_2 is for the most part very much smaller than k_{-1} (and k_1). *This is the equilibrium state condition.*
(4) After a very short time, the rates of formation and decomposition of the complex ES become, and then remain, much smaller than the rate of variation of $[S]$ and $[P]$. *This is the stationary state condition:* k_2 is no longer negligible compared with k_{-1} and k_1.

Let us first consider the conditions for the state of equilibrium. Equation (2.3(3)), with condition (3) ($k_2 \ll k_{-1}$), implies the equilibrium state:

$$\frac{d[ES]}{dt} = 0 \tag{2.5}$$

since:

$$K_S = \frac{[ES]}{[E][S]} = \frac{k_1}{k_{-1}}$$

which is the equilibrium condition for the reaction, following equation (2.3(3)), leads to:

$$\frac{d}{dt}[ES] = 0 \tag{2.6}$$

which, taking into account the law of conservation: $[E] + [ES] = E_{total}$ gives:

$$[ES] = \frac{E_{total}[S]}{k_{-1}/k_1 + [S]}. \tag{2.7}$$

At equilibrium, this formula gives the concentration of ES in terms of total concentrations of enzyme and substrate. The variation of the concentration of product is given by:

$$v = \frac{d[P]}{dt} = k_2[ES]$$

or following (2.7):

$$v = \frac{k_2 E_{total}[S]}{k_{-1}/k_1 + [S]}$$

$$\boxed{v = \frac{v_{max}[S]}{K_m + [S]}} \tag{2.8}$$

with, using (1) and (2): $[S] \approx [S_0]$, writing $v_{max} = k_2 E_{total}$ and:

$$K_m = \frac{k_{-1}}{k_1}, \qquad k_2 \ll k_{-1}. \tag{2.9}$$

v_{max} is the *limit* or the *maximum* rate corresponding to an infinite substrate concentration, i.e. to a *saturation* of enzymic sites bound to the substrate; K_m is the Michaelis constant (the equilibrium dissociation constant). Figure 2.2 shows the $v(s)$ curve, with $s = [S]$, which directly gives the measurable parameters: v, v_{max} and K_m such that $v(K_m) = v_{max}/2$.

2. Briggs–Haldane equations: quasi-stationary state conditions

If condition (3) of the equilibrium state is replaced by the hypothesis of quasi-stationary state (4), we obtain the Briggs–Haldane formulation. Thus in the quasi-stationary state:

$$\frac{d[ES]}{dt} \approx 0.$$

A calculation identical to that above gives:

$$\frac{d[ES]}{dt} = 0 = k_1[E][S] - (k_{-1} + k_2)[ES]$$

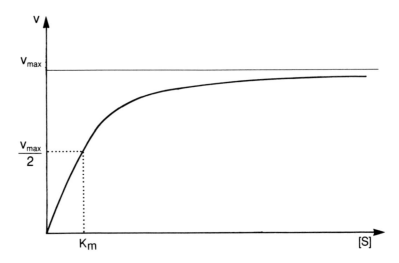

Fig. 2.2. Rate of reaction, v (Eq. 2.8), in terms of the concentration of the ligand $[S]$ in Henri–Michaelis–Menten kinetics.

$$[E] = E_{total} - [ES].$$

Thus:

$$[ES] = \frac{E_{total}[S]}{\dfrac{k_{-1}+k_2}{k_1} + [S]} \qquad (2.10)$$

finally:

$$v = \frac{d[P]}{dt} = k_2[ES] \qquad (2.11)$$

and

$$\boxed{v = \frac{v_{max}[S]}{K_M + [S]}} \qquad \text{where} \qquad \left. \begin{array}{l} v_{max} = k_2 E_{total} \\[2mm] K_M = \dfrac{k_{-1} + k_2}{k_1} \end{array} \right\} \qquad (2.12)$$

with $[S] \approx [S_0]$.

The new constant K_M is the *generalised Michaelis constant*.

In conclusion, whether we use the hypothesis of the equilibrium state or the hypothesis of the quasi-stationary state, the rate of the enzymic reaction is expressed in terms of two values: the maximum rate v_{max} and the Michaelis constants K_m or K_M. The enzyme–substrate system is characterised by these two values.

An interesting case is that of proteolytic enzymes synthesized as inactive precursors. Before producing their catalytic action, these precursors undergo an activating process which consists of a limited proteolytic step catalysed by a protease. Examples of this are the pancreatic zymogens, the phenomenon of blood-clotting that involves proteolytic activation of prothrombin and plasminogen, the activation of protyrosinase to tyrosinase, of proinsulin to insulin, and so on. A model for zymogen activation has been proposed by Varon *et al.* (1988):

$$E + E_i \underset{k_{-1}}{\overset{k_1}{\rightleftharpoons}} (EE_i) \overset{k_{-1}}{\longrightarrow} (EE_a) \overset{k_2}{\longrightarrow} E + E_a$$
$$\searrow$$
$$W$$

$$(2.13)$$

$$E_a + A \underset{k'_{-1}}{\overset{k'_1}{\rightleftharpoons}} (E_aA) \overset{k'_2}{\longrightarrow} (E_aY) \overset{k'_2}{\longrightarrow} E_a + Y$$
$$\searrow$$
$$X$$

In this scheme, there are four intermediate substrates: EE_i, EE_a, E_aA, and E_aY; X and Y are the products of the E_a reaction on A, and W is a related peptide from E_i during EE_a formation. The reader will find the solutions of the kinetic system, deduced from the general equations, in the original article by Varon *et al.* (1988) and Havsteen *et al.* (1993).

3.   The mathematical viewpoint: the pseudo-steady state hypothesis

Equations (2.1) can be rendered dimensionless by putting:

$$\tau = k_1[E](0)t, \quad u(\tau) = \frac{[S](t)}{[S](0)}, \quad v(\tau) = \frac{[ES](t)}{[E](0)}$$

$$\lambda = \frac{k_2}{k_1[S](0)}, \quad K = \frac{k_{-1} + k_2}{k_1[S](0)}, \quad \varepsilon = \frac{[E](0)}{[S](0)}.$$

$$(2.14)$$

They become:

$$\frac{du}{dt} = -u + (u + K - \lambda)v$$

$$\varepsilon \frac{dv}{dt} = u - (u + K)v$$

$$(2.15)$$

with initial conditions:

$$u(0) = 1, \quad v(0) = 0.$$

$$(2.16)$$

From these equations, it is clear that the final steady-state is $u = 0$, $v = 0$. In the Henri–Michaelis–Menten model, the reaction kinetics are such that $\varepsilon \ll 1$, i.e. a very small concentration of enzyme is needed for a large concentration of substrate. The constant $0 < \varepsilon \ll 1$ specifies a *singular perturbation problem* because it multiplies

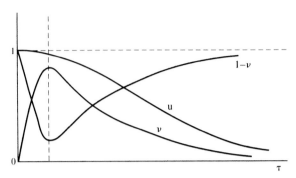

Fig. 2.3.   Solutions of Eq. (2.15) where the substrate is u, the substrate–enzyme complex is v and the free enzyme is $1-v$. Singular solutions appear on the left (after J. D. Murray, *Mathematical Biology*, Springer-Verlag, Berlin (1989)).

a derivative in Eq. (2.15). In this case, there are very powerful methods available for determining asymptotic solutions (Murray, 1989). The principal reason for this kind of problem is that if $\varepsilon = 0$, the order of the system is reduced and the reduced system cannot, in general, satisfy all the initial conditions. It is therefore important to know when the ε-terms can be neglected in practical situations. Murray gives a detailed description of the method. The result is that in many experimental situations, the singular solution for $u(\tau)$ and $v(\tau)$ is never observed: the solution is obtained from Eqs (2.15) by setting $\varepsilon = 0$ and satisfying only the initial condition on the concentration $u(\tau)$ of substrate. Mathematically, $\varepsilon \, dv/d\tau \approx 0$, and we may say that the reaction for the complex $[ES](\tau) = v(\tau)$ is in the pseudo-steady state because the reaction is so fast that it is more or less in equilibrium at all times. This *pseudo-steady state hypothesis* reflects the difference of time scales between kinetic reactions. It will often be used in the course of our work.

The *rate of reaction*, also called the rate of uptake, is $du/d\tau$, which is given by the non-singular or outer equation (Murray, 1989) (Fig. 2.3). It is shown to be equal to $\lambda/(1 + K)$, which corresponds, in dimensional terms, to Eq. (2.3).

We have dwelt upon the subject of enzyme kinetics because of its importance in biology. Under normal conditions, cell function in living organisms depends on a great variety of enzymes. It is therefore useful to recall the theoretical concepts which are the bases of a formulation now unanimously adopted.

II.   Regulatory function of the enzymic reaction

What happens when a molecule has a structure sufficiently similar to the substrate but different enough not to produce a catalytic action? In such a case, the molecule–enzyme complex formed may produce different kinds of inhibition:

— *Simple intersecting linear competitive inhibition*

In this case, the classical Henri–Michaelis–Menten reaction is supplemented by:

$$E + S \rightleftharpoons ES \rightarrow E + P$$

$$+$$

$$I$$

$$k'_{-1} \updownarrow k'_1$$

$$EI$$

$$K_I = \frac{[E][I]}{[EI]} = \frac{k'_{-1}}{k'_1} \qquad (2.17)$$

whence the equations:

$$E_{tot} = [E] + [ES] + [EI] \quad \text{(according to the law of conservation)}$$

$$\frac{d}{dt}[EI] = k'_1[E][I] - k'_{-1}[EI].$$

With the steady state condition for the Henri–Michaelis reaction and the equilibrium condition for EI ($K_I = k'_{-1}/k'_1$), and by using the same technique as above, we find:

$$v = \frac{v_{max}[S]}{K_M\left(1 + \dfrac{[I]}{K_I}\right) + [S]} \qquad \text{where} \qquad \left.\begin{array}{l} v_{max} = k_2 E_{tot} \\[4pt] K_I = \dfrac{k'_{-1}}{k'_1} \\[6pt] K_M = \dfrac{k_{-1} + k_2}{k_1} \end{array}\right\} . \qquad (2.18)$$

The inhibition of the oxidation of succinic acid by malonic acid is an example of competitive inhibition. This system, with its three components (E, S and I), behaves as if the Michaelis constant K_M had been modified to:

$$K'_M = K_M\left(1 + \frac{[I]}{K_I}\right) \qquad (2.19)$$

in other words, the presence of I leads to an increase of K_M without altering the rate of reaction v_{max} (Fig. 2.4).

Although it has been generally held that there is an analogy between substrate and inhibitor, competitive inhibition can exist without any obvious structural relationship between substrate and inhibitor (Segel, 1975). Similarly, in the cases of noncompetitive and uncompetitive inhibition which we shall now consider, there is often no structural relationship between substrate and inhibitor which bind on different sites.

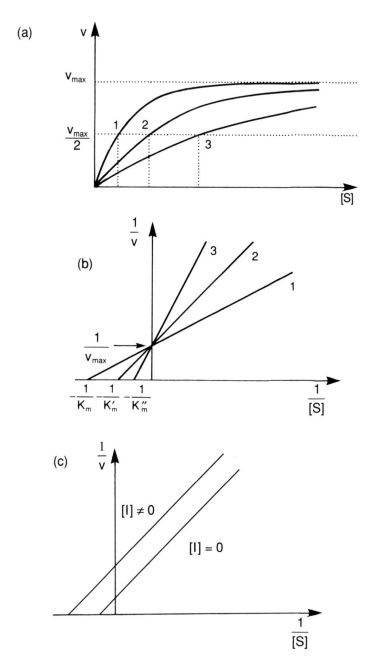

Fig. 2.4.   (a) Rate of reaction v as function of ligand concentration $[S]$ in the case of competitive inhibition. (b) The same, using coordinates $1/v$ and $1/[S]$. The curves 1, 2 and 3 correspond to increasing quantities of inhibitor. (c) The case of a strictly uncompetitive inhibitor.

— 'Classical' non-competitive inhibition (simple intersecting linear non-competitive inhibition)

Here the inhibitor has no effect on the binding of the substrate and vice versa; *ESI* is inactive and we then have:

$$
\begin{aligned}
E + S &\underset{k_{-1}}{\overset{k_1}{\rightleftharpoons}} ES \xrightarrow{k_2} E + P \\
&+ \qquad\qquad + \\
&I \qquad\qquad\ I \\
&k'_{-1}\Big\Updownarrow k'_1 \quad k'_{-1}\Big\Updownarrow k'_1 \\
EI + S &\underset{k_{-1}}{\overset{k_1}{\rightleftharpoons}} ES\,I
\end{aligned}
\qquad\qquad (2.20)
$$

where we may write:

$$
\begin{aligned}
K_I &= \frac{k'_{-1}}{k'_1} \\[2mm]
K_m &= \frac{k_{-1}}{k_1} \\[2mm]
K_M &= \frac{k_{-1} + k_2}{k_1}.
\end{aligned}
\qquad\qquad (2.21)
$$

The constant K_M, defined in the steady state conditions, has here been replaced by the constant K_m. However, the passage to the steady state conditions is not quite so simple. Indeed, if we suppose that k_2 is not negligible compared with k_{-1}, we have:

$$
\frac{d[ES]}{dt} = k_1[S][E] + k'_{-1}[ESI] - k'_1[ES][I] - (k_{-1} + k_2)[ES] = 0
$$

with:

$$
[E] = E_{tot} - [ES] - [EI] - [ESI].
$$

Similarly

$$
\frac{d[ESI]}{dt} = k'_1[I][ES] + k_1[S][EI] - (k_{-1} + k'_{-1})[ESI] = 0
$$

$$
\frac{d[EI]}{dt} = k'_1[I][E] - (k'_{-1} + k_1[S])[EI] + k_{-1}[ESI] = 0.
$$

This system of three equations with three unknowns can be solved by the classical methods of linear algebra. In particular, [ES] can be explicitly written as:

$$
[ES] = E_{tot} \cdot \frac{k'_{-1}k_1[S]\{k_1[S] + k'_1[I] + k'_{-1} + k_{-1}\}}{(k'_1[I] + k'_{-1})\{(k_1[S] + k_{-1})^2 + (k_1[S] + k_{-1})(k'_1[I] + k'_{-1} + k_2) + k'_{-1}k_2\}}
$$

with $v = k_2 [ES]$. We thus obtain non-Michaelis kinetics with terms in $[S]^2$. It is only when $k_2 = 0$ that we have the classical equation of non-competitive inhibition:

$$[ES] = E_{tot} \cdot \frac{k'_{-1} k_1 [S]}{(k'_1 [I] + k'_{-1})(k_1 [S] + k_{-1})}$$

whence

$$[ES] = \frac{E_{tot}}{\left(1 + \dfrac{[I]}{K_I}\right)} \cdot \frac{[S]}{(K_m + [S])} \tag{2.22}$$

which leads to the classical equation:

$$v = \frac{\dfrac{v_{max}}{\left(1 + \dfrac{[I]}{K_I}\right)} \cdot [S]}{K_m + [S]} \cdot \tag{2.23}$$

Thus, under the hypothesis of the steady state, 'classical' non-competitive inhibition involves non-Michaelis kinetics except when $k_2 = 0$, in which case we have quasi-equilibrium conditions. Here, K_m remains unaffected by the presence of the inhibitor but v_{max} is reduced and the system behaves as if there had been a decrease in k_2. We observe that this behaviour is quite opposite to that of the competitive inhibitor where v_{max} was not modified whereas K_M changed to K'_m.

— *Uncompetitive inhibition (simple linear uncompetitive inhibition)*

In this case, the inhibitor binds exclusively to the *ES*:

$$E + S \rightleftharpoons ES \rightarrow E + P$$
$$+$$
$$I$$
$$k''_{-1} \updownarrow k''_1 \tag{2.24}$$
$$ESI$$

We again find:

$$v = \frac{v_{max} [S]}{K_M + [S] \left(1 + \dfrac{[I]}{K_{IC}}\right)} \qquad \text{where} \qquad \left.\begin{array}{l} K_M = \dfrac{k_{-1} + k_2}{k_1} \\[2ex] K_{IC} = \dfrac{k''_{-1}}{k''_1} \end{array}\right\} \tag{2.25}$$

— *Linear mixed-type inhibition*

Here, the inhibitor binds not only to the free enzyme E but *also to the complex ES* with two different dissociation constants K_{IE} and K_{IC}. The reaction scheme may then be represented as follows:

$$E + S \rightleftharpoons ES \rightarrow E + P$$

$$K_m$$

with: $K_{IC} = \alpha K_{IE}$, the case $\alpha = 1$ corresponding to Eq. (2.23), obtained using the same hypotheses as above:

$$
v = \frac{v_{max}[S]}{K_m\left(1 + \dfrac{[I]}{K_{IE}}\right) + [S]\left(1 + \dfrac{[I]}{K_{IC}}\right)}
\qquad \text{where} \qquad
\left.
\begin{aligned}
K_{IE} &= \frac{k'_{-1}}{k'_1}\\[4pt]
K_{IC} &= \frac{k''_{-1}}{k''_1}
\end{aligned}
\right\}
\quad (2.26)
$$

We again find the results above according to the limit values of K_{IC} and K_{IE}:

$$
\left.
\begin{aligned}
&K_{IC} \rightarrow +\infty \text{ competitive inhibition Eq. (2.18)}\\[4pt]
&K_{IC} = K_{IE} \text{ 'classical' non-competitive inhibition } (\alpha = 1) \text{ Eq. (2.23)}\\[4pt]
&K_{IE} \rightarrow +\infty \text{ strictly uncompetitive inhibition Eq. (2.25)}
\end{aligned}
\right\}
\quad (2.27)
$$

The phenomena of enzyme activation and inhibition may intervene in the regulation of the enzyme itself. This is the first example of a two-factor system in biology (see below). In particular, a system of this kind has been discovered in nucleotide synthesis.

Criteria for the validity of the steady state hypothesis applied to an enzyme–substrate–inhibitor system have been determined by recent investigations: Segel (1988) has found a simple new condition by estimating relevant time scales; Segel and Martin (1988) have discussed conditions for the reduction of the steady state equation with a general modifier for a unireactant enzyme mechanism; Sen (1988) has developed mathematical methods giving approximate solutions for the time course of the reversible Michaelis–Menten reaction; Frenzen and Maini (1988) and Topham (1988) have examined the validity of the quasi-steady state hypothesis.

Topham (1988), for instance, represents irreversible enzyme inactivation due to an unstable inhibitor by the following set of reactions:

$$E + I \xrightarrow{k_i} E^- \tag{2.28(1)}$$

$$E + I \overset{k_1}{\rightleftharpoons} EI \xrightarrow{k_2} E^- \tag{2.28(2)}$$

$$I \xrightarrow{k'} R. \tag{2.28(3)}$$

In these reactions, two basic mechanisms of enzyme inactivation are assumed depending on the type of reaction between the enzyme (E) and the unstable inhibitor (I): (i) a simple bimolecular reaction to form E^- (Eq. 2.28(1)), an inactive species; (ii) the formation of an enzyme through an enzyme inhibitor complex (EI) (Eq. 2.28(2)) where the inhibitor can be decomposed into a product R (Eq. 2.28(3)). Recent research on irreversible enzyme activity kinetics (Wang, 1990a; Wang and Tsou, 1990; Wang, 1991) has determined the range of validity for the kinetic equations of substrate reaction in the presence of the modifier.

The fractal approach to enzyme kinetics

Equations (2.8) or (2.12) are useful in analysing kinetic data by using the least-squares fitting of the experimental results to the rate equation. However, the Michaelis–Menten expression of the kinetics is an approximation of the complex mechanisms that occur in the regulation of enzyme catalysis. As mentioned above, other equations have been proposed, which could be generalized as the ratio of two polynomials:

$$V = \frac{\alpha_1[S] + \alpha_2[S]^2 + \ldots + \alpha_n[S]^n}{\beta_0 + \beta_1[S] + \beta_2[S]^2 + \ldots + \beta_m[S]^m}$$

where α_i and β_j are the appropriate parameters. Of course, the difficulty that arises concerns the physical interpretation that may be given to these parameters. To avoid the risk of possible subjectivism, a new model, based on the fractal concept, has been proposed by Lopez-Quintela and Casado (1989). An object is said to be fractal (Mandelbrot, 1983) when one of its properties, say P, obeys the auto-similarity law:

$$P(\varepsilon) = \varepsilon^{1-D}$$

where ε represents a certain measurement scale of the property P, and D is the fractal dimension. The latter may be a fractional number such that:

$$d_t < D < d$$

where d_t is the topological dimension (see Appendix D, Volume III), and d is the Euclidean dimension of the medium in which the object is included. A classical example is given by three-dimensional Brownian motion which depends on the scale of observation and satisfies the expression: 1 $(d_t) < 2$ $(D) < 3$ (d).

Since complex chemical reactions may involve several intermediate substrates, the idea is to describe them using a schema which is formally equivalent to that of Michaelis–Menten:

$$E + S \underset{k_{-1}^{eff}}{\overset{k_1^{eff}}{\rightleftharpoons}} (ES)^{eff} \xrightarrow{k_2^{eff}} \text{Products} \tag{2.29}$$

Table 2.1. Comparison between polynomial and fractal models for various enzyme reactions (López-Quintela and Casada, 1989).

Enzyme	Substrate	Polynomial method (degree)	Fractal method (dimension)
Human thrombin	Tos–Gly–Pro–Arg–pNa	1:1	1.00
Human thrombin	CBZ–Gly–Pro–Arg–pNa	2:2	0.73
Human thrombin	H–D–Phe–Pip–Arg–pNa	2:2	0.91
Alkaline phosphatase of *E. coli*	*p*-nitrophenyl phosphate	2:2	0.87
NADPH cytochrome-c (P-450) reductase	cytochrome-c	1:1	1.00
NADPH cytochrome-c (P-450) reductase	cytochrome-c	1:1	0.98

Depending on the observation scale, the potential energy curve, for example, may not have the same profile in terms of the reaction coordinate. In this case, the property P is represented by the rate k^{eff} and may show fractal behaviour:

$$k^{eff} = \varepsilon^{1-D}. \tag{2.30}$$

Of course, this approach can be used only if the assumed complex mechanisms are such that the resolving power (according to the observation scale) increases as the working concentration decreases, i.e. for a step i of the sequence of reactions, we have:

$$k_i^{eff} = A_i[S]^{1-D}. \tag{2.31}$$

The concentration has been chosen as the observation scale because it is generally the physical aspect observable in chemical kinetics. Equation (2.31), which is the basis of the method proposed by Lopez-Quintela and Casado (1989), leads to a new formulation of the reaction rate (2.12):

$$v = \frac{v_{max}^{eff}[S]^{2-D}}{K_M^{eff} + [S]}.$$

In this expression, the coefficients v_{max}^{eff} and K_M^{eff}, which describe the global mechanisms, are defined in terms of the individual rate constants in Eq. (2.29). When $D = 1$, the Michaelian situation is re-established, and the effective constants of the chemical reaction (2.29) are identified with the usual constants of the Michaelis–Menten equation. Thus, the fractal dimension D deviates from unity as the mechanisms become more complex. This corresponds to a greater irregularity in the free energy curve, i.e. it is far from being a Michaelian curve. Table 2.1 shows a comparison between the polynomial and fractal models for different enzyme reactions. This method of identifying parameters for a given reaction is useful when we want to replace a sequence of unknown reactions by a representative one. Of course, it complements other classical methods by calculating the new coefficients v_{max}^{eff} and K_M^{eff}.

III.   Molecular interactions and the active site concept

As we have just seen, the physical phenomenon of the destabilisation of chemical bonds, and the consequent activation of the enzyme–substrate complex, occur at a specific site. The concept of the *active site* is of course very suggestive but immediately raises the question of the mechanism leading to the encounter between enzyme and substrate. The simplest idea is that of a collision, as conceived in the statistical mechanics of gas kinetics. But here the difficulty is obvious: according to probability theory, collision between two molecules is considered to be a rare phenomenon, so what would be the chances of a random encounter involving not only enzyme and substrate but a host of other molecules such as inhibitors, co-factors and other cooperative sites?

Let an enzyme molecule in its quaternary structure be ideally represented by a sphere of radius r_1, and the active site on the sphere by a small surface σ_1. Then the probability of a collision with a substrate molecule, similarly defined by (r_2, σ_2), is given by: $Pr = Pr_1 \times Pr_2$, where Pr_1 is the probability of an encounter between the two spheres, and Pr_2, under the same conditions as above, is the probability of an encounter between the two surfaces.

The calculation of Pr is likely to be complicated because of the several factors that have to be considered: the distribution of molecule orientations, the distribution of shape and position of the active sites, and so on. Finally, the rate constant k of the chemical reaction, which depends only on the reactivity of the atomic groups, will have to be multiplied by the probability Pr to obtain the corrected constant:

$$K_C = k \times Pr. \tag{2.32}$$

This simple expression clearly shows the first consequence of the concept of the active site: a full knowledge of protein and substrate surfaces is necessary for a description of the mechanism of adjustment between the molecules. The probability Pr concerns the 'static' adjustment between enzyme and substrate.

The second consequence relates to the fundamental role played by *induced adjustment*, the theory of which was initially proposed by Koshland. This is inseparable from the concept of the allosteric effect discovered by Monod. In fact, not only is there a definite relationship between the molecular structure of the surface of the protein and that of the ligand, but there also exists a dynamic reciprocity between the conformational changes of the molecules involved.

But can the surface of a protein be completely known? Some authors have attempted to solve the problem, which may be reduced to the determination of the stable conformational states of a tertiary proteinic structure (Chapter 1, Section II). Calculations run on today's powerful computers may be expected to 'map' protein surfaces, but several questions arise. Which exactly are the atomic groups that make up the protein 'surface'? How are these structures modified during a conformational change? We know that the polar and non-polar groups on the protein surface strongly influence molecular interactions. We also know that associations between protein subunits depend closely on the nature and the geometry of the surfaces in contact.

Thus, the problem of *static* structural adjustment during molecular interaction is

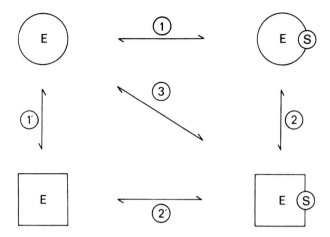

Fig. 2.5. Chemical reactions involved in induced adjustment: 1' shows the change in conformation (after Yon, 1969).

compounded by that of *induced* adjustment. The dynamic theory of enzyme–substrate association may be summed up, taking into account the allosteric molecular interactions discussed in Chapter 1, Section III, by the following postulates:

(i) association between enzyme and substrate leads to modification of the geometrical structure of the enzyme;

(ii) enzymic activity occurs only if the atomic groups on the enzyme are perfectly oriented with respect to the corresponding groups on the substrate; and

(iii) the orientation of the substrate undergoes a change induced by the modified geometry of the enzyme. Thus, there is a *dynamic reciprocity.*

Postulates (i) and (iii) are the bases of the theory of molecular interactions and allosteric effects. Postulate (ii) is formally expressed in Eq. (2.3). It is encouraging to find that X-ray diffraction analysis has successfully demonstrated conformational changes localised in the neighbourhood of the active sites.

The mechanism of induced adjustment at the active site of the enzyme is no more than the three-dimensional image of an elementary system of chemical reactions (Fig. 2.5). An elementary system of this type would be one of the items of the set of reactions leading to Eq. (1.9).

Let us close this discussion of the fundamental notions of enzymic catalysis and regulation with two examples that introduce new chapters in biology through the basic concepts developed above. The first example concerns the phenomenon of antigen–antibody interaction. It would of course, have been just as interesting to discuss hormonal interactions which use similar processes, but we shall examine these in Chapter 3, Volume III. The example chosen here will show how the corrected equilibrium constant of the reaction may be calculated. The result is of fundamental importance in current theoretical work on immunology (see for example De Lisi, 1976).

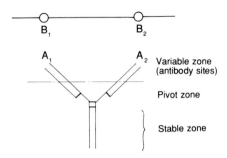

Schematic representation of immunoglobin

Fig. 2.6. Highly simplified antibody structure with two sites A_1 and A_2 interacting with the epitopes B_1 and B_2 on an antigen.

The second example illustrates the notion of an enzymic system with independent fixation sites, i.e. with the allosteric effects suppressed. The interesting point here is that it shows the effect of coupling of enzymic reactions, which is indeed a permanent feature of biological systems. Moreover, this example will allow us to introduce the mathematical formalism of the reaction–diffusion equations used in the description of chemical reactions.

Example 1: Antigen–antibody interaction

The immunological system of an organism is based on two mechanisms: first, the *recognition*, by specialised cells, of the molecular configurations of a foreign substance, the *antigen*, and, secondly, the *induced response* that appears during the highly discriminating process of recognition which leads to the formation of *antibodies*, molecules which belong to the class of immunoglobulins. Before going into the details of immunological defence, let us schematically consider antibody structure to be a two-part structure: one part variable V, and the other constant C (Fig. 2.6).

The two antibody sites, A_1 and A_2, fix on to the corresponding antigen sites, B_1 and B_2 (*epitopes*). We may then write the following chemical reactions:

$$B_1B_2 + A_1A_2 \underset{k_-^1}{\overset{k_+^1}{\rightleftharpoons}} A_1B_1 \qquad (1)$$

$$k_-^2 \big\Vert k_+^2 \qquad\qquad (2.33)$$

$$A_1B_1 + A_2B_2 \qquad (2)$$

The first reaction is bimolecular as it creates a bond between the two molecules (A) and (B) at sites A_1 and B_2. The second is intramolecular, being in fact another bond between the sites A_2 and B_2 of the *same* molecule.

Here we find the conditions for the recognition of active sites:

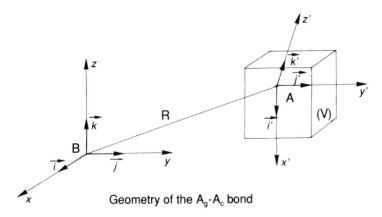

Geometry of the A_g-A_c bond

Fig. 2.7. Antigen–antibody reaction between two molecules A and B in volume V, centred respectively in frames (i', j', k') and (i, j, k) (after De Lisi, 1976).

- collision between (A) and (B), followed by
- recognition between A_1 and B_1, and then by
- recognition between A_2 and B_2.

We will now seek the geometrical conditions for the formation of a bond between (A) and (B). Let A be the centre of mass of (A), B that of (B) and R the distance AB, as in Fig. 2.7.

If we suppose that A is located in a volume V at the extremity of R, (x, y, z) will be the coordinates of A in the reference frame $\mathcal{R}_B(\mathbf{i}, \mathbf{j}, \mathbf{k})$. The system (A) being in a certain configuration, its orientation must be specified with respect to R_B: the system (A) in the frame $\mathcal{R}_A(\mathbf{i'}, \mathbf{j'}, \mathbf{k'})$ will have to be oriented according to the solid angle $\Delta\Omega$ and $\Delta\alpha$ about $\mathbf{k'}$.

The distribution function of the probability Pr is thus, per unit volume, the product of two probabilities:

- the probability P_R of finding the system (A) in the volume ΔV, and
- the probability P_Ω of finding (A) oriented according to θ_1, θ_2, θ_3 in the domain $\Delta\Omega$, $\Delta\alpha$.

Hence:

$$\text{Pr} = P_R(R)\Delta V \cdot P_\Omega(\theta_1, \theta_2, \theta_3)\Delta\Omega\Delta\alpha.$$

Following Eq. (2.32):

$$K_C = P_R(R) \cdot P_\Omega(\theta_1, \theta_2, \theta_3)k\Delta V\Delta\Omega\Delta\alpha.$$

The probability Pr is of the same form for the two types of reaction (2.33) (1) and (2) for which the reaction constants are:

$$K_{C_1} = \frac{k_+^1}{k_-^1} \qquad K_{C_2} = \frac{k_+^2}{k_-^2}$$

and

$$\frac{K_{C_1}}{K_{C_2}} = \frac{P_R^1 P_\Omega^1}{P_R^2 P_\Omega^2} \cdot \frac{k_1}{k_2}$$

with $k_1 = k_2$, since the epitopes and combination sites of the antibody may be considered chemically identical.

Here, the functions P_R and P_Ω are easily determined if they are supposed constant in all space. Thus:

$$\int_V P_R \, dV = n \Rightarrow P_R = \frac{n}{V}$$

as the extremity of R may be at any of the points volume V which contains n antibodies, and:

$$\int_{\text{half-space}} P_\Omega \, d\omega \, d\alpha = \frac{1}{2} \int_0^{4\pi} d\omega \int_0^{2\pi} P_\Omega \, d\alpha = 1; \, P_\Omega = \frac{1}{4\pi^2}$$

for molecule (A) can only be oriented in the complementary half-space occupied by system (B). Finally, the reaction constant for n molecules (A) in the volume V may be written:

$$K_{C_1} = \frac{n}{4\pi^2 V} k_1 \Delta V \Delta \Omega \Delta \alpha$$

so that the ratio $\dfrac{K_{C_1}}{K_{C_2}}$ is given by:

$$\boxed{\frac{K_{C_1}}{K_{C_2}} = \frac{n}{4\pi^2 V P_R^2 P_\Omega^2}} \qquad (2.34)$$

This is the fundamental equation of the theory of antigen–antibody recognition (De Lisi, 1976).

Example 2: An enzymic system with n *independent sites*

The idealised case of an enzyme with n independent active sites has been analysed by Botts and Morales (1953). They generalise the Henri–Michaelis–Menten equation in the form of a system of coupled enzymic reactions, with $i = 1, 2, \ldots, n$:

$$E + S \underset{k_{-}^{(1)}}{\overset{k_1^{(1)}}{\rightleftharpoons}} ES_1 \overset{k_2^{(1)}}{\longrightarrow} E + P$$

$$\vdots$$

$$ES_{i-1} + S \underset{k_2^{(i)}}{\overset{k_1^{(i)}}{\rightleftharpoons}} ES_i \overset{k_2^{(i)}}{\longrightarrow} ES_{i-1} + P$$

The notation here is the same as in Section I of this chapter: the complex ES_i has i sites occupied by i molecules of the substrate. Using the hypothesis (Eq. (2.10)) of the quasi-stationary state and Eq. (2.3(3)) we obtain:

$$\frac{d[ES_i]}{dt} = k_1^{(i)}[ES_{i-1}][S] - (k_-^{(i)} + k_2^{(i)})[ES_i] = 0$$

and by writing:

$$[ES_i] = \{k_1^{(i)}/(k_-^{(i)} + k_2^{(i)})\}[ES_{i-1}][S] = k^{(i)}[ES_{i-1}][S]$$

we get:

$$k^{(i)} = k_1^{(i)}/(k_-^{(i)} + k_2^{(i)})$$

From this we deduce the rate v of the production of P:

$$v = \sum_{i=1}^{n} v_i \qquad \text{where} \qquad v_i = k_2^{(i)}[ES_i]$$

thus:

$$v = \sum_{i=1}^{n} k_2^{(i)} k^{(i)} [ES_{i-1}][S]$$

By recurrence we find:

$$v = [E] \sum_{i=1}^{n} \left\{ k_2^{(i)}[S]^i \prod_{l=1}^{i} k^{(l)} \right\}$$

Finally:

$$[E] = E_{tot} - \sum_{i=1}^{n} [ES_i] = E_{tot} - \sum_{i=1}^{n} \left\{ [S]^i [E] \prod_{l=1}^{i} k^{(l)} \right\}$$

$$= E_{tot} \left[1 + \sum_{i=1}^{n} \left(\prod_{l=1}^{i} k^{(l)} \right) [S]^i \right]^{-1}$$

Then by writing:

$$O_i = \prod_{l=1}^{i} k^{(l)} [S]^i$$

we have:

$$v = \frac{E_{tot}}{1 + \sum_{1}^{n} O_i} \sum_{i=1}^{n} k_2^{(i)} O_i .$$

(2.35)

In this calculation we have supposed that, at a given instant t, only one molecule of the substrate may bind to, or dissociate from, an active enzymic site.

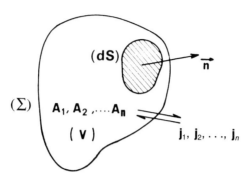

Fig. 2.8. Volume V with r reactions taking place and n fluxes of matter across its surface.

However, for enzymic processes of this kind it should be observed that a probabilistic description may be more appropriate. The main results obtained will be found elsewhere (Darvey and Staff, 1967; Bartholomay, 1972).

This example naturally leads us to a study of coupled chemical reactions, the results of which are of prime importance in biology. From the discussion above, trimolecular reactions would appear to be unlikely, and reactions involving four or more molecules even more so. It may therefore be supposed that complex enzymic reactions, such as those envisaged in biology, are the result of a set of coupled bimolecular chemical reactions.

IV.   Coupled chemical reactions: reaction–diffusion equations

1.   *General equation for a chemical transformation*

All the results presented hitherto are based on the general fact that a transformation of a chemical nature is merely a phenomenon of mass transport, i.e. the spatiotemporal variation of molecular concentration by the displacement of molecules. In particular, in the state of thermodynamic equilibrium or in the stationary state of rates of transformation, there is continuous renewal and destruction of molecular species through random collisions, which are most frequently bimolecular.

If we examine such transformations more closely, we find that the *local* creation of substance, and thus the local increase in concentration, i.e. in the heterogeneity of the medium, is necessarily accompanied by mass transport towards the zone of lower concentration. This can be mathematically demonstrated.

Let us consider a volume V containing the chemical species $A_1, A_2, \ldots, A_n$ reacting according to r chemical reactions (Fig. 2.8). This open system allows an exchange of matter with the outside across the surface (Σ). The mass flux corresponding to the chemical species is written $\mathbf{j}_1, \mathbf{j}_2, \ldots, \mathbf{j}_n$ and the concentrations inside V are $[A_1], [A_2], \ldots, [A_n]$.

The quantity of a chemical species A_i, for example, may vary during unit time according to two processes:

(1) Transport of mass exchanged with the outside: $(dm_i/dt)_{ext}$
(2) Creation of substance by chemical reaction, inside volume V: $(dm_i/dt)_{int}$.

For species i, the total variation is evidently:

$$\frac{dm_i}{dt} = \left(\frac{dm_i}{dt}\right)_{ext} + \left(\frac{dm_i}{dt}\right)_{int}$$

which is the sum of two terms:

- Contribution (2), per unit volume, is given by:

$$\left(\frac{d[A_i]}{dt}\right)_{int} = \sum_{j=1}^{r} \nu_{ji}\upsilon_j$$

where υ_j is the rate of reaction j per unit volume (Eq. (1.1)). For example, for the reaction:

$$\upsilon_{1j}A_1 + \upsilon_{2j}A_2 + \ldots + \upsilon_{pj}A_p \rightleftharpoons \upsilon_{p+1,j}A_{p+1} + \ldots + \upsilon_{qj}A_q$$

we would have:

$$\upsilon_j = k_j^+ \prod_{i=1}^{p} [A_i]\nu_{ij} - k_j^- \prod_{i=p+1}^{q} [A_i]\nu_{ij}$$

where $q \neq r$ is the number of chemical species involved in reaction j.
- Contribution (1) is calculated as follows:

$$\left(\frac{dm_i}{dt}\right)_{ext} = -\int_{(\Sigma)} \mathbf{j}_i \cdot d\mathbf{S}$$

where $d\mathbf{S}$ is the vector with a modulus equal to the surface element and normal with respect to the surface. Then, using Green's theorem:

$$\left(\frac{dm_i}{dt}\right)_{ext} = -\int_V \operatorname{div} \boldsymbol{j}_i \, dV$$

and per unit volume:

$$\left(\frac{d[A_i]}{dt}\right)_{ext} = -\operatorname{div} \boldsymbol{j}_i$$

Finally we deduce the *local equation of mass balance*:

$$\frac{\partial[A_i]}{\partial t} = \sum_{j=1}^{r} \nu_{ij}\upsilon_j - \operatorname{div} \boldsymbol{j}_i, \qquad i = 1 \text{ to } n \tag{2.36}$$

of which the interpretation is clear: in this *equation of conservation of mass*, the first term represents the productive mass reaction and the second term the displacement

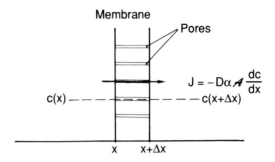

Fig. 2.9. Passage of molecules across a porous membrane from a compartment with concentration $c(x)$ to a compartment with concentration $c(x + \Delta)x$.

of mass between points of different concentration. In biological systems, convection phenomena are hardly ever observed. The commonest form of displacement is thus a diffusion flux. Equation (2.36) is therefore called the *reaction–diffusion equation*.

2. *The physics of diffusion: Fick's equation*

The local variation of substance may be physically represented by two compartments, (1) and (2), separated by a permeable membrane of thickness Δx with a pore density α. If A is the area of the membrane, the total pore surface is αA. Let $c(x)$ be the concentration at x of a chemical species in compartment (1), and $c(x + \Delta x)$ the concentration in compartment (2) (Fig. 2.9).

Experimental observation shows that the flow across the membrane is proportional to $c(x)$ and αA and inversely proportional to the membrane thickness Δx. If D is the constant of proportionality, the flow at x may be written:

$$Q(x) = D \frac{\alpha A c(x)}{\Delta x}$$

and similarly at $x + \Delta x$:

$$Q(x + \Delta x) = D \frac{\alpha A c(x + \Delta x)}{\Delta x}.$$

The net flow between x and $x + \Delta x$ is thus:

$$\Delta Q = Q(x) - Q(x + \Delta x) = -D \alpha A \frac{\Delta c}{\Delta x}$$

and for $\alpha = 1$ we have the equation describing the phenomenon of free diffusion:

$$\Delta Q = -D A \frac{\partial c}{\partial x}$$

The molecular flow in the direction $x \rightarrow x + \Delta x$ is such that the gradient $\Delta c = c(x + \Delta x) - c(x)$ is of a sign opposite to that of Δx. In the case shown in Fig. 2.9, where $\Delta x > 0$, the flow ΔQ occurs from x towards $x + \Delta x$ if $c(x + \Delta x) < c(x)$. In differential form we have:

$$\Delta Q = -D\alpha A \frac{dc}{dx} \qquad (2.37)$$

The dimensions of this equation are: moles·[time^{-1}] = $[D] \cdot L^2 \cdot \text{moles} \cdot L^{-3}/L$, or: [time^{-1}] = $[D] \cdot L^{-2}$. The diffusion 'constant' will therefore be expressed in terms of surface per unit time.

This equation, known as *Fick's first law of diffusion*, applies in particular to mass transport across cell membranes and is thus of capital importance in physiology.

Fick's second law expresses the conservation of matter in a given volume. In the case of a permeable membrane, it reflects the fact that the quantity of matter undergoing variation during time Δt is equal to that undergoing variation in space Δx. More precisely, according to Fick's first law, with dc developed to the second order:

$$\Delta Q(x + \Delta x) = -D\alpha A \left(\frac{\partial c}{\partial x} + \frac{\partial^2 c}{\partial x^2} dx \right) = \Delta Q(x) - D\alpha A \frac{\partial^2 c}{\partial x^2} dx$$

thus:

$$\Delta Q_{x \rightarrow x + \Delta x} = D\alpha A \frac{\partial^2 c}{\partial x^2} dx$$

is the spatial variation of the quantity of matter. The temporal variation is expressed by the derivative of the quantity of matter:

$$\frac{\partial}{\partial t}(\alpha A c \, dx) = \alpha A \frac{\partial c}{\partial t} dx$$

from which, by equating these two expressions, we obtain the equation of conservation, known as *Fick's diffusion equation*:

$$\frac{\partial c}{\partial t} = D \frac{\partial^2 c}{\partial x^2} \qquad (2.38)$$

Thus we retrieve the formula in Eq. (2.36), since **j** expresses a flux, i.e. the number of molecules per unit surface per unit time. In the one-dimensional space x, we have:

$$\Delta Q = -\alpha A D \frac{\partial c}{\partial x} = \alpha A j$$

so that, in general:

$$\frac{\partial c}{\partial t} = -\operatorname{div} \mathbf{j}.$$

It can be shown that the diffusion constant D may be written:

$$D = \omega RT$$

where ω is the mobility of the particle in the environment, R the perfect gas constant and T the temperature.

3. The mathematics of a reactional–diffusional system

Fick's law (2.38) combined with the equation of mass balance (2.34) gives the general form of the reaction–diffusion equation for a chemical species i contributing to r reactions j:

$$\frac{\partial}{\partial t}[A_i] = \sum_{j=1}^{r} v_{ij} v_j + D_i \nabla^2 [A_i] \quad i = 1, 2, \ldots, n$$

where ∇ is the gradient operator and $\nabla \cdot \nabla = \nabla^2 \equiv \Delta$ is the Laplace operator.

The reaction term being in general a *non-linear* function of the $[A_i]$ values in the mathematical sense (see, for example, Eq. (2.8) for the Henri–Michaelis–Menten kinetic rate), we may write:

$$\frac{\partial}{\partial t}[A_i] = f_i([A_1], \ldots, [A_n]) + D_i \nabla^2 [A_i] \quad i = 1, 2, \ldots, n \quad . \tag{2.39}$$

The variation of the $[A_i]$ values in time and in space is thus governed by the solution of this non-linear system of partial differential equations, for given initial conditions and limits.

However, the problem is extremely complex and, in the absence of appropriate general mathematical methods, we have to be satisfied with the study of the *qualitative* behaviour of the system.

But what exactly does qualitative behaviour mean? Essentially, it is the study of the *stability* of the system. It may be found useful to consult the literature on the subject (see for example Sattinger (1973), on the theory of bifurcations; La Salle and Lefchetz (1961), for stability in the Lyapunov sense; and Nicolis and Prigogine (1977), for a general review). Murray (1989) gives many examples together with a detailed description of the mathematical methods used for the resolution of reaction–diffusion equations. Let us now consider briefly the three forms of stability that may be attained by a system of the type (2.39), rewritten below with $X_i = [A_i]$:

$$\frac{\partial X_i}{\partial t} = f_i(X_1, X_2, \ldots, X_n) + D_i \nabla^2 X_i, \quad i = 1, 2, \ldots, n. \tag{2.40}$$

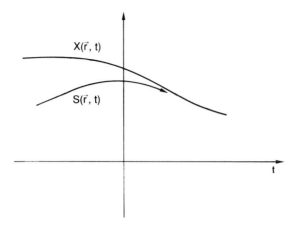

Fig. 2.10. Asymptotic stability in the Lyapunov sense: S tends asymptotically towards X as t tends towards infinity.

- *Lyapunov stability*

Let $\mathbf{X} = (X_i(\mathbf{r}, t))_{i=1,n}$ be a solution of the system (2.40) physically and mathematically defined (limited and positive), for given initial conditions and limits. Then:

(i) $\mathbf{X}(\mathbf{r}, t)$ is *stable* in the Lyapunov sense if: $\forall \varepsilon > 0$ and for $t = t_0$, there exists $\eta(\varepsilon, t_0)$ such that $\forall \mathbf{S}(r, t)$, solution of (2.40), we have:

$$\| \mathbf{X}(\mathbf{r}, t_0) - \mathbf{S}(\mathbf{r}, t_0) \| < \eta \Rightarrow \| \mathbf{X}(\mathbf{r}, t) - \mathbf{S}(\mathbf{r}, t) \| < \varepsilon \text{ for } t \geq t_0.$$

This mathematical definition simply means that the two solutions $\mathbf{X}$ and $\mathbf{S}$ will not be distant from one another for $t \geq t_0$ if for a value t_0 the two solutions are separated by a distance less than η (Fig. 2.10).

(ii) $\mathbf{X}(\mathbf{r}, t)$ is *asymptotically stable* if $\mathbf{X}(\mathbf{r}, t)$ is stable and:

$$\lim_{t \to \infty} \| \mathbf{X}(\mathbf{r}, t) - \mathbf{S}(\mathbf{r}, t) \| = 0.$$

- *Orbital stability*

Let us consider systems with the property of invariance by translation in time, i.e. such that if $\mathbf{X}(\mathbf{r}, t)$ is a solution of the system, then $\mathbf{X}(\mathbf{r}, t + T)$ is also a solution. This property has been demonstrated by Arnold (1974): given an autonomous equation (or system)

$$\frac{d\mathbf{X}}{dt} = \mathbf{v}(\mathbf{X})$$

it is readily seen that, if $\mathbf{X} = \boldsymbol{\phi}(t)$ is a solution, then $\mathbf{X} = \boldsymbol{\phi}(t + T)$ is also a solution as:

$$\frac{d\boldsymbol{\phi}(t + T)}{dt}\bigg|_{t=t_0} = \frac{d\boldsymbol{\phi}(t)}{dt}\bigg|_{t=t_0+T} = \mathbf{v}(\boldsymbol{\phi}(t_0 + T)) = \mathbf{v}(\boldsymbol{\phi}(t + T))\big|_{t=t_0}.$$

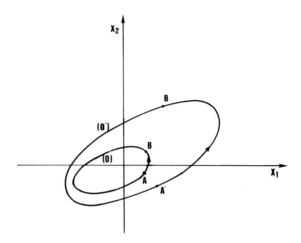

Fig. 2.11. Orbital stability. The points B and B' move away from points A and A', but B' moves further than B when t increases, even though the two orbits (O) and (O') remain close together in space. There is thus no Lyapunov stability.

The corresponding orbits are not necessarily closed. However, if $\boldsymbol{\phi}(t_1) = \boldsymbol{\phi}(t_2)$ for $t_1 \neq t_2$, then the orbit will be closed (or we shall have a stationary point) and a new property will appear:

$$\forall t \ \boldsymbol{\phi}(t + T) = \boldsymbol{\phi}(t) \quad \text{with} \quad T = t_2 - t_1.$$

Each phase thus defines a trajectory or orbit of the $[A_i] = \mathbf{X}_i$ in space. Let (O) be such an orbit. (O) is said to be orbitally stable if:

$$\forall \varepsilon > 0 \qquad \exists \eta > 0$$

such that if $d(\mathbf{X}(\mathbf{r}, t_0), (O)) < \eta$ at t_0, then:

$$d(\mathbf{X}(\mathbf{r}, t), (O)) < \varepsilon \qquad \text{for } t \geq t_0$$

where d is the distance between the trajectories.

Similarly, (O) is asymptotically stable in the orbital sense if $d(\mathbf{X}, (O)) \to 0$ when $t \to \infty$ (Fig. 2.11).

● *Structural stability*

Let us suppose that the physical system described by Eq. (2.40) depends on a parameter λ. For example:

$$\frac{\partial X_i}{\partial t} = f_i^{\lambda}(X_1, X_2, \ldots, X_n) + D_i \nabla^2 X_i \quad i = 1, 2, \ldots, n. \tag{2.41}$$

Then, if the parameter λ varies while $\mathbf{X}$ undergoes a change, the problem of the variation of the 'structure' of these equations and consequently of their stability will arise.

The system will be said to be *structurally stable* if a variation ε of λ leads only to a variation of the order of ε, i.e. $o(\varepsilon)$, of the solution of the 'undisturbed' system, and to an invariance of the topological properties of the trajectories.

Now let us go back to system (2.41) and examine its asymptotic stability. A method often used is that of linearisation. In the neighbourhood of the 'non-perturbed' solution $X^{(0)}(\mathbf{r}, t)$, the 'perturbed' solution may be written:

$$X(\mathbf{r}, t) = X^{(0)}(\mathbf{r}, t) + \boldsymbol{\varepsilon}(\mathbf{r}, t) \tag{2.42}$$

whence the system (2.41) in ε_i, for $i = 1, 2, \ldots, n$:

$$\frac{\partial \varepsilon_i}{\partial t} = f_i^\lambda(X_1^{(0)} + \varepsilon_1, \ldots, X_n^{(0)} + \varepsilon_n) - f_i^\lambda(X_1^{(0)}, \ldots, X_n^{(0)}) + D_i \nabla^2 \varepsilon_i$$

or:

$$\frac{\partial \varepsilon_i}{\partial t} = g_i^\lambda(\varepsilon_1, \ldots, \varepsilon_n) + D_i \nabla^2 \varepsilon_i, \qquad i = 1, 2, \ldots, n. \tag{2.43}$$

The transformation of the variable (2.42) gives an equation in ε_i, the stability of which may now be studied in the neighbourhood of 0. Moreover, we may write ($i = 1, 2, \ldots, n$):

$$g_i^\lambda(\varepsilon_1, \ldots, \varepsilon_n) = \sum_{j=1}^n \left(\frac{\partial f_i^\lambda}{\partial \varepsilon_j} \right)_0 \varepsilon_j + \text{second order terms.}$$

In the first approximation, the linearised system (2.43) may be written:

$$\frac{\partial \varepsilon_i}{\partial t} = \sum_{j=1}^n \left(\frac{\partial f_i^\lambda}{\partial \varepsilon_j} \right)_0 \varepsilon_j + D_i \nabla^2 \varepsilon_i \qquad i = 1, 2, \ldots, n. \tag{2.44}$$

The interest of the linearisation is two-fold: first, it renders the system (2.41) solvable by the classical methods of linear algebra; and secondly, it allows the application of the following theorem of Lyapunov:

<u>Theorem</u>: *If the solution of (2.44) is asymptotically stable (resp. unstable) in the neighbourhood of 0, then* $(X_i^{(0)})_{i=1,n}$ *is an asymptotically stable (resp. unstable) solution of (2.41).*

4.   An autocatalytic reaction

The chemical processes studied above are all supposed to occur in the neighbourhood of thermodynamical equilibrium. The chemical reactants A_i being placed in a volume V (the reactor), the kinetic study for a given environment bears on the formation of products and, if the reaction is reversible, the destruction of these products.

But if we suppose the reaction to be maintained, i.e. by continuous removal of products and appropriate input to the chemical reactor, then the system will be in a stationary state which may be far removed from thermodynamic equilibrium. A thermodynamic interpretation of this process is discussed in the next chapter.

In the meantime, the mathematical methods presented above allow us to determine the stability of the system.

To illustrate this, let us consider a very particular reactional system, called the 'Brusselator' (Prigogine and Lefever, 1968), which is the basis of several recent models. This class of systems has two essential characteristics:

- The systems comprise *autocatalytic reactions* of the type:

$$A + X \rightarrow 2X$$

  i.e. the molecule X serves as catalyser to its own chemical reaction. This is why such reactions are called autocatalytic.
- They have a behaviour frequently observed in nature, that of a *stable limit cycle* (Appendix B). Thus, a good number of phenomena, such as morphogenesis and cell division, may be interpreted in terms of limit cycles.

Under which conditions does a reactional system possess these properties? Hanusse (1972) and Tyson (1976) have demonstrated the following theorem:

Theorem: *If a reaction consists only of monomolecular or bimolecular steps, it will be impossible to obtain a limit cycle surrounding a stable node in a set of reactions comprising two intermediate variables.*

This is the case, for example, in the trimolecular model, or 'Brusselator', which is written as follows:

$$
\begin{align}
&(1) \quad A \rightarrow X \\
&(2) \quad B + X \rightarrow Y + D \\
&(3) \quad 2X + Y \rightarrow 3X \\
&(4) \quad X \rightarrow E.
\end{align}
\tag{2.45}
$$

As explained above, *non-equilibrium* is a necessary condition: for example, the final products D and E are removed from the reactor as fast as they are produced. There are two chemical reactants, A and B, and two intermediate products, X and Y, respectively in concentrations of: a, b, X_1 and X_2. Thus, A, B, D and E have fixed values and only X_1 and X_2 are variable. Expression (2.45(3)) represents a trimolecular autocatalytic step, but the improbability of such a reaction suggests that it should be replaced by a series of bimolecular reactions:

$$Y + X \rightarrow Z \quad \text{and} \quad Z + X \rightarrow 3X$$

with the condition that this step occurs very rapidly and that Z is adiabatically (see Appendix B) eliminated.

Using (2.41) with appropriate units (Nicolis and Prigogine, 1977), the system with two variables X_1 and X_2 may be written:

$$\frac{\partial X_1}{\partial t} = a - (b + 1)X_1 + X_1^2 X_2 + D_1 \frac{\partial^2 X_1}{\partial x^2}$$
$$\frac{\partial X_2}{\partial t} = bX_1 - X_1^2 X_2 + D_2 \frac{\partial^2 X_2}{\partial x^2} \qquad\qquad (2.46)$$

with initial and limit conditions (Dirichlet):

$$X_1(0, t) = X_1(1, t) = a$$
$$X_2(0, t) = X_2(1, t) = \frac{b}{a}.$$

In keeping with the principle of linearisation, the stability of the reaction is analysed about the *stationary state*:

$$X_1^{(0)} = a \qquad\qquad X_2^{(0)} = \frac{b}{a}$$

then with:

$$X_1 = X_1^{(0)} + \varepsilon_1 \qquad\qquad X_2 = X_2^{(0)} + \varepsilon_2$$

we obtain the system (Eq. 2.44):

$$\frac{\partial \varepsilon_1}{\partial t} = (b - 1)\varepsilon_1 + a^2 \varepsilon_2 + D_1 \frac{\partial^2 \varepsilon_1}{\partial x^2}$$
$$\frac{\partial \varepsilon_2}{\partial t} = -b\varepsilon_1 - a^2 \varepsilon_2 + D_2 \frac{\partial^2 \varepsilon_2}{\partial x^2} \qquad\qquad (2.47)$$

with limit conditions:

$$\varepsilon_1(0, t) = \varepsilon_1(1, t) = \varepsilon_2(0, t) = \varepsilon_2(1, t) = 0.$$

We thus have a system for which the change of the *homogeneous* state (with zero-value space derivatives) can be easily sought. The complete resolution of this model lies beyond the scope of this work, but the detailed account give by Nicolis and Prigogine (1977) should be consulted as it is the basis of a whole class of systems represented by ordinary differential equations. Recently, Li and Goldbeter (1989) have made an exhaustive study of coupled biochemical oscillators. Reaction–diffusion systems allow the interpretation of many biological phenomena, examples of which will be found all through this work. For example, we shall see how time change (with zero diffusion) and space change (with zero time-variation) are characteristic of biological oscillations and three-dimensional structuration (Fig. 2.12). This also applies to cell division and differentiation, as well as to many other fascinating chapters of biology. This being the case, we can hardly resist the temptation of seeking the far more fundamental 'reality' underlying the *self-organisation* of all living organisms.

The major problem is that, at the theoretical level, the study of non-linear dynamic systems, including the class of reaction–diffusion systems, is a difficult chapter of

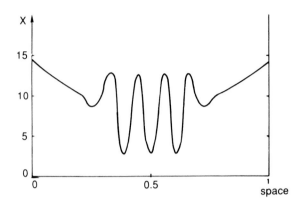

Fig. 2.12. A solution of the system of differential equations (2.46) (called 'Brusselator') showing space structuration with arbitrary units of space (after Nicolis and Prigogine, 1977).

mathematics. Thus, it has perhaps been useful to recall the more important theorems, allowing recognition of the *stability* of phenomena in the neighbourhood of stationary states and the observable concept of the limit cycle. Indeed, seen in a highly simplified way, cell division may well be considered as a macroscopic expression of a limit cycle followed by the concentration of specific chemical substances. Similarly, a chain of enzymic reactions of the type:

$$S_1 \xrightarrow{E_1} S_2 \xrightarrow{E_2} \ldots \ldots \to S_n$$

where the enzymes E_i may be allosteric, follows a course corresponding to the solution of a dynamic non-linear system. For certain values of concentrations and rate constants, and with certain conditions we shall specify later, oscillations of the limit cycle type can be demonstrated. *Glycolysis*, an essential pathway of energetic metabolism, is an example of this (Volume II, Chapter 6). It is difficult to study a combination of reactions such as that represented above. In addition to the fractal method already discussed, Savageau has proposed another approach in a series of articles (Savageau, 1991a,b; Sorribas and Savageau, 1989a,b; Savageau, 1979). This method, now known as the formalism of S-systems (Savageau and Sands, 1991), will be presented in Chapter 8.

Some of the concepts presented in this chapter are quite ancient, stemming directly from experimental biochemistry, while others are recent, deduced from the mathematical theory of stability. In the rest of this work we shall examine the impact of the non-linearity of dynamic systems on the evolution of the phenomena of life. The fact that today non-linear systems are no longer studied merely qualitatively is a sign of the considerable progress that has been made. Indeed, while the theoretical scientist is attracted by the behavioural aspect of a reactional system with a given topology, and the experimental scientist strives to determine the constants of such systems in given conditions, it must be agreed that both lines of research converge to a finer understanding of the phenomena observed.

However, the organism cannot be equated to a homogeneous set of chemical reactions. Each cell is composed of a complex set of substructures within which occur numerous enzymic reactions corresponding to specific metabolic processes. There is in fact a great diversity of local micro-environments. Current debate on the phenomena of 'channelling' and 'pooling' is an illustration of this at the cellular level. The classical results discussed above require to be modified so as to take into account the spatial enzymic and metabolic organization. Thus, even at the molecular level, the chemical reactions are organized in space to constitute structural units capable of interaction. This is an example of a functional interaction occuring at a distance, i.e. a non-local functional interaction, and as we shall see in Chapter 4, a non-symmetric functional interaction. In short, the internal chemistry of cells should be viewed as a vast network of non-local functional interactions acting from a region where the local chemical reactions occur towards other regions where other local chemical reactions occur. The following property, which sums up the ideas presented in Chapters 1 and 2, will serve as a basis for further development:

If an organism can be equated to a highly regulated set of local chemical reactions, then local reactions must exist between structures which, in physical space, are necessarily situated at a distance.

This problem will be fully treated in Chapter 4.

3

Methods in Biological Dynamics

I. Biology and complexity

Any theoretical approach to biology will first have to deal with the choice of the best mathematical instruments for describing the great complexity of biological dynamics. The working of even the most elementary living cell involves the control and regulation of thousands of different molecular interactions, so the study of such a system will obviously require much simplification. One way of doing this would be by breaking down the original system into simpler subsystems according to rules as precise as possible such as, for example, the rules determined by equivalence relations. We could thus proceed *from one level of description to another*, using a method often found satisfactory in explaining the working of complex mechanisms through a study of the component parts. This process of decomposition could evidently be pushed to extreme limits, but where is one to stop when confronted by the complexity of biological phenomena? Will the reductionist method, so useful in physics, turn out to be equally successful in biology?

Clearly, one of the main problems of modern theoretical biology concerns the *use of mathematical tools* that may be as varied as the phenomena under investigation. Several authors have worked out theories offering a combined descriptive, explicative and predictive approach. The more important of these are: Rosen's relational theory; Delattre's theory of transformation systems which describes relations between the component parts of a system; Prigogine's thermodynamic theory of irreversible processes; Katchalsky's network theory which establishes, by analogy with thermodynamics, a generalised phenomenological description of biological systems; and Thom's theory of catastrophes which, based on methods of qualitative dynamics, describes phenomena at particular points for the relevant parametric values. All of these theories will be briefly reviewed below.

Another problem is raised by the *nature of the central concepts* used in these theories. Often specific to biological phenomena, such concepts need to be thought out differently from those in physics. This applies to the more important concepts such as those of the fundamental level of description, of time and of space. It also applies to abstract concepts such as complexity, autonomy and self-organisation, which appear in their true form only in biology. This question will be formally treated in the last chapter of Volume III, but a few words here may be useful.

We shall show that the choice of the *fundamental level description* is one of the more important concepts in biology because of the intricacy of the physiological subsystems involved, in other words because of the *functional organisation* of the biological system. Biological organisation being essentially a *hierarchical* functional system, we have to look for mathematical tools capable of producing a dynamic description.

Prigogine (1980) recalls the distinction made by Eddington (1958) between 'primary' laws controlling individual behaviour and 'secondary' laws governing group behaviour. However, it is difficult to say exactly what these fundamental laws might be. While examining the significance of the second law of thermodynamics at the microscopic level, Prigogine and his co-workers were obliged to make a total re-assessment of the concepts of classical as well as of quantum dynamics. One of the conclusions is that *the future is not included in the past*, so that a new concept of time emerges from the theory. Ordinary time is associated with movement, and to introduce irreversibility at the microscopic level the dynamic equations will have to be expressed in terms of the microscopic entropy operator and the time operator, the eigenvalues of which would give the possible ages of the system. This would then be the *internal* time—a highly revolutionary concept! Yet one cannot help being struck by the analogy between physical systems, as studied by Prigogine, and biological systems.

Defining the complexity of a biological system is admittedly a difficult task, but at least it should be possible to give a precise definition within a restricted framework. While it is easy to compare the complexity of a living organism with that of an inanimate object, it is difficult to appreciate differences in the degree of complexity between different species. For instance, could a mammal be held to be more complex than a fish? May not the difference be simply due to the greater *autonomy* or the better *self-organisation* of mammals in a generally larger environment? The concepts of complexity, autonomy and self-organisation will be developed in Volume III, Chapter 6. Closely linked to the notions of control and regulation, these concepts are rooted in the evolutionary past of living organisms. According to Dobzhansky and Boesiger (1968) nothing will make sense in biology if it is not examined in the light of the theory of evolution. We shall therefore try to present a dynamic view of functional organisation. Just as a physicist studies the structure of matter, the physiologist may be said to be interested mainly in the functional aspects of life. With this in mind, let us propose the following definition of functional complexity or the potential of functional organisation:

Let an element of a given structure in its functional context, e.g. a molecule, an organelle or a cell, be called a *structural unit*. A *sink* is a structural unit that accepts a chemical substance, i.e. a physiological product, that it lacks. A *source* is a structural element that produces this substance. It will be shown that the *functional complexity C* of a biological system may be expressed as a *potential of functional organisation* (see Chapter 12), following Chauvet (1982, 1993b), as follows:

$$C = \sum_{i\text{-levels}} \sum_{\alpha\text{-functions}} n_\alpha^i \ln(v - n_\alpha^i) \qquad (3.1)$$

where n_α^i is the number of sinks at level i and the physiological function is assimilated to product P_α. v is the number of structural units in the system. Functional complexity may be expressed in a manner better adapted to the construction of a field theory as follows:

$$C = \sum_{\text{levels}} \sum_{\text{functions}} [\text{sinks}] \ln [\text{sources}]. \qquad (3.2)$$

This expression takes into account the conservation of sinks and sources at different levels and for different functions, with control relations evidently existing between the structures.

One of the characteristics of living organisms appears to be the *permanent transformation towards increasing organisation*. How and why does a living system evolve from the simple to the complex and not the other way about? It is probably this characteristic of life that convinced P.-P. Grassé (1973) of the existence of an 'Evolutionary Project'. The evidence seems to indicate that the evolution of the living world, all species considered, is the result of an irreversible super-organisation analogous to a self-organising system constantly seeking greater complexity. This type of behaviour is clearly the opposite of that of the physical world in a state of equilibrium. However, as we shall see later, when a system is far from the equilibrium state, new spatiotemporal structures may appear even in the case of the physical world. We shall also see how information theory as applied to the living world, considered in its totality as a self-organising system, partly answers this fundamental question. For the present, we suggest that a very close analogy may be made between a self-organising system and a non-conservative, irreversible system. According to Prigogine, it is the structure of the equations of movement at the microscopic level with their random element that manifests itself as irreversibility at the macroscopic level. Is the increasing, irreversible self-organisation the result of a fundamental law of living matter? From this point of view, the equation of 'movement' could be that of the density of species in the phase space where each species is subject to random mutagenesis. The result would be the *irreversibility of transformation at the macroscopic level*. It is obvious that 'movement' in this context has little to do with the usual sense of the term, and that the age of the species, taken as a specific value of the time operator, will depend directly on the distribution function of the species. Thus, we shall have to tackle the fundamental problem of expressing the '*permanent*

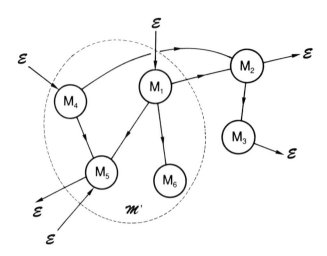

Fig. 3.1.   Rosen's ($\mathcal{M}$–$\mathcal{R}$) system. $\mathcal{M}'$ is a subsystem of $\mathcal{M}$ as M_5 has an output towards $\mathcal{E}$, and M_1, M_4, M_5 are the only inputs. M_1, M_4, and M_5 are the origin components, and M_2 and M_3 are terminal components. M_5 is also a terminal component.

super-organisation' compatible with the physiological functions of the species. All the methods presented below will have to be considered in the light of this objective.

The main purpose of this introduction has been to stress the coherence of concepts necessary to theoretical physiology.

II.   Relational theories

1.   *Rosen's theory of abstract biological systems*

A system $\mathcal{M}$ is described by a set of components M_α and by component–component or component–environment relations (Rosen, 1958). A molecule or, in more general terms, a substance produced by a component M_α and considered as its output may serve as an input to a component M_β. If the description is restricted to this level, a component is simply a structureless 'black box'. Such a system is conveniently represented by a mathematical graph (Fig. 3.1), on which each component corresponds to a *vertex* and each relation, as described above, to an oriented *edge* or *arc*. Let us recall some classical definitions of graph theory:

- a *path* is a succession of arcs, with the end of one arc corresponding to the same vertex as the origin of the next arc;
- a *component* is a set of vertices and arcs between vertices;
- a component is said to be *strongly connected* when, between any two of its vertices, E_j and E_k, it is possible to find a path from E_j to E_k as well as one, not necessarily identical, from E_k to E_j.

A component of $\mathcal{M}$ that receives an input from the environment $\mathcal{E}$ is called the *origin* component. A component of $\mathcal{M}$ with an output to the environment $\mathcal{E}$ is called the *terminal* component. A subsystem $\mathcal{M}' \subset \mathcal{M}$ is a subset of components of $\mathcal{M}$ connected as in $\mathcal{M}$ and such that:

(i)   the set of outputs $\mathcal{M}' \to \mathcal{E}$ contains a subset of outputs $\mathcal{M} \to \mathcal{E}$;
(ii) $\mathcal{M}'$ receives no inputs from a component of $\mathcal{M}$ not included in $\mathcal{M}'$. In other words, $\mathcal{M}'$ can receive inputs from no external component except the environment (Fig. 3.1).

Another characteristic of biological systems is the time lag introduced by a component during its operation. Rosen takes into account two types of time lag: that which is intrinsic to the operation of the component — the *operational lag* — and that due to the finite time of propagation of inputs from one component to another — the *transport lag*. We may ask how the relative importance of one component could be determined with respect to the others. In a biological system it would appear natural to consider the number of environmental outputs that are inhibited when a given component is itself inhibited. Let Θ be the set of environmental outputs of a system and $S_\alpha \subset \Theta$ the subset of outputs for a component M_α which cease to exist when M_α is inhibited. S_α is called the *dependent set* relative to M_α.

To be applicable to a living cell, such a structure must satisfy at least two conditions:

(i)   it must represent a metabolising system (anabolic and catabolic); and
(ii)   it must be capable of auto-reproduction.

These conditions are formalised below:

(i)   The components are supposed to have the property of *non-contractibility*, i.e. no component of the system can produce an output unless *all* the inputs directed towards the component have reached it. In addition, it is assumed that each component $M_\alpha \in \mathcal{M}$ has a finite lifetime τ_α, i.e. it can only function for a time τ_α. Consequently, the system will cease to function unless its components are renewed before the end of their lifetime.
(ii)   To $\mathcal{M}$ is associated a new system $\mathcal{R}$ containing components R_α, the function of which is the replication of components M_α. It is supposed that $\mathcal{R}$ is defined by inputs which are a subset $\Theta_\alpha \subset \Theta$ of outputs $\mathcal{M} \to \mathcal{E}$, that the output of R_α which is the replica of M_α has no transport lag, i.e. it operates as soon as it is produced, and that each R_α is non-contracting. A condition is imposed on the formation of the components R_α: Θ_α, which is the set of inputs R_α, is constructed on Θ, the set of outputs $\mathcal{M}$, such that if M is a terminal component of $\mathcal{M}$, and T_M represents the set of outputs to the environment of $\mathcal{M}$ produced by M, we have:

$$\forall M \in \mathcal{M}: \qquad T_M \subseteq S_M$$

where S_M is the dependent set relative to M and:

$$\forall M \in \mathcal{M}: \qquad T_M \cap \left(\bigcup_\alpha \Theta_\alpha \right) \neq \emptyset$$

In other words, this condition requires that *each terminal component M of $\mathcal{M}$ produce at least one output to the environment serving as an input to R_α* (Fig. 3.2a). Rosen considers this to be weakest condition allowing satisfactory results. Such a system is called the ($\mathcal{M}$–$\mathcal{R}$) system and roughly represents, on the one hand, the biochemical mechanism of cell metabolism and, on the other, the system of nucleic acid replication. To be of value, this abstract model would have to be able to explain certain biological phenomena.

The consequences of Rosen's theory may be summed up as follows:

(1)   In a metabolic system, there must always exist a structure that cannot be re-established by the system following inhibition or destruction of the structure.
(2)   Under certain imposed conditions there exists a biological structure the inhibition or destruction of which destroys the metabolic activity of the system.
(3)   Certain components M_α of a metabolic system may be re-established by supplying the corresponding components R_α with appropriate inputs from the environment.
(4)   Most metabolic systems possess a nucleus containing the substances necessary to the conservation of the genetic potential.

These results are based on the theorems that follow.

Theorem 1: Dependence structure of an ($\mathcal{M}$–$\mathcal{R}$) system. *If a component $M_{\alpha 0}$ of an ($\mathcal{M}$–$\mathcal{R}$) system is inhibited, then:*

- *either the system dies completely and $M_{\alpha 0}$ is called the central component,*
- *or there exists a subsystem $\mathcal{M}' \subset \mathcal{M}$ such that $\mathcal{M}$ is unaffected by the inhibition if we consider components R_α corresponding to $M_\alpha \in \mathcal{M}'$, $\forall \alpha$. This maximal unique subsystem is called the subsystem $\mathcal{M}'_{\alpha 0}$ corresponding to $M_{\alpha 0}$. The set of all the outputs from $\mathcal{M} \to \mathcal{E}$, not produced by $\mathcal{M}'_{\alpha 0}$, is $\overline{S}_{\alpha 0}$, the augmented dependent set relative to $M_{\alpha 0}$ (Fig. 3.2b).*

This theorem describes the behaviour of the ($\mathcal{M}$–$\mathcal{R}$) system when one of its components is inhibited. We have seen that, if $S_{\alpha 0}$ is the set of outputs that cease to exist when $M_{\alpha 0}$ is inhibited, then, because of the property of non-contractibility, each output from $\mathcal{M}$ in $S_{\alpha 0}$ will be suppressed. Moreover, Θ_α being constructed on Θ, each output belonging to $S_{\alpha 0}$ is an input for a component R_α. Thus, the inhibition of $M_{\alpha 0}$ will lead, again because of the hypothesis of non-contractibility, to the extinction of all the R_α components having inputs from the outputs of $\mathcal{M}$ belonging to $S_{\alpha 0}$. Then, as their lifetime is finite, the components M_α corresponding to R_α will finally cease to function, the components S_α dependent on M_α will in turn be no longer productive, and so on.

Theorem 2: Re-establishment of an ($\mathcal{M}$–$\mathcal{R}$) system. *A component M_α of an ($\mathcal{M}$–$\mathcal{R}$) system is said to be re-establishable if, after inhibition or destruction, it can only be re-established with none of the outputs from the* augmented dependent set $\overline{S}_\alpha$ of M_α *being used as an input to the components R_α corresponding to M_α, i.e. $\Theta_\alpha \cap \overline{S}_\alpha = \varnothing$. Then, in an ($\mathcal{M}$–$\mathcal{R}$) system in which the graph of $\mathcal{M}$ is connected, it is impossible for all components of $\mathcal{M}$ to be re-establishable.*

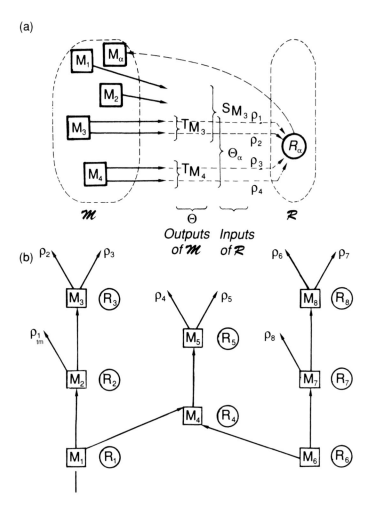

Fig. 3.2. $\Theta_1 = \{\rho_5\}$; $\Theta_2 = \{\rho_4, \rho_5\}$; $\Theta_3 = \{\rho_2, \rho_7\}$; $\Theta_4 = \{\rho_5\}$; $\Theta_5 = \{\rho_4, \rho_5\}$; $\Theta_6 = \{\rho_4\}$; $\Theta_7 = \{\rho_1, \rho_6\}$; $\Theta_8 = \{\rho_2, \rho_5, \rho_8\}$. Example of an ($\mathcal{M}$–$\mathcal{R}$) system (after Rosen, 1958), illustrating a *central component* M_1. $S_1 = \{\rho_1, \rho_2, \rho_3, \rho_4, \rho_5\}$ and the inhibition of M_1 leads to the inhibition of R_6 as $\Theta_6 = \{\rho_4\}$ whence the inhibition of M_6. As $S_6 = \{\rho_6, \rho_7, \rho_8\}$, we deduce that all the outputs are finally inhibited: $\bar{S}_1 = \{\rho_1, \rho_2, \rho_3, \rho_4, \rho_5, \rho_6, \rho_7, \rho_8\}$ and $\mathcal{M}'_1 = \varnothing$. M_1 is therefore a central component. Similarly, $\mathcal{M}'_2 = \{M_1, M_4, M_5, M_6\}$ as the inhibition of M_2 leads to the inhibition of $S_2 = \{\rho_2, \rho_5, \rho_8\}$ and that of M_3. So R_7 and R_8 are inhibited followed by the inhibition of M_7 and M_8 as $\Theta_7 = \{\rho_1, \rho_6\}$ and $\Theta_8 = \{\rho_2, \rho_5, \rho_8\}$. We deduce that $\bar{S}_2 = \{\rho_1, \rho_2, \rho_3, \rho_6, \rho_7, \rho_8\}$. This is the *augmented dependent set* relative to M_2 which indicates that ρ_6, ρ_7 and ρ_8 are also affected (indirectly) by the inhibitor of M_2. Similarly, $\mathcal{M}'_2$ is the complementary set of the set of components affected by the inhibition of M_2 (i.e. M_3, M_7 and M_8).

Thus, in a metabolic system there must always exist a structure that cannot be re-established by the system following the inhibition or destruction of the structure.

Theorem 3: Existence of central components. *If in an ($\mathcal{M}$–$\mathcal{R}$) system in which the graph of $\mathcal{M}$ is connected there exists an origin component M of $\mathcal{M}$ which is not re-establishable while all the other components are re-establishable, then this origin component is central to the system.*

In other words, under certain conditions the metabolic activity of a system may be completely destroyed by the inhibition or suppression of a particular biological structure.

We see that an ($\mathcal{M}$–$\mathcal{R}$) system can describe some of the essential properties of living organisms. However, this description covers only the coarse structure of the organisms but not their intimate working. This, of course, is only to be expected with a highly general approach to biological systems based more on powerful mathematical methods than on the interpretation of biological facts which, unfortunately, still remain poorly established. Indeed we may wonder whether this theory is not too ambitious. Since Berthalanffy's work (1973), the question has been much discussed and some authors have indeed been sharply critical of such theories. Perhaps, as Delattre (1981a) has suggested, it is as yet too early to make a general synthesis of fundamental physiological phenomena. In fact it seems likely that a 'grand', unifying approach will only be possible after partial theoretical interpretations have been successfully made in various fields of biology. However, in the present context, Rosen's ($\mathcal{M}$–$\mathcal{R}$) system is exemplary. This essentially relational theory models the structure — the $\mathcal{M}$ and $\mathcal{R}$ subsystems — on the two principles of life considered to be fundamental: *metabolism* and *self-reproduction*. But what if these properties were consecutive to an even more general principle of organisation? There can be no doubt that such a principle, if it exists, must eventually emerge from a study of the evolution of the species. Nevertheless, Rosen's theory as well as the work of Rashevsky (1961) brings to light a set of highly important concepts emphasising the *functional* aspects of biology. We could hardly afford to ignore such a capital contribution in our search for a logical approach to the functional organisation of life.

Let us now introduce Delattre's theory of transformation systems (Delattre, 1971a, b). This relational theory stems from a close analysis of molecular systems and is likely to be of considerable importance as it is a generalisation based on a vast number of observations on the transformation of natural systems.

2.   *Delattre's theory of transformation systems*

Delattre (1971a) proposes a very general formalism using canonical language to describe complex systems within a clearly specified framework. Here, we shall merely recall the principal axioms of the theory, referring the reader to Delattre's work for further details. Based on extensive experimental evidence as well as on observations of concrete situations, the theory is of interest because of its rational, coherent approach to universality.

A transformation system is considered to be involved whenever a set of individual
elements, through interactions between elements or because of some external action,
undergoes a change with time of at least some of the functional characteristics that
define the elements in the set. The definition of elements by their functional characteristics appears to be fundamental in the case of living organisms and this in itself
would amply justify the interest in relational theories.

The formalism of transformation systems is based on axioms, some or all of which
may be needed according to the nature of the systems, which may be linear (multiplicative or non-multiplicative) or non-linear.

a. Axioms

Axiom 1: *The objects under study are assumed to be distributed in classes of
functional equivalence* E_i, *i = 1 to n, according to the identity of their properties.*

Axiom 2: *Between any two classes,* E_j *and* E_k, *the possibility of transformation may
be either non-existent or it may exist one-way or both ways.*

These two axioms define the topological support of the descriptive mode. The next
three axioms analytically characterise the possible transformations admitting that
three, and only three, fundamental processes are at the origin of the transformations:
a *spontaneous* transformation of an element, a transformation involving *only one
element* under the action of a force such as an external field, and, lastly, a transformation simultaneously involving *several elements* of the same class or of different
classes. In the axioms below, *F* represents the number of elementary transformations
per unit time:

Axiom 3: *For a spontaneous transformation from* E_j *to* E_k *we have:*

$$F_{kj} = \eta_{kj} N_j.$$

Axiom 4: *For a transformation from* E_j *to* E_k, *caused by a field of intensity* ϕ, *we
have:*

$$F_{kj} = \sigma_{kj} \phi N_j.$$

Axiom 5: *For a transformation involving* α_j *elements of* E_j, ..., α_{j+p} *elements of* E_{j+p},
we have:

$$F = K N_j^{\alpha j} N_{j+1}^{\alpha_{j+1}}, ..., N_{j+p}^{\alpha_{j+p}} \tag{3.3}$$

where η_{kj}, σ_{kj} *and K are factors of proportionality depending on the classes considered, and* N_j *is the number of elements in class j.*

These basic axioms express the fact that a transformation from a given class
occurs at a time rate that depends on the number of elements in the class.

In fact, all the elements of a given class are equally likely to participate in the transformation. This is indeed a very general law of nature. It is found, for example, in the law of mass action for chemical concentrations (van't Hoff) or in the law of ecological encounters between species (Volterra):

$$\sum_i \alpha_i A_i \xrightarrow{k} B$$

$$\frac{dB}{dt} = k \prod_i [A_i]^{\alpha_i}.$$

(3.4)

Axioms 3, 4 and 5 thus allow the quantitative definition of the number of transformations per unit time. The difficulty in application arises from Axiom 1 which requires that the classes of functional equivalence be determined. While in some cases this may be easy enough — see below the example of a chemical reaction in a homogeneous solution — in others the situation may be very complex indeed. Some examples will be considered further on.

Let us add two more axioms:

Axiom 6: *A transformation from* E_j *to* E_k *may, in a general way, subtract* α_{kj} *elements from* E_j*, and add* β_{kj} *elements to* E_k *(α, $\beta \geq 1$, integers).*

Axiom 7: *Some restrictions, particularly of a geometrical nature, may require that, according as the reaction kinetics are more or less rapid, certain relations be respected between the contents of some of the classes. These relations are of the type:*

$$\xi(N_j, N_k, \ldots) = 0.$$

(3.5)

Axiom 6 would apply to such phenomena as dissociation or polymerisation in chemistry, to cell multiplication or reproduction in biology, and so on. Axiom 7 takes into account possible restrictions in class content, for example in the case of a constant flux from a class where the flux is independent of class content.

The balanced equation for each class j is given by:

$$\frac{dN_j}{dt} = -(\text{outputs from } E_j)$$

$$+ (\text{inputs to } E_j \text{ from the other classes } E_k \ (k \neq j))$$

$$+ (\text{direct external inputs}).$$

Formally,

$$\frac{dN_j}{dt} = -S_j + Q_j + \varepsilon_j$$

$$\boxed{\frac{dN_j}{dt} = -\left(\sum_k \alpha_{kj} F_{kj} + \alpha_{ej} F_{ej} \right) + \left(\sum_k \beta_{jk} F_{jk} \right) + \varepsilon_j}$$

(3.6)

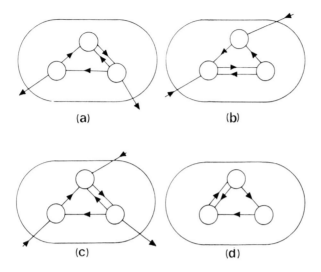

Fig. 3.3. Transformation graphs (adapted from Delattre, 1971a): (a) a true exotropic graph; (b) a true endotropic graph; (c) a true mixed graph; (d) a true closed graph.

where $k \neq j$, and F_{ej} represents outputs from class E_j towards the exterior of the system. The F_{kj} values corresponding to Axioms 3, 4 and 5 are expressed as non-linear functions of N_j, N_k, ... The terms ε_j represent the inputs ($\varepsilon_j > 0$) or the outputs ($\varepsilon_j < 0$) of the elements E_j, independently of class content.

b. Graphic representation

Here, as in the case of Rosen's relational theory, it is convenient to use the idea of a graph to represent the passage of elements from one class to another. The vertices of the graph correspond to equivalence classes; the oriented arcs between the vertices correspond to the relations between the classes. As before, a path is a continuous succession of arcs. A component of the graph, i.e. a set of vertices and arcs between the vertices, is said to be a *strongly connected* component if it is possible to find a path from E_k to E_j and a path, not necessarily identical, from E_j to E_k. This is by definition a *maximum* component if by the addition of neighbouring components it remains impossible to constitute a new, strongly connected, component. According to Delattre (1971a), the graph is a 'true graph' when the arcs crossing the frontier are added to it. He then introduces the following notions:

- a *true exotropic graph* is a true graph with its frontier crossed only by divergent arcs (Fig. 3.3a);
- a *true endotropic graph* is a true graph with its frontier crossed only by convergent arcs (Fig. 3.3b);

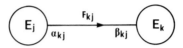

Fig. 3.4. An oriented graph representing transformations from E_j to E_k.

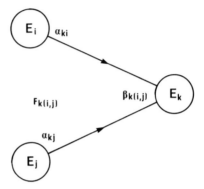

Fig. 3.5. A graph showing transformations from E_i and E_j to E_k. $F_{k(i,j)}$ is the number of transformations.

- a *true mixed graph* is a true graph with its frontier crossed by divergent as well as convergent arcs (Fig. 3.3c); and
- a *true closed graph* is a true graph with no arcs crossing its frontiers (Fig. 3.3d).

Thus, we have the following property of *transformation systems*:

> *Any system that is not strongly connected can be broken down into four types of true graph: exotropic, endotropic, mixed and closed true graphs. This decomposition constitutes a partition of the system showing the affiliation of the elements, thus revealing structural stability or change in the system.*

What do these graphs contribute to the equations above? Let us consider the simplest graph (Fig. 3.4) where the symbols F_{kj}, α_{kj} and β_{kj} have the same meaning as before: F_{kj} is the number of transformations from E_j to E_k per unit time, α_{kj} the number of elements of E_j participating in this transformation, and β_{kj} the number of elements that result from the transformation.

As a slightly more complex case (Fig. 3.5), we may consider a transformation involving two classes E_i and E_j, and ending up in a class E_k. Following Axiom 5, the number of transformations $F_{k(i,j)}$ may be written:

$$F_{k(i,j)} = K N_j^{\alpha_{kj}} N_i^{\alpha_{ki}} \tag{3.7}$$

This term appears in the equations of $N_i(t)$, $N_j(t)$ and $N_k(t)$. Here we have two *simultaneous arcs* which are interdependent because of stoichiometric restrictions between the two classes from which the arcs arise. Moreover, in some cases F may depend on internal fields ϕ_i or on external fields ϕ_e. Autocatalytic reactions may also

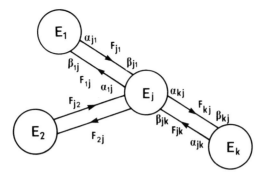

Fig. 3.6. A three-class transformation system. Such a system is physically linear if F_{kj} is proportional to N_j.

occur but in this case the criteria of rationality, as expressed by the formalism, make it clear that no equation of N_j will contain a positive term including N_j. In fact, as Delattre puts it: that which ends up in E_j cannot *directly* depend on E_j (inversion of the principle of causality); such dependence may obviously exist in an *indirect* form, i.e. through elementary looped processes involving only the classes from which they arise. Thus, if we write:

$$\frac{dN_j}{dt} = aN_j + \dots \quad \text{with } a > 0$$

we have to bear in mind the approximation due to the higher dimension considered. For example, in a classical chemical reaction with a kinetic constant K we would have: $a \equiv K(N_i)$ with $i \neq j$. Here, another class (E_i) will have to be considered. The usefulness of this axiomatico-deductive approach will be appreciated in the application to Eigen's theory of macromolecular transformation (Chapter 6).

Another consequence for the behaviour of the system is immediate if $\alpha_{kj} = 1 \forall (k, j)$ and if, as above, there are no simultaneous arcs. With these conditions, $F_{kj} = K_{kj}N_j$, where K_{kj} is the constant of proportionality. The function F_{kj} is proportional to the cardinal N_j of the class from which the arc arises, and the system is said to be *physically linear*. If necessary, it is also assumed to be linearly field-dependent through the coefficients K (cf. Axiom 4). In a particular case, the representative graph would be as in Fig. 3.6. The general equation of the linear system is:

$$\frac{dN_j}{dt} = -\left(\underbrace{\sum_{\substack{k \\ k \neq j}} K_{kj} N_j + K_{ej}N_j}_{\text{output}} \right) + \underbrace{\sum_{\substack{k \\ k \neq j}} K_{jk} N_k + \varepsilon_j}_{\text{input}} \tag{3.8}$$

$$j = 1 \text{ to } n$$

where $K_{ej}N_j$ is the number of transformations E_j towards the exterior of the system. It is convenient to write the term:

$$-\left(\sum_k K_{kj} + K_{ej}\right)$$

where $k \neq j$, as K_{jj} so that the differential equation relative to class E_j may be expressed as:

$$\frac{\mathrm{d}N_j}{\mathrm{d}t} = \sum_k K_{jk} N_k + \varepsilon_j, \qquad K_{jj} \leq 0, \qquad K_{jk} \ (k \neq j) \geq 0 \qquad (3.9)$$

or, in matrix form, as

$$\dot{N} = KN + \varepsilon.$$

It is obviously more difficult to use the formalism in the case of non-linear transformations. However, we may study the local behaviour of such systems by linear transformation in the neighbourhood of stationary points (Chapter 2) and approach the trajectories leading to these points by methods of simulation. Some of the more important applications of this are:

- discrimination between possible models of a given system, for example the ergosterol–vitamin D system;
- determination of relevant variables for transformation systems subject to external action as in radiobiology or toxicology;
- examination of internal coherence of certain types of model, for example the Leslie matrices in population theory (Chapter 11);
- determination of necessary or sufficient conditions for the existence of certain phenomena, for example direct and inverse regulation (Chapter 8); and
- choice of procedure in the search for models.

c. *An application showing the impossibility of spontaneous, undamped oscillations in physically linear systems and the existence of oscillations in non-catalytic systems*

Using only the axioms of the theory of transformation systems, Hyver (1972) showed that a physically linear system as described above, with $K_{ij} \geq 0$ for $i \neq j$, will not oscillate. It can be shown that the complete solution of Eq. (3.9), with $j = 1$ to n, may be written:

$$N_j(t) = C_j + \sum_{k=1}^{n} V_{jk} \, \mathrm{e}^{\lambda_k t} \qquad (3.10)$$

where λ_k are the eigenvalues of the matrix $\mathbf{K}$. To this matrix may be added a matrix containing only positive or zero elements: $\mathbf{A} = a\mathbf{I} + \mathbf{K}$ where $a > |K_{jj}|$ for all j, and $\mathbf{I}$ is an n-order unit matrix. But, being non-negative, $\mathbf{A}$ will always have a non-negative, real eigenvalue r such that the modulus of all the other eigenvalues of $\mathbf{A}$ is not greater than r. Therefore, λ_{max}, the eigenvalue of $\mathbf{K}$ with the greatest real part, is real. Let us here recall a result that will also be used in Part II: λ_{max}, the eigenvalue with the greatest real part, controls the long-term behaviour of the species. For a species j, the quantity: $V_{jmax} e^{\lambda_{max} t}$ becomes preponderant when t is great, so that in the long term it would be enough to consider just this quantity. In the present case, the system (Eq. (3.9)), in which λ_{max} is real, will have a monotonous, asymptotical behaviour. Thus any autonomous or endogenous oscillation will be impossible. Given that the results of chemical kinetics and thermodynamics coincide only if the principle of micro-reversibility is admitted, this result will be generally valid, even beyond thermodynamic equilibrium. An interpretation of this result will be found in the next section, but it may be recalled here that, for a system in equilibrium, the principle of micro-reversibility requires that each elementary reaction be in equilibrium with the corresponding inverse reaction.

Another equally important result follows. The Michaelis–Menten bimolecular mechanism was deduced from the Lindemann–Hinshelwood monomolecular mechanism and then related to the formation of an activated complex. Using the same process, we can break down, at least in mathematical terms, a trimolecular reaction or one of a higher order into a series of first and second order reactions. If the principle of micro-reversibility is assumed for these reactions, we can show that the eigenvalues of the matrix of coefficients $\mathbf{K}$ are all real and that it is impossible for the system to oscillate about its equilibrium state. If the system is an open one, it will be necessary to suppose all the reversible reactions to be simultaneously in equilibrium.

It is interesting to see that this result corresponds to a very special kind of graph, called 'tree-graph' by Hyver (1973), which is applicable to a fairly large class of systems. A loop being defined as a finite chain which starts and ends at the same vertex, a tree graph is classically a finite, connected graph free of loops. This graph can be conveniently used to illustrate any system of chemical reactions. Thus Fig. 3.7a represents, according to Delattre's method, the set of reactions (Eq. (3.11)) in the form of a graph of subsystems of interactions:

$$
\begin{aligned}
&(1) \quad A + B \rightleftharpoons C + D \\
&(2) \quad D + E \rightleftharpoons F + G \\
&(3) \quad B + U \rightleftharpoons H + I \\
&(4) \quad F + G \rightleftharpoons J + K \\
&(5) \qquad\;\; I \rightleftharpoons L \\
&(6) \quad I + Q \rightleftharpoons R + S + T \\
&(7) \quad P + G \rightleftharpoons M \\
&(8) \quad D + X \rightleftharpoons Y + Z \\
&(9) \quad E + V \rightleftharpoons W
\end{aligned}
\qquad (3.11)
$$

For example, it may be seen that the chemical species E intervenes in two reactions: E reacts with D (Eq. (3.11(2))) as well as with V (Eq. (3.11(9))). The figure is a tree graph and the set of reactions represented corresponds to a non-linear tree.

Hyver (1973) demonstrates the following results:

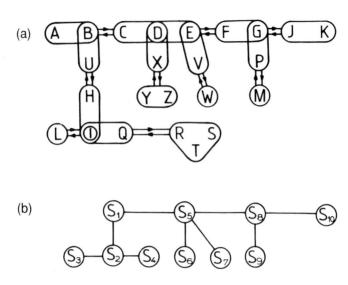

Fig. 3.7. Graphical representation of the system of chemical reactions in Eq. (3.11) (adapted from Hyver, 1973): (a) graph of the subsystem of interactions; (b) S-graph obtained by concentrating in a single class the subsystems with intersections containing common elements.

(i)   if a linear graph is a tree graph with or without loops, then all the eigenvalues are real;

(ii) the graph must contain at least one loop to have a complex eigenvalue; and

(iii) in the non-linear case represented in Fig. 3.7a, an S-graph may be deduced from the graph of the subsystems of interactions as follows: a vertex represents the union of all the subsystems of which the intersection is not an empty set. The edges correspond to previously existing relations. Then, if the S-graph is free from loops, we have a new graph as in Fig. 3.7b. In this case, all the eigenvalues are real and the system cannot oscillate.

In conclusion, although the method described is quite general in its axiomatic context, it still remains very powerful as it takes into account all the external and internal fields. *The representation of the subsystems of interactions by a simple graph immediately indicates whether or not the biological system under investigation is an oscillator.*

The method also demonstrates that it is not necessary for a system to have autocatalytic or crossed catalytic reactions for it to be an oscillating system. This is a most important result, as some studies based on the early work of Lotka and Volterra on population growth had led to the conjecture that at least one autocatalytic reaction was necessary for free oscillations and that an oscillating system must therefore have one or more autocatalytic or crossed catalytic reactions. To negate this hypothesis, Hyver (1972) gives a satisfactory counter-example which the reader may wish to consult.

Using the notion of *functionally equivalent classes* for chemical species, other applications of Delattre's formalism will be considered further on, for example in

Eigen's theory of macromolecular transformations (Chapter 6) and, in particular, in the concept of oscillators and limit cycles (Chapter 5, Volume III). Before going on to the next section, which deals mainly with the structural organisation of matter, let us stress two points:

(i)   The formalism of transformation systems, as in the case of the ($\mathcal{M}$–$\mathcal{R}$) systems, deals with the functional aspect of phenomena particularly by introducing the concept of functionally equivalent classes, as defined by Delattre, within a rigorous axiomatic context. Prior to Delattre's work, Rashevsky (1961) had already paid considerable attention to the topological aspect of living organisms. This is the approach we intend to adopt and develop in Chapters 4 and 12, and in Volumes II and III, where we shall describe a functional logic of life.

(ii) The formalism of transformation systems is a good example of the axiomatico-deductive method as opposed to the more usual analogical or hypothetico-deductive methods. Indeed, as Delattre emphasised:

> In the usual development of a model, the choice of elementary processes and the way these processes are combined depend, on the one hand, on what is recognised as acquired knowledge and, on the other, on the ideas that the creator of the model may have concerning the possible origin of the phenomena — in other words, on his imagination and on his insight. Thus, by definition the model proposed is hypothetical. The fact that the model fits experimental data is too insufficient a criterion to justify and hence validate the hypothesis ... To try to remedy these defects, we first need a canonical language of description. In principle, this should constitute a well-constructed reference framework, i.e. a framework which is rational, or at least as rational as we can make it.

It is along these lines that we shall describe an axiomatic framework of *functional biological organisation* with the corresponding dynamic processes. But for the present let us continue with our review of the methods used in biological dynamics with the theory of thermodynamics which, being very general, is mainly applicable to the overall behaviour of matter.

III.   Thermodynamic theory

Can the origin of biological structures be explained by thermodynamic theory? If not, do the major principles of thermodynamics — the first and second law (Boltzmann's principle of order) — need to be further developed to attain this goal? Indeed, recent work in this field, stimulated by results obtained by Prigogine, establishes relationships between the equations of chemical kinetics and ordered structures generated in space and in time. Let us now present this comprehensive thermodynamic interpretation.

A system of chemical reactions of the type studied in Chapter 1 is generally non-linear, with several solutions according to the parametric value. The solution corresponding to the thermodynamic equilibrium state is called the '*thermodynamic branch*' by Prigogine.

In other words, it should be possible to interpret the stability of the solutions of a chemical system through the stability of the thermodynamic branches in close relationship with non-equilibrium behaviour. This description is of interest because of the possibility of the appearance of ordered structures different from those found in the case of crystals. These structures are called *dissipative structures* as they are maintained in the stationary state (of non-equilibrium) by a permanent external supply of matter (or energy).

The theory of non-equilibrium thermodynamics, which is an extension of the classical theory, studies three problems (see Nicolis and Prigogine, 1977):

(1)   stability of systems beyond the thermodynamic branch;
(2)   growth kinetics of dissipative structures; and
(3)   application of the concepts of dissipative structure and order through fluctuation.

1.   *Principles of equilibrium thermodynamics*

(i)   A closed physicochemical system, with no exchange of matter or of energy with the outside, tends towards an equilibrium state which is totally disorganised at the molecular level. This is the second law of thermodynamics. The transformation is characterised by a function of state S, the *entropy*, such that:

$$\frac{dS}{dt} \geq 0 \qquad (3.12)$$

which means that S increases continuously to a maximum value defining the state of thermodynamic equilibrium.

(ii)   If the system is open, with exchange both of matter and of energy with the outside, the variation of entropy may be decomposed according to the external (e) and internal (i) contributions to the system, i.e. respectively, the flow of entropy due to exchanges with the outside and the production of entropy due to irreversible processes inside the *closed* system:

$$\frac{dS}{dt} = \frac{d_e S}{dt} + \frac{d_i S}{dt}.$$

From thermodynamic theory we have:

$$\frac{d_i S}{dt} \geq 0. \qquad (3.13)$$

We may note a fundamental consequence due to the flux term: this may compensate the production of entropy so that a *stationary state* may be reached:

$$\frac{dS}{dt} = 0 \quad \text{with} \quad d_e S = -d_i S \leq 0.$$

When the system is not in equilibrium, this may be written:

$$d_e S = -d_i S < 0$$

and in the equilibrium state:

$$d_i S = d_e S = 0 .$$

Thus a new structure may be obtained by putting negative entropy into the system.

(iii) If the system is closed, allowing exchange only of energy with the outside, the transformation is described by the Helmholtz *free energy* function F:

$$F = E - TS \tag{3.14}$$

so that, following Eq. (3.13), with:

$$T \, d_i S = T d S - dQ \quad \text{and} \quad dQ = dE + p dV \quad \text{(conservation of energy)}$$

where dQ is the quantity of heat received by the system during time dt, p the pressure and V the volume, we have at constant T and V:

$$\frac{dF}{dt} \leq 0.$$

A biological system may be considered as a system within which entropy is produced through coupled and autocatalytic chemical reactions leading to the term $d_i S$ which is characteristic of an irreversible process.

2. The entropy production term. Consequence in the linear field: non-equilibrium stationary states in the neighbourhood of equilibrium

Let us find an expression for the entropy production term $T \, d_i S$ in the non-equilibrium state. The reasoning is based on Prigogine's hypothesis for *local equilibrium: the expression for S in non-equilibrium conditions has the same functional form as that of thermodynamic equilibrium, except that it should be interpreted locally rather than for the whole system.*

If s is the density of entropy defined by

$$S = \int_V s dV$$

the expression of the local entropy function for concentrations $c_1, c_2, \ldots, c_n$, we may apply the known results where μ_i is the chemical potential of species i for a set of r reactions of the type (1.1):

$$\frac{\partial s}{\partial t} = \sum_{i=1}^{n} \frac{\partial s}{\partial c_i} \frac{\partial c_i}{\partial t} = \sum_{i=1}^{n} -\frac{\mu_i}{T} \frac{\partial c_i}{\partial t}.$$

As the variation of concentration with time is known (Eq. (2.36)), the equation above may be written:

$$\frac{\partial s}{\partial t} = -\sum_{i=1}^{n} \frac{\mu_i}{T} \left(\sum_{j=1}^{r} \nu_{ij} v_j - \mathrm{div}\, \mathbf{j}_i \right)$$

$$= \sum_{j=1}^{r} \left(-\sum_{i=1}^{n} \frac{\mu_i}{T} \nu_{ij} \right) v_j + \mathrm{div} \left(\sum_{i=1}^{n} \frac{\mu_i}{T} \mathbf{j}_i \right) - \sum_{i=1}^{n} \mathbf{j}_i \cdot \nabla \left(\frac{\mu_i}{T} \right)$$

$$\frac{\partial s}{\partial t} = \sum_{j=1}^{r} \frac{a_j}{T} v_j + \mathrm{div} \left(\sum_{i=1}^{n} \frac{\mu_i}{T} \mathbf{j}_i \right) - \sum_{i=1}^{n} \mathbf{j}_i \cdot \nabla \left(\frac{\mu_i}{T} \right)$$

where a_j is the chemical affinity of reaction j:

$$\alpha_j = -\sum_{i=1}^{n} \nu_{ij} \mu_i. \tag{3.15}$$

The expression $\partial s / \partial t$ may be transformed:

$$\frac{\partial s}{\partial t} = \mathrm{div}\, \mathbf{J}_s + \sigma \tag{3.16}$$

with:

$$\mathbf{J}_s = \sum_{i=1}^{n} \frac{\mu_i}{T} \mathbf{J}_i \tag{3.17}$$

$$\sigma = \sum_{j=1}^{r} \frac{\alpha_j}{T} v_j - \sum_{i=1}^{n} \mathbf{j}_i \cdot \nabla \left(\frac{\mu_i}{T} \right) \tag{3.18}$$

which, by integrating over volume V limited by surface $\mathcal{A}$, remarkably produces the decomposition:

$$\frac{\partial S}{\partial t} = \int \frac{\partial s}{\partial t}\, dV = \int_{\text{volume}} \mathrm{div}\, \mathbf{J}_s \, dV + \int \sigma\, dV.$$

Then using Green's theorem (Appendix A):

$$\frac{\partial S}{\partial t} = \int_{\text{surface}} \mathbf{n} \cdot \mathbf{J}_s \, d\mathcal{A} + \int \sigma\, dV \equiv \frac{\partial_e S}{\partial t} + \frac{\partial_i S}{\partial t} \tag{3.19}$$

where $\mathbf{n}$ is the vector normal to the surface.

Thus, we again have the terms for the flux $\mathbf{J}$ and the production of entropy σ which may be written:

$$\sigma = \sum_{\alpha} J_\alpha X_\alpha \tag{3.20}$$

where J_α is the generalised current associated with the irreversible process, either of diffusion (**J**) or of reaction (v), and X_α is the generalised force, either of diffusion $\left(-\nabla \dfrac{\mu}{T}\right)$ or of reaction (a/T).

The condition $\partial_i S/\partial t \geq 0$ implies that $\sigma \geq 0$.

At thermodynamic equilibrium, the forces have zero values:

$$\nabla\left(\frac{\mu_\alpha}{T}\right) = 0 \quad \text{and} \quad a_\alpha = 0$$

while the currents also have zero values. It is natural to use Taylor's approximation in the *neighbourhood of this singular point*. Let us write:

$$J_\alpha = \sum_\beta L_{\alpha\beta} X_\beta \tag{3.21}$$

keeping only the first-order terms

$$L_{\alpha\beta} = \left(\frac{\partial J_\alpha}{\partial X_\beta}\right)_0$$

in Taylor's development of the current J_α.

The coefficients $L_{\alpha\beta}$ are called the *Onsager coefficients*. In the region of validity of this linear approximation, the term σ becomes:

$$\boxed{\sigma = \sum_\alpha J_\alpha X_\alpha = \sum_{\alpha,\beta} L_{\alpha\beta} X_\alpha X_\beta} . \tag{3.22}$$

This is a quadratic form in X where the coefficients characterise the coupling between the various irreversible processes of the *same* scalar or vectorial nature, or between a force of a scalar nature and a vectorial flux, for example active transport in an *anisotropic* medium. More precisely, in an isotropic medium or one with central symmetry, a scalar force cannot produce a vectorial flux, and vice versa, since the Onsager coefficients for a medium with central symmetry will all be odd-order tensors which will cancel out.

It is of the greatest importance to be able to determine the variation of σ with time and thus characterise the *non-equilibrium stationary states in the neighbourhood of equilibrium*. If P is the *total production of entropy* in volume V, then:

$$P = \int_V \sigma \, dV. \tag{3.23}$$

It can be shown that:

$$P \geq 0$$

and that:

$$\frac{dP}{dt} < 0 \tag{3.24}$$

in the non-stationary state, with a zero value in the stationary state.

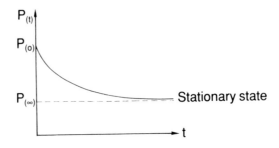

Fig. 3.8.   Non-equilibrium stationary state defined by a minimum production of entropy P.

A linear system thus obeys the very general expression of inequality above, expressing the fact that a non-equilibrium stationary state presents a minimal production of entropy (Fig. 3.8) in the neighbourhood of equilibrium. As P is a Lyapunov function (Appendix B), the stability of the system is ensured.

In other words, a perturbed stationary state returns to a reference state, which is a non-equilibrium state because of given limit conditions, so that the production of entropy has a minimal but non-zero value. In a way, entropy production plays the role of thermodynamic potential in equilibrium thermodynamics. The fundamental question now is whether these results can be extended to the non-linear case.

3.   Non-linear thermodynamics of chemical reactions: non-equilibrium stationary states far from equilibrium

Let us consider a simple chemical reaction:

$$A \underset{k_-}{\overset{k_+}{\rightleftharpoons}} B$$

characterised by the affinity:

$$a = k_B T \ln \frac{k_+ A}{k_- B}$$

where k_B is the Boltzmann constant.

This reaction will be considered to be linear if it is possible to write the reaction rate v in linear form:

$$J = LX \quad \text{where} \quad v = La$$

as:

$$v = k_+ A - k_- B = k_+ A \left(1 - \exp\left[-\frac{a}{k_B T} \right] \right)$$

so that we have:

$$v = \frac{k_+ A}{k_B T} a = La \quad \text{and} \quad L = \frac{k_+ A}{k_B T}. \tag{3.25}$$

This development is possible only if $a \ll k_B T$. *Consequently, as soon as* $a \approx k_B T$, *the chemical reaction will be non-linear.* In practice, this is the most frequent case and, as we shall see, particularly so in biology.

a. The universal criterion of evolution

Let us rewrite the expression for the total production of entropy in the form:

$$P = \int \sigma \, dV = \int (\Sigma J_\alpha X_\alpha) \, dV.$$

Using the same method as for the linear case, we calculate dP/dt, which describes the evolution of P towards the stationary state. The technique used by Glansdorff and Prigogine (1971) consists of writing:

$$\frac{dP}{dt} = \frac{d_x P}{dt} + \frac{d_J P}{dt}. \tag{3.26}$$

This separates the variations of entropy production due to forces and to currents. Then, a fairly simple calculation leads to the result:

$$\boxed{\frac{d_x P}{dt} \le 0} \tag{3.27}$$

and as in the linear region

$$\frac{d_x P}{dt} = \frac{1}{2} \frac{dP}{dt}$$

we again have the same criterion as before for the production of minimal entropy.

This last result, because of its general nature, is called the *universal criterion of evolution. In particular, it establishes the behaviour of non-equilibrium states in the neighbourhood of the stationary state.* But the deduction of stability is not immediate because, in the nonlinear region, P is no longer a Lyapunov function. Thus, a new function, $\delta^2 S$, will be required for the non-linear case.

b. The Glansdorff–Prigogine functional

If the differences between the stationary state and the functions of entropy and of entropy production, respectively $\Delta S = S - S_{st}$ and $\Delta P = P - P_{st}$, are considered to be small as they describe random perturbances of the system, then they may be expressed simply in terms of first and second order contributions:

$$\Delta S = \delta S + \tfrac{1}{2}\delta^2 S + \dots$$

$$\Delta P = \delta P + \tfrac{1}{2}\delta^2 P + \dots \; . \tag{3.28}$$

Glansdorff and Prigogine demonstrate the following results:

- in the neighbourhood of equilibrium (stationary state = equilibrium state):

$$\begin{cases} (\delta^2 S)_{eq} \leq 0 \\ \dfrac{1}{2}\dfrac{\partial}{\partial t}(\delta^2 S)_{eq} = \delta_x P = \displaystyle\int \sum_\alpha J_\alpha X_\alpha dV = P \geq 0 \end{cases}. \tag{3.29}$$

This is the definition of a Lyapunov function for $(\delta^2 S)_{eq}$: it ensures the stability of the system, and the result obtained above is confirmed.

- far from equilibrium (stationary state $\neq$ equilibrium state):

$$\begin{cases} \delta^2 S \leq 0 \\ \dfrac{1}{2}\dfrac{\partial}{\partial t}(\delta^2 S) = \delta_x P = \displaystyle\int_V \sum_\alpha \delta J_\alpha \delta X_\alpha dV \end{cases}. \tag{3.30}$$

The last term corresponds to the *production of entropy due to the perturbation.* Here we again come upon the difficulty that the universal criterion of evolution applies, in the non-linear region, to $d_x P$, in the sense that the sign of $\delta_x P$ is undefined and depends on the system. It may thus be shown that:

$$d_x P = \frac{1}{2} d(\delta_x P) + Q_e^{(a)}$$

where $Q_e^{(a)}$ is the anti-symmetric contribution of the expression: $\delta J = \Sigma(l)\delta X$ which is of undefined sign, so that a condition imposed on $\delta_x P$ cannot in general lead to $d_x P \leq 0$.

c. *Physical interpretation*

Figure 3.9 represents the physical interpretation of the results above. It is an observed fact that the equilibrium state of a system is characterised by an absence of spatiotemporal order. And the inequalities (3.29) show precisely that, *in the neighbourhood of equilibrium*, the system moves towards an unorganised spatiotemporal state as long as the perturbations remain weak, with the inequality $\delta_x P \geq 0$ holding good.

- Thus, the thermodynamic interpretation of the kinetics of chemical reactions leads to an important result: a system of chemical reactions in equilibrium cannot

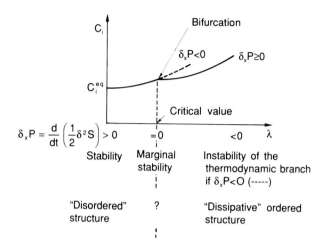

Fig. 3.9. Passage from a disordered structure to an ordered, dissipative structure, i.e. from the 'thermodynamic branch' of solutions to a 'new branch' (after Glansdorff and Prigogine, 1971). Abscissae: parametric value (constraint); ordinates: concentration.

create spatiotemporal structures if it is subject only to random perturbations of its parameters such as, for example, the values of the limit concentrations.

- Another result is obtained in the neighbourhood of the equilibrium state if a constraint applied to the parameters of the system, for example λ, leads to a systematic divergence from equilibrium. We know that the production of entropy will then be minimised, with $d_x P \leq 0$ and $\delta_x P \geq 0$. The system will be stable, with the mathematical solution belonging to the *thermodynamic branch*. By continuity, the solution follows this thermodynamic branch when the values of the parameters λ vary.

- The third result is fundamental: *beyond a certain critical value (or bifurcation value) λ_c of λ there exists a new branch of solutions corresponding to an entirely new ordered structure, such that $\delta_x P \geq 0$, while the thermodynamic branch becomes unstable if $\delta_x P < 0$.*

The thermodynamic theory of irreversible processes is based on the production of entropy and, in particular, on the excess of entropy production in cases far from equilibrium, $\int_V \sum \delta J \delta X dV = \frac{1}{2} \partial_t \delta^2 S$. Biological systems of course involve a great number of chemical reactions, and here this theory clearly plays a fundamental role, bringing to light the surprising existence of new structures beyond the thermodynamic branch, at least at the molecular level, even if it is not quite clear what happens at the critical value λ_c (the point of *marginal stability*). We have also seen that the functional $\delta^2 S$ is a Lyapunov function which, by its physical expression, shows the link between the theory of stability and thermodynamics. However, it must be admitted that although these thermodynamical concepts lead to an entirely new physical interpretation, they are not necessary for the rigorous study of chemical systems. Thus Haken (1978), working on the synergetics of chemical systems, found the same results (see also Landauer, 1975). Hyver (1973), using Delattre's formalism of transformation systems, also found similar results, especially in cases to which it is difficult to apply thermodynamic theory, such as the inexistence of oscillations in linear systems.

In any case, the thermodynamic theory of irreversible processes far from equilibrium has allowed the clear and rigorous formalisation of phenomena, the importance of which has long been recognised in biology, for instance chemical diffusion and reaction, i.e. transport of matter and chemical transformation by the combination of reactants (Turing, 1952). Thus, the notions of chemical current J and chemical force X have been generalised. These dynamic quantities are essential, the system being characterised by the current of chemical reaction $J^{(r)}$ due to the associated conjugated force $X^{(r)}$. The product of these quantities gives the energy dissipated in the system during the reaction, so that production of entropy is:

$$\sigma = J^{(r)}X^{(r)} = v\,\frac{a}{T}. \tag{3.31}$$

But $v a$, the product of the reaction rate v and the affinity $a = -\Sigma v_i \mu_i$, i.e. the difference of free energy between the reactants and the products, is homogeneous to a power. Moreover, even with the linear approximation, it remains difficult to make the equation $J = LX$ explicit. For example, in the study of the 'Brusselator' (Glansdorff and Prigogine, 1971), the behaviour of a relatively small number of equations requires quite complicated calculations. It is therefore important to find a formalism capable of representing a large number of coupled enzymic reactions, some of which may be in a non-homogeneous phase, in convenient equation form. What is sought is a highly figurative language allowing almost automatic writing of a complex system of chemical reactions. This is the aim of *network thermodynamics*, which generalises the notions presented above.

d.   Network thermodynamics

This method for the calculation and representation of a system of coupled chemical reactions was developed by Oster *et al.* (1973). The authors generalise the notions of current J and force X by seeking the profound analogy between the power dissipated, expressed by the product JX, in various fields (Table 3.1), such as:

Electricity	Current I	Voltage U
Fluid mechanics	Flow Q	Pressure P
Translation	Velocity v	Force F
Rotation	Angular velocity ω	Couple T
Diffusion	Current of matter J	Chemical potential μ
Chemical reaction	Reaction rate v	Chemical affinity $-\Sigma\mu v$

Here we have written the force of chemical reaction $X = a$ instead of a/T.

Table 3.1. Classification of conceptual analogies between current, force, displacement, impulse and power for various fields of physics. The constitutive relationship between generalised elements and their electrical analogue is also shown (following Oster *et al.* (1973) and Atlan (1976)).

Energy field	Current $J = \dfrac{\partial q}{\partial t}$	Force $X = \dfrac{\partial p}{\partial t}$	Displacement $q = \int J dt + q_0$	Impulse $p = \int X dt + p_0$	Power $P = JX$
Electricity	Current I	Voltage U	Charge q	Flux Φ	IU
Fluid mechanics	Flow Q	Pressure P	Volume V	Pressure impulse Γ	QP
Translation	Velocity v	Force F	Displacement x	Impulse p	vF
Rotation	Angular velocity ω	Couple T	Angular displacement θ	Angular moment c	ωT
Diffusion	Flow of matter J	Chemical potential μ	Mass m (nmoles)	—	$J\mu$
Chemical reaction	Rate of reaction v or chemical current J_r	Chemical affinity $a = -\Sigma\mu v$	Degree of reaction ζ	—	$J_r a$

Constitutive relationship / Generalised elements:

$$R = \frac{\partial X}{\partial J}$$

$$C = \frac{\partial q}{\partial X}$$

$$L = \frac{\partial p}{\partial J}$$

$\psi_R(X, J) = 0$

$\psi_C(X, q) = 0$

$\psi_L(J, p) = 0$

Electrical analogy:

Resistance $R = \dfrac{\partial U}{\partial I}$

Capacity $C = \dfrac{\partial q}{\partial U}$

Self-induction $L = \dfrac{\partial \phi}{\partial I}$

α.   *Characteristics of an element in a chemical network.*   With the generalisation of the *dissipation function* to the various fields in which there is dissipation of energy, we may use knowledge from electricity or mechanics to determine relationships between chemical quantities (Thoma, 1971). The aim of network thermodynamics is thus to:

(i)   to define for each element of the network: the current J going through the element; and the conjugated force X causing the current;

(ii)   to determine the relations, called *constitutive* relations:

$$f_R(X, J) = 0, \qquad f_C\left(X, \int_0^t J\,dt\right) = 0 \qquad f_L\left(\int_0^t X\,dt, J\right) = 0$$

(iii)   to describe the network and the characteristics of the network elements using simple rules.

We thus define the notions of resistance $R = \partial X/\partial J$, of capacitance $C = \partial q/\partial X$ (with $q = \int_0^t J\,dt + q_0$) and of self-inductance $L = \partial p/\partial J$ (with $p = \int_0^t X\,dt + p_0$) which generalise the classical definitions in electricity: $R = \partial U/\partial I$, $C = \partial q/\partial U$ and $L = \partial \phi/\partial I$ where the quantities U, I, q and ϕ represent respectively the voltage, the current, the quantity of electricity and the inductive flux.

The analogy with the well-known electrical characteristics, R, C and L allows a generalisation of the notions of a resistive, capacitive or inductive element to a network of chemical reactions.

We can thus demonstrate the existence of:

(1) Energy dissipating *chemical resistances*. The power dissipated is:
$P = aJ$.

The relation between a and J may be determined in the same way as for the simple example leading to Eq. (3.25). The law of mass action for a reaction of the type:

$$\sum v_i A_i \underset{k_-}{\overset{k_+}{\rightleftharpoons}} \sum \bar{v}_k B_k \qquad (3.32)$$

is written according to the relations (3.3) and those that follow:

$$J = v_+ - v_- = k_+ \prod_{\substack{i \\ \text{reactants}}} [A_i]^{|v_i|} - k_- \prod_{\substack{k \\ \text{reactants}}} [B_k]^{\bar{v}_k}$$

$$= k_+ \prod_{\substack{i \\ \text{reactants}}} [A_i]^{|v_i|} \left(1 - \frac{k_-}{k_+} \frac{\prod [B_k]^{\bar{v}_k}}{\prod [A_i]^{|v_i|}} \right)$$

$$= k_+ \prod_{\substack{i \\ \text{reactants}}} [A_i]^{|v_i|} \left(1 - e^{-\alpha/RT} \right)$$

in the form

$$J = v_+(1 - e^{-a/RT})$$

using the expression for the equilibrium constant (Glansdorff and Prigogine, 1971) in terms of the concentrations c_j:

$$RT \ln K(T) = -\Sigma v_j(\mu_j(T) - RT \ln c_j)$$

and the affinity:

$$a = -\Sigma v_j \mu_j.$$

Then:

$$a = RT \ln \frac{K(T)}{\prod_j c_j^{v_j}}.$$

Similarly:

$$J = k_- \prod_k [B_k]^{\bar{v}_k} (e^{\alpha/RT} - 1) = v_- (e^{\alpha/RT} - 1).$$

According to the theory of the activated complex, the chemical reaction (3.32) can be written in the form:

$$\sum_{i=1}^{n} v_i A_i \underset{1}{\rightleftharpoons} C \underset{2}{\rightleftharpoons} \sum_{k=1}^{m} \bar{v}_k B_k \qquad (3.33)$$

which is composed of two reactions, 1 and 2, of fluxes J_1 and J_2 respectively. In the stationary state: $J = J_1 = J_2$. The affinities of each of these reactions depend on the partial affinities:

$$a_1 = \left(\sum_{i=1}^{n} |v_i| \mu_i \right) - \mu_c = a^+ - \mu_c$$

$$a_2 = \mu_c - \sum_{k=1}^{m} \bar{v}_k \mu_k = \mu_c - a^-.$$

Thus the overall power dissipated is:

$$P = a_1 J_1 + a_2 J_2.$$

and in the stationary state:

$$P = J(a_1 + a_2) = J(a^+ - a^-) = Ja$$

which is coherent with the following expression for J, in which k is the common mean value, and $\bar{v}_+$ and $\bar{v}_-$ are the mean direct and inverse reaction rates:

$$J = \bar{v}_+ - \bar{v}_- = ke^{\alpha^+/RT} - ke^{\alpha^-/RT}. \qquad (3.34)$$

Energy dissipation is then expressed in terms of two resistances R_1 and R_2 defined for each of the two reactions 1 and 2 by:

$$\frac{\partial J_1}{\partial a_1} = \frac{1}{R_1} = \frac{\partial J}{\partial a_1} = \frac{\partial}{\partial a_1}\left[k\,\exp\left(\frac{a_1 + \mu_c}{RT}\right) - k\,\exp\left(\frac{-a_2 + \mu_c}{RT}\right)\right]$$

whence:

$$\frac{1}{R_1} = \frac{k}{RT}\,\exp\left(\frac{a_1 + \mu_c}{RT}\right) = \frac{k}{RT}\,e^{a^+/RT} = \frac{\bar{v}_+}{RT} \tag{3.35}$$

and similarly:

$$\frac{1}{R_2} = \frac{\partial J}{\partial a_2} = \frac{\bar{v}_-}{RT}.$$

When the reaction tends towards equilibrium, the rates $\bar{v}_+$ and $\bar{v}_-$ tend towards a common value $\bar{v}_+^{eq} = \bar{v}_-^{eq} = \bar{v}^{eq}$ so that the two chemical resistances R_1 and R_2 become equal to $RT/\bar{v}^{eq}$. *Thus the resistances are not constant but vary with concentration during the reaction.* In the same way, in non-linear regions, i.e. far from thermodynamic equilibrium, only one of the two resistances R_1 or R_2 will subsist.

(2) *Chemical capacitance*: This concept is easier to demonstrate by starting from the usual expression of chemical potential μ and the standard chemical potential μ_0:

$$\mu = \mu_0 + RT\ln[A]. \tag{3.36}$$

We seek a relationship between a and the integral of the current J. We know that the chemical force is the variation of chemical potential. But what is the current? It is obviously the variation of the concentration of chemical species *with time*, i.e. $d[A]/dt$.

From this we deduce the capacitance C:

$$C = \frac{\partial}{\partial X}\,(\textstyle\int J dt) = \frac{\partial[A]}{\partial\mu} = \frac{[A]}{RT}. \tag{3.37}$$

Here again *the chemical capacitance varies with the concentration.*

(3) *Chemical self-inductance*: We seek a relationship between the current J and the integral of the chemical force, $\int X dt$ + constant, i.e. a coefficient of self-inductance:

$$L = \frac{\partial}{\partial J}\,(\textstyle\int X dt). \tag{3.38}$$

The existence of time lags in autocatalytic reactions seems to favour the notion of inductive effects similar to those in electricity (Atlan, 1976).

It thus appears reasonable to represent a chemical reaction by characteristics analogous to the electrical element R, L and C. This generalisation has the advantage of extending the notion of an electrical network to that of a chemical network. If the laws governing complex electrical networks (Kirchoff's laws) hold good for chemical networks, this highly elaborate technique would be of considerable importance

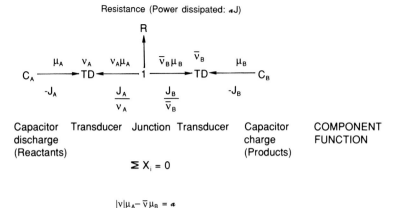

Fig. 3.10. Relation graph of a simple chemical reaction: $|v|A \rightleftharpoons \bar{v}B$, indicating analogous electrical components and functions (after Atlan, 1976).

in the study of complex biochemical reactions and might even herald the advent of the chemical computer.

β. Bond graphs.   Here, as for electrical networks, the main object is the visual description of circuits representing chemical systems. *Bond graphs* serve this purpose. With Delattre's formalism of transformation systems, we have already seen how valuable graphical techniques can be for complex systems. Another example of the use of such graphs will be found in the physiology of respiration (Volume II, Chapter 7).

   In bond graphs for electrical networks, the arcs represent the pathways of the circulating element, i.e. the energy, while in the other cases, for graphs in the proper sense of the term, the circulating element is matter.

   Let us first consider a simple chemical reaction of the type:

$$|v| A \rightleftharpoons \bar{v}B$$

close to equilibrium where v and $\bar{v}$ are the stoichiometric coefficients. We shall show that the representative graph is as in Fig. 3.10.

   The rules for constructing the graphs are as follows:

(R_1)   The arrows indicate the *energy pathway.*

(R_2)   The numbers 1 and 0 represent the *type of junction*: 1 represents a *series junction* (as in Fig. 3.10); 0 represents a *parallel junction* (coupled chemical reactions). These junctions allow a convenient extension of Kirchoff's laws. Thus for a series junction (1) we have: $\Sigma X = 0$ (mesh law): and for a parallel junction (0) we have: $\Sigma J = 0$ (point law).

(R_3)   The elements of the network are connected by their *exchanges of energy*, in other words: the capacitor *C stores* chemical energy; the resistor *R dissipates*

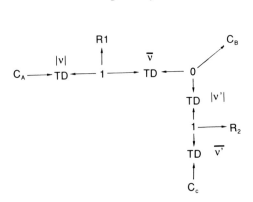

Fig. 3.11. Illustration of a series-parallel junction in a system of two coupled chemical reactions (after Atlan, 1976): $|v| \rightleftharpoons \bar{v}B \rightleftharpoons \bar{v}C$.

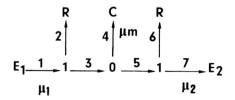

Fig. 3.12. Bond graph of the diffusion of a chemical substance across a membrane (after Atlan, 1976).

chemical energy; and the transducer *TD* maintains the power *XJ* by modification of the factors *X* and *J*.

The example shows that the force relative to *A* is multiplied by v_A while the current is simultaneously divided by v_A so that the product remains constant. Thus the role of the transducer is to adjust the reaction rates (currents *J*) taking into account the stoichiometry of the reaction.

Figure 3.10 clearly shows the equivalence between a chemical reaction and the transformation of a chemical substance of which the energy is stored in capacitors, and the energy of reaction is dissipated in a resistor. Let us consider two examples.

Example 1: Coupled chemical reactions

$$|v| A \rightleftharpoons \bar{v}B$$

$$|v'| B \rightleftharpoons \bar{v}'C$$

The representative graph is the set of the two bond graphs of the type shown in Fig. 3.10 connected by a junction 0 (Fig. 3.11).

Example 2: Diffusion across a membrane

The simplest model for the diffusion of a chemical substance (*E*) across a membrane may be represented by the graph in Fig. 3.12. We may seek the equation of state for the system in the form:

$$\frac{\mathrm{d}\mu_m}{\mathrm{d}t} = f(\mu_m; \mu_1, \mu_2) \qquad (3.39)$$

where μ_m is the chemical potential of E within the membrane itself, μ_1 and μ_2 are the chemical potentials on one side and the other of the membrane. Let $i = 1$ to 7 be the relations as on the graph in Fig. 3.12.

Starting from the properties of the junctions 0 and 1 of the relations between R and C, a simple calculation (see Atlan (1976)) shows that:

$$RC\,\frac{\mathrm{d}\mu_m}{\mathrm{d}t} = -2\mu_m + (\mu_1 + \mu_2). \qquad (3.40)$$

This *equation of state of the diffusive system* determines the variation of the membrane load μ_m for given inputs μ_1 and μ_2 and the R and C values characteristic of the membrane.

Thus, non-linear thermodynamics is based on the unification of various scientific disciplines. Although only a few applications of this technique have been found up to now, it is quite likely to turn out to be a powerful tool in the near future for quantitative studies of complex chemical systems. For example, Mikulecky *et al.* (1977) have already applied the theory to the mass transport of tissues without, however, the use of bond graphs (see Volume II, Chapter 2). Just as the thermodynamics of irreversible processes, far from equilibrium, has extended the ideas of classical thermodynamics to reveal the qualitative elaboration of new structures of order, we may say that network thermodynamics, coupling automatic calculations to powerful interpretation, allows the construction of physicochemical models for several biological phenomena.

Recently, Mikulecky (1993) has remarkably demonstrated the theoretical advantages of network thermodynamics as well as its practical applications, for example through the use of the SPICE (Simulating Program with Integrated Circuit Emphasis) simulator which permits the numerical resolution of highly complex problems. As Mikulecky points out, the problem of the compartmentation of chemical reactions has often been neglected. One of the aims of our work is to develop a general theory of functional organisation (see Chapter 4) taking into account the spatial organisation of the organism (Chauvet, 1993a).

We have considered thermodynamic theory at some length because it appears to be the only comprehensive theory capable of unifying several very different sciences. It provides a multidisciplinary approach based on relatively well-known laws of matter. Another advantage of thermodynamics is that it gives a macroscopic description of phenomena occurring at a much lower level. It is thus an example of a theory capable of linking two levels of description, the global level and the local level.

Nevertheless, thermodynamic theory cannot be considered to be a general theory of systems. But once again, we may wonder whether such a general theory could actually exist.

IV.   Thom's theory of elementary catastrophes

The various phenomena studied above can be interpreted, at least at the molecular level, by a dynamic system. For example, this is the case for the reaction–diffusion system, of the type (2.40) or (2.41), in chemical kinetics:

$$\frac{\partial x_i}{\partial t} = f_i^\lambda(x_1, x_2, \ldots, x_n) + D_i \nabla^2 x_i \qquad i = 1, 2, \ldots, n. \tag{3.41}$$

Some systems, for instance that of Herschkovitz-Kaufman (1975), (cf Eq. (2.46)), have been completely solved by analytical and numerical methods. When the constraints A and B are greater than a certain critical value that can be calculated, the system follows a new branch of the curve, beyond the thermodynamic branch, associated with the presence of a new ordered structure (Fig. 3.9).

Thus, knowledge of the parameters of a non-linear dynamic system is essential to a study of its behaviour. In the absence of the diffusion term $D_i \nabla^2 x_i$ (negligible in the first approximation), at the stationary state, Eq. (3.41) becomes:

$$f_i^\lambda(x_1, x_2, \ldots, x_n) \equiv f_i(x_1, x_2, \ldots, x_n; \lambda_1, \lambda_2, \ldots, \lambda_p) = 0 \qquad i = 1, 2, \ldots, n.$$

This expression clearly shows the *internal variables* of the system, $x_1, x_2, \ldots, x_n$, as well as the so-called '*external*' parameters: $\lambda_1, \lambda_2, \ldots, \lambda_p$.

The process observed takes place in the space–substrate (D) which, in general, is part of the space–time $\mathbb{R}^4$. $(\mathbf{r}, t)$ is thus a point of (D) at which the state of the system is defined by $X = (x_1(\mathbf{r}, t), x_2(\mathbf{r}, t), \ldots, x_n(\mathbf{r}, t)) \in V$ where V is a compact differentiable variety (Appendix B). The variety which contains $\mathbf{X}(\mathbf{r}, t)$ is written $V(\mathbf{r}, t)$. The parameters $\lambda_1, \lambda_2, \ldots, \lambda_p$ describe space Λ. It is convenient to separate the internal variables of state of the system from the parameters that act from the outside, for in biological systems n may be large while p may be small or at least manageable.

The study of the stationary system may be included in the theory of catastrophes. This theory takes into account *physical conventions external to the formulation of problems* to examine the behaviour of the system at precisely defined points when the parameters vary, these points corresponding in fact to stationary states. The critical values of the parameters at which the behaviour of the system changes are defined by Thom as the *catastrophic set*. Much work has now been published on the theory of catastrophes and it may be useful to consult, for example, Ekeland (1977), Bruter (1974, 1982) and Saunders (1980). Here we shall deal only with the general aspects of the theory, but we may remind the reader that it refers strictly to phenomena of the physical world and so the application to any other field would be purely metaphorical.

When $(\mathbf{r}, t)$ gives a regular description of the region (D) of the space–substrate, the vector of state $\mathbf{X}$ also gives a regular description of the variety $V(\mathbf{r}, t)$. However, there may be a discontinuity of $\mathbf{X}$ at certain points of (D). This set is noted A_D. If Γ is the trajectory of $\mathbf{X}$ when $(\mathbf{r}, t)$ undergoes a change, it is convenient (but not indispensable, as we shall see later) to suppose that it is defined by a field of vectors deriving from a potential F:

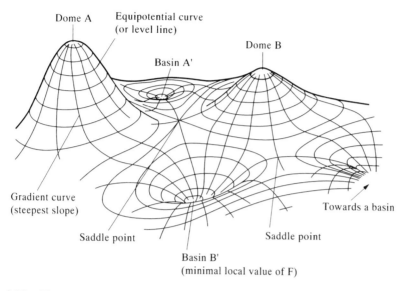

Dome A

Equipotential curve
(or level line)

Dome B

Basin A'

Gradient curve
(steepest slope)

Towards a basin

Saddle point

Saddle point

Basin B'
(minimal local value of F)

Fig. 3.13. The particles at the bottom are in the equilibrium state defined by Eq. (3.43), i.e. for a minimal local value of F. When λ varies, the contours undergo a change and the local minima are displaced (see Fig. 3.14).

$$\frac{\mathrm{d}x_i}{\mathrm{d}t} = -\frac{\partial}{\partial x_i} F(x_1, x_2, ..., x_n; \lambda_1, \lambda_2, ..., \lambda_p) \qquad i = 1, 2, ..., n. \qquad (3.42)$$

A suggestive illustration is found in the behaviour of a torrent rushing down the slopes of two mountains separated by a valley (Fig. 3.13). The values of the parameters $\lambda_1, \lambda_2, ... \lambda_p$, together with the analytical form of F, define the geometry of the equipotential curves, i.e. the contours. The continuous variation of the parameters is equivalent to that of the contours. Each particle of water on the surface of the earth, being subject to gravitational potential, will move along the gradient lines until it reaches equilibrium, defined by:

$$\frac{\mathrm{d}x_i}{\mathrm{d}t} = -\frac{\partial F}{\partial x_i} = 0 \qquad i = 1, 2, ..., n. \qquad (3.43)$$

A lake situated at the bottom of the valley corresponds to a state of equilibrium for the system (3.42), the equilibrium state being defined by Eq. (3.43). Suppose that a phenomenon of erosion, especially active on the geological structure of B, modifies the relief shown in a sectional view in Fig. 3.14 from the situation as in diagram (1) to that in diagram (4). For critical values of $\lambda = (\lambda_1, \lambda_2, ..., \lambda_p) \in \mathbb{R}^p$, lake C disappears after having emptied itself into lake D. This illustration allows the presentation of several concepts.

The concept of the *local variety of equilibria* corresponds to stationary states and consequently to states that are statistically the most easily observable. At these points, we have:

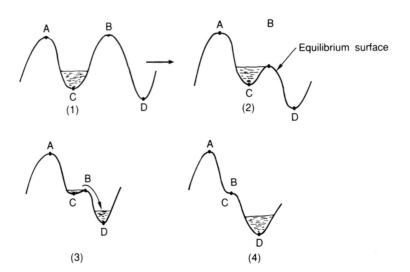

Fig. 3.14.   Illustration of critical points when the contours undergo a deformation from step (1) to step (4).

$$\frac{\partial F}{\partial x_i}(x_1, x_2, \ldots, x_n; \lambda_1, \lambda_2, \ldots, \lambda_p) = 0 \qquad i = 1, 2, \ldots, n$$

or in the global form:

$$\boxed{F'_x(X, \lambda) = 0} . \tag{3.44}$$

In the hydraulic example above, the local variety consists of four points A, B, C and D. When $\lambda = (\lambda_1, \lambda_2, \ldots \lambda_p)$ varies, the equilibrium states $X(\lambda) = (\hat{x}_1(\lambda), \hat{x}_2(\lambda), \ldots \hat{x}_n(\lambda)) \in \mathbb{R}^n$ will be displaced. The term *variety of equilibria* applies to the locus of the equilibrium states in the space $\mathbb{R}^n \times \mathbb{R}^p$ when λ moves over $\mathbb{R}^p$. The equilibrium points $X(\lambda)$ are also called *critical points* or *singularities*. The singular points can be classified: the largest classes are the *attractors*, such as C and D towards which the trajectories lead, and the *repulsors*, such as A and B from which the trajectories start. These notions may be generalised to more varied sets. When dealing with points, the attractors are called *sinks* and the repulsors *sources*, but to avoid confusion it may be better to use the terms *receivers* and *emitters* instead of sinks and sources for *singular* points.

Let us now consider formally a variety as shown in Fig. 3.15 that corresponds to the potential:

$$F(x; \lambda_1, \lambda_2) = x^4 + \lambda_1 x^2 + \lambda_2 x. \tag{3.45}$$

This is the locus of the singularities of F, solutions of the equation:

$$F'_x(x; \lambda_1, \lambda_2) = 4x^3 + 2\lambda_1 x + \lambda_2 = 0.$$

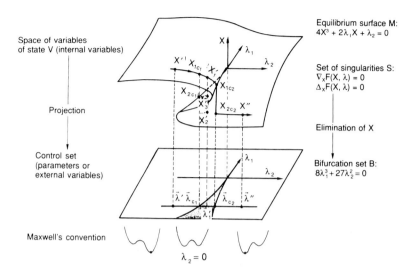

Space of variables of state V (internal variables)

Projection

Control set (parameters or external variables)

Maxwell's convention

Equilibrium surface M:
$4X^3 + 2\lambda_1 X + \lambda_2 = 0$

Set of singularities S:
$\nabla_x F(X, \lambda) = 0$
$\Delta_x F(X, \lambda) = 0$

Elimination of X

Bifurcation set B:
$8\lambda_1^3 + 27\lambda_2^2 = 0$

$\lambda_2 = 0$

Fig. 3.15. Equilibrium surface, singularity set S and bifurcation set B (which is here identical to the catastrophic set K for the potential $x^4 + \lambda_1 x^2 + \lambda_2 x$.

Let us seek the values of λ for which two singularities become identical. In this case:

$$F_x'' = 0$$

so that:

$$F_x'' = 12x^2 + 2\lambda_1 = 0.$$

By eliminating x between the equations $F_x' = F_x'' = 0$ we find the equation in λ_1 and λ_2, which determines the existence of the double singularities. From the equation above we have:

$$x^2 = -\frac{\lambda_1}{6}$$

then:

$$F_x' = x(4x^2 + 2\lambda_1) + \lambda_2 = \frac{4\lambda_1}{3}x + \lambda_2$$

from which, by substituting for x in $F_x'' = 0$, we get:

$$\delta = 8\lambda_1^3 + 27\lambda_2^2 = 0. \tag{3.46}$$

In the plane (λ_1, λ_2) (Fig. 3.15), the locus of the values of λ for which $\delta = 0$ is called the *cusp* or the *semicubical parabola*.

Let $\lambda(\lambda_1, \lambda_2)$ move along a straight line parallel to the axis of the λ_2 values, in the plane containing negative λ_1 values (cf. Fig. 3.15). We see that, when λ is outside the shaded zone, the local variety of equilibrium reduces to a point X′ or X″; and, when λ is on the semicubical parabola, the variety of equilibrium has two points:

(X_{1c_1}, X_{2c_1}) or (X_{1c_2}, X_{2c_2}). And when λ is inside the shaded zone, the local variety of equilibrium consists of three points (X'_1, X'_2, X'_3). The theory of catastrophes *introduces* the *physical conventions* which allow the determination of the singular point corresponding to the *observable* state.

With Maxwell's convention, for example, the observable state is that which corresponds to the singular point ensuring the absolute minimum of the potential function. In this case, when $\lambda = (\lambda_1, \lambda_2)$ reaches a value where $\lambda_2 = 0$, the absolute minimum, which corresponds to a point $X(\lambda)$ moving over the upper layer of the Riemann–Hugoniot surface $(\lambda_2 < 0)$, will suddenly undergo a change of position, since for $\lambda_2 > 0$ the absolute minimum is associated with a point $X(\lambda)$ moving over the lower layer of the Riemann–Hugoniot surface. When passing through $\lambda_2 = 0$, $X(\lambda_1, -\varepsilon)$ passes through $X(\lambda_1, +\varepsilon) = -X(\lambda_1, -\varepsilon)$.

Thus, with the variation of $\lambda_1, \lambda_2, \ldots, \lambda_p$, the state of the system undergoes continuous deformation with sudden variations of potential for certain critical values $\lambda_1^c, \lambda_2^c, \ldots \lambda_p^c$ of $\lambda_1, \lambda_2, \ldots \lambda_p$. The critical values form a set K called the *catastrophic set* of the system. The set A_D of the points of (D) associated with the catastrophic points of K, in which the variables of state of the system $x_1, x_2, \ldots x_n$ undergo a sudden change in values, is also the locus of the points where a change in the *morphology* appears.

More generally, the space Λ containing the parameters is called the *control space*. As we shall see in Volume III, Chapter 7, using the multiple field theory, this space also contains the external variables corresponding to a different level of functional organisation. When λ varies continuously in Λ, then $F(X, \lambda) \equiv F_\lambda(X)$ is a family of functions (with p parameters λ_i) which behave in one of two possible ways: (i) $F_{\lambda'}$ for the value λ' of the parameters is of the *same type* as F_λ, i.e. F_λ and $F_{\lambda'}$ have the same number of singular points; or (ii) the structure is different and the function no longer has the same qualitative behaviour. In the first case of stability, F_λ is said to be a *structurally stable* function. The corresponding points with coordinates $\lambda_1, \lambda_2, \ldots, \lambda_p$ are called the *generic points*. In the second case, these points constitute the *bifurcation set K*. To sum up, for a potential F (cf. Fig. 3.15), we have the following definitions:

- the *variety of equilibrium*, the subset M of $V \times \Lambda$ of the *critical points* (or singularities) of F, is defined by:

$$\nabla_X F = 0 \; ; \tag{3.47}$$

- the *variety of singular equilibria*, the subset S of M of the *degenerate critical points*, is defined by:

$$\nabla_X F = 0 \quad \text{and} \quad \Delta = \det(\mathbf{H}(F)) = 0 \tag{3.48}$$

where $\mathbf{H}$ is the matrix of the second derivatives and Δ is the determinant of the matrix;

- the *bifurcating set B* is the subset of Λ obtained by the projection of S in Λ by eliminating the variables of state $x_1, x_2, \ldots x_n$ between Eqs (3.48);
- the *catastrophic set* is the set K of the parameters $\lambda_1, \lambda_2, \ldots, \lambda_p$, for which a *maximum coincides with a minimum*; and
- the *conflicting set* is the set C of the parameters for which *two minima have identical values*.

Obviously, $K \subset B$. In the case of the fold (Fig. 3.15), the bifurcating and the catastrophic sets are identical.

The major problem of the mathematical theory of catastrophes is the determination of the classes of structurally stable potentials. These classes are sometimes incorrectly called 'catastrophes'. Thom suggested that, if the number of parameters p was four or less than four, then there could be only seven 'elementary' catastrophes, involving no more than two variables of state. Let us present the problem in simple terms.

The potential governing the dynamic course of a system must be of a regular nature. But how does this potential vary when λ varies? Let $F(X, \lambda)$ be written as:

$$F(X, \lambda) = f(X) + \sum_{i=1}^{p} \lambda_i g_i(X). \tag{3.49}$$

It can then be shown that under certain conditions $F(X, \lambda')$, another application of the same type, is stable with respect to $F(X, \lambda)$. The set of F thus constructed is said to be a *versal unfolding* of f and $F(X, \lambda)$ is the realisation of the *unfolding*. A most important result is that all the unfoldings can be reduced to one of them, called the *universal unfolding*, which has the minimum number of parameters. It would therefore be natural, according to Bruter (1974), to represent the various forms in which the physical phenomenon may appear by the universal *unfolding* $F_\lambda(X)$ as the potential function of a dynamic system. Then by supposing that the number of parameters λ_i is less than four for reasons of 'natural' simplicity (the space of the parameters being identified to space–time), we find Thom's seven elementary potentials. Finally, let us add two more definitions: the *co-dimension* is the number of parameters $p \leq 4$ of the unfolding; and the *co-rank* is the number of internal variables in the phase space (Appendix B). Table 3.2 gives the list of the potentials, called *archetypes*. The last column indicates the numbers of the figures representing the equilibrium surfaces and the catastrophic sets (when representable in $\mathbb{R}^3$) (Figs 3.16.1 to 3.16.7).

The fundamental results of Thom's theory may be summed up as follows:

(i)   There exists a limited number of geometrical figures representing the catastrophic sets associated with a system, the state of which is governed by a potential: five figures if the potential depends on three parameters, and seven figures if the potential depends on four parameters.

(ii) If p is the number of the parameters λ, the catastrophic sets are as follows: a fold for $p = 1$; a crease for $p = 2$; a swallowtail, a hyperbolic umbilic, or an elliptic umbilic for $p = 3$; a butterfly, or a parabolic umbilic for $p = 4$.

Table 3.2. Thom's theory of elementary catastrophes: for a given number of variables (co-rank) and parameters (co-dimension), the table shows the unfolding polynomial in x and the denomination of the corresponding catastrophe illustrated in Fig. 3.16. The coefficient of the monomial of the highest degree has been given a unit value in each case.

Number of state variables = co-rank	Number of parameters of deployment = co-dimension	Unfolding	Denomination	Figure 3.16
	0	x^2	minimum simple	
1	1	$x^3 + \lambda_1 x$	fold	1
	2	$x^4 + \lambda_1 x^2 + \lambda_2 x$	pleat	2
	3	$x^5 + \lambda_1 x^3 + \lambda_2 x^2 + \lambda_3 x$	swallowtail	3
	4	$x^6 + \lambda_1 x^4 + \lambda_2 x^3 + \lambda_3 x^2 + \lambda_4 x$	butterfly	4
2	3	$x^3 + y^3 + \lambda_1 xy - \lambda_2 x - \lambda_3 y$	hyperbolic umbilic	5
	3	$x^3 - 3xy^2 + \lambda_1(x^2 + y^2) - \lambda_2 x - \lambda_3 y$	elliptic umbilic	6
	4	$x^2 y^2 + y^4 + \lambda_1 x^2 + \lambda_2 y^2 - \lambda_3 x - \lambda_4 y$	parabolic umbilic	7

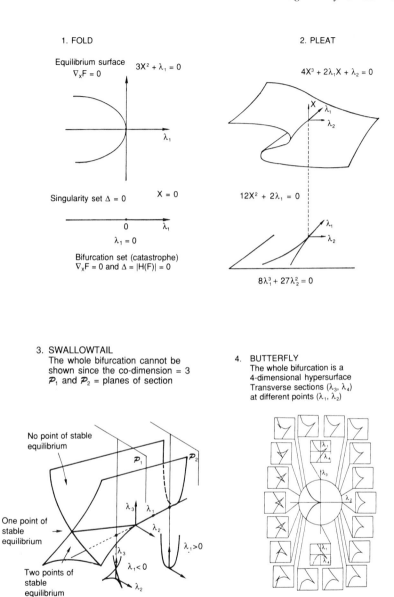

Fig. 3.16.   Bifurcation sets corresponding to Table 3.2.

5. ELLIPTIC UMBILIC
   𝒫 = plane of section
   Co-rank = 2; Co-dimension = 3

6. HYPERBOLIC UMBILIC
   𝒫 = plane of section
   Co-rank = 2; Co-dimension = 3

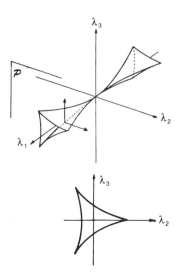

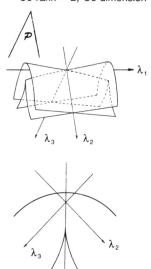

7. PARABOLIC UMBILIC: MUSHROOM
   Co-rank = 2; Co-dimension = 4
   Transverse sections (λ_3, λ_4)
   at different points (λ_1, λ_2)

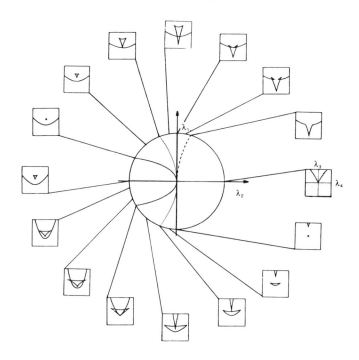

Fig. 3.16. (*cont.*)

According to Thom, various combinations of these seven catastrophes are fundamental to the morphogenesis of living organisms which maintain their structural stability during their development: *a basic logic thus appears to be subjacent to the permanent renewal of forms in Nature.*

The number of phenomenological variables relevant to morphological definition in the neighbourhood of a singularity is estimated to be of the order of 30,000. But it has been suggested that, if the analytical form of the potential were known, the number of variables playing an important role in environmental development could be reduced to as few as three. However, it must be admitted that the experimental determination of the form of the potential, at least for the present, seems far from simple.

Summary of Part I

All living organisms possess two main classes of organic macormolecules: *nucleic acids* (*DNA* and *RNA*) and *proteins*. The former are polymers of nucleotides in which the nitrogenous bases may be purines (adenine and guanine) or pyrimidines (cytosine and thymine). In RNA, thymine is replaced by uracil. DNA is a bicatenary, helicoidal molecule, the *double helix* of Watson and Crick. *Proteins* are made up of amino acids which differ by the radical R. There are 20 different species of amino acids of the type $A–R_i$.

Nucleic acids are generated from an alphabet of four letters A, T, G, C and proteins from an alphabet of 20 elements $A–R_i$ with $i = 1$ to 20. Nucleic acids transmit hereditary information, and proteins ensure catalytic activity and regulation.

A polypeptide chain may have a *primary* structure (a sequence of amino acids), a *secondary* structure (due to hydrogen bonds between oxygen of the carboxyl groups and hydrogen of the amino group on the peptides) forming a right α-helix or a folded β-sheet, a *tertiary* structure (caused by the folding of the polypeptide chain on itself), or a *quaternary* structure formed by the association of several, more or less independent, polypeptide chains, called subunits. The quaternary structure is mainly responsible for biological activity through the phenomenon of *cooperativity*, or dynamic association between subunits, which is one of the fundamental mechanisms of cell *regulation*.

Cooperativity depends on the conformation of biological macromolecules. Conformational states may be studied by thermodynamics and statistical mechanics. In particular, it is possible to calculate configuration integrals and thence the free energy variations during changes of state. A conformational change corresponds to an *allosteric* transition. The *allosteric effect* is the *indirect regulation of an enzymic reaction through an effector acting at another site and without any direct interaction*.

Protein–ligand interactions, essential to molecular biology, have been formalised by Wyman. The concept of the binding polynomial allows a systematic study of interactions by introducing the notion of the binding matrix, an abstract image of the protein–ligand chemical reactions. Two allosteric models, the Monod–Wyman–Changeux and the Koshland–Nemethy–Filmer models, are presented. The former is examined in detail and the saturation function $\overline{Y}_s$ is considered in relation to the general case.

The chemistry of living organisms is based on enzymic reactions described by the Henri–Michaelis–Menten equations. Equilibrium state conditions allow deduction of the reaction rate (variation of product concentration) in terms of two parameters: maximum rate and the Michaelis constant. Quasi-stationary state conditions (the Briggs–Haldane equations) give the reaction rate in terms of the generalised Michaelis constant. The regulatory function of the enzyme is demonstrated not only by its allosteric properties but also by the phenomena of enzymic activation and inhibition. The existence of active molecular sites is fundamental to the dynamic theory of enzyme–substrate associations. The coupling necessary to enzymic reactions is theoretically studied by systems of reaction–diffusion equations, the behaviour of which depends on thermodynamic constraints applied to the reactor. The 'Brusselator' is an example of historical and methodological importance.

The methods of biological dynamics use recent physical and mathematical theories that may be classified as relational, thermodynamic and qualitative. Relational theories consider classes of functional equivalence: Rosen's theory is very general (perhaps too much so) but establishes a theoretical framework for the study of living organisms through the fundamental activities of metabolism and self-reproduction. Delattre's theory — the formalism of transformation systems — appears better adapted to the great variety of phenomena observed in biological systems and has the merit of introducing a very powerful *axiomatico-inductive method* and the notion of *true graphs*. The thermodynamic theory of irreversible processes describes the course of chemical reactions far from equilibrium and demonstrates the existence of *new stable structures*, corresponding to structural self-organisation in living organisms, which may be an answer to Schrödinger's famous question: 'What is Life?' Thermodynamic formalism is also useful for the description of huge sets of chemical reactions, called networks, which then benefit from techniques used for the analysis of electrical networks. These methods have opened up the new field of *network thermodynamics*, which has already been successfully applied to the transport of substances across epithelial linings and seems well adapted to automatic, computerised processing of chemical networks. Finally, Thom's theory of catastrophes, developed from qualitative dynamics, represents a subtle approach to the study of very large systems involving numerous variables and parameters. The theory predicts the course of a system at certain points of interest, the *critical points*, the set of which is called the *catastrophic set*. In fact, it is precisely at the points of the space–substrate associated with the catastrophic points that morphological changes occur. If the number of parameters is less than or equal to four (as is often the case when only the relevant control variables are considered), there will be only seven catastrophes, called *elementary catastrophes*, involving no more than two state variables. These seven archetypical potentials of Thom's theory may be considered to govern the dynamic evolution of biological systems.

Part II: The Molecular Organisation of Living Matter

Rien n'a de sens en biologie, si ce n'est à la lumière de l'évolution.

Nothing makes sense in biology unless seen in the light of evolution.

Dobzhansky and Boesiger, *Essais sur l'évolution* (1968)

Introduction to Part II

A living system is *dynamically organised*, but, much more than this, it is dynamically *self-organised*. Whether we consider a simple virus or a complex multicellular creature, a living organism is fundamentally different from any of the other structures found in nature. With this problem we touch upon one of the great mysteries of life. Indeed, no solution appears possible except within the framework of the evolution of the species: self-organisation and perpetuation being the two undissociable characteristics of all forms of life. It was in the seventeenth century that Harvey's work on blood circulation first introduced the idea of a true *mechanism* in living organisms, the organisation representing nothing but the complexity of the visible structure, the normal extension of the sum of knowledge, constantly increasing, of the physical nature of life. The all-important role attributed to physical processes, even primary ones, originated in the eighteenth century idea of the 'vital force', the *organising principle* that was considered to be an integral part of the organism. The seventeenth century theory of 'animal-machines' continued to flourish during the eighteenth century and was finally consecrated by Lavoisier: 'The animal-machine is principally governed by three main regulators: *respiration*, which consumes oxygen and carbon to produce caloric; *transpiration*, which increases or decreases according to the need for transporting more or less caloric; and finally, *digestion*, which renders to the blood that which is lost by respiration or transpiration'. So that, at the beginning of the nineteenth century, there was every reason to believe that living beings were *organised* on the basis of components obeying the laws of physiology.

Thus, we first need to define, as rigorously as possible, exactly what we mean by the organisation of biological systems, or rather by the self-organisation, either from the point of view of the theory of dynamic systems or from that of the theory of information (Chapter 4). As heredity is the key to the invariance of the organisation, we shall undertake a formal study of the replication–translation apparatus (Chapter

119

5). Finally, we shall discuss two fundamental problems: (i) the organisation of macromolecules during the course of evolution (Chapter 6); and (ii) the conservation, often accompanied by the complexification of physiological processes, in particular the process of control and regulation (Chapter 7).

4

Organisation of Biological Systems

In the chapters above we have described living matter from a physical point of view, in terms of statics and dynamics, as being made up of atoms assembled in the form of molecules grouped in structural units of increasing complexity. In fact, this structural complexity is related to functional complexity. But how can functional complexity be measured? This essential problem, which in reality is that of the organisation of living beings, is now one of the major branches of research in biology, and a new approach will be found here and in Volume III. The organisation and the self-organisation of biological systems are obviously crucial to the understanding of evolution. Before going on to discuss current work in this field, let us consider some definitions. Intuitively it may be admitted that an organised system is capable of responding to its environment by means of internal reorganisation. Such a system would therefore be easier to comprehend through a study of its own transformation. For instance, an object falling in a gravitational field does not undergo a structural change whereas an animal subjected to the same field will 'reorganise' itself so as to fall on its feet. What is the nature of the reorganisation? Here, we clearly recognise a reorganisation of position in space. As we shall see in Chapter 6, the idea of the structure of a system changing under the effect of external causes may be applied to molecular organisation.

Haken (1978), with a physicist's interest in generalisation, extended these notions from inert matter to living matter, defining organisation in terms of simple dynamic systems. While the notion of organisation is quite easy to understand intuitively, no precise and rigorous definition seems to be available. Thus, the definition of organisation within the context of generalised physical systems introduces an original concept. Indeed, we have become quite used to recognising the specificity of biological systems by the organisation at the atomic, molecular and supramolecular levels, i.e. as mentioned above, by the double complexity — structural and functional — arising from organ–function relationships.

Yet, if we define a function as a relationship between the elements or subsystems of a system, the concept of functional complexity, as we saw in Chapter 3, goes a good deal further. A description of functional organisation will be found in Section III and in Chapter 12. *It would appear to be of fundamental importance to formalise the development of the relations between elements, or functional interactions, rather than the development of the elements themselves.*

A very different approach, using information theory, has led to the construction of a model of biological organisation. It is true that this theory, which is in fact a theory of communication, describes the functional rather than the structural aspect of organisation but, as we shall see, the method has its limits. However, these descriptions of functional organisation have produced an idea that Haken (1978) has satisfactorily formulated in his work on synergetics: the concept of the dynamic evolution of a system under the effect of external causes. We shall therefore begin with a formal presentation of Haken's mathematical definition of an organised system, from which the definition of a self-organised system may be immediately deduced.

I.   A formal definition of self-organisation

1.   *Organisation*

A system will be considered to be organised if the behaviour of its elements changes under the effect of an external cause, the disappearance of which cancels the effect. According to this definition, the organisation expresses a *cause–effect relationship*.

In mathematical terms, Haken (1978) defines organisation as the response of a class of systems to external causes with the *adiabatic approximation* below. More precisely, let $\mathbf{X}$ be a vector that describes the state of the system, and F the set of forces acting on the system. The dynamic change in the system is given by the matrix equation:

$$\dot{\mathbf{X}}(t) = \mathbf{A}\mathbf{X}(t) + \mathbf{B}(F)\mathbf{X}(t) + \mathbf{C}(F) \tag{4.1}$$

where $\mathbf{A}$ and $\mathbf{B}$ are matrices independent of $\mathbf{X}$, and where $\mathbf{B}$ and $\mathbf{C}$ tend to zero with F. Stability is imposed on the system by the condition of the eigenvalues λ of the real parts: $\mathrm{Re}(\lambda) < 0$.

Under the adiabatic approximation, the variation of F is supposed to be very slow compared with that of the free system $\mathbf{X}$, so that $\dot{\mathbf{X}} \approx 0$ and the solution is then immediate:

$$\mathbf{X} = -(\mathbf{A} + \mathbf{B}(F))^{-1}\mathbf{C}(F). \tag{4.2}$$

For example, in the simplest case where $\mathbf{X}$ reduces to a single component x, we have: $\dot{x} = -\lambda x + F(t)$ with $\lambda > 0$, $B = 0$ and $C(F) = F$, which is the kinetic equation of a chemical reaction where F is the external force depending on the concentrations, in other words the diffusion term (Chapter 2, Eq. (2.39)). When the force F cancels out, the system returns to the stable state. Moreover, the adiabatic approximation itself expresses stationariness.

Thus, according to Haken, an organised system will move towards the stable stationary state following a certain curve of phase space depending on external forces, i.e. the stable stationary state is an attractor for the dynamic system. But the important hypothesis here is that of the adiabatic approximation, which leads to the solution of systems that would otherwise remain unsolved. As we shall see in the course of this work, this hypothesis is quite frequently used. More than just a convenient mathematical tool, it appears to be a *natural hierarchical construction* basic to the dynamics of physiological phenomena. If the organisation were measurable, it would be possible to compare different organisms in terms of their complexity since the information content of a given system is related to the complexity and the organisation of the system. The concepts of *complexity, self-organisation* and *autonomy* will be formally defined within the framework of our theory of functional organisation in Volume III.

2. Self-organisation

Again, according to Haken (1978), a self-organising process is naturally defined from the organisational point of view as the *evolution of a system in which the organising forces, instead of being external as in the case above, are internal to the system.* Self-organisation thus results from the internal forces of the system, which in turn are modified along with the system, i.e. there is reciprocal action between effect and cause. We may therefore consider a system to be composed of subsystems, at least one of which controls or rather organises the whole system. In other words, the system is *regulated* in view of a fundamental teleonomical project, or invariant reproduction as described by Monod (1970).

A simple example will help us to understand Haken's ideas. The force F may be identified to the variable x_1 and variable of state x to the second variable x_2. Thus, compared with the organised system as described above, in which the evolution is defined by the vector of state x subject to an external force F, we now have a self-organised system (x_1, x_2). Let us now consider the dynamic system in the form:

$$(1) \quad \dot{x}_1 = -a_1 x_1 + g_1 x_1 x_2$$
$$(2) \quad \dot{x}_2 = -a_2 x_2 + g_2 x_1^2, \quad a_2 > 0 \tag{4.3}$$

and suppose a temporal hierarchy defined by $a_2 \gg |a_1|$. Then:

$$\dot{x}_2 \approx 0 = -a_2 x_2 + g_2 x_1^2$$

so that:

$$(1) \quad \dot{x}_2(t) \approx \frac{g_2}{a_2} x_1^2(t)$$

$$(2) \quad \dot{x}_1(t) = -a_1 x_1 + \frac{g_1 g_2}{a_2} x_1^3(t). \tag{4.4}$$

This mathematical formulation, which is a consequence of the temporal hierarchy imposed, leads to the following interpretation: $x_2(t)$ is obtained from $x_1(t)$ by Eq. (4.4(1))) so that the subsystem (x_2) may be considered to be a slave to (x_1), its reaction on (x_1) being represented by Eq. (4.4(2)). The solution of this equation leads to a positive or zero value for x_2. If $x_1 = 0$ then $x_2 = 0$, and if $x_1 \neq 0$ then $x_2 \neq 0$. In other words, the effect x_2 is cancelled if the cause (internal) x_1 is cancelled. For this reason, x_1 is called the *parameter of action or of order*.

In Chapter 2, we found this kind of temporal hierarchy for the reduced system of the Henri–Michaelis–Menten equations (2.15). Equations (4.4) are then similar to the 2(1) equations obtained for $n = 0$ by solving the *singular perturbation problem*, i.e. when the general form of the ordinary differential equations is:

$$\frac{du}{d\tau} = f(u, v), \quad \frac{dv}{d\tau} = \frac{1}{\varepsilon} g(u, v), \quad 0 < \varepsilon \ll 1$$

with the following development of the solution in terms of the powers of ε:

$$u(\tau; \varepsilon) = \sum_{n=0} \varepsilon^n u_n(\tau), \quad v(\tau; \varepsilon) = \sum_{n=0} \varepsilon^n v_n(\tau).$$

The most general system may be written in a form identical to (4.3):

$$\dot{x}_1 = -a_1 x_1 + g_1(x_1, \ldots, x_n)$$
$$\vdots$$
$$\dot{x}_n = -a_n x_n + g_n(x_1, \ldots, x_n)$$

or in matrix form:

$$\dot{\mathbf{X}} = -\mathbf{A}\mathbf{X} + \mathbf{G} \qquad (4.5)$$

where $\mathbf{A}$ is a diagonal matrix made up of $a_i \equiv a_{ii}$ and $\mathbf{G}$ is a column matrix where each g_i is a function of x_i, $i = 1$ to n.

A certain subset $\{x_1, \ldots, x_i, \ldots, x_s\}$ of $\mathbf{X}$ corresponds to weak negative damping (unstable mode with $a < 0$), while the other $\{x_{s+1}, \ldots, x_j, \ldots, x_n\}$ corresponds to positive damping (stable mode with $a > 0$).

The hypothesis of adiabatic approximation is expressed by:

$$|a_j| \gg a_i \quad \begin{array}{l} 1 \leq i \leq s \\ s + 1 \leq j \leq n \end{array} \qquad (4.6)$$

so that, as in the particular case above, we may write:

$$\dot{x}_j \approx 0$$

and, following (4.5):

$$x_j \approx \frac{1}{a_j} g_j(x_1, \ldots, x_n) \quad s + 1 \leq j \leq n \qquad (4.7)$$

from which we deduce the *master* equations:

$$\dot{x}_i = -a_i x_i + g_i(x_1, \ldots, x_s; x_{s+1}(x_i) \ldots) \quad 1 \leq i \leq s \tag{4.8}$$

where the variables x_i are the parameters of order of the system. These variables control the system since, if x_i is zero for all i, then the g_j values are zero for all j. *The behaviour of the entire system is governed by the behaviour of the parameters of order.*

Under these conditions, the variables x_j of the second group in the stable mode depend directly on the variables of the first group in the unstable mode. The second group is a slave to the first since it is controlled by the variables x_i, $i = 1, \ldots, s$. From this point of view, the variables x_i, $i = 1, \ldots, s$, describe the order of the system, or rather express the *degree of organisation*. Finally, the dynamic course of the system, considered as being made up of two intimately linked subsystems $(1, s)$ and $(s + 1, n)$, is determined by the values of the first subsystem which turns the second into a slave subsystem.

The dynamic systems (4.3) and (4.5) are based on an important hypothesis which allows the introduction of the damping terms $-a_i x_i$ for all i. It can be shown that the general non-linear dynamic system:

$$\dot{x}_i = h_i(x_1, \ldots, x_n) \quad i = 1, 2, \ldots, n \tag{4.9}$$

can be expressed in this form if there exists a stable stationary solution:

$$\mathbf{X}^{(0)} = (x_1^{(0)}, x_2^{(0)}, \ldots, x_n^{(0)}).$$

After analysis of the stability of the system (Eq. (2.42)), we may write:

$$x_i(t) = x_i^{(0)} + \varepsilon_i(t) \quad i = 1, 2, \ldots, n \tag{4.10}$$

and linearise the system (4.9) in the matrix form:

$$\dot{\boldsymbol{\varepsilon}} = \mathbf{L}\boldsymbol{\varepsilon} \tag{4.11}$$

where the non-linear contribution to the equation, $\mathbf{Q}(\boldsymbol{\varepsilon})$, has been suppressed (this notation signifies that the elements of the matrix $\mathbf{Q}$ are functions of $\varepsilon_1, \varepsilon_2, \ldots, \varepsilon_n$):

$$\dot{\boldsymbol{\varepsilon}} = \mathbf{L}\boldsymbol{\varepsilon} + \mathbf{Q}(\boldsymbol{\varepsilon}). \tag{4.12}$$

$\mathbf{L}$ is the Jacobian matrix (Appendix B). The general solution is known to be given by:

$$\boldsymbol{\varepsilon} = \sum_\alpha \eta_\alpha e^{\lambda_\alpha t} \boldsymbol{\varepsilon}_\alpha(0) \tag{4.13}$$

where the λ_α are the eigenvalues obtained by the diagonalisation of $\mathbf{L}$, i.e. they are the solutions of:

$$\lambda_\alpha \boldsymbol{\varepsilon}_\alpha(0) = \mathbf{L}\boldsymbol{\varepsilon}_\alpha(0) \tag{4.14}$$

with $\boldsymbol{\varepsilon}_\alpha(0)$ as eigenvectors to the right. Substituting the *function of time*, η_α, in (4.12), it may be shown that $\eta_\alpha(t)$ satisfies the equation:

$$\dot{\eta}_\alpha(t) = \lambda_\alpha \eta_\alpha(t) + g_\alpha(\eta_1, \eta_2, \ldots, \eta_n) \quad \alpha = 1, 2, \ldots, n \tag{4.15}$$

where g_α is a function from the non-linear term $\mathbf{Q}(\varepsilon)$. This new system of equations deduced from (4.9) in the neighbourhood of the stable stationary state $\mathbf{X}^{(0)}$ is analogous to the system (4.8) discussed above. Thus, it will suffice to arrange the variables η_α in two groups. The variables with $\mathrm{Re}\,(\lambda_\alpha) > 0$ correspond to unstable modes and are the parameters of order which control the adiabatically stable slave modes. These two groups of variables are distinct and under the control of other parameters.

Finally, we see that Haken's definition of self-organisation is based on the mathematical structure of equations in the neighbourhood of stationary states and on the existence of a *temporal hierarchy* allowing adiabatic elimination. It reveals two types of variables, the *parameters of order* which control the dynamic course of the system, analogous to the external forces in the definition of organised systems, and the *slave variables* analogous to the variables of state of the subsystem.

II.   Biological organisation and information theory

1.   *Von Foerster's self-organising system*

The theory of information in its most elaborate form, as proposed for example by Weiner (1948), Shannon and Weaver (1972), and Brillouin (1959), is essentially a theory of *communication*. It attempts to express the quantity of information contained in a message sent from transmitter to receiver over a single channel. The message is transmitted by means of a given arbitrary number N_α of symbols α_i, $i = 1, 2, \ldots, N_\alpha$. If $P_i = \mathrm{Pr}[\alpha_i]$ or the probability of the appearance of the symbol α_i in the message (measured asymptotically by the relative statistical frequency of appearance according to the law of large numbers), the *mean quantity of information per symbol* contained in the message x is given by the Shannon function $H(x)$:

$$H(x) = -\sum_{i=1}^{N_\alpha} P_i \log_2 P_i . \tag{4.16}$$

This is the mathematical expectation (Appendix D) of $-\log_2 P_i$, which represents the quantity of information of the symbol. If $P_i = p = \dfrac{1}{N\alpha}$ $\forall i$, we evidently have:

$$H(x) = -\log_2 p = \log_2 N_\alpha$$

which is the mean quantity of information if the symbols are assumed to be equiprobable. The function H is thus characteristic of the symbols used in the message. In the case of an alphabet of two symbols 0 and 1, as used by computers, we have:

$$H(x) = -P_1 \log_2 P_1 - P_2 \log_2 P_2$$

and if

$$P_1 = P_2 = \tfrac{1}{2}, \text{ then } H(x) = 1.$$

This value of H, or rather the definition given (in terms of logarithms to the base 2), is the unit of measurement of the quantity of information, called the *bit* (binary digit). *Generally, the function* H *passes through a maximum when the symbols are equiprobable.*

Compared with Haken's theory above, Shannon and Weaver's theory of communication (1948) uses a totally different approach to a theory of biological organisation. But what kind of organisation does it describe?

We know there are two kinds of complexity: *structural* complexity defining what may be called a '*static*' organisation, and *functional* complexity defining a '*dynamic*' organisation. The first type of organisation has been described by Dancoff and Quastler (1953). The authors calculate the quantity of information needed to create a structure, without considering its function, by using Shannon's expression for H (1948). The description of the second type of organisation involves a process varying with time. Von Foerster (1960) has defined this by the function $R(t)$, which is the variation with time of the interdependence of the symbols in the message, i.e. the elements of the system (see also Eq. (5.2)). According to Haken (1978), there exists an external force acting on the interactions between the elements, causing R to vary between total disorder ($R = 0$) and maximum order ($R = 1$).

Von Foerster defines a self-organising system by:

$$\frac{dR}{dt} > 0 \tag{4.17}$$

so that the order of the system in its environment increases with time. As $R = 1 - \dfrac{H}{H_{max}}$, it follows that:

$$\frac{dR}{dt} = \frac{-H_{max}\dot{H} + H\dot{H}_{max}}{H_{max}^2} > 0$$

whence the condition:

$$\boxed{H\dot{H}_{max} > H_{max}\dot{H}} \tag{4.18}$$

which expresses the increase of R with time. But what do the functions H and H_{max} represent? We have seen that H is the degree of constraint within the system or the dependence between the elements, while H_{max} measures potential information, i.e. that which exists in the case of total disorder. R may therefore increase through two causes, an internal cause which increases the constraint in the system, and an external cause which increases potential information. During the change with time, the condition above must be valid.

Von Foerster gives us an intuitive idea of the organisation within a system with the following analogy: he considers a box containing a large number of small, cubical magnets, magnetised in a given direction for three faces with a common vertex and in the opposite direction for the other three faces. After the box is shaken for a time, the chaotic mass of cubes is transformed into an apparently highly organised,

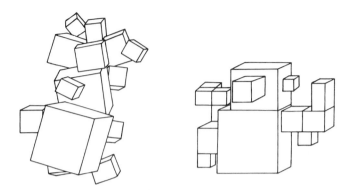

Fig. 4.1. Von Foerster's representation of structural organisation. A set of magnetised cubes (left) appears to be organised (right) after being shaken together.

ordered structure (Fig. 4.1). A disordered agitation is thus seen to be capable of producing order, according to Von Foerster's principle of 'order arising from noise'. But it must be admitted that the organisation here is structural and not functional. Of course, the principle is novel compared with the earlier idea proposed by Schrödinger (1945), that of 'order arising from order', which explains the increase in the order of a system by an import of order from the environment. It is interesting to compare Von Foerster's principle with other theories. For instance, we have seen how a new spatiotemporal order may be created far from thermodynamic equilibrium under the condition that the system be supplied with negative entropy. We may also observe that Haken's formal definition of self-organising systems holds good if the system considered is a global, isolated system (structure together with the environment). *But the originality of Von Foerster's work lies in the idea of* noise *as the additional source of order.* Noise allows the introduction of potential order in relation to the properties of the elements of the system. More recently, Eigen (1971) has demonstrated that a system can only develop through random errors, i.e. through noise. Glansdorff and Prigogine (1970) have also shown that an isolated system may develop into a new structure beyond the thermodynamic branch for random fluctuations of the parameters in the neighbourhood of a bifurcation point. We shall return to this subject later, but for the moment let us consider the concept of self-organisation from the point of view of information theory as currently proposed by several authors (see Atlan, 1972).

The logic of the organisation of a system must be based on the interactions between the system and the environment so that:

(i)   the system cannot be conceived as being isolated from the environment; and

(ii)   the process of organisation and adaptation depends, at least in part, on random or unplanned effects of the environment on the system.

Using Yockey's equation (1958), Atlan (1972) gives the mathematical formulation below for the 'creation of order from noise', i.e. the increase in the quantity of information of a system under the influence of random fluctuations.

2.   The Yockey–Atlan theory of self-organisation

Let us first clarify the notion of the organisation of a system in terms of information theory. *If a system (S) is organised, there will necessarily be a transfer of information between its subsystems (E$_i$),* which, as we have seen, reflects the probabilities of transition between the (E$_i$) subsystems. A transfer of information of this kind is called a *constraint*.

There are thus two extreme cases:

- *total dependence* between the subsystems, so that any 'movement' in one of the subsystems implies the same 'movement' in all the others. There is no more information in (S) than there is in any one of the subsystems (E$_i$):

$$\text{total constraint} \Leftrightarrow H_S = H_{E_i}; \tag{4.19}$$

- total independence between the subsystems, so that the disorganisation between the parts is complete:

$$\text{zero constraint} \Leftrightarrow H_S = \sum_i H_{E_i}.$$

The organisation of a system thus corresponds to an *intermediate state*, called *ambiguity* (for it expresses the incertitude existing in the subsystem E_i when E_j is known), such that:

$$H_{E_i/E_j} \neq 0 \quad \forall i, j.$$

It represents the constraint between the subsystems E_i and E_j, zero organisation being characterised by:

$$H_{E_i/E_j} = H_{E_i} \quad \forall i.$$

We shall see that H may be written:

$$\boxed{H = H_{E_i} + H_{E_j/E_i} - E_{E_i/x}}. \tag{4.20}$$

The first term H_{E_i} is the total information of the system when the transmission on the channel $E_i \rightarrow E_j$ occurs without ambiguity, i.e. such that $H_{E_i} = H_{E_j}$. As above, the noise acting on the system produces an *ambiguity–autonomy* which increases the information by the quantity H_{E_j/E_i}. This is the second term. The third term similarly describes the influence of noise on the transmission channel from the system to the observer, i.e. the *destructive ambiguity*. There is a loss of information $H_{E_i/x}$ where x represents the set of input elements to the system, the only ones known to the observer. Let us look more closely at the transmission channel between E_i and E_j. If there is noise on this channel, the quantity of information of the system *perceived* by the observer increases. Paradoxically, there is an increase in autonomy and thus a reduction of constraints and of R, which is why Atlan (1972) calls the term *ambiguity–autonomy*. In fact, although R decreases, showing that the structural disorder increases, $H = H_{max}(1 - R)$ increases. Similarly, the quantity of information in the

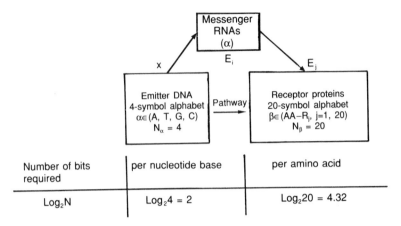

Fig. 4.2.   Information transfer along a single channel of communication in the case of protein biosynthesis, $x \equiv$ DNA, $E_i \equiv$ mRNA and $E_j \equiv$ proteins.

systems E_i and E_j increases as soon as E_i and E_j become autonomous or as soon as the redundance of E_i and E_j decreases. We may therefore express the principle of order arising from noise as follows: the total quantity of information of an organised system is equal to the sum of the information of the different substructures to which should be added a quantity of information corresponding to the ambiguity between the substructures. *Thus, noise reduces redundance.* The order 'created' from noise, as considered by Von Foerster (1960), differs in origin from that due to decreased redundance. However, noise on the transmission channel between the system and the observer will cause a decrease in the information transmitted to the observer, because of the *destructive ambiguity.*

In the case of protein biosynthesis, represented schematically in Fig. 4.2, the noise corresponds to copying errors. The entire system consists of only two substructures, (E_i) and (E_j), with just one transmission channel from E_i to E_j. The total information perceived by the observer, in the particular case where the elements x are at the input and E_j is at the output, is (Eq. (4.20)):

$$H = H_{E_i} + H_{E_j/E_i} - H_{E_i/x}.$$

Let us now calculate this expression explicitly. If x (or E_i) and E_j are messages written with the aid of alphabets (α) and (β) containing respectively N_α and N_β elements ($N_\alpha = 4$, $N_\beta = 20$), we may write (see Appendix D):

$$P_{\beta/\alpha} = \Pr[E_j = \beta/E_i = \alpha].$$

This is the probability of finding the symbol β in the outgoing message E_j when the symbol α is in the message E_i, in the place corresponding to β, for the channel $E_i \to E_j$. Following (Eq. (D-3)):

$$H_{E_j/\alpha} = -\sum_{\beta} P_{\beta/\alpha} \log_2 P_{\beta/\alpha}$$

where $P_{\beta/\alpha}$ is the conditional probability of the symbol β being in privileged correspondence with the symbol α (of the message E_i). We deduce that the mean incertitude on E_j, when E_i known, is given by:

$$H_{E_j/E_i} = \left\langle H_{E_j/\alpha} \right\rangle_{\Pr[E_i]} = \sum_\alpha P_\alpha H_{E_j/\alpha} \tag{4.21}$$

where $\left\langle H_{E_j/\alpha} \right\rangle_{\Pr[E_i]}$ is the mean value weighted by the probabilities $\Pr[E_i]$. These expressions are deduced from the conditional probabilities in Shannon's theory. The second term of (4.20) is given by (4.21), i.e. the mean value weighted by the distribution of the probabilities for the symbols α (Eq. (D-4)):

$$H_{E_j/E_i} = -\sum_{\alpha,\beta} P_\alpha P_{\beta/\alpha} \log_2 P_{\beta/\alpha}.$$

Similarly, we may calculate $H_{E_i/x}$:

$$H_{E_i/x} = \sum_\sigma P_\sigma H_{E_i/\sigma} = -\sum_{\alpha,\sigma} P_\sigma P_{\alpha/\sigma} \log_2 P_{\alpha/\sigma}.$$

It is supposed that the noise factor varies with time on the correspondence between the symbols (and thus, as we shall see, on the rules of transcription and translation), i.e. on the conditional probabilities $P_{\alpha/\sigma}$ and $P_{\beta/\alpha}$, according to the following kinetics:

$$\frac{\mathrm{d}}{\mathrm{d}t} P_{\beta/\alpha} = -J_{\alpha\to\beta}(t) \cdot P_{\beta/\alpha} + C_{\alpha\to\beta}(t). \tag{4.22}$$

In this differential equation, due to Yockey, the second member evaluates the copying error: the first term represents the decrease of $P_{\beta/\alpha}$ with the 'rate constant' $J_{\alpha\to\beta}(t)$; the second term represents a favourable error (e.g. another symbol is replaced by β) so that it is independent of $P_{\beta/\alpha}$.

Summing over β, with $\sum_\beta P_{\beta/\alpha} = 1$, leads to:

$$\frac{\mathrm{d}}{\mathrm{d}t} \sum_\beta P_{\beta/\alpha} = -\sum_\beta J_{\alpha\to\beta} P_{\beta/\alpha} + \sum_\beta C_{\alpha\to\beta} = 0. \tag{4.23}$$

If we suppose the values $J_{\alpha\to\beta}$ and $C_{\alpha\to\beta}$ to be replaced by their mean values $J(t)$ and $C(t)$, the equation above may be expressed in *average* terms as follows:

$$J(t) \sum_\beta P_{\beta/\alpha} = \sum_\beta C(t) = N_\beta C(t)$$

where N_β is the number of symbols in the alphabet (E_j). We then deduce:

$$\boxed{\frac{\mathrm{d}}{\mathrm{d}t} P_{\beta/\alpha} = -J(t) P_{\beta/\alpha} + \frac{1}{N_\beta} J(t)}. \tag{4.24}$$

This fundamental equation, due to Yockey (1958), expresses the dynamic change of $P_{\beta/\alpha}$ in terms of the noise factor represented by $J(t)$ on the channel $E_i \rightarrow E_j$. Similarly, on the channel $x \rightarrow E_i$, we have:

$$\frac{d}{dt}P_{\alpha/\sigma} = -J(t)P_{\alpha/\sigma} + \frac{1}{N_\alpha}J(t). \tag{4.24'}$$

In the example of protein biosynthesis, we have $N_\alpha = 4$ and $N_\beta = 20$. The transition $x \rightarrow \alpha$ represents the *transcription* and the transition $\alpha \rightarrow \beta$ represents the *translation* (Fig. 4.2). We may deduce the equation of H in terms of time (4.20):

$$H = H_{E_i} + H_{E_j/E_i} - H_{E_i/x}$$

where H_{E_i} represents the quantity of information of the message of mRNA, $H_{E_i/x}$ the *destructive ambiguity* on this channel and H_{E_j/E_i} the *ambiguity–autonomy*. By deriving with respect to t, and supposing (i) that the probabilities P_α are independent of time since the errors affect *only the transition probabilities* during transcription and translation, and (ii) that $J(t) = J_0 = $ constant in the phases of transcription and translation, as appears to be experimentally verified, we have:

$$\frac{dH}{dt} = \frac{dH_{E_j/E_i}}{dt} - \frac{dH_{E_i/x}}{dt}.$$

A fairly long calculation, based on Eqs (4.21) and (4.24), then leads to the following expressions for the terms of the second member:

$$\frac{d}{dt}H_{E_j/E_i} = -J_0 H_{E_j/E_i} - J_0 \sum_{\alpha,\beta}\frac{1}{N_\beta}P_\alpha \log_2 P_{\beta/\alpha}$$

and

$$\frac{d}{dt}H_{E_i/x} = -J_0 H_{E_i/x} - J_0 \sum_{\alpha,\sigma}\frac{1}{N_\alpha}P_\sigma \log_2 P_{\alpha/\sigma}.$$

Finally, since:

$$J_0 H = J_0 H_{E_i} + J_0 H_{E_j/E_i} - J_0 H_{E_i/x}$$

we have:

$$\boxed{\frac{dH}{dt} + J_0 H = J_0 H_{E_i} + J_0 \sum_{\sigma,\alpha}\frac{1}{N_\alpha}P_\sigma \log_2 P_{\alpha/\sigma} - J_0 \sum_{\alpha,\beta}\frac{1}{N_\beta}P_\alpha \log_2 P_{\beta/\alpha}} \tag{4.25}$$

This is Atlan's equation for the quantity of information in a system generating protein biosynthesis under the influence of noise. By integrating this equation, we find the solution in the following form:

$$H(t) = H_{E_i} + \left(1 - \frac{1}{N_\alpha}\right)\left(\log_2 \frac{N_\beta}{N_\alpha}\right)(1 - e^{-J_0 t}).$$

In effect, we easily obtain the solution of the equation in $P_{\alpha/\sigma}$ $((4.24'))$ analogous to (4.24) in $P_{\beta/\sigma}$) as

$$P_{\alpha/\sigma}(t) = \frac{1}{N_\alpha} + \left[(P_{\alpha/\sigma})_{t=0} - \frac{1}{N_\alpha} \right] e^{-J_0 t}.$$

Neglecting the correct transition $x \rightarrow E_i$, and retaining only the average of the 'incorrect' transitions (i.e. those *with errors*), the P_σ values can be approached by the mean values $\dfrac{1}{N_\alpha}$, and the $\log_2 P_\alpha/\sigma$ values by the values for the incorrect transitions alone. With this approximation (Atlan, 1972), and writing $N_\alpha = N_\sigma$, we find:

$$\sum_{\alpha,\sigma} P_\sigma \log_2 P_{\alpha/\sigma} \approx (N_\alpha - 1) \log_2(1 - e^{-J_0 t}) - (N_\alpha - 1) \log_2 N_\alpha.$$

Thus, the fundamental equation (4.25), without the term $J_0 H_{E_j/E_i}$, is written:

$$\frac{\mathrm{d}H}{\mathrm{d}t} + J_0 H = K_1 J_0 \ln(1 - e^{-J_0 t}) + (H_{E_1} - K_2)J_0$$

where:

$$K_1 = \left(1 - \frac{1}{N_\alpha} \right) \log_2 e$$

and

$$K_2 = \left(1 - \frac{1}{N_\alpha} \right) \log_2 N_\alpha.$$

Integrating this equation (see Atlan, 1972) and taking the term H_{E_j/E_i} into account, we obtain:

$$H(t) = H_{E_i} + \left(1 - \frac{1}{N_\alpha} \right) \left(\log_2 \frac{N_\beta}{N_\alpha} \right) (1 - e^{-J_0 t}).$$

This expression gives the variation of H under the cumulative effects of the noise factors considering the *two types of ambiguity*. We can then have $\mathrm{d}H/\mathrm{d}t > 0$.
But from this equation in $H(t)$ we deduce that $\mathrm{d}H/\mathrm{d}t$ is positive only if:

$$\boxed{\frac{N_\beta}{N_\alpha} > 1} . \tag{4.26}$$

Thus, protein biosynthesis requires a change of alphabet such that the number of symbols in the message E_j *(the amino acid sequence of the protein) is greater than the number of symbols in the message* E_i *(the nucleotide sequence of mRNA). This* remarkable result embraces the very existence of the genetic code which thus appears to be the consequence of the necessary increase in the quantity of information within the system *through the increase of what Atlan calls the ambiguity–autonomy.*

As $H = H_{max}(1 - R)$, we may write:

$$\frac{dH}{dt} = H_{max} f(t) + (1 - R)g(t) \tag{4.27}$$

with:

$$f(t) = -\frac{dR}{dt} \quad \text{and} \quad g(t) = \frac{dH_{max}}{dt}.$$

The process of organisation may then be defined by the two functions $f(t)$ and $g(t)$ which are such that (at least for small values of t):

$$f(t) = -\frac{dR}{dt} > 0 \quad \text{as } R \text{ decreases}$$

which corresponds to the ambiguity–autonomy, i.e. the creation of order from noise on the internal channel by the decrease of redundance; and

$$g(t) = -\frac{dH_{max}}{dt} < 0 \quad \text{as } H_{max} \text{ decreases}$$

which corresponds to the destructive ambiguity, i.e. the decrease in the total quantity of information on the channel from the system to the observer. *The reciprocal action of the influence of redundance and of maximum information determines the self-organising behaviour of the system.* Indeed, we have the following result:

$$\boxed{dH/dt > 0 \quad \text{with} \quad H_{max} f(t) > (1 - R) \,|g(t)| \text{ for } t < t_M} \tag{4.28}$$

This phase corresponds to an increase in complexity. Then, for a certain pair of values $(f(t_M), g(t_M))$, $H(t)$ passes through a maximum $(dH/dt = 0)$ and then decreases. The expression: $H = H_{max}(1 - R)$ shows that $(1 - R)$ increases under the effect of noise, thus compensating the decrease of H_{max} which allows H to increase. But from the time t_M, H_{max} decreases more than $(1 - R)$ increases. The interpretation is that noise leads to a change of structure without loss of functionality for the system as long as the decrease in redundance is enough to offset the total loss of potential information H_{max}. Therefore, the initial redundance $R(0)$ must be high. Figure 4.3 illustrates this type of behaviour: t_M represents a measure of the *reliability* of the system, i.e. the degree of the functional organisation of the system which would be self-organising only for *positive* values of reliability.

This description is interesting in that it clearly shows how a system, initially self-organising, may undergo an ageing process under the effect of noise factors. *Thus, the self-organising character of a system appears to be determined by two quantities:*

(1) the value of R_0, the initial redundance, which determines the self-organising potential with increasing $H(t)$; and

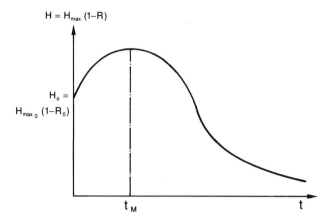

$$H = H_{max}(1-R)$$

$$H_o = H_{max\,0}(1-R_0)$$

t_M

t

Fig. 4.3. Shannon's $H(t)$ function for a single transmission channel between two substructures (after Atlan, 1972). The function passes through a maximum $t = t_M$, the reliability of the system.

(2) the reliability, which expresses the rapidity with which $H(t)$ attains its maximum value, and characterises the duration of self-organisation.

Moreover, as well shall see in Chapter 7, these two quantities could play an important role in an explanation of the evolution of the species.

III. A theory of the functional organisation of formal biological systems: some concepts and definitions

1. *Introduction to the functional organisation of biological systems*

How are biological systems organised? In the sections above, we have presented two approaches, based respectively on physical intuition and on the concept of information. However, these approaches are extremely general and therefore difficult to apply to the physiological functions of the organism.

Many authors have discussed biological organisation from points of view based on a well-established mathematical or physical theory. Thom (1972), with his catastrophe theory based on qualitative dynamics (Chapter 3), conceived a theory of morphogenesis (Chapter 10), which was then extended by Zeeman (1977); Prigogine and his co-workers developed a theory of structural self-organisation (Chapter 3) based on the principles of the thermodynamics of irreversible processes (Nicolis and Prigogine, 1977; Prigogine *et al.*, 1977); structural pattern-forming, including the mechanochemical approach to morphogenesis (Chapter 12) was investigated by Oster *et al.* (1983), Murray and Oster (1984a, b); Eigen (1971), and Eigen and Schuster (1979) applied neo-Darwinian principles to macromolecular self-organisation. Other types of formalism have also been extensively used: transformation systems (Delattre, 1971b), compartmental analysis (Jacquez, 1985; Walter, 1980, 1983a), general and hierarchical systems (Arbib, 1972; Pattee, 1970), automaton theory (Kauffman, 1985), graph theory (Rosen, 1958a, b; Rashevsky, 1961; Levins, 1970), graph theory for neural networks (Von Foerster, 1967; Hopfield, 1982), and statistical mechanics (Demetrius, 1983) (Chapter 11).

The objective of this book is to develop our own point of view of the functional organisation of biological systems. As shown in Chapter 3, and recalled above, the relational aspect of biological systems has already been suggested but, curiously enough, has not been extended to the functional organisation of physiological functions. Our aim is to show how some concepts and definitions deduced from the nature of systems in numerous fields of biology may lead to the formalisation of realistic properties common to the functional organisation of a general physiological system (Chauvet, 1993a,b,c). The 'bottom-up' approach represented by the contents of the first two volumes finds its counterpart in the 'top-down' approach with which we attempt to establish the integrative theory of biological systems presented in Volume III. Let us here introduce some of the concepts and definitions that will be used in several of the following chapters.

The two new concepts of *'functional interaction'* and *'structural discontinuity'* have been demonstrated to be fundamental to biology. As a consequence, biological systems have been shown to be driven by specific criteria of evolution that are different from those of physical systems. Three specific properties: *non-symmetry, non-locality,* and *non-instantaneity,* are related to these basic concepts. Using the specific properties of functional interactions, we have shown that a biological system can be considered to be composed of two hierarchical systems: (i) the observed functional biological system (O-FBS) that describes the topology of the FBS, i.e. the functional organisation, with a hierarchical directed graph; (ii) the dynamic functional biological system (D-FBS) that describes the continuous non-linear dynamics of the FBS in terms of a field. In the framework of this theory, the problem of the relation between structure and function is considered to be due to the distinction between the structural and functional organisations.

Although structure and function appear to be indissociable since a biological function cannot be conceived without a structure to support it, the formalisation of a functional organisation will be shown to involve hierarchical systems that do not necessarily coincide with the corresponding structural systems. Epistemologists have put forward definitions of structure and function that are difficult to formalize within a self-coherent theory. The point of view of mathematical biologists, such as Rashevsky (1961), often concerns the topological nature of biological systems. Although the topological description seems to approach the idea of a set of relations between the elements of a system, the principles that underlie its origin have to be determined in order to answer to following questions: How does a functional organisation evolve? Can its behaviour be expressed with a minimum number of hypotheses? What is a physiological function?

Physical systems at all levels of description are characterised by their structure, i.e. a combination of structural interactions or forces between elements of matter. Physical laws specify how the stability of this set of elements may be established.

Similarly, biological systems are composed of material elements, and therefore satisfy these physical laws. However, being physiological systems, they possess specific properties. As each substructure acts at a distance on another substructure, it can be shown that *functional interactions*, playing the role of forces in physical systems, exist between all substructures in the physiological system. Functional interactions have three specific properties: *non-symmetry, non-locality*, and *non-instantaneity*, which give biological systems their unique characteristics. Since, in terms of functional interactions, the observed functional organisation has to be a stable combination of these interactions, two problems arise. The first concerns the conditions of stability of the functional organisation; and the second concerns a criterion of organisation. Ultimately, the solution to these problems may be expected to lead to a general principle of the evolution of the biological system.

Thus, *in the biochemical pathways of living organisms, non-locality is a rule*. For example, an organ such as the thyroid, located in the neck, can act chemically on a muscle cell, situated for instance in the foot, only through a *non-local* interaction involving the production of a thyroxine molecule capable of triggering the necessary chemical reaction at a distance. Similarly, when there is an excess of glucose in the blood, the insulin secreted by the pancreas increases the transport of glucose across the membranes of liver and muscle cells. This phenomenon is perfectly general and may be observed in all structures of the organism.

We believe that the fundamental cause of the non-locality lies in the *structural discontinuities* existing between structural units and, consequently, in the global regulation of the organism which must work as a whole. This important property corresponds to a major difference between the biological and the physicochemical systems, introducing what may now be called a *function*. Thus, the function of the thyroid is to produce an action on a cell situated elsewhere in physical space, this action being coordinated with other phenomena of chemical origin in the organism.

2. *The problem of representation in biology*

In a remarkable epistemological analysis, Delattre (1971a) has given very general definitions of the notions of *system, structure, function* and *evolution* that are equally useful in the mathematical and the biological sciences. Although the definitions adopted by Delattre are not the only ones possible, they contribute to a clear theoretical presentation. These notions are closely related to the formalism of transformation systems (Chapter 3), the usefulness of which becomes evident after further development from an epistemological point of view.

Let us now try to extend Delattre's epistemological and mathematical analysis to the problem of representation in theoretical physiology. We shall see that, for practical reasons, the formalisation of physiological functions requires different definitions. After recalling and discussing these definitions, we shall introduce the concepts that appear to be necessary to the representation of physiological phenomena.

a. Notions of system, structure, function and evolution

α. *System: general considerations.*   Since each *system* is a set of interacting elements, it is necessary to define an element. A functional definition of an element should contain all the qualitative and quantitative characteristics required to account for the role and the behaviour of the element with respect to the other elements of the system. The interactions within the system are expressed by relationships between the characteristics of the elements constituting the system. These interactions produce two types of effect as far as the elements are concerned (Delattre, 1971a):

— a modification of the quantitative characteristics, in which case we have a system with *modifiable elements*; or

— the appearance or disappearance of qualitative characteristics, in which case the system is said to have *transformable elements.*

Evidently, these effects may occur simultaneously.

With the introduction of the notion of *reference states*, i.e. the states in which the real elements of the system may be found, the interaction between the elements expresses the relationship between the states or the conditions that allow the elements to pass from one reference state to another. The elements that are in the same reference state at a given instant may then be considered to constitute a *reference class*. Since the definition of the states involves the characteristics of the elements, the classes possess the same characteristics. This immediately raises the problem of the definition of classes on the basis of continuous quantitative characteristics. The best solution is obtained by taking the mean values of the dimensions relative to the elements involved in the definition of the reference states.

β. *Structure and system.*   Various meanings have been attributed to the term '*structure*'. Formerly, the word was used mainly in its restricted sense to refer to the *spatial disposition of the elements* of a system, whereas today it is commonly used in a very general sense to describe the relationship between the elements of a system, which may be abstract or material. More precisely, the term 'structure' will here be used with reference to the arrangement of the elements of a system within the framework of the theoretical model chosen to represent the system. Thus, we may distinguish between the terms 'structure' and 'system': the former concerns the static arrangement of the elements whereas the latter refers to the dynamic interaction between the elements.

Delattre's general definition of a system may be stated as follows: a system is a set of interacting elements which, by the application of equivalence rules, may be classified into a finite (or infinite) number of reference classes. The interactions between the elements may be finally represented by relations of three fundamental types between the reference classes. These may be topological relations, or relations of order or of transfer. Specific axioms concerning the existence conditions of the elements may be applied to each system. A system may be defined by a set of eight items:

(1) the affirmation of the existence of reference classes;
(2) the number of reference classes;
(3) the definition of each reference class, i.e. the set of corresponding intrinsic qualitative and quantitative characteristics;
(4) the number of elements in each reference class;
(5) the axioms of existence for the elements in the reference classes;
(6) the affirmation of the existence of relations between the reference classes;
(7) the fundamental type of the relations; and
(8) the definition of the relations between the reference classes, i.e. the sets of intrinsic and extrinsic characteristics that express the relations, and eventually the corresponding numerical values.

The *total structure* of the system may be deduced from items 1, 2, 3, 6 and 7, and the *relational structure* from items 1, 6 and 7.

γ. *Function.* The function of an element depends upon its role and its behaviour in the environment. It is therefore necessary to define the characteristics of the elements of the environment. However, it is difficult to determine the *elementary functions* which have to be distinguished so as to ensure the partition of the elements. This raises major problems, especially in a biological context in which it is difficult to consider that only one element or a small group of elements is involved. Applied to material systems, this definition would give magnetic interaction, for example, the status of an elementary function since it represents the action of one element on another. The difficulty lies in defining the characteristics of the element undergoing the interaction since it is precisely these characteristics which define the elementary function.

If we consider a *system of systems*, the function expresses the set of properties manifested by a given system in the presence of the other systems. In these conditions, we shall have to take into account the structure–function relationship. However, there are two types of structure: total structure and relational structure. As Delattre points out, a discussion of the structure–function relationship of a material system must first of all specify the environment in which the system is situated. Only then can the notion of function in its most general sense have any real significance. Moreover, it should be borne in mind that the term 'structure' considered in relation to function, and applied to the various systems observed, refers to the total structure and not merely to the relational structure.

δ. *Evolution.* Without going into details of the analysis of the influence of the time factor on the notions outlined above, we may consider that the structure corresponds to a qualitative notion since the variation of the total structure is finally reducible to the appearance or the disappearance of quantitative classes situated at the level of a quantitatively stable subsystem. In contrast, the function has a qualitative aspect due to the nature of the interactions, as well a quantitative aspect due to the intensity of the interactions.

This essential difference in the nature of the concepts leads to the rejection of the couple 'structure–function' in favour of the notion of a *system* which offers a simultaneous approach to the qualitative and quantitative aspects of the function and its time-variation.

We have here briefly presented the main points of Delattre's epistemological ideas concerning some fundamental notions of systems. Let us now see if it is possible to use these definitions to represent a coordinated set of physiological processes.

b.   A representation of physiological systems: the hypothesis of associative functional self-organisation

α.   Biological structure and systems.   Can Delattre's very general definitions, valid for material as well as for abstract systems, be applied directly to biological systems or do we need to extend the definitions so as to incorporate specific physiological characteristics? Biological systems appear to have three essential characteristics:

(i)   The definition of a biological structure may be deduced from the experimental constraints: *a structure is an object that may be defined by criteria that are morphological or at least quantitative.* Intuitively, the function may be defined as *the set of activities or transformations subtended by the structure.*

(ii)   A major difficulty due to the nature of biological or physiological systems arises from the hierarchy of the constitutive objects. The *anatomical hierarchy*, in terms of molecules, organelles, cells, tissues, organs and apparatuses, implies a *functional hierarchy* which does not necessarily coincide with the anatomical hierarchy.

(iii)   Finally, the higher we go in the hierarchy (anatomical or functional), the smaller is the number of elements involved, until it reaches the order of unity.

These are the three major characteristics that distinguish physiological systems from other material systems such as physical systems. Let us now discuss these characteristics taking the physiological function of respiration as an example.

β.   The hypothesis of associative self-organisation.   When an organism is represented by a set of functional interactions between structural units belonging to several hierarchical systems, each corresponding to a physiological function, we may wonder how a given interaction is conserved in spite of unfavourable gene micromutations or other incidents leading to dysfunction. In other words, *how does the organisation maintain its invariance?* The nature of this problem will be touched upon and illustrated by an evolutionary model in Chapter 7. *The hypothesis of associative self-organisation* may be stated as follows:

> *If, at a given instant a structural unit does not produce the elementary physiological function (for example the product P_j in Fig. 4.4) necessary to its survival, then, in order to survive, it must receive this function from another unit, in which case a new functional unit is created.*

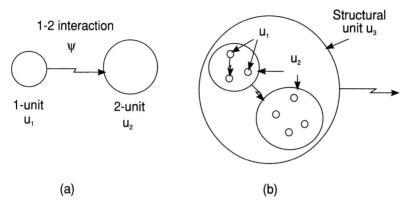

Fig. 4.4. Representation of the action (called functional interaction), i.e. the physiological function (called elementary physiological function), between structural units: (a) functional interaction; (b) hierarchical system based on the association of structural units u_1, u_2, u_3 at three successive levels.

This hypothesis may serve as a paradigm for the creation of functional inter-actions, as we shall show in Section IV below which deals with the problems of spatial organisation in the cell.

γ. The physiological system as a hierarchical system. Functional interactions, levels of organisation.   Whereas a biological system may be readily conceived in terms of structural levels following its anatomical description and its structural or-ganisation from the cell level to that of the organism, it is quite another matter to describe an organism on the basis of its levels of functional organisation. It appears natural to distinguish between the *structural association* (the set of units involved in a given structure) and the *functional association* (the set of units involved in a given physiological function). *We thus conceive an organisational level in terms of the elementary physiological function identified with the collective behaviour of a given set (L) of structural units. Then (L) defines an organisational level, the links at this level being termed organic links.* From the physiological point of view, it is first clearly necessary to consider the *function*. As Delattre put it, starting from the ob-servation of a given function, we may inductively attribute a certain number of characteristics to the elements, i.e. arrive at a definition of the elements.

For example, the respiratory function ensures the elimination of the carbon diox-ide produced by the cells and supplies the oxygen needed for the metabolism (Vol-ume II, Chapter 7). More precisely, the respiratory function must maintain the right levels of partial gas pressure for carbon dioxide, oxygen and H^+ ions in the blood. This involves the action of several organs and tissues which may be considered as subsystems: certain parts of the central nervous system participate in the regulation and control of respiration; the lungs ensure the transport of gases; the pulmonary alveoli and capillaries allow the transfer from the gaseous to the liquid phase; the heart and the blood vessels distribute the blood to the tissues; the kidneys cleanse the blood; and the haemoglobin transports oxygen and carbon dioxide molecules. Now, should the acid-base equilibrium be considered as the essential function of which the respiratory function is an element, or vice versa?

Among the subsystems of the respiratory function, let us consider, as a second example, the subsystem that ensures blood transport, i.e. the heart and the blood vessels. This subsystem, like the others, is evidently hierarchical from the anatomical point of view, being made up of sets of capillaries, arterioles, arteries and so on. The cardiovascular system may thus be considered to be a set of hierarchical systems, all of which are under the control of the central nervous system. Similarly, the ventilatory 'subfunction' of respiration involves several hierarchical subsytems such as muscles, bronchi and alveoli. In the normal physiological state, alveoli are classified according to their bronchial generation. In the pathological state, alveoli are distinguished according to the structural modifications observed, for instance, during inflammation. Thus, there exist different *classes* of alveoli according to the anatomical localisation on one hand, and the histological status on the other.

Clearly, *the histoanatomical hierarchy does not coincide with that of the physiological functions or subfunctions.* These two types of organisation do not generally coincide, thus making the representation of biological systems rather difficult. For example, the tubule of the nephron, which is part of the hierarchical renal system, interacts with the CO_2 receptor controlled by the nervous system *via* a buffer system, since the function of certain parts of the tubule (the localisation of which should therefore be specified) is to secrete H^+ ions (Chapter 9, Volume II).

This important finding is due to the mode of *construction* of the hierarchical system which *creates* the physiological function. We can hardly do otherwise than begin by isolating a given function, seek the anatomical structures underlying the function, and then, for each subsystem, determine the subfunctions, the sub-subsystems, the sub-subfunctions, and so on. This procedure finally leads to a set of parallel hierarchical systems, raising the problem of the definition of the levels of organisation.

The terminology that follows (Chauvet, 1993a) may appear to be less rigorous than that proposed by Delattre from an epistemological point of view but, applied to complex physiological phenomena, it has the advantage of being closer to that used in experimental physiology. Thus, *the structure will be treated as the 'physical' arrangement of the elements of the system in space,* and will not identified with the relational structure of the system. The structure coincides with the histo-anatomy of the element. The *structural hierarchy* thus corresponds to the hierarchy of the material system *observed.* The structure being defined as a physico-anatomical arrangement, *the functional hierarchy will be defined as the set of interactions between the elements of the material system 'useful' for the function considered.* This functional organisation satisfies the definition of the relational structure. The total structure of the corresponding system would require the addition of the interacting elements. Obviously, this new organisation will not coincide with the histo-anatomical structures. Although it may be possible to describe the organisation of the system in terms of the total structure, this does not seem to be particularly attractive, at least for the time being, considering the complexity involved.

3. *Mathematical representation of the functional organisation*

Definition I: Functional interaction and structural unit

Let us now give a mathematical definition of the functional interaction. Latin subscripts $i, j, \ldots$ represent units $u_i, u_j, \ldots$, and Greek subscripts correspond to products. Thus $P_{\alpha,i} \equiv P_{\alpha,u_i}$ denotes an α-product synthesised in the i-unit u_i. The functional interaction (α) from the i to the j-unit is written as ψ_{ij}^α. Levels of organisation are represented by Latin superscripts. For example, with an elementary di-graph:

(i)   Each element u_i or u_j (nodes i and j) represents a structural unit with an elementary function ψ_{ij}^α from u_i to u_j.

(ii) However, the result of this interaction is a product which may be either the direct value of the elementary function:

$$P_{\alpha,j} = \psi_{ij}^\alpha(P_{\alpha,i}; r) \tag{4.29}$$

or the transformed value:

$$P'_{\alpha,j} = \Phi_{ij}^\alpha(P_{\alpha,i}; r) = \phi_j^\alpha \cdot \psi_{ij}^\alpha(P_{\alpha,i}; r) \tag{4.30}$$

inside the unit localised in r. The variables $P_{\alpha,j}$ or $P'_{\alpha,j}$ will be identified with elementary physiological functions. More generally, μ products $P_{\alpha,i}$, $1 \le \alpha \le \mu$ in the i-unit could occur during the execution of the elementary physiological function.

(iii) A physiological function will result from a set of hierarchically organised and functionally interacting elements. The physiological function will be identified with the collective behavior of the elements whose product is denoted by F:

$$F = f(F^1, F^2, \ldots, F^n) \tag{4.31}$$

where F^l ($l = 1, \ldots, n$) is an elementary physiological function. A system in which $F = 0$, or a constant, is self-controlled.

Finally, these concepts and definitions should allow us to express the stability of the physiological function as well as that of the representative hierarchical system. *Thus, the fundamental problem may be stated, on one hand, as that of the topological stability of the functional interactions and, on the other, as that of the stability of the dynamic processes associated with these interactions.* These two aspects of the problem will be taken up in Chapter 12 of this volume and in Chapter 6 of Volume III.

Definition II: Level of organisation

A level of organisation (L^l), *as an elementary physiological function* F^l, *is identified by the collective behaviour, i.e. the dynamics, of a given set of* L *elementary functions between structural units.*

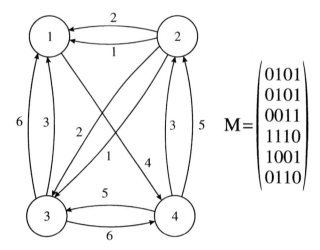

Fig. 4.5. An example of a graph and its incidence matrix to describe, together, an organisation whose degree equals 4. Numbers inside nodes of the graph represent the structural unit, i.e. the column of the matrix (on the right). Numbers along the arrows represent the product acting from the source to the sink, i.e. the rows of the matrix. Therefore, a row α is constituted by a zero in column j if the structural unit u_j is a sink for the product $P_{\alpha,j}$.

Thus, a physiological function is the collective product of a set of elementary physiological functions such as F^l, and because (L^l) uniquely defines one level of organisation and an elementary physiological function, then a physiological function is a set of L levels such as (L^l), i.e. a hierarchical system that produces F. Most often, the dynamics is specified for a given time scale of the process and therefore defines the level of organisation.

Definition III: Degree of organisation

> *The degree of the functional organisation of an FBS at level 1 is the number v^1 of structural units (structural equivalence classes) that constitute the subsystem at this level.*

Thus, the creation of a functional link between one structural unit and the subsystem is the consequence of the association of this structural unit with the subsystem. The association increases the degree of functional organisation.

Definition IV: Graph and matrix of the functional organisation of an FBS: $S^{(1)}$ (G, T)

> *A biological subsystem $S^{(1)}$ at level 1 is described by two elements (Fig. 4.5): the graph **G** of the functional interactions, and some parameters characteristic of the dynamics of the system:*

(1) The graph **G** *specifies the elementary functions (edges) between structural units (vertices). A matrix* **M** *of elements 0 and 1 is associated with his
graph (incidence matrix of* **G***).
(2) The parameters (for example the time scale* T^1*) for level* 1*, are defined
by the dynamical processes that describe the collective behaviour at this level.*

M has μ rows, i.e. the number of elementary functions such as P_α ($\alpha = 1, \ldots, \mu$),
and v^l columns, i.e. the number of structural units u_j ($j = 1, \ldots, v^l$) that are included
in the collective function at level l:

$$\mathbf{M} = (a_{\alpha j}) \begin{cases} a_{\alpha j} = 1 \Leftrightarrow P_{\alpha,j} \in^* u_j & (4.32.1) \\ a_{\alpha j} = 1 \Leftrightarrow P_{\alpha,j} \notin^* u_j. & (4.32.2) \end{cases}$$

The symbol $\in^*$ denotes the fact that the product $P_{\alpha,j}$ is emitted by the structural unit
u_j which, in this case, is called a *source*. All structural units that do not possess P_α
are called *sinks*. An elementary function is created from a source to a sink:

$$\psi^\alpha_{jk} \neq 0 \Leftrightarrow P_{\alpha,j} \in^* u_j \ \ P_{\alpha,k} \notin^* u_k. \tag{4.33}$$

Relations (4.29) and (4.30) may now be written in a more general form:

$$P_{\alpha,k} = \psi^\alpha_{jk}(P_{\alpha,j})$$
$$F^k_\alpha = \Phi^\alpha_{jk}(P_{\alpha,i}) = \phi^\alpha_k \circ \psi^\alpha_{jk}(P_{\alpha,i}). \tag{4.34}$$

We thus have a description of the topology of the system, i.e. the relational aspect
between its elements, and the dynamical process at the level considered. The time
scale will be shown to be important for the construction of the functional organisation. Moreover, it implies a close connection between structure and function.

Definition V:

The functional organisation at level 1 *is defined by the distribution* $(n^{(1)}_\alpha)_{\alpha=1,\mu^{(1)}}$
of functional links between structural units at this level. Then $n^{(1)}_\alpha$ *is also the
number of zeros in the row* α *of the matrix* **M***, i.e. the number of sinks for
the function* P_α *of the system.*

When there are L levels of organisation within a given physiological function, i.e.
when a set of elementary physiological functions F^k, $k = 1, L$, constitutes a *physiological function* F, we have the relation $F = f(F^1, F^2, \ldots, F^L)$. This equation expresses
an implicit control, or an intrinsic regulation, between the individual F^ks. Its relation
to equation (4.31) is obvious: F^k represents the collective product created by
the elementary functions ψ^α_{ij} at level k, and F is the collective product of all the
levels that constitute the *hierarchical biological system*. Of course, this relation is a
condensed form of several equations such as:

$$\dot{\psi}^{\alpha,kl}_{ij} = f^{\alpha,kl}_{ij}(\psi^{\alpha,11}_{11}, \ldots, \psi^{\alpha,LL}_{v^k v^l}) \tag{4.35}$$

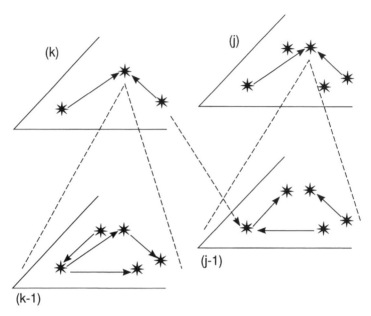

Fig. 4.6. Functional interactions within and between levels of parallel hierarchical systems. In terms of functional interactions, a given unit at a level (*k*) is composed of a set of units at level (*k*–1), i.e. represents the collective behaviour of these units, and is a physiological function. This physiological function is therefore a hierarchical system. The same structural unit can be included in the collective behavior of level (*j*–1) in another hierarchical system. An example is given by a system that controls another system, such as the nervous and the respiratory systems.

each of which describes the dynamics of elementary functions between an *i*-structural unit at level *k* (L^k) and a *j*-structural unit at level *l* (L^l): $i = 1,\ldots, v^k$, $j = 1,\ldots, v^l$, and $k, l = 1, L$. When $\psi_{ij}^{\alpha,kl} \neq 0$ with $k \neq l$, the corresponding link is called an inter-level link because it implies equation (4.31). When two physiological functions with an interaction between them (such as the respiratory and the cardiovascular functions, represented respectively by airflow and cardiac flow) are considered, two *parallel* hierarchical systems are obtained (Fig. 4.6). An important property as regards the practical consequences, is the '*relativism*' of levels in the functional organisation. Relativism is involved when one variable, at level *l* for the first system, is at level *k* for the second. For example, a group of neurons, which are organised in a hierarchical system, can be connected to a group of neurons organised in several subgroups of neurons, where the groups, and the levels of organisation, are defined by their collective behaviour.

   Although it may be easy to think of a biological system in terms of structural levels, due to its anatomical description and organisation (from cellular to organismal structure), it is considerably more difficult to describe an organism in terms of its levels of functional organisation. However, at least because of the property of relativism, it appears to be necessary to distinguish between the *structural association* of the units involved in a given structure and the *functional association* among units involved in a given physiological function. In terms of functional interactions, an organism is identified as a set of parallel hierarchical systems, one for each physiological function, and can be modelised and implemented on a parallel computer. Thus, one major aim will be to investigate the laws of conservation applicable to a given interaction and to a set of interactions, i.e. a physiological function, and to determine which aspects of the related organisation of an organism are kept invariant.

All these concepts and definitions are useful to conceive the biological functional organisation of some systems. We have studied a particular system of metabolic and self-replicative units with two levels of organisation: the so-called 'Eigen–Goodwin' system (Chauvet and Girou, 1983), which will be considered below as an example of an evolutionary process. Such a formalised description has additional interesting consequences regarding the concept of the *potential of functional organisation*, and the ability to define dynamics in the (ψ, ρ) representation (see Section 4 below). Systems with a time hierarchy have been considered in the past, for example in the case of metabolic systems. These are usually analysed by means of Tikhonov's theorem, which has been extended to the case of fast, oscillating variables in metabolic systems with complex slow and fast dynamics (Dvorak and Siska, 1989).

4. Dual representations: (N, a) and (ψ, ρ)

In Delattre's formalism of transformation systems (Chapter 3), the partition of the elements is such that the system remains a statistical one, i.e. a system in which all the elements of a class have exactly the same functions with respect to the other elements of the system. The probability that a given reaction occurs per unit time is proportional to the product of the cardinals of the species involved in the reaction, each cardinal being raised to a power equal to the number of elements of the corresponding species. The statistical nature of the system implies that the probability of reaction of all the elements of a class is uniform. Moreover, if the number of elements is sufficiently great, the balance of the appearance and the disappearance of the elements of a class (E_j) is given by the differential equation:

$$\frac{dN_j}{dt} = -\left(\sum_k a_{kj} F_{kj} + a_{ej} F_{ej} \right) + \left(\sum_k b_{jk} F_{jk} \right) + \varepsilon_j \tag{4.36}$$

i.e. Equation (3.6) where N_j is the number of elements in the class E_j, F_{kj} is the number of elementary transformations per unit time from class E_j to class E_k, and F_{ej} is the number of elementary transformations towards the environment ε. Since the system is of a statistical nature, we may write the general term as (3.7):

$$F = K N_j^{\alpha_j} N_{j+1}^{\alpha_{j+1}} \dots N_{j+p}^{\alpha_{j+p}} \tag{4.37}$$

where the transformation concerns a_j elements of E_j, ..., a_{j+p} elements of E_{j+p}. ε_j describes the inputs $(\varepsilon_j \geq 0)$ or the outputs $(\varepsilon \leq 0)$ of the elements of E_j, which are *independent* of the occupation of the classes of the system. *As N represents the occupation number of the classes, and as a represents the 'rate constant' of the transformation, this representation may be called* (N, a).

Delattre's formalism of transformation systems may be applied when the partition of the elements of a system allows the construction of functionally equivalent classes, and when the number of elements per class is sufficient to establish a statistical system. The definitions given above for a physiological system, in particular with respect to its double hierarchy: histo-anatomical and functional, lead to a new representation since these two conditions are not satisfied. Indeed, if ψ_{ij} is the functional interaction between two structural units u_i and u_j, which may be at the same level or not, with $i, j = 1$ to p, then the system is usually governed by a system of partial derivative equations (4-35) where we must include specific physical or geometrical parameters ρ_i of the structure. The dynamics of the system may be considered to be represented by (ψ, ρ). The two representations (N, a) and (ψ, ρ), constructed respectively on the classes of functional and structural equivalence, may be considered as a *dual representation*. In (N, a), the dynamics of the system is that of the number of elements N in a class relative to a reference state in which a, the characteristic of the interaction, is a constant. In (ψ, ρ), the dynamics of the system is that of the functional interaction ψ between two classes of structural equivalence, where ρ characterizes the geometry of the structure.

Finally, whereas the (N, a) representation is adapted to systems with a large number of elements, such as chemical or ecological systems, we shall see that the (ψ, ρ) representation may be preferable in the case of systems of physiological control. In general, a complex system with multiple levels of organisation (in terms of Definition II) may be described by a *mixed* representation in which the lowest levels are made up of classes with a large number of elements (molecules), and in which the highest levels are made up of classes with a small number of elements (organs and tissues). For example, Fig 4.7a illustrates the (ψ, ρ) representation of the respiratory function versus the (N, a) representation. The more advanced schematic diagram representing the respiratory function in terms of functional interactions is given in Fig. 4.7b, and its relations with other physiological systems in the organism in Fig 4.7c. Illustrations such as these will be further discussed in Volume III, Chapter 7.

The search for new representations of phenomena involving a large number of levels of organisation can only be useful if the notion of organisation is correctly defined. Biological systems in general, and physiological systems in particular, have multiple levels of structural as well as functional organisation. Delattre's epistemological approach, offering insight into the notions of structure, system, function and evolution, and introducing the concepts of total structure and relational structure, has laid down a sound foundation for this investigation. The notion of the (ψ, ρ) representation appears to be neccessary for the case of systems of physiological regulation. The nervous system affords a good example, in which ψ could represent the field variable (functional interaction varying with time and space), allowing the use of the field theory in the description of certain phenomena. We shall use this approach to calculate synaptic efficacy involving at least two levels of organisation (Volume III, Chapter 2). However, a mixed representation $(N, \alpha) \otimes (\psi, \rho)$ may be usefully applied at the molecular level.

IV.   Spatial organisation in the cell: concept of structural discontinuity

In the conceptual framework we propose, the non-local functional interaction is produced at a well-defined region called the *active site*. It is convenient to view this functional interaction in terms of an application, in the mathematical sense, with the space of definition E being applied to a space of values F, i.e. $f: E \rightarrow F$. This active site (space F from the mathematical point of view) is the seat of two highly complex mechanisms stemming from the chemistry of living organisms: the allosteric interaction (Chapter 1) and enzymatic catalysis (Chapter 2). Each chemical reaction in the cell is catalysed by a protein — an enzyme — which is generally specific, so that there are as many different enzymes as there are chemical reactions. A problem that immediately arises concerns the regulation of the 'local' set of reactions that may be treated, in simple terms, as a homogeneous medium. However, such a local description is rather coarse since the cell is in fact composed of local micro-environments within which the chemical reactions occur and because non-local interactions are at play between these micro-environments.

In Chapter 2, we examined the phenomena involved in the adjustment of reactions at a given site. Let us now suppress the usual biochemical constraint of a homogeneous, uniform and isotrope medium, that applies under *in vitro* experimental conditions, and consider the spatial organisation of the cellular micro-regions in which the chemical reactions actually occur.

1.   *On the existence of non-local interactions: the concept of the active site re-examined*

As presented in the sections above, the organism appears to be an immense network of chemical reactions. Indeed, even the cell consists of an extremely complex set of substructures within which specific metabolic processes occur. As mentioned above, the medium is not homogeneous and isotrope. Local micro-environments, such as those displaying the phenomena of 'channelling' and 'pooling', imply a true spatial metabolic organisation of the cell. Let us recall the property of non-locality as presented in Chapter 2:

> *If an organism can be equated to a highly regulated set of local chemical reactions, then it is necessary that non-local reactions exist between structures which, in physical space, must be situated at a distance.*

Sets of this kind may be represented as in Fig. 4.8. The question of how the organism manages to make the 'best' organisation of all its functions is indeed difficult to answer. However, we shall attempt to tackle the problem in the last chapter of this work. As we have demonstrated (Chauvet, 1993c), the non-locality and hierarchy of biolological systems are closely linked. We shall use this idea to deduce that the 'best' organisation (which is 'best' in a sense we shall define later) may be a hierarchical organisation. Several specific examples illustrating this point of view will be found in the course of this work.

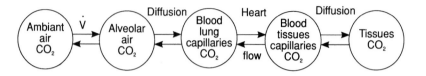

COMPARTMENTS

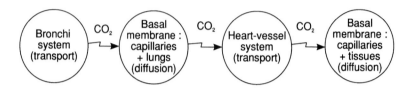

FUNCTIONAL INTERACTIONS

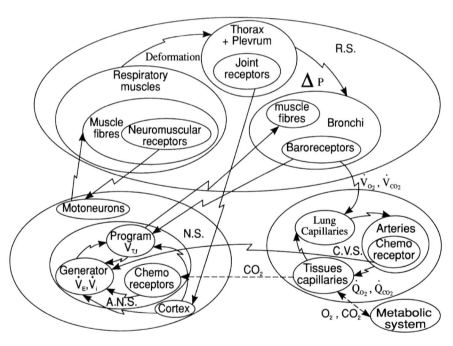

Fig. 4.7.   Representations (ψ, ρ) and (N, a) in the case of the respiratory function. (a) Ventilatory function represented (above) in representation (N, a) (below) in representation (ψ, ρ). (b) The same function represented as a hierarchical system connected with nervous and cardiovascular functions according to functional interactions. R.S. = Respiratory System; C.V.S. = cardiovascular system; E.S. = Endocrine System; N.S = Nervous System; A.N.S. = Autonomic Nervous System; f = frequency; ΔP = Pressure gradient. (c) Functional interactions for the respiratory system. $f^{(\ )}$, $\hat{\imath}$, $i = 1, 8$ = Control functional interactions; p = hydrostatic pressure; $\dot{Q}$ = fluid flow: $\dot{V}$ = ventilation; $\dot{q}$ = metabolic flow, Inputs: Molecules, O_2, CO_2. Output: H^+, H_2O. (*See table for nomenclature.*)

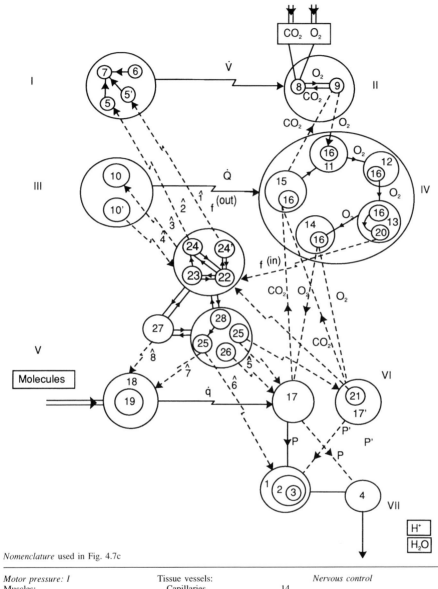

Nomenclature used in Fig. 4.7c

Motor pressure: I		Tissue vessels:		*Nervous control*	
Muscles:		Capillaries	14		
Inspiratory	5	Veins	15	— *Respiratory control*	
Expiratory	5′	Haemoglobin	16	Neuronal pools	22
Diaphragm	6	Chemo-receptors	20	Pneumotaxic centres	23
Plevrum	7	*Digestive system: V*		Bulb centres:	
Ventilatory mechanics: II		Intestines	18	Inspiratory	24
Bronchi	8	Mitochondria	19	Expiratory	24′
Alveoli	9	*All cells and receptors: VI*			
Heart: III		All cells	17	— *Hormonal control*	
Myocard	10	Mecano-receptors	21	Pancreatic cells	25
Vessels: IV		*Kidney: VII*		Thyroid cells	26
Pulmonary vessels:		Tubules	1	Post-hypophysis	27
Capillaries	11	Tubular cells	2		
Veins	12	Membrane	3	Hypothalamus	28
Arteries	13	Collector tube	4		

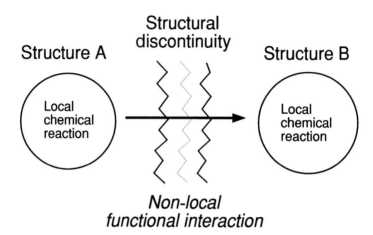

Fig. 4.8. Diagrammatic representation of a structural discontinuity between two sets of chemical reactions in two different structures A and B. The non-local functional interaction acts from A to B.

2.   Enzyme organisation: microcompartmentation and the example of 'channelling'

a.   Introduction to the problem of enzyme organisation

A much-discussed question concerns the existence of 'channelling', i.e. the phenomenon involved when a metabolite produced by an enzyme reaches the enzyme at the next step of the metabolic chain without getting lost in the surrounding cellular medium, as if it actually moved through a channel. The product of an enzymatic reaction is transferred in the form of a substrate towards the next enzyme in the metabolic chain *without being equilibrated* with the solution in which it travels. Thus, there would appear to be a *direct transfer* of metabolites in a molecular pathway. This idea is now widely accepted (Friedrich, 1991; Ovádi, 1991; Westerhoff and Welch, 1992). However, the term 'metabolite channelling' is not restricted to a specific molecular mechanism such as the direct transfer of an intermediate substrate but, in a more general sense, it includes the phenomenon of segregation or *microcompartmentation* (Spivey, 1991; Ovádi, 1991). The phenomenon of microcompartmentation corresponds to an isolation of intermediate substrates from the rest of the compartment resulting in a high local concentration of these substrates. Ovádi reviews the mechanisms of metabolite channelling, including the elementary steps of the transfer of intermediate substrates in interacting and non-interacting enzyme systems.

Kempner (1980) defines a metabolite compartment as a subcellular region of biochemical reactions kinetically isolated from other cellular processes. Friedrich (1990) suggests that this kinetic isolation may be due to a diffusional barrier, to the distance, to segregation of a metabolite pool in a micro-environment at membrane surfaces or to binding at macromolecular sites. However, the concept of channelling has been questioned in some cases such as that of the activity of glycolytic enzymes (Gutfreund and Chock, 1991), or that of coupled enzyme kinetics (Salerno, 1991). According to Friedrich (1991), a substance may be kinetically confined in a restricted intracellular region by enzymes, called 'quenching enzymes', that degrade or eliminate the given substance. An example of this kind of specific confinement is the behaviour of second messengers, such as cyclic AMP, that are produced at the submembrane level. In the case of Ca^{++}, the quenching enzymes being the various Ca^{++}-transport ATPases.

A major difficulty in the interpretation of channelling is due to dynamic enzyme associations. The physiological relevance of covalent and tight non-covalent enzyme clusters in direct metabolite transfer seems to be established. In fact, enzyme–enzyme aggregates *in situ* or at least in the spatial vicinity of enzymes catalysing consecutive reactions has not been completely proved. Batke (1991) points out an apparent discrepancy between the high concentration of glycolytic enzymes in the cytoplasm and the assumption of intermediate channelling, and comments on the difficulty of identifying the specific channelling complex of enzymes in the cytoplasm as well as the problem raised by the extrapolation of *in vitro* experimental results to *in vivo* conditions.

Pettersson (1991) feels that the idea of metabolite channelling between enzymes forming dynamic complexes is still highly speculative, and that theoretical calculations would be required to transform the speculations into falsifiable hypotheses. Several questions require to be answered: What would be the expected local metabolite concentration gradients in the neighbourhood of a producing enzyme? What effect would these gradients have on the interaction of the metabolite if the consuming enzyme formed a complex with the producing one? How large is the assumed increase in catalytic efficiency of the coupled reaction due to this effect, and how does it vary with the geometry of the space between the sites of the interacting enzymes?

b. Relation with physiology and physiological consequences

From a physiological point of view, metabolite channelling by enzyme complexes would provide several catalytic advantages. Ovádi (1991) suggests that it would at least: (i) prevent loss of intermediate substrates by diffusion; (ii) decrease the transit time required for an intermediate substrate to reach the active site of the next enzyme; (iii) decrease the transit time for the system to reach the new steady state; and (iv) protect chemically labile intermediate substrates.

Compartmentation would of course be useful for maintaining a high local concentration of intermediate substrates, which in turn would improve enzyme efficiency. The formation of a pool of intermediate substrates requires that the diffusion of intermediates in the solution be limited either by steric hindrance or by the juxtaposition of active sites. The geometry of the space between the active sites should therefore be such as to limit the loss of intermediate substrates.

Another advantage offered by compartmentation is the maintenance of high local concentrations of metabolites precisely in the regions in which the specific active sites are located so that the overall solubility of the cytoplasm remains within reasonable limits. It may be recalled that the cell contains thousands of enzymes and that the mean distance between enzymes is much smaller than the diameter of an average-sized protein. There must therefore exist a very high local concentration of intermediate substrates. *Can the phenomenon of diffusion alone ensure the passage of a metabolite from one enzyme to another?* The time required for diffusion can be readily estimated at about r^2/D, where r is the mean distance between enzymes and D is the coefficient of diffusion of the substrate. When r is small and D is large with respect to enzyme turnover time, metabolic channelling may not be necessary since bulk diffusion could be an effective means of material transport of small substrates in the cytoplasm especially if the medium closely resembles water. However, the cellular medium is more like a viscous matrix than a buffer solution. Although the high protein concentration (100–300 mg/l) would suggest a high viscosity, measurements indicate that the viscosity of the medium is only 3 to 5 times that of water. The only explanation of this anomaly is that most of the proteins are not in solution but adsorbed by cell membranes and the cytoskeleton, so that the cytosol remains fairly fluid. This suggests the presence of cytoplasmic barriers, which we have called *structural discontinuities*. If there were no compartments *in vivo*, the transfer of metabolites, substrates and allosteric effectors by the classical diffusion process would be hindered by the viscosity of the matrix. Metabolite channelling would contribute to increasing the rate of transfer by limiting diffusion.

The idea of a '*coordinative effect*' was introduced by Welch (1977) in order to explain how multi-enzymatic complexes might produce the intracellular compartmentation of intermediate substrates, common to two or more metabolic pathways, without the need for membranes or other barriers as in the case of cellular organelles. The metabolic pathways could be coordinated by channelling mechanisms, the terminal products being carried to their destination by specific couplings between the biochemical pathways. We shall re-examine this interesting effect further on and propose another interpretation.

From a theoretical point of view, there are at least two possible approaches: the estimation of the rates of transfer along metabolic pathways, i.e. the expression of biochemical events in terms of the physicochemical context, and the search for an explanation in the general physiological context. One approach has been explored in biophysical terms by Welch *et al.* (1988), Westerhoff and Welch (1992), and Ricard *et al.* (1992). Another approach, based on the functional organisation of a biological system, was introduced by us (Chauvet and Girou, 1983; Chauvet, 1993a).

c. Definitions of channelling on biophysical bases

α. *Metabolic organisation.*   The metabolic organisation may appear under various forms such as:

(i)   A static assembly, as in the interaction of glycolytic enzymes with muscle cells or in the working of the Krebs cycle in the internal mitochondrial membrane

(Srere, 1991), where the intermediate substrates are protein-bound. Such systems are found in the anabolic pathways in which there is an initial and a final product.

(ii) Weakly interacting protein–protein networks, as in the metabolic pathways of protein biosynthesis (Welch, 1977), in which the intermediate substrates are grouped in local pools. Such systems are generally found in the amphibolic pathways characterised by numerous bifurcations at which the metabolites share several routes.

Each of these forms exhibits two properties: 1) a part of the metabolic flux is sequestered from the cytosol; and 2) there is a relation between the state of organisation and the functioning of the metabolic process, i.e. between the structural organisation and the functional organisation.

β. Application of Curie's principle of symmetry:

The phenomenon of channelling is characterised by the existence of a local anisotropic reaction–diffusion coupling under certain physical conditions.

In an *in vitro* system, the uniform concentration of metabolites leads to the formation of an isotropic medium. A trivial case of macroscopic heterogeneity in an isotropic medium would be that produced by membranes or even by a set of chemical reactions involving limited diffusion. This would also be the case for *dissipative structures*, far removed from thermodynamic equilibrium (Prigogine and Nicolis, 1971).

Curie's principle states that, in an isotropic medium, it is physically impossible to couple a scalar chemical reaction with a vectorial diffusion of matter. Given the conditions of homogeneity and isotropy, this principle would of course be applicable *in vitro*. However, Westerhoff and Welch (1992) have demonstrated the validity of the principle in the case of an anisotropic medium such as may be found *in vivo*. In fact, contrary to a widespread physical intuition, Curie's principle imposes a reaction–diffusion coupling in a structured anisotropic medium, i.e. an ordered *in vivo* metabolism, called the mechanism of channelling. Thus, in an anisotropic medium, a symmetric cause may induce an asymmetric effect.

Enzymatic catalysis has very non-symmetric qualities. Locally, the catalytic process is of a vectorial nature. In these conditions, how does the locally anisotropic character of the events occurring at the active site actually generate a directed flow of matter? Directed reactions, associated with membrane structures, are known to occur in numerous metabolic processes. However, a structural localisation is not necessary for a vectorial flow of matter coupled to an enzymatic reaction. Directed reactions are also found in enzymatic reaction sequences involving a multi-enzymatic complex which represents the transfer of metabolites from one site to another, i.e. a phenomenon rather like that of channelling. A multi-enzymatic complex can produce a directed flow of matter if it is associated with a structural matrix. This is precisely what occurs on the internal face of cell membranes. Thus, directed metabolic flow does not always involve molecular channelling of metabolites.

On the basis of biophysical mechanisms at different levels of structural organisation, we may propose several definitions:

At a macroscopical level, channelling may be defined as the selective inclusion of metabolically related enzymes within an enclosed region, bounded by a semi-permeable membrane which serves to confine intermediate substrates or products to a specific metabolic process. Such a process involves a metabolic pathway with several steps for which all the enzymes constitute a *pool*. The intermediate metabolites P_i can be pooled and used by any enzyme E_i as follows:

$$S \ldots \to P_1 \xrightarrow{E_1} P_2 \xrightarrow{E_2} P_3 \xrightarrow{E_3} P_4 \xrightarrow{E_4} \ldots P$$

$$S \ldots \to P_1 \underset{E_1}{\rightrightarrows} P_2 \underset{E_2}{\rightrightarrows} P_3 \underset{E_3}{\rightrightarrows} P_4 \underset{E_4}{\rightrightarrows} \ldots P \qquad (4.38(1))$$

At the molecular level, channelling implies the site-to-site transfer, or the 'direct transfer', of intermediate metabolic substrates between individual, sequentially-acting enzyme molecules, without equilibration with the surrounding medium. The definition is based on an elementary enzymic step of the pathways that are totally independent:

$$S \ldots \to P_1 \xrightarrow{E_1} P_2 \xrightarrow{E_2} P_3 \xrightarrow{E_3} P_4 \xrightarrow{E_4} \ldots P$$

$$S \ldots \to P_1 \underset{E_1}{\rightrightarrows} P_2 \underset{E_2}{\rightrightarrows} P_3 \underset{E_3}{\rightrightarrows} P_4 \underset{E_4}{\rightrightarrows} \ldots P \qquad (4.38(2))$$

Thus, in the case of perfect channelling, there is no pooling of free intermediate substrates. In another type of channelling, the intermediate product is dissociated from the enzyme that produces it but remains confined to its micro-environment. It may be added that the microcompartment that initiates the channelling is not necessarily a permanent structure.

γ. The respective roles of diffusion and reaction.   Can the transfer of metabolites from one enzyme to another be accounted for by the diffusion process alone? We may attempt to answer this question, posed above in connection with the phenomenon of compartmentation, following the reasoning proposed by Westerhoff and Welch. Since enzymatic action combines individual processes whose influence on the transport properties of the solvent is difficult to study, these authors use the transit time τ which corresponds to the time required for the accumulation of intermediate substrate to reach the steady state. In the case of a simple system composed of two enzymatic reactions, the transit time may be obtained from the relation $\tau = \tau_d + \tau_r$ which clearly shows the breakdown of the transit time into two parts, one corresponding to the diffusion (d), and the other to the reaction (r). The quantities $1/\tau_d$ and $1/\tau_r$ are pseudo first-order rate constants. The first constant, estimated at 5 μsec, characterizes the rate at which a molecule of the substrate encounters the recognition volume of an active site; and the second constant characterizes the rate of formation of the complex (ES), once the encounter has taken place, with τ_r being estimated at 1 msec. The kinetic theory of fluids gives another expression for $1/\tau_r$ involving a factor of steric probability which, however, is difficult to quantify since it represents the local approach of a molecule of substrate towards the active site. With this theory, we obtain a value for τ_r of the order of 1 msec, which confirms the validity of the estimation made from the individual rate constants. Thus, the reaction–diffusion coupling in the enzymatic reaction is in agreement with the approximations of the kinetic theory of fluids.

These results show that simple bulk diffusion is quite sufficient to explain the kinetics of the metabolic flow. However, the two examples considered are exceptions to the rule. In most of the pathways of intermediate metabolism, the enzyme and reactant levels are much lower than in the cases of biosynthesis in *E. coli* and glycolysis in yeast. Thus, supposing the mean distance between enzymes to be 0.1 μm, and the turnover time less than the usual average value, limited diffusion would occur in the absence of channelling. However, the transit time would be very long, *except if subcellular compartmentation existed.*

δ. How should the diffusion process be characterised in order to account for the existence of molecular channels in the absence of a physical limit such as a membrane? This question may be answered by determining the escape rate that allows limited diffusion in the absence of natural physical barriers. Westerhoff and Welch use the Smoluchowski constant k_{-d}, estimated to be $\approx 10^9$ sec^{-1}. If we compare this value of the escape rate of the substrate molecule with the catalytic turnover time (1 msec), we see that *the electrostatic interaction is insufficient, kinetically, for the recapture of metabolites in a channelling mode.* This leads to the important finding that the existence of a barrier is necessary to prevent the escape of a metabolite by diffusion in the case of channelling.

Some other aspects concerning channelling may also be considered. For instance, the physical nature of the micro-environment has considerable influence on the kinetic parameters of localised enzymatic processes. Similarly, as we have mentioned above, the viscosity of the medium plays a role in enzyme–substrate dissociation. In the case of channelling, the compartment tends to become macroscopic, and this naturally affects the thermodynamic analysis of the phenomenon (see Chapter 3).

In conclusion, we may say that the heterogeneous and anisotropic conditions in which enzymatic phenomena at the subcellular level occur make it difficult to apply the usual theoretical methods. Enzymatic organisation varies from cell to cell, as well as with time, so that the dynamics of the organisation is highly complex. It can scarcely be separated from the underlying functional metabolic organisation. As Westerhoff and Welch (1992) put it, the real problem here is that of the relationship between the structural organisation and the causal properties of cell metabolism.

3. On the functional organisation in a biological structure: the example of enzyme organisation

The question that we propose to address in this section is that of the physiological significance of the cellular enzymatic organisation (Chauvet and Costalat, 1995). Since the cell, far from being a mere bag containing a homogeneous cytosol, is in fact a highly structured medium, we may wonder what functional advantages the enzymatic organisation could offer. As seen in the previous section, this point has been discussed by several authors (Welch, 1977; Ovádi, 1991; Westerhoff and Welch, 1992), and some functional consequences of enzymatic organisation have been suggested: (i) the segregation of related metabolic pathways; (ii) the increased production of metabolites, i.e. the improved efficacy of the biochemical pathways because of the proximity between substrates and active sites; and (iii) the modification of the intrinsic catalytic properties of enzymes through enzyme–enzyme interactions or interactions between enzymes and cellular substructures. Marmillot *et al.* (1992) have investigated a model in which phosphofructokinase is assumed to be distributed between the bulk phase and the cellular substructures. The study reveals the possibility of a polarised wave phenomenon which might explain the transmission of signals in the larger eucaryote cells.

However, some of the functional consequences of enzymatic organisation have remained unexplored, in particular the effects that may be produced on the enzymatic kinetics of the metabolic pathways, especially on the stability of the stationary state. This state, distinct from that of thermodynamic equilibrium (Nicolis and Prigogine, 1977), is known to be of importance in living organisms. In fact, the homeostatic condition of a living system is closely approximated by a steady state far removed from equilibrium, and most cells are in this type of state for relatively long periods of time (Mikulecky, 1993). However, the homeostatic condition can be satisfied only if the stationary state is sufficiently stable, i.e. if the biological system, after moving away from the stationary state under the effect of a perturbation, returns to the stationary state within a 'reasonable' period of time. We shall examine this notion more closely in the subsection below. The stability of a metabolic pathway in a structureless, homogeneous medium has been discussed by several authors (Walter, 1970; Palsson and Lightfoot, 1985).

In our study of the consequences of enzymatic organisation on the stability of the stationary state of metabolic systems (Chauvet, 1993; Chauvet and Girou, 1983), we propose to establish a link between the structural properties, i.e. the distribution of enzymes in the cell, on the one hand, and the dynamics of cellular processes, on the other. We shall consider models of increasing complexity, ranging from the Henri–Michaelis–Menten type reactions to a metabolic pathway with an allosteric feedback loop of the Yates–Pardee type. This approach allows the representation of the functional organisation of metabolic pathways in terms of the stability of the dynamics of the system, which in fact correspond to the biochemical kinetics. The relationship between the structural and functional properties in the context of enzymatic organisation leads to a definition of the concept of structural unit and, because of the stability, the related concept of the physiological function.

a.   Definition of a metabolic pathway as a structural unit

α.   Michaelis enzymatic reactions: metabolic flux and transport between the local medium and the bulk phase.   Let us consider a metabolic pathway:

$$\rightarrow P_1 \xrightarrow{E_1} P_2 \xrightarrow{E_2} P_3 \xrightarrow{E_3} P_4 \xrightarrow{E_4} \dots \qquad (4.39(1))$$

in which P_i represents a metabolite and E_i the enzyme catalysing the reaction at the step of rank i. Let us first assume that all the enzymes are of the Michaelis type so that the step of rank i may be represented by a Henri–Michaelis–Menten enzymatic reaction:

$$E_i + P_i \rightleftharpoons E_i - P_i \rightarrow E_i + P_{i+1} \qquad (4.39(2))$$

which may be written:

$$E + S \rightleftharpoons E - S \rightarrow E + P$$

by replacing E_i, P_i, and P_{i+1} by the classical notations E, S and P. The different terms of this kinetic equation may be successively interpreted as: (i) E–S complex formation and dissociation with the rate constants k_{+s} and k_{-s}, and (ii) a catalysis with a rate constant k_{cat} (Westerhoff and Welch, 1992). Let us now suppose that the metabolites are present in two phases: (i) the local medium close to the enzymes, or local phase; and (ii) the core of the solution or bulk phase. Writing k_{+t} and k_{-t} for the kinetic constants of metabolite transport from the core of the solution towards the local medium and *vice versa*, we obtain the following schema:

$$S \quad \underset{k_{-t}}{\overset{k_{+t}}{\rightleftharpoons}} \quad S + E \quad \underset{k_{-s}}{\overset{k_{+s}}{\rightleftharpoons}} \quad E - S \quad \overset{k_{cat}}{\longrightarrow} \quad E + P.$$

(core of solution)   (local medium)

transport   binding   catalysis

This schema may be compared with the molecular enzyme kinetic model proposed by Damjanovitch and Somogyi (see Welch (1977)), although there are some differences. In our model: (i) the enzymes are considered to be present mainly in the local medium; (ii) no particular hypothesis is made concerning the mechanisms of transport between the two phases, for instance, there may be diffusion and/or migration in an electric field; and (iii) the approach is 'kinetic' rather than 'molecular'.

The net flux of metabolite transport between the bulk phase and the local phase is given by:

$$J_t = k_{+t} c_{bulk} - k_{-t} c_{local} \qquad (4.40)$$

where c_{bulk} is the concentration of the substrate S in the bulk phase, and c_{local} the concentration of S in the local medium. The transport flux J_t is expressed as the number of moles per unit time, and the rate constants of transport k_{+t} and k_{-t}, which are expressed in terms of volume per unit time, are assumed to be proportional to the surface of exchange between the two phases. The flux of the chemical reaction is described by the classical Michaelis expression:

$$J_m = n_E \frac{k_{cat} c_{local}}{\dfrac{(k_{-s} + k_{cat})}{k_{+s}} + c_{local}} = n_E \frac{k_{cat} c_{local}}{K_M + c_{local}} \qquad (4.41)$$

n_E being the number of moles of the enzyme E in the local medium, J_m the metabolic flux (the number of moles of substance produced per unit time), and K_M the Michaelis constant. In the stationary state, the transport flux J_t is equal to the metabolic flux J_m in the particular case of a unique enzymatic reaction (4.39) (2). With the hypothesis of a weak concentration of substrate, such that $c_{local} \ll K_M$, the metabolic flux in the stationary state may be written:

$$J_m = n_E \frac{\dfrac{k_{cat}}{K_M} k_{+t}}{\dfrac{k_{cat}}{K_M} k_{-t}} c_{bulk} \tag{4.42}$$

Equations (4.40–42) show that the metabolic flux in the stationary state depends on the two constants k_{+t} and k_{-t}, which may be equal (as in the case of simple diffusion) or different (because of a difference of electric potential between the two phases, as illustrated by Ussing's equation for a two-compartment system separated by a diffusion barrier (Giebisch *et al.*, 1978). We shall now investigate the stability of the stationary state by modelling the transitory states that appear when a metabolic state moves away from the stationary state.

β. Stability of the dynamics of a step in the metabolic chain.   Let us consider an element of the metabolic pathway (4.39.1) supposing that two of its enzymes: E_{i-1} and E_i, are associated with a cellular substructure, i.e. the local medium, and that the metabolite P_i may be exchanged between the local medium and the bulk phase, according to the schema:

$$\rightarrow P_{i-1,local} \xrightarrow{E_{i-1}} P_{i,local} \xrightarrow{E_i} P_{i+1,local} \rightarrow$$

$$\tag{4.43}$$

where V_{local} and V_{bulk} represent respectively the volumes of the local medium and the bulk phase. The variation of the concentrations $P_{i,local}$ and $P_{i,bulk}$ (moles per unit volume) may be described by the following system of differential equations:

$$\frac{dP_{i,local}}{dt} = -\frac{V_{max,i} P_{i,local}}{K_{M,i} + P_{i,local}} + \frac{J_{i-1} + k_{+t,i} P_{i,bulk} - k_{-t,i} P_{i,local}}{V_{local}}$$

$$\frac{dP_{i,bulk}}{dt} = \frac{k_{-t,i} P_{i,local} - k_{+t,i} P_{i,bulk}}{V_{bulk}} \tag{4.44(1)}$$

where $V_{max,i}$ (moles per unit volume per unit time) and $K_{M,i}$ (moles per unit volume) are respectively the maximum rate and Michaelis constant of the reaction catalysed by the ith enzyme, $k_{+t,i}$ and $k_{-t,i}$ (volume per unit time) are defined as above, and J_{i-1} is the number of moles of P_i produced per unit time during $(i-1)$th reaction

[Pi,local]

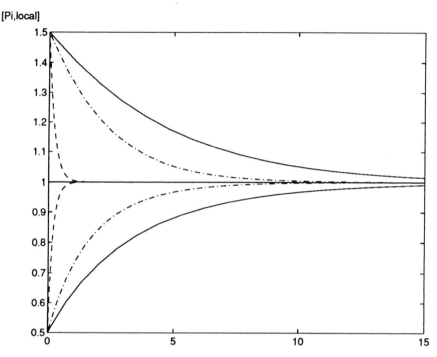

Fig. 4.9a. Influence of the ratio V_{local}/V_{bulk} on the variation in time of $[P_{i,local}]$ in case of channelling without pooling (confinement in the local phase): $V_{bulk} = 0$ (full lines); V_{local}/V_{bulk} = 1 (-.-); $V_{local}/V_{bulk} = 0.05$ (---). In the upper curves (resp. bottom), the initial perturbation is $[P_{i,local}]$ $(t = 0) = 1.5P_{i,local}{}^{ss}$ (resp. $[P_{i,local}](t = 0) = 0.5P_{i,local}{}^{ss}$), where $P_{i,local}{}^{ss}$ is the steady state value. The degree of stability is increased when the local volume decreases (after Chauvet and Costalat, 1995).

(J_{i-1} being supposed constant for the present analysis). We may observe that: $V_{max,i}$ = $k_{cat,i}n_E/V_{local}$, n_{E_i} being the number of moles of the enzymes E_i, and $k_{cat,i}$ the kinetic constant as defined above.

The numerical resolution of equations (4.44.1) shows how the metabolite constants vary after a perturbation leads to a displacement from the stationary state. Over the entire domain of the parameters considered, the stationary state is asymptotically stable. Intuitively, we may say that, after a perturbation, the concentrations return indefinitely to their stationary state values. However, the resilience or the degree of stability, i.e. the rate at which the perturbed solution converges towards the stationary state, depends mainly on the parameters that characterise the enzymatic organisation (Fig. 4.9). Firstly, the confinement of the element of the metabolic pathway to the local medium tends to increase the degree of stability (Fig. 4.9a), all the more that the ratio V_{local}/V_{bulk} is small. Secondly, the degree of stability depends on the constants of exchange between the local medium and the bulk phase, i.e. $k_{+t,i}$ and $k_{-t,i}$: (i) in the case of simple diffusion, the increase of $k_{+t,i} \equiv k_{-t,i}$ augments the rate of convergence towards the stationary state until the concentrations are relatively close to those of the stationary state; after this the rate of convergence decreases (Fig.

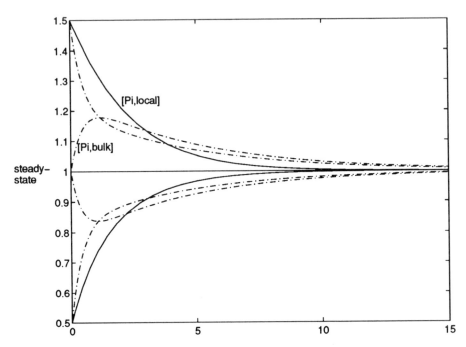

Fig. 4.9b.   Effect of exchange constants $k_{+t,i} = k_{-t,i}$ on the stability of the steady state after perturbation of $[P_{i,local}]$: variation in time of $[P_{i,local}]$ for $k_{+t,i} = k_{-t,i} = 0$ (full lines) and $k_{+t,i} = k_{-t,i} = 1$ (-.-); simultaneous variation in time of $[P_{i,bulk}]$ for $k_{+t,i} = k_{-t,i} = 1$ (-.-).

4.9b); and (ii) when $k_{+t,i} \neq k_{-t,i}$ (in the presence of an electric field for example), the increase in $k_{+t,i}$ alone diminishes the rate of convergence when $P_{i,local}$ is perturbed, and augments it when $P_{i,bulk}$ is perturbed (see Table 4.1).

Thus, for a single step in a metabolic pathway confined to a limited subcellular volume, the stability of the dynamics depends on the eventual exchanges between the metabolites in the local volume (or phase) and the bulk phase. It also depends on the volume of the local phase for a given total volume. Let us now consider the more general case of a chain of reactions and see how the enzymatic organisation influences the stability of the stationary state. This will enable us to introduce the definition of a structural unit.

γ. Stability of a metabolic pathway with allosteric control of production.   Let us now consider a common metabolic pathway of the Yates–Pardee type in which the last substance produced inhibits the first reaction by means of an allosteric mechanism. If there is no enzymatic organisation, the variation of the concentration P_i of each of the metabolites may be written as follows (Fig. 4.10):

$$\frac{dP_1}{dt} = -\alpha_1 P_1 + f_1(P_n; \tilde{\omega}, \kappa, \alpha_0)$$

$$(4.44(2))$$

$$\frac{dP_i}{dt} = -\alpha_i P_i + \alpha_{i-1} P_{i-1} \quad i = 2, 3, \ldots, n$$

Table 4.1.   Stability of a metabolic network according to the associations between the metabolic pathways

Model	Results	Reference
Element of the metabolic pathway with spatial localisation (Eq. (4.41(1)) and Fig. 4.9))	$\searrow$ from $V_{local}/V_{bulk} \Rightarrow \nearrow$ in the rate of convergence towards the stationary state $\nearrow$ of exchanges by diffusion between local and bulk phases $(k_{+t,i} \equiv k_{-t,i} \nearrow)$ $\Rightarrow \nearrow$ in the rate of convergence towards the stationary state	This section (Chauvet and Costalat, 1995)
Pathway of the Yates– Pardee type without spatial localisation (Eq. (4.44(2)))	$\nearrow$ in the length of the metabolic pathway $\Rightarrow \searrow$ in the domain of stability	Walter (1970) (Rapp and Berridge, 1977)
Pathway of the Yates– Pardee type with spatial localisation (Eq. (4.44(3)) and Fig. 4.14a)	$\nearrow$ in exchanges by diffusion between local and bulk phases (under certain conditions) $\Rightarrow$ $\searrow$ in amplitude of oscillations. A periodic solution may be transformed into a stationary solution	Doubabi *et al.* (1994) This section (Chauvet and Costalat, 1995)
Association in parallel of two pathways of the Yates-Pardee type (Eq. (4.63) and Fig. 4.14b)	$\nearrow$ in exchanges by diffusion between local and bulk phases (under certain conditions) $\Rightarrow$ $\nearrow$ in the domain of stability except in the case of very high Hill coefficients, e.g. $\tilde{\omega} \sim 100$	*General case:* This section (Chauvet and Costalat, 1995) *Specific cases:* Chauvet (1993); Chauvet and Girou (1983); Morillon *et al.* (1994)

where the coefficients $\alpha_i = (k_{cat,i}/K_{M,i}) E_i$, $i = 2, \ldots, n$ represent the rate constants of the non-allosteric reactions (which is the case when the metabolite concentration is much smaller than the Michaelis constant, i.e. when $P_i \ll K_{M,i}$); f_1 has the form: $f_1(P_n; \tilde{\omega}, \kappa, \alpha_0) = \alpha_0/(1 + \kappa P_n^{\tilde{\omega}})$, $\tilde{\omega}$ being Hill's coefficient. Walter (1974), and Rapp and Berridge (1977) have determined the conditions of stability for which this type of system remains asymptotically stable. The method used is similar to that described above except for the complication introduced by the control term f_1. The essential result of this study of stability is related to the number of steps in the metabolic chain: when the length of the metabolic pathway increases, the domain of stability of its unique stationary state decreases. We may deduce that a metabolic pathway is functional if the conditions above are satisfied.

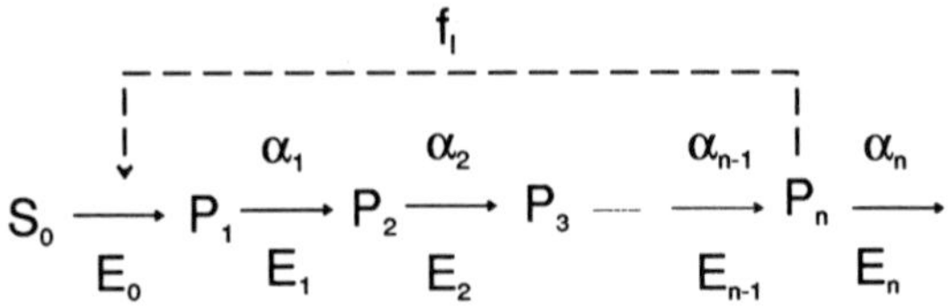

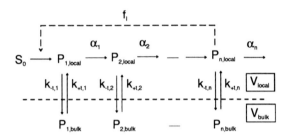

Fig. 4.10. Metabolic pathway of the Yates–Pardee type.

Fig. 4.11. Metabolic pathway of the Yates–Pardee type with enzymatic organisation.

The effect of enzymatic organisation on the dynamics of a metabolic pathway of the Yates–Purdee type has been studied by Doubabi *et al.* (1994) using a harmonic balancing technique. Let us consider the system shown in Fig. 4.11:

$$\frac{dP_{1,local}}{dt} = -\alpha_1 P_{1,local} + f_1(P_{n,local}; \tilde{\omega}, \kappa, \alpha_0) + \frac{k_{+t,1}P_{1,bulk} - k_{-t,1}P_{1,local}}{V_{local}}$$

$$\frac{dP_{i,local}}{dt} = -\alpha_i P_{i,local} + \alpha_{i-1}P_{i-1,local} + \frac{k_{+t,1}P_{i,bulk} - k_{-t,i}P_{i,local}}{V_{local}} \qquad (4.44(3))$$

$$\frac{dP_{i,bulk}}{dt} = \frac{k_{-t,i}P_{i,local} - k_{+t,i}P_{i,bulk}}{V_{bulk}}, \quad i = 2, 3, \ldots n$$

where

$$\alpha_i = \frac{V_{max,i}}{K_{M,i} + P_{i,local}}.$$

The general study of the system (4.44.3) is rather difficult. However, when $V_{local} = V_{bulk}$, and $k_{-t,i} = k_{+t,i}$ (simple diffusion), it can be analytically shown, and numerically verified in the non-linear case above, that the exchange of metabolites between the local medium and the bulk phase is accompanied by: (i) a decrease in the amplitudes and frequencies of the periodic solutions of the system (4.44(3)) (see Fig. 4.14a); and (ii) the eventual replacement of a periodic solution by a stationary state. This confirms the hypothesis that a metabolic pathway tends to a stabilisation through exchanges with metabolite pools present in the bulk phase.

The two results presented may be summed up as follows: (i) the stability of the dynamics of a step in the metabolic chain within a spatial subcellular compartment depends on the exchanges between the metabolites of the local volume (or phase) and the core of the solution; and (ii) the domain of stability of the unique stationary state of the metabolic pathway decreases when the length of the metabolic pathway increases. These properties suggest that the metabolic pathway may be considered as a structural unit, i.e. a sequential organisation of type (4.39.2) steps, the function of which is the production of a terminal metabolite P. More generally, we may use the following definitions:

A structural unit is a structural equivalence class constituted by units that are identical with regard to their structure, and independent with regard to their function (according to certain criteria). An elementary physiological function is the collective behaviour (cooperation in some tasks) of at least one functional interaction.

A physiological function (a biological system) is the collective product of a set of structural units which can be hierarchically classified according to their elementary interactions.

Let us now consider how the stability may be affected by the presence of two metabolic pathways associated in the production of a physiological function, i.e. two identical structural units working in parallel, but each being in a different phase or compartment.

4. *A paradigm for the creation of functional interactions: the self-association hypothesis*

a. *The structural unit and the physiological function*

In the preceding section we have considered a certain number of concepts and definitions which form the basis of our integrative approach to biological systems. According to the definitions of the structural unit and the physiological function, the structural organisation, which constitutes the micro-environment in which enzymatic catalysis occurs, is the structural unit channelling towards a given metabolic pathway. This structural unit is related to the physical environment by the input (the substrate or initial product) and the output (the final product) of the metabolic pathway. In the case of bifurcations, or transfers from one metabolic pathway to another, one structural unit, say u_1, acts on another structural unit u_2.

Let us now formalize this problem. As mentioned in Section 3 above, the functional interaction, denoted ψ_{ij}^{α}, is mathematically defined from the i-unit to the j-unit. The result of this interaction is a product which may be either the direct value of the elementary function:

$$P_{\alpha,j} = \psi_{ij}^{\alpha}(P_{\alpha,i}; r) \tag{4.45}$$

or the transformed value:

$$P_{\alpha,j}' = \Phi_{ij}^{\alpha}(P_{\alpha,i}; r) = \phi_j^{\alpha} \cdot \psi_{ij}^{\alpha}(P_{\alpha,i}; r) \tag{4.46}$$

inside the unit localised in r. The variables $P_{\alpha,j}$ or $P_{\alpha,j}'$ are identified as elementary physiological functions. More generally, μ products $P_{\alpha,j}$, $1 \le \alpha \le \mu$ in the i-unit may appear during the execution of the elementary physiological function. A physiological function results from a set of hierarchically organised and functionally interacting elements. With these definitions, a general schema for a metabolic pathway system constituting a unit u_i may be written from Eq. (4.38) in terms of compartments:

$$\rightarrow P_1 \xrightarrow{\alpha_1} P_2 \xrightarrow{\alpha_2} P_3 \xrightarrow{\alpha_3} P_4 \xrightarrow{\alpha_4} \dots \qquad (4.47)$$

When each product can diffuse from one unit u_i to another u_j according to the transport function $g_i(P_j, P_i)$ between two different locations, we have:

$$\begin{array}{c}
\rightarrow P_1 \xrightarrow{\alpha_1} P_2 \xrightarrow{\alpha_2} P_3 \xrightarrow{\alpha_3} P_4 \xrightarrow{\alpha_4} \dots \\
g_1 \Updownarrow \quad g_2 \Updownarrow \quad g_3 \Updownarrow \quad g_4 \Updownarrow \\
\rightarrow P_1 \xrightarrow[\alpha_1]{} P_2 \xrightarrow[\alpha_2]{} P_3 \xrightarrow[\alpha_3]{} P_4 \xrightarrow[\alpha_4]{} \dots
\end{array} \qquad (4.48)$$

Under specific circumstances, e.g. in phenomena related to the mechanism of regulation, micromutation, and so on, the switchover from one metabolic pathway to another takes place in a favoured direction according to one of the transport functions g_i, i.e. the metabolic flow is redirected, creating a functional interaction from a structural unit channelling u_i onto a structural unit channelling u_j:

$$\begin{array}{c}
(u_i) \\
\downarrow g_{ij}. \\
(u_j)
\end{array} \qquad (4.49)$$

We shall now develop this idea to examine the association between structural units and the consequence of an increase in the stability of the dynamics of the system (Chauvet, 1993a). From the point of view of global dynamics, the biological system appears to tend towards an increase in stability. In other words:

> *The existence of structural units, i.e. of structural discontinuities, is a necessary condition for the functioning of the system and, in the case of channelling, the metabolic systems are structured such that their functional interactions increase the stability of the functioning.*

b.   The self-association hypothesis

To carry the investigation further, we shall use a paradigm based on the relation between physiology and evolution (more details of this will be found in Chapter 7). We introduced the hypothesis of self-association (Chauvet and Girou, 1983) to justify the concept of the functional interaction. The study of the physiological function seen as a combination of functional interactions led us to formulate the property of invariance, called the 'vital coherence' of the function, which is in fact a generalisation of the hypothesis of self-association. This paradigm will also be found useful in the study of the growth and development of organisms and in that of the evolution of the species. It represents a general, abstract schema of the possible functional relations between structural units:

> *A unit that has undergone a redirection of flow may either:*
> (i)   *stop functioning, i.e. the unit 'dies', if* $[P_{\alpha,j} < P_{\alpha,j}(0)]$ *where* $P_{a,j}(0)$ *is a threshold; or*

(ii) associate with other units whose properties are not necessarily identical, but which still possess the missing product, i.e. a $P_{a,i}$, such that $P_{a,j} = \psi_{ij}^{\alpha}(P_{\alpha,i})$. Thus, an α-functional interaction would be created from the i-unit to the j-unit.
This is the paradigm we propose to use.

The population level (U) is constituted by the association of metabolical units, such as the u_i mentioned above, in a network. This network is submitted to variations in its functioning following internal regulations and/or environmental modifications that can stop the synthesis of $P_{a,j}$, for example when certain g_i are null, and redirect the metabolic flow in another direction.

c. On the nature of a break in the self-association

α. *A break in the functional interaction of the metabolic pathway.*   Let us assume that for some reason a break occurs at a lower level in a sequence of reactions in the metabolic system, for example from the product $P_i \equiv P_{i,u_1} \in^* u_1$ to the product $P_1^* \equiv P_{i,u^*_1} \in^* u_1^*$, u_1^* being the unit at which the break occurs. If u_1^* needs P_{i+1} in order to 'survive', then according to the above paradigm, an elementary function from u_1 to another unit u_1^* has to be created. According to the notations defined in equations (4.45) or (4.46), let $\psi^i_{u_1u_1^*}$ be this interaction, i.e.:

$$P_{i,u_1^*} = \psi^i_{u_1u_1^*}(P_{i,u_1}) \qquad P_{i,u_1^*} \in^* u_1^* \qquad P_{i,u_1} \in^* u_1. \qquad (4.50)$$

In the present case, with only one functional interaction represented by the product P_i emitted by u_1 acting on u_1^*, we can simplify the notations as follows:

$$P_i^* = \psi(P_i) \qquad P_i^* \in^* u_1^* \qquad P_i \in^* u_1. \qquad (4.50')$$

Because of the micromutation that has disrupted the biochemical pathway (the enzyme E_{i-1} is suppressed), the product P_i no longer exists in u_1^*. Therefore, P_{i+1} which is obtained from P_i in this unit disappears, as does the unit, except if the product P_i can be captured from another unit that possesses it (according to the principle of vital coherence). In this case, *an association between the metabolic pathways is obtained at a higher level.* We shall now discuss the general formulation of the possible mechanisms of this association, and then study the stability of the process.

β. *Basic mechanisms of the association.*   As mentioned above, various mechanisms could be involved in the creation of this association, leading to a relation such as (4.50). For example, $P_1 \in^* u_1$ may diffuse passively towards the 'pathological' unit u_1^* and, when it arrives in u_1^* (we shall then call it P_1^*, all the metabolites in this pathological unit being indicated by an asterisk (*), and assume subscript i to be 1), it could initiate the transformation leading to P_2^*. Such a sequence of transport–transformation may be represented by the diagram:

$$P_1 \xrightarrow{\Phi} P_2^*$$
$$\psi \searrow \quad \nearrow \phi \qquad\qquad (4.51)$$
$$P_1^*$$

where the left part is non-local and the right part is local. Then:

$$P_2^* = \Phi(P_1) = \phi \circ \psi(P_1)$$

i.e. equation (4.45). Various types of system may be used to describe these transformations by considering different mechanisms for Φ, ϕ, ψ.

(i)   *The simplest mechanism would be a linear transformation* from P_1 to P_2^* including both transport and chemical reactions. It is similar to a classical chemical reaction, i.e. a transfer from the P_1-compartment to the P_2^*-compartment:

$$P_1 \xrightarrow{\Phi} P_2^* \qquad (4.52)$$

where the direct transformation is indicated by Φ. This case, which is indeed the simplest, is studied here with:

$$P_2^* = \Phi(P_1).$$

The direct transformation will be expressed below in terms of a rate constant $\bar{\alpha}$.

(ii)   *A passive diffusion* of the product P_1 can be explicitly included in the transformation above:

$$P_1 \xrightarrow{\Phi} P_2^*$$
$$g_1 \searrow \quad \nearrow \alpha_1 \qquad (4.53)$$
$$P_1^*$$

where ψ and ϕ are replaced by linear transformations:

(i) $g_1(P_1, P_1^*) = \beta(P_1 - P_1^*)$ to describe a simple passive diffusion with coefficient β; and

(ii) $\phi(P_1^*) = P_2^*$ (Eq. (4.51)) given by the kinetic equation:

$$\frac{dP_2^*}{dt} = -\alpha_2 P_2^* + \alpha_1 P_1^* \qquad (4.54)$$

to describe the chemical transformations with the rate constants α_1, α_2.

5.   *Functional association between two metabolic pathways defined as structural units*

a.   A general and generative schema of the association

On the basis of these simple mechanisms, the general schema for a metabolic pathway system similar to (4.48):

$$\rightarrow P_1^* \xrightarrow{\alpha_1} P_2^* \xrightarrow{\alpha_2} P_3^* \xrightarrow{\alpha_3} P_4^* \xrightarrow{\alpha_4} \dots$$
$$g_1 \Updownarrow \quad g_2 \Updownarrow \quad g_3 \Updownarrow \quad g_4 \Updownarrow$$
$$\rightarrow P_1 \xrightarrow[\alpha_1]{} P_2 \xrightarrow[\alpha_2]{} P_3 \xrightarrow[\alpha_3]{} P_4 \xrightarrow[\alpha_4]{} \dots$$

corresponding to the dynamics (4.54) may be described by the dynamical system:

$$\frac{dP_i}{dt} = \alpha_{i-1}P_{i-1} - \alpha_i P_i - g_i(P_i, P_i^*) \tag{4.55(1)}$$

$$\frac{dP_i^*}{dt} = \alpha_{i-1}P_{i-1}^* - \alpha_i P_i^* + g_i(P_i, P_i^*) \tag{4.55(2)}$$

$$i = 2, 3, \dots .$$

In the present case of association between the metabolic systems (4.47) where the number of steps is $n = 4$, let us assume, for example, that α_2 becomes null in a given unit u_1, leading to the appearance of a pathological unit u_1^*. If u_1^* receives P_3 from u_1, and if P_1^* can also diffuse towards u_1, we may deduce the following schema from (4.48):

$$
\begin{array}{c}
\downarrow \hspace{5em} | \\[0.5em]
P_1^* \xrightarrow{\alpha_1} P_2^* \xrightarrow{\alpha_2} P_3^* \xrightarrow{\alpha_3} P_4^* \xrightarrow{\alpha_4} \\[0.3em]
g_1 \Updownarrow \hspace{4em} g_3 \Updownarrow \\[0.3em]
P_1 \xrightarrow{\alpha_1} P_2 \xrightarrow{\alpha_2} P_3 \xrightarrow{\alpha_3} P_4 \xrightarrow{\alpha_4} \\[0.5em]
| \hspace{6em} \uparrow
\end{array}
\tag{4.56}
$$

Similarly, to obtain the dynamical system (4.54), we may write:

$$\frac{dP_1}{dt} = -\alpha_1 P_1 + f_{1,4}(P_4; \tilde{\omega}, \kappa, \alpha_0) - g_1(P_1, P_1^*)$$

$$\frac{dP_2}{dt} = -\alpha_2 P_2 + \alpha_1 P_1$$

$$\frac{dP_3}{dt} = -\alpha_3 P_3 + \alpha_2 P_2 - g_3(P_3, P_3^*)$$

$$\frac{dP_4}{dt} = -\alpha_4 P_4 + \alpha_3 P_3$$

$$\frac{dP_1^*}{dt} = -\alpha_1 P_1^* + f_{1,4}(P_4^*; \tilde{\omega}, \kappa, \alpha_0) + g_1(P_1, P_1^*) \tag{4.57}$$

$$\frac{dP_2^*}{dt} = \alpha_1 P_1^*$$

$$\frac{dP_3^*}{dt} = -\alpha_3 P_3^* + g_3(P_3, P_3^*)$$

$$\frac{dP_4^*}{dt} = -\alpha_4 P_4^* + \alpha_3 P_3^*.$$

The case of a simple passive diffusion for P_1 and P_2 is obtained, as described above, by putting:

$$g_1(P_1, P_1^*) = \beta_1(P_1 - P_1^*)$$
$$g_3(P_3, P_3^*) = \beta_3(P_3 - P_3^*).$$

(4.58)

We can simplify this kinetic system by using the schema (4.53) rather than (4.52), i.e. by introducing the constant $\bar{\alpha}$. With the non-direct feed-back of P_4^* on P_1, the following system of equations is obtained:

$$\frac{dP_1}{dt} = -\alpha_1 P_1 + f_{1,4}(P_4; \tilde{\omega}, \kappa, \alpha_0) + f_{1,4}(P_4^*; \tilde{\omega}, \kappa, \alpha_0)$$

$$\frac{dP_2}{dt} = -\alpha_2 P_2 + \alpha_1 P_1$$

$$\frac{dP_3}{dt} = -(\alpha_3 + \bar{\alpha})P_3 + \alpha_2 P_2$$

(4.59)

$$\frac{dP_4}{dt} = -\alpha_4 P_4 + \alpha_3 P_3$$

$$\frac{dP_4^*}{dt} = -\alpha_4 P_4^* + \bar{\alpha}P_3$$

where $\bar{\alpha}$ is a positive constant, included in Φ as explained in paragraph 4.c.β above, which simply describes the non-local contribution of product $P_3 \in^* u_1$ to the production of $P_4^* \in^* u_1^*$. It is assumed here that P_4^* can modify the synthesis of P_1 in an additive manner, in the same way as P_4, and that the coefficients for the degradation of P_4 and P_4^* are the same. Such a system represents the dynamics of a new unit noted $u_2 \equiv (u_1, u_1^*)$.

In reality, structural units are located at different points in the physical space. Thus, the variation in time of the product satisfies the partial differential equations which describe the dynamics in $u(r_0)$ and in $u^*(r)$ at two points r_0 and r. We shall extend our study to these cases in Volume III.

b.   Mathematical study of the dynamics in a u_2*-unit: a specific system*

The system (4.59) can be made dimensionless by using a transformation given by Walter (see Rapp, 1976) where:

$$\xi = (\alpha_0 \alpha_X \alpha_Y \alpha_1 \kappa^{1/\tilde{\omega}})^{1/4} \qquad t^* = \xi t \qquad b_1 = \frac{\alpha_X}{\xi} \qquad b_2 = \frac{\alpha_Y}{\xi} \qquad b_{i+2} = \frac{\alpha_i}{\xi}, i = 1, 2. \quad (4.60)$$

A new system of equations is then obtained:

$$(1) \quad \frac{dx_1}{dt} = -b_1 x_1 + \frac{1}{1 + x_4^{\tilde{\omega}}} + \frac{1}{\left(1 + \left(\dfrac{b_5}{b_3} x_5\right)^{\tilde{\omega}}\right)}$$

$$(2) \quad \frac{\mathrm{d}x_2}{\mathrm{d}t} = -b_2 x_2 + x_1$$

$$(3) \quad \frac{\mathrm{d}x_3}{\mathrm{d}t} = -(b_3 + b_5)x_3 + x_2 \tag{4.61}$$

$$(4) \quad \frac{\mathrm{d}x_4}{\mathrm{d}t} = -b_4 x_4 + x_3$$

$$(5) \quad \frac{\mathrm{d}x_5}{\mathrm{d}t} = -b_4 x_5 + x_3$$

in terms of new state variables:

$$t^* = a_0 t \qquad x_1(t^*) = a_1 X(t) \qquad x_2(t^*) = a_2 Y(t)$$
$$x_3(t^*) = a_3 P_1(t) \qquad x_4(t^*) = a_4 P_2(t) \qquad x_5(t^*) = a_5 P_2^*(t) \tag{4.62}$$

with dimensionless coefficients a_i and b_i instead of the dimensional ones α_i, κ and $\tilde{\omega}$. In this example, the functional interaction (4.50), created according to the proposed paradigm, is mathematically expressed by Eq. (4.45) (or (4.34)), and corresponds to the general equation (4.35), where $k = l$, $i = j$. This is a very simple case of an organic link at a given level of organisation, which is here the metabolical level.

Walter (1969a, b), Viniegra-Gonzalez (1973), and Rapp (1976), have studied the stability of the system (4.55.1), which describes the time-variation of units u_1, determining the sufficient conditions for the system to be asymptotically stable. We have deduced the stability of system (4.61), which represents $u_2 \equiv (u_1, u^*)$, from an analysis of the linearised systems (Chauvet and Girou, 1983). Around steady states, the domain of stability of the new system (4.61) of units $u_2 \equiv (u_1, u_1^*)$ can be shown to be larger than the domain of stability of the preceding system (4.55.1) with $n = 4$ for units u_1, i.e. the new system is *more stable*.

The results are shown in Fig. 4.12: (i) For a given set $(\alpha_0, \alpha_1, \alpha_2, \tilde{\omega})$ of parameters, the condition of stability of system (4.55.1) (with $n = 4$) is studied in the plane (α_3, α_4). The domain of stability of the 1-unit metabolic system is *outside* the solid line. The same study is repeated with the system (4.61) (corresponding to the 2-unit metabolic system (u, u^*)) for various values of $\bar{\alpha}$ (only one value of $\bar{\alpha}$ is represented in Fig. 4.12a), and for increasing values of parameter κ, both of which describe the existence of the association. *We can see that the area of stability (represented by the space outside the closed lines) increases when the formal biological system is complexified.* In other words, unit $u_2 \equiv (u_1, u_1^*)$ is metabolically 'more stable' than unit u_1. With this statement, we postulate that the likelihood of the existence of a system is all the greater that the system is more complex.

More recently, Machbub et al. (1992) have used a similar method to investigate the stability of the units $u_j \equiv (u_{j-1}, u_1^*)$ obtained with the same self-association process as $u_2 \equiv (u_1, u_1^*)$. The linear part of the system is found to be exponentially stable. Moreover, the stationary states of u_j are asymptotically stable through a balance between the linear and non-linear terms of the equation that describes the time-variation of u_j. An important result has been obtained for the domains of stability of

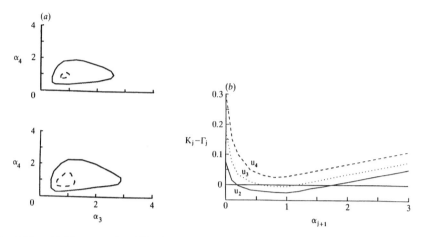

Fig. 4.12. (a) The area of stability increases when the system (full line) is rendered complex (dashed line). For each system, i.e. each closed curve, the corresponding domain of stability lies outside the internal space: $\kappa = 0.1$ in the upper figure, and $\kappa = 0.01$ in the lower figure. These values express the 'intensity' of the molecular binding. (b) The stability of an association between a unit u_{j-1}, of degree $j - 1$, and a unit u_1. The association is represented by the factor $K_j - \Gamma_j$ (these terms come respectively from the linear and non-linear parts of the generalised dynamic system), and α_{j+1}, for $j = 1, 2, 3$, is the added parameter (after Machbub *et al.* (1992)).

successive units u_j: locally, unit u_j is more stable than u_{j-1}. All these results have been confirmed numerically (Machbub *et al.*, 1992), and it has been shown that successive associations increase the domain of stability (Fig. 4.12b). Thus, building an association by creating a functional interaction appears to lead to a better stability at the level (M). This mathematical property could be generalised to a class of dynamical systems such as (4.57).

c.   Numerical study of the dynamics in a u_2-*unit: a general dynamical system*

The same method has been used to study the general dynamical system (4.57), which is not mathematically equivalent to the specific system (4.59). Numerical simulations (Burger *et al.*, 1993) have shown that, in this case too, *association increases the domain of stability.* The values of the parameters are: $\alpha_1 = \alpha_2 = 1$, $\alpha_0 = 50$, $\kappa = 1$. Coupling between the two units is realised by simple passive transport: $g_i (P_i, P_i^*) = \beta_i (P_i - P_i^*)$, $\forall_i$, and the study is made in the plane α_3, α_4. An increase in stability can be demonstrated even when all the coefficients β_i are assumed to be unequal, and, in this case, *the higher the value of* β_i, *i.e. the 'intensity' of coupling, the wider is the region of stability.*

The existence and unicity of the steady state have been proved and it has been shown that no bifurcation exists. In some particular cases, it can be proved that: (i) periodic solutions with coupling have lower amplitude and lower frequencies;

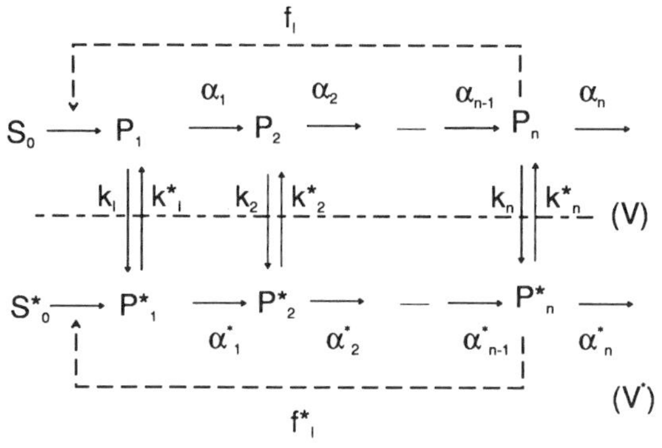

Fig. 4.13. Two associated metabolic pathways.

(ii) coupling can give rise to a steady state instead of a periodic solution (Morillon *et al.*, 1994).

6. The paradigm of self-association applied to the enzyme organisation: role of local and bulk phase

Let us apply the paradigm of self-association to the model of two metabolic pathways, such as those given by (4.44(3)) (Fig. 4.13), which are identical and in parallel but situated in different places: the molecules P_i and P_i^* being the same in chemical terms. The two pathways have the same substrates and the same intermediate products. The association between the two pathways u_1 is represented by the coefficients of exchange k_i which characterise the passive transport between the two pathways. If all the coefficients are null, then the two units are *independent*. If one of the coefficients is non-null, then the two units are associated forming a new structural unit we shall call u_2. The same notations apply to the two associated pathways, one of the pathways being identified by an asterisk (*). The kinetics of the two associated metabolic pathways may be described by the following dynamic system:

$$\frac{dP_1}{dt} = -\alpha_1(P_1) + f_1(P_n; \tilde{\omega}, \kappa, \alpha_0) + \frac{k_1^* P_1^* - k_1 P_1}{V}$$

$$\frac{dP_i}{dt} = \alpha_{i-1}(P_{i-1}) - \alpha_i(P_i) + \frac{k_i^* P_i^* - k_i P_i}{V}, \quad i = 2, ..., n$$

$$\frac{dP_1^*}{dt} = -\alpha_1^*(P_1^*) + f_1^*(P_n^*; \tilde{\omega}^*, \kappa^*, \alpha_0^*) + \frac{k_1 P_1 - k_1^* P_1^*}{V^*}$$

$$\frac{dP_i^*}{dt} = \alpha_{i-1}^*(P_{i-1}^*) - \alpha_i^*(P_i^*) + \frac{k_i P_i - k_i^* P_i^*}{V^*}, \quad i = 2, ..., n$$

(4.63)

where f_1^* is of the form: $f_1^*(P_n^*) = \alpha_0^*/[1 + \kappa^* (P_n^*)^{\tilde{\omega}^*}]$, $\alpha_i(P_i) = -V_{max,i} P_i/(K_{M,i} + P_i)$ and V and V^* are the respective volumes of each of the phases (compartments);

the constants of exchange k_i and k_i^* are analogous to $k_{-t,i}$ and $k_{+t,i}$ in section 3 and the constants α_i and α_i^* take into account the volumes, for example: $\alpha_i = V_{max,i}/K_{M,i}$ $= k_{cat,i} n_{E_i}/(K_{M,i}V)$, $i = 1, 2, ..., n$, assuming that $P_i \ll K_{M,i}$.

The domain of stability of this dynamic system, which is more general than the systems described above since it includes the non-linear terms α_i, may be investigated, for example, in the case of a chain of length $n = 4$, in terms of the two coefficients α_3 and α_4. When the two metabolic pathways are associated by coefficients of exchange k_i (non-null), the corresponding domain of stability has been shown to be increased compared to that of a single pathway in the linear case (Chauvet, 1993) and, here more generally, for the non-linear case (Fig. 4.14b). Morillon *et al.* (1994) have demonstrated this result mathematically. Moreover, the dynamic system possesses a unique, stable stationary solution.

We have considered how the spatial localisation of enzymatic reactions, ranging from the elementary type (one step reaction) to that of one or several metabolic pathways, may affect the functional stability of the organisation (Table 4.1). The spatial localisation of molecules involved in the reaction depends on the physics of the medium and on the structure of the molecules, i.e. the spatial arrangement of the molecules with respect to the surrounding pool. In fact, it depends on the exchanges that occur either through the diffusion of metabolites, or through a molecular mechanism in the neighbourhood of the active enzymatic site, or through an exchange between two phases, one of which corresponds to a cellular substructure. The organisation may therefore be considered to be *structural*. We may also study the enzymatic kinetics of the metabolic pathway which leads to the synthesis of an element of the metabolic network: the product P, by investigating the properties of stability. The nature of this stability, which involves the products of the metabolic network, is *functional*.

As shown in Table 4.1, an increase in the exchanges by diffusion between the pools of metabolites, or a confinement of molecules in the neighbourhood of an active site, leads to an increase in the functional stability about the stationary state. This holds good not only for a single reaction in the metabolic pathway but also for the whole metabolic pathway. These results demonstrate that enzymatic organisation produces an increase in functional stability when: (i) a metabolic pathway is confined to a reduced space; (ii) the metabolic pathway exchanges metabolites with external pools; and (iii) the same metabolic pathway is present in two phases or adjacent compartments. We may therefore consider this molecular enzyme–substrate ensemble, together with sequential organisation which leads to the production of a terminal metabolite, as a *structural unit*.

We have seen that the association of identical metabolic pathways interpreted as structural units increases the stability of the functional dynamics of the set of pathways. We suggest that this result may be generalised so that if a functional organisation is correctly represented, the graph of the metabolic networks, constructed along functional lines, may be expected to reveal the same property, i.e. an increase in the stability of the system. In this framework, using the definitions given above for the *structural unit* and the *physiological function*, the bifurcations in the network would correspond to the association between structural units.

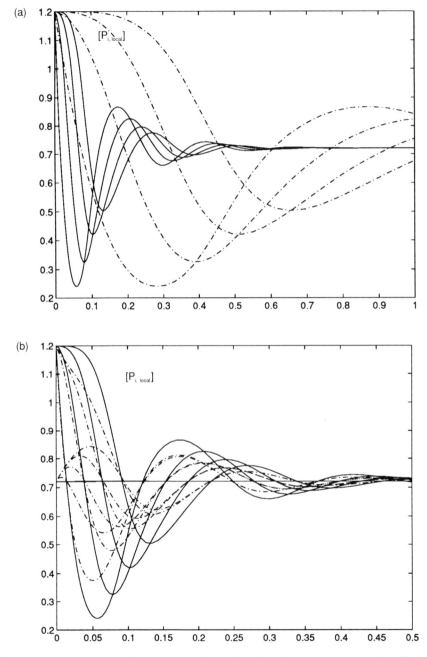

Fig. 4.14. Bundle of solutions [$P_{i,\,local}$]: (a) for Eq. (4.44(3)): the amplitude of oscillations decreases when the volume of compartment V_{local} decreases (effect of confinement), for two values of volumes $V_{local} = 1$ (full lines), and $V_{local} = 5$ (dashed lines); (b) for Eq. (4.63): the amplitude of oscillations decreases when coefficients of exchange k_i increase: *non-null exchange coefficients* (association) lead faster to stability ($k_i = 0$ (full lines) and $k_i = 10$ (dashed lines).

In other words, the approach proposed here, based on the investigation of stability, may be considered to be a '*global*' method since it integrates the functional behaviour of the set of metabolites into a unit which itself may be considered to be a structural unit. Although we do not yet know whether the stability produced by functional association is conserved in the case of the association of non-identical units, we may suggest, on the bases of the examples considered above, that the functional association of metabolic pathways tends to promote stability. This property is fundamental to our theory of functional organisation in living organisms (Chauvet, 1993).

As far as the metabolic organisation is concerned, the significance of these results is evident: *a considerable functional advantage results from spatial compartmentation in the form of a structural channelling unit.* All redirection of matter in a metabolic network corresponds to the creation of a functional interaction due to an increase in the domain of stability of the dynamics. Thus, the existence of channelling, as defined here, corresponds to a necessity in the functional organisation of the cell. The exchange of matter between compartments, e.g. cells, organelles, may be a source of stability for the cell metabolism.

5

The Replication–Translation Apparatus

Let us now see how the structures described in Chapter 1 may be transmitted during the evolution of an individual or a group of individuals. Behind the extreme diversity of biological macromolecules lies a chemical unity related to the chemistry of heredity, which is the foundation of molecular biology. In this connection, several questions arise. How are genetic characteristics transmitted? What gives them their stability? Can we establish a molecular theory explaining the selection and the evolution of the species? Finally, how can we deduce the physiological function of the hereditary material?

We shall discuss what has been called the 'central dogma' of molecular biology and, following Zimmerman and Simha (1965), give a dynamic representation of the principal phenomena observed: DNA replication, genetic transcription and translation, and the biosynthesis of peptide chains. Unfortunately, this formal approach to gene dynamics has been somewhat abandoned of late. However, it still remains of interest from the methodological point of view and is likely to prove useful again as soon as sufficient experimental kinetic data become available. Recent ideas concerning the structure of *nucleosomes* on the one hand and the role of *introns* and *exons* on the other will of course need to be included. Finally, the theory of information deduced from the work of Shannon and Weaver (1972) will be used to describe the transfer of information from the gene to the synthesised product. Information theory can be expected to quantify the genetic message and, as we shall see, it actually lays down the basis for a theory of biological organisation.

I. The 'central dogma' of molecular biology

One of the fundamental discoveries of modern biology is that DNA is the support for the hereditary characteristics of the cell. We also know that the nucleotide sequence

Table 5.1. The genetic code with the amino acids indicated by their symbols as in Table 1.2. Each protein chain ends with the codon 'STOP'.

1st position (5′ end)	2nd position				3rd position (3′ end)
	U	*C*	*A*	*G*	
	Phe	Ser	Tyr	Cys	*U*
U	Phe	Ser	Tyr	Cys	*C*
	Leu	Ser	STOP	STOP	*A*
	Leu	Ser	STOP	Trp	*G*
	Leu	Pro	His	Arg	*U*
C	Leu	Pro	His	Arg	*C*
	Leu	Pro	Gln	Arg	*A*
	Leu	Pro	Gln	Arg	*G*
	Ile	Thr	Asn	Ser	*U*
A	Ile	Thr	Asn	Ser	*C*
	Ile	Thr	Lys	Arg	*A*
	Met	Thr	Lys	Arg	*G*
	Val	Ala	Asp	Gly	*U*
G	Val	Ala	Asp	Gly	*C*
	Val	Ala	Glu	Gly	*A*
	Val	Ala	Glu	Gly	*G*

of DNA determines the structure of cell proteins (according to the genetic code shown in Table 5.1) and, by extension, the physiological function. The description of the several subtle and beautiful experiments that have led to these conclusions will be found elsewhere; here we intend to develop a constructive, formalised synthesis of the above results within the framework of the physicochemical sciences. See, for example, the calculation of enthalpies and entropies for the reaction between 18 N'-hydroxysuccinimide esters of N-protected proteinaceous amino acids and p-anisidine (Siemion and Stephanowicz, 1992) which reveals the correlation between the free enthalpies of activation and the corresponding amino acid codons. In Chapter 1 we saw the structure of the elementary 'building blocks' and their assembly in the form of macromolecules: the nucleotides composing DNA or RNA, and the amino acids forming proteinic units, an example of which is myoglobin. The functional role of the proteinic units was identified as that of catalytic support regulating the multitude of chemical reactions taking place within the organism. Briefly, the functional role of DNA may be said to be the delivery of the genetic message. In this sense, its role is essentially informational.

1. *Genes and chromosomes*

When a dividing eucaryotic cell is looked at through a microscope, filaments of varying thickness may be seen forming rod-like structures which then seem to disappear. These are the *chromosomes*, visible as entities of extraordinary diversity of

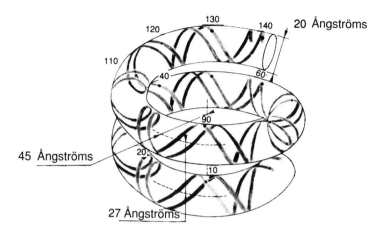

Fig. 5.1.   The nucleosome: geometrical details (after Kornberg and Klug, 1981).

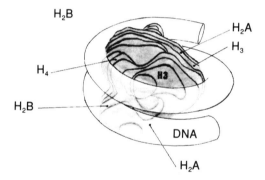

Fig. 5.2.   The nucleosome: DNA coiled around a core made up of four types of histone: H_2A, H_2B, H_3 and H_4 (after Kornberg and Klug, 1981).

form and structure. They contain the genetic information, i.e. a sequence of units called *genes* which mainly determine the structure of the proteins basic to the physiological function of the cell. Thus, *the chromosome is the polystructural support of genes which are the functional units of information.*

It has long been known that the chromosome is made up of a substance called *chromatin* which consists of proteins and nucleic acids, so that DNA is only part of the chromosome although it contains the entire genetic information. It is only recently that the role of the five different types of protein, the *histones*, has been identified to some extent. We have seen that DNA is composed of two strands of nucleotide units coiled about each other in a helix, forming a fairly rigid structure. But when DNA is associated with histones in the chromatin, a 10 nm periodic structure is observed. This new structural organisation corresponds to elementary units called *nucleosomes* (Fig. 5.1). The nucleosome has been shown to be a DNA superhelix coiled around a histone core (Fig. 5.2). This superstructure may explain

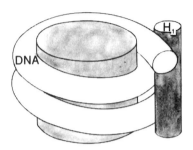

Fig. 5.3. The nucleosome: a fifth type of histone, H_1, marks the entry and the exit of the DNA chain (after Kornberg and Klug, 1981).

the behaviour of chromosomes during the cell cycle. For example, the 'disappearance' of chromosomes between *mitoses* (cell divisions) in reality corresponds to the decondensation of chromosomes into very fine, invisible filaments, followed by a phenomenon of diffusion.

The fundamental fibre of chromatin is thus a chain of nucleosomes like beads on a string. The *core* of the nucleosome is a DNA superhelix (146 nucleotide base pairs) coiled around a histone octamer: the tetramer $(H_3)_2$ $(H_4)_2$ defines the central spiral of the DNA superhelix; H_2A and H_2B are in the form of two dimers (H_2A) (H_2B), one on each face of the tetramer; each dimer is bound to the rest of the DNA to complete the superhelix. The remaining type of histone, H_1, is attached to sites at which the DNA enters and leaves the nucleosome (Fig. 5.3). The role of H_1 is not only to close off the nucleosome but also to favour the organisation of a new helical superstructure with nucleosomes as elements (Fig. 5.4). The aggregation of the histones H_1, which have high ionic forces, leads to the formation of a helicoidal polymer with an irregular structure containing six nucleosomes per spiral turn. Thus, the two types of chromatin fibre may be explained by the spiralling of the 10-nm fibre, the chain of elementary nucleosomes, into a 30-nm superhelix by an increase in the ionic force.

According to Kornberg and Klug (1974), the problem of DNA condensation is not that of compacting a stiff ribbon of DNA, which is a formidable problem, but that of winding the ribbon around spools and then arranging the spools, which is much easier. Histones are not directly involved in the expression of the genetic message, but they may facilitate modifications of chromosome structure by allowing selective gene expression during induction. At the end of this chapter we shall see how enzymes carry out the mathematical operations implicit to the 'manipulation' involved in the super-coiling. It is also known that proteins other than histones may influence the active chromosome, which suggests that, in spite of the progress achieved, several mysterious points remain to be cleared up. Here we shall restrict ourselves to the formalised methodological aspect of only such information as is widely accepted currently.

However, we must mention a major difference that has been found in gene distribution between higher and lower organisms (Chambon, 1981). Indeed, while the DNA of procaryotic cells such as *Escherichia coli* raises no major problem as the

Fig. 5.4.   Helicoidal superstructure formed by nucleosomes (after Kornberg and Klug, 1981).

three million base-pairs code for all of the approximately three thousand proteins found, the same is not true of the chromosomes in eucaryotic cells. Thus, it appears that although the 46 chromosomes in humans could actually code for more than three million proteins, the estimated number of proteins is less than 150,000. How is this excess of DNA to be explained?

Delicate experiments have shown that the excision of non-coding gene sequences after transcription is necessary for the formation of mature mRNA which may then be translated into a protein (see below). Indeed, such *split genes* appear to be a general characteristic of higher organisms.

2.   *Replication, transcription and translation*

The process of *replication* corresponds to the 'exact' reproduction, errors excepted, of the structure of the DNA molecule during cell division and growth. It is based on the complementarity of the nucleotides (A–T and G–C). The biochemical mechanism is evidently enzymic: an enzyme allows the formation of phosphodiester bonds between the DNA strand and the nucleotides. The double strand thus formed is undone by breaking the hydrogen bonds, each strand then serving as a template for the formation of a new double strand, and so on. This process was immediately understood as soon as the DNA structure was determined. The discovery of Watson and Crick was indeed the crowning achievement of a great body of scientific work by Pauling, Avery and many others.

Let us mention some of the models proposed to describe DNA replication. Genetic and physiological approaches have shown that DNA replication is initiated by a protein called DnaA acting on a unique sequence which is the origin of replication, termed *ori*C. The role of DnaA as the principal controller of the initiation of replication has been investigated in *Escherichia coli* by means of a mathematical model (Mahaffy and Zyskind, 1989). Stochastic models have been used to describe gene regulation in RNA replication (Fernandez, 1988), as well as DNA replication during growth (Herrick and Bensimon, 1991). Several models have been conceived to elucidate the essential features of overall plasmid replication kinetics (Keasling and Palsson, 1989). Parvoviral DNA replication has been modelled by Tyson *et al.* (1990) using some of the specificities of parvoviruses, a group of small animal viruses with a single-stranded linear DNA genome. The relative abundance of the terminal sequence orientations of the DNA molecules has been measured to obtain more information about the replication process. A kinetic 'hairpin transfer' model, based on differential rates of hairpin formation and inversion processes, has been developed (Tyson *et al.*, 1990). It has also been shown that the interconversion of replication and recombination structures can generate concatemers through an initial single-strand DNA inversion to a duplex, as in the case of the Herpes Simplex Virus DNA (Morgan and Severini, 1990). Takahashi (1989) has developed a fractal model for chromosome and chromosomal DNA replication based on the assumption that a part of a chromosome, in the form of a radial loop, is similar in shape to the whole chromosome, and thus satisfies an assumption of self-similarity. This model allows the calculation of a fractal dimension of a representative metaphase chromosome, and the prediction of the double-duplex mode of replication. The role of DNA methylation in DNA replication and transcription through molecular changes at the electron level has been recently demonstrated by means of a mathematical model (Liu *et al.*, 1991).

From our point of view, it is interesting to analyse the dynamics underlying DNA replication. Much remains to be learned about the enzymological phenomena of *in vivo* replication: certain *replicases* have been identified, but these are certainly not the only enzymes playing a part. In particular, it is not known whether replication takes place in a single, continuous operation rather than in separate, simultaneous operations over several distinct regions. The latter hypothesis is currently favoured, for it has been found that special enzymes, the *ligases*, allow the fragments to associate in a single strand. This immediately raises the question: how is the topological structure of the uncoiled circular double strand maintained when replication begins simultaneously in its different parts? Clearly, a precise description of the phenomena involved will require the definition of the global, topological conditions compatible with a theoretical model of spatiotemporal kinetics.

However, a realistic hypothesis has been proposed to explain the enzymatic self-replication of short RNA molecules (Kanavarioti, 1992). The author illustrates this hypothesis by the self-replication of an oligopyrimidine strand, involving the following steps: (i) the template-directed polynucleotide synthesis is based on Watson–Crick base-pairing; (ii) the double helix, acting as template for the synthesis of a second oligopyrimidine strand from activated pyrimidine monomers, produces a new free strand; (iii) a triple helix is then formed from the double helix and the free strand in dynamic equilibrium; and (iv) the unassociated free strand directs the synthesis of

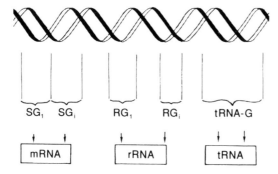

Fig. 5.5. The three classes of RNA, necessary for protein biosynthesis, transcribed from a DNA chain. For convenience, the drawing is not to scale; in reality, tRNA is very much shorter than rRNA, and mRNA can be very long indeed. RG, regulatory genes; SG, structural genes; tRNA-G, tRNA genes.

the complementary oligopurine strand to form a new double helix which will in turn direct the synthesis of an oligopyrimidine strand identical to the first strand that started the whole sequence. Clearly, this approach, under certain conditions, could lead to a satisfactory explanation of replication. It constitutes a good example of current research aimed at understanding the phenomenon of non-enzymatic self-replication.

Transcription involves copying the DNA nucleotide sequence by a corresponding RNA nucleotide sequence, in the antiparallel form, i.e. $3' \rightarrow 5'$ for DNA and $5' \rightarrow 3'$ for RNA. As seen in Chapter 1, three different types of RNA may be formed: mRNA, tRNA and rRNA, according to the gene copied from DNA (Fig. 5.5). The copying process follows the same rules as replication. The enzyme that allows transcription is RNA polymerase. Whereas DNA has two strands (the double helix), i.e. it is bicatenary, RNA is monocatenary. This is an essential difference since only one of the strands is transcribed.

This is the type of transcription observed in procaryotes. As mentioned above, transcription in higher organisms is much more complex as it involves elimination of the non-coding sequences copied off DNA to produce 'mature' RNA (Fig. 5.6). Chambon (1981) has studied the gene coding for ovalbumin, a protein found in birds' eggs. The transcription of the gene gives rise to a *primary transcript*, which is perfectly complementary to the corresponding DNA segment. Then, *gene splicing* eliminates the non-coding sequences, producing a *mature* mRNA containing only the useful coding sequences. The coding sequences are called *exons* and the non-coding sequences are called *introns*. The different steps are shown in Fig. 5.6. The precision of the enzymic machinery during the splicing operations is astounding. A splicing enzyme must at the least be capable of recognising the sites to be snipped to the accuracy of a single nucleotide. In fact, all intron transcripts begin with the sequence *GU* and end with the sequence *AG*. The specificity of the splicing could also be due to a splicer RNA. The importance of this research is obvious when we recall the extraordinary disparity between the size of the genome and the number of genes. But even more exciting is the idea that splicing may actually play a regulatory role in

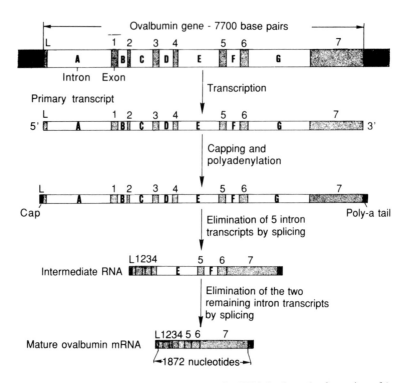

Fig. 5.6. Elimination of non-coding sequences of mRNA leads to the formation of 'mature' mRNA (after Chambon, 1981).

gene expression. Indeed, splicing may be necessary for the stabilisation of RNA and its transfer from nucleus to cytoplasm. We shall return to these fundamental notions in the following chapters.

The regulation of transcription is thought to be due in part to the non-linear solitary wave that is propagated along the DNA double helix due to the transmission of conformational changes (Polozov and Yakushevich, 1988). The total number of transcriptional regions in human genomic DNA has been estimated (Koziol, 1991). Since DNA-dependent conformational transitions of the transcription complex occur during transcription, models have been constructed to analyse the kinetics showing the different types of transition with characteristic catalytic properties. For example, potential memory effects and hysteresis have been investigated by Job *et al.* (1988), and Ko (1991) has constructed a stochastic model of gene induction based on the transcriptional complex.

Translation corresponds to a set of complex processes leading to the biosynthesis of proteins according to the nucleotide sequence of mRNA. Translation is considered to take place in two steps (Fig. 5.7a):

Step 1: The formation of aminoacyl-tRNA or the activation of amino acids (AA_i):

$$E_i \sim AA_i + tRNA_i \rightarrow tRNA_i\text{-}E_i\text{-}AA_i$$

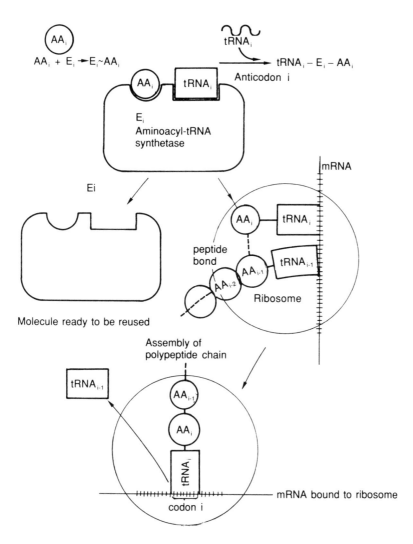

Fig. 5.7a. Steps in the genetic translation of the tRNA$_i$ into amino acids (*AA$_i$*).

Step 2: The assembly of the polypeptide chain from the aminoacyl-tRNAs trans-ported to the mRNA–ribosomal complexes, which corresponds to the *transfer* of amino acids. Schematically, the activation is carried out by an enzyme E_i, called aminoacyl-RNA$_i$ synthetase, which has two specific sites, one for the amino acid AA_i and the other for tRNA$_i$ (Fig. 5.7a). The formation of the polypeptide chain follows upon the installation of the AA_i–tRNA$_i$ on the corresponding triplet of mRNA through a bijective spatial application.

$$AA_i\text{–}tRNA_i \rightarrow Triplet_i(mRNA)$$

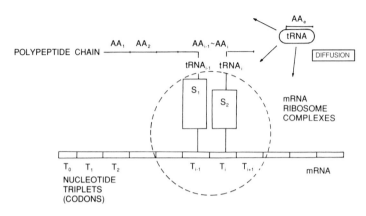

Fig. 5.7b. The mechanism of translation on the ribosome. The elongation of the polypeptide chain formed by the AA_i is carried out on the two ribosomal subunits, the big subunit S_1 and the small subunit S_2.

which constitutes the *genetic code* shown in Table 5.1. However, it should be mentioned that some results, such as those due to the wobble hypothesis (see below), do not admit this bijection.

A triplet of mRNA is called a *codon* and the corresponding part of tRNA is called the *anticodon*. The degeneracy of the third base of the codon seems to be responsible for the lesser specificity of a message translated from mRNA to polypeptide. The wobble hypothesis, due to Crick (1966), implies that there is a non-unique correspondence in recognition between the 3'-base of the codon and the 5'-base of the anticodon. Variations of wobble pairing, which appears to be informationally more complex than hitherto believed, have been formalised by using the cooperative symmetry of the genetic code (Shcherbak, 1989a, c). In the context of the cooperative symmetry of the genetic code, Rumer (1966) suggested an alternative form of writing the genetic code by separating the 5'-primary doublet of codon bases (the root) from the 3'-codon base (the ending). Using this form, Rumer in fact implemented some of the rules of the systematisation principle of the cooperative symmetry of the genetic code (Shcherbak, 1989b). The spatial alignment of the AA_i through the codon–anticodon complex is carried out sequentially (for successive values of i) on the ribosome, which works like a biochemical factory. Each ribosome has two sites: the P (peptidyl) site and the A (amino-acyl) site which are situated at the junction of the 30 S and 50 S subunits, and the specific mRNA codon. All the AA_i–$tRNA_i$ are first fixed on the A-site, but it is not yet known if the initiator, f-met-tRNA, is similarly first fixed on this site. Then a peptide bond links the successive AA_i to the polypeptide chain under construction (Fig. 5.7b).

The chemical phenomena basic to protein biosynthesis are the reaction–diffusion processes described in Chapter 4. Let us briefly recall the three main steps:

(1) *Initiation*:
    (1.1) The codon T_0 is always the same (AUG which codes for *methionine*).
    (1.2) N-formyl-methionine–tRNA (AA_0–$tRNA_0$) is formed.

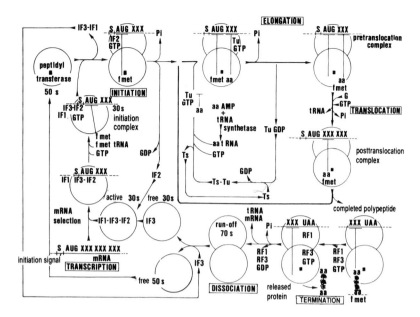

Fig. 5.8. A detailed diagram of protein biosynthesis (Revel *et al.*, 1972).

(1.3) Then the *N*-formyl part is eliminated leaving a methionine residue (*the most frequent case*) which may sometimes be removed by a specific amino-peptidase, so that all proteins do not necessarily have a structure beginning with methionine.

(1.4) The coupling is due to an initiating factor.

(1.5) The translation sequence can then begin.

(2) *Elongation*: increase in the length of the polypeptide chain involves several enzymes, in particular peptidyl-transferase.

(3) *Termination*: the signal to end the polypeptide chain is given by one of the three codons T_N: UAA, AGA or UAG, called *stop* codons (Table 5.1). A succession of two or three stop codons appears to be required to terminate a polypeptide chain.

All this suggests the extreme complexity of the chemical processes of protein biosynthesis. Figure 5.8, taken from Revel *et al.* (1972), shows the intricate molecular relationships involved. The authors give a fairly complete description of protein biosynthesis, with the different signals for reading the genetic code, the punctuation, and so on. In short, they describe the informational language used by living organisms to ensure adequate functioning. This model provides a good illustration of the powerful technique of network thermodynamics. In fact, the cycle of operations may be represented quite simply by a set of elementary biochemical reactions associated with a Paynter graph (Atlan, 1976).

The technique used actually corresponds to information processing as in computer science with data capture at the DNA level, initialisation by AUG, codon–anticodon

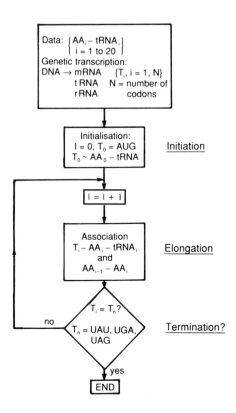

Fig. 5.9.   An algorithm for protein biosynthesis.

loops, and, finally, the end of the procedure by UAA, UGA or UAG. The algorithm corresponding to this is shown in Fig. 5.9.

Is this kind of qualitative description of the phenomena of replication and protein synthesis sufficient? The main question that arises is whether a quantitative description will have a real advantage. For the theoretician, the study of the kinetics of these reactions is of the greatest importance for it is necessary to understand exactly how the processes occur in order to be able to control them. For instance, deeper knowledge of the intimate chemistry of biosynthesis should be of help in the artificial synthesis of indispensable proteins (of which insulin is an example) or again be useful in solving some of the problems of molecular pathology. But in our opinion, and we shall try to demonstrate this all through the book, quantitative description should yield a formalism useful at all levels of description, and perhaps even lead to a unification of the processes of control and regulation.

In connection with the dynamic analysis of protein biosynthesis, another problem, fundamental at the conceptual level, has been raised by Crick (1976) and other authors (Stettler *et al.*, 1979): *Is all DNA really in the form of a double helix?* This question involves topological considerations and will be examined at the end of this chapter.

3.   *Computational methods of determining the sequencing properties of DNA*

An extremely interesting method of investigation is the analysis of nucleic acid and protein sequences. If unity exists at the gene level — and from what we have seen there is little doubt about the spatiotemporal dynamics of genes — we should be able to study it through the intra- or inter-sequence homologies of a nucleic acid or a protein chain by using data banks. The enormous task of compiling all the known protein sequences and structures has been undertaken by Dayhoff (1976), who has already produced several atlases, with new volumes being added each year. The information is available on magnetic tape so that a computer can be conveniently used for the programmed exploration of the data. The most immediate application of this data bank naturally concerns evolution and, more precisely, we may hope that the comparison of genetic material from different species will enable us to deduce the sequences of proteins that have actually disappeared. For example, Ninio (1979) gives the principle of such comparisons applied to haemoglobin. Degrees of dissemblance (or resemblance) between sequences can be defined by means of a distance, calculated by an appropriate technique. The map of the distances can then be used to construct a genealogical tree. However, such a phyletic tree is not the result of direct deduction but involves arbitrary choices. The largely empirical nature of this approach has encouraged further research towards the determination of local or global homologies.

The problem of determining the greatest similitude between nucleic acid or protein sequences is a difficult one since the solution is not unique and depends on the algorithm used. Indeed, the differences between the sequences are not limited to the substitution of one nucleotide base (or an amino acid) for another. Indeed, there may be deletions or insertions; the bases may not be functionally equivalent; certain contradictions between bases may be taken for agreements; and so on. In short, each possible operation, coincidence/non-coincidence or insertion/deletion, will have to be weighted so as to determine possible alignments which will then be classified according to specified optimum conditions. *Aligning two sequences* means putting them into correspondence through the shortest mutational pathway allowing deduction of one from the other.

Needleman and Wunsch (1970) have constructed an algorithm that will seek optimum alignment in terms of the greatest number of coincidences while taking into account the jumps due to insertions/deletions. This intuitive algorithm, working by induction, is perfectly adapted for computer use. Moreover, Sankoff (1972) has shown that it satisfies the principle of optimum alignment so that the alignment corresponds to the minimum value of the measurement. One problem concerns the definition of the measurement of the global distance between sequences or the *global alignment*. Another problem, very different and even more delicate, appears in the search for *local alignments*, i.e. possible correspondence between fragments of sequences that may be identical or different. Sellers (1980) has extended the algorithm due to Needleman and Wunsch, now called the NWS algorithm, by defining the local optimum of the distance as *the set of the local minima of a function of which the global minimum gives the global alignment*.

To illustrate this, let us consider an example taken from Kanehisa and Goad (1982). After arranging two sequences in matrix form, a 'path' is constructed such that the element (i, j) be defined by a diagonal line if there is a coincidence between the bases of line i and column j, by a horizontal line if there is a deletion of a base in the horizontal sequence, and by a vertical line if there is a deletion of a base in the vertical sequence. We thus obtain a matrix, such as in Fig. 5.10a, enabling us to visualise a particular path corresponding to the alignment:

$$\begin{array}{llllllllllllll} \text{G} & \text{T} & \text{T} & \text{A} & \text{A} & \text{G} & \text{G} & \text{C} & \text{G} & & \text{G} & \text{G} & \text{A} & \text{A} & \text{A} \\ \vdots & \vdots & \vdots & & & & & & & & \vdots & \vdots & \vdots & \vdots & \vdots \\ \text{G} & \text{T} & \text{T} & & & & & \text{G} & \text{A} & \text{G} & \text{A} & \text{G} & \text{G} & \text{A} & \text{A} & \text{A} \end{array}$$

Obviously, other paths, corresponding to other alignments, could have been constructed. The definition of such paths, in particular of an optimum path, enforces the choice of a distance. For instance, we may choose a *cumulative* distance D such that:

$$D(i, j) = \max\{D(i - 1, j) - \beta, D(i, j - 1) - \beta, D(i - 1, j - 1) + \delta\}$$

where β is the weight assigned to the deletion of a base i or j, and $\delta = 1$ if base i coincides with base j and $\delta = -\alpha$ if it does not. Let us then suppose we seek the homology between the two subsequences beginning in (i_1, j_1). If $D(i, j) = 0$, with $\alpha = \beta = 3$ for example, we may say that there is one base substitution or one insertion/deletion for three coincidences, thus defining the best local alignment. This is the principle of the method based on the intuitive NWS algorithm. Several variations and extensions of this approach have been constructed and, in particular, Goad and Kanehisa (1982) show how to determine regions in which the density of coincidences is greater than a selected threshold value. Figure 5.10b shows the results we have obtained using this technique.

The repetition structure of several large human genes has been investigated using similar methods, for example that of fragmentation sequencing (Smillie and Bains, 1990) which has allowed the detection of cryptically simple DNA sequences, i.e. short repeated sequences which are near but not necessarily adjacent to each other. Incidentally, the study confirms that cryptically simple DNA sequences are over-represented in the genome.

Such methods (Dumas and Ninio, 1982; Wilbur and Lipman, 1983) are certainly full of promise. A dynamic programming algorithm has been used to align two sequences when the alignment is constrained to lie between two arbitrary boundary lines in the dynamic programming matrix (Chao *et al.*, 1993). One of the difficulties involved in aligning sequences is that of the reduction of space requirements. The use of spaces proportional to the sum of the sequence lengths allows the imposition of appropriate penalties. The comparison of multiple macromolecular sequences may be based on a hierarchical sequence synthesising procedure (Wong *et al.*, 1993), or on a method of inductive inference known as *minimum message length encoding* (Allison and Yee, 1990). A problem very similar to that of aligning sequences, i.e. finding consensus sequences such as the longest common subsequence in a set of molecular sequences, can be formulated as follows: finding a consensus of k aligned molecular sequences, in which n aligned positions have been identified, can be viewed as a set of n simpler problems, each of which consists in finding a consensus of k alternatives at an aligned position (Day and McMorris, 1992). This problem has been solved on the basis of plurality rules. Other methods have been extensively studied in the field of sequence alignments: the 'skeletal' representation (Gotoh, 1990), the Shannon information calculations (Hariri *et al.*, 1990), the number of matches (Rinsma *et al.*, 1990; Myers and Miller, 1989), the reduction of the number of residue pair comparisons performed between the two structures (Orengo and Taylor, 1990), dynamic programming algorithms (Zuker and Somorjai, 1989). When no sequence-homology to proteins with established spatial structure is known, the prediction of protein structures can be made by methods of artificial intelligence (Kaden *et al.*, 1990).

The recognition of sequences depends on some peculiar feature of a DNA sequence, e.g. dinucleotide asymmetry, prevalence of RNY codons, CG content, or the information value of the protein chains. Some methods measure the bias in the base composition and in the codon usage tables. Another method for characterising nucleotidic sequences, taking into account the length of the sequences, is based on maximum entropy techniques (Cosmi *et al.*, 1990).

Obviously, the problem of aligning sequences and the correlation of nucleotides in protein-coding DNA sequences (Luo and Li, 1991) is related to the determination of the genetic code, and consequently to its dynamics. For example, long-range two-body correlations in a DNA sequence exhibit oscillations that persist for very large correlation lengths. Oscillations have been shown to be three-point cycles related to the coding regions in the DNA (Mani, 1992a, b). The solution of the problem of aligning sequences may provide the key to the interpretation of the genetic code regarding its evolutionary properties (Luo and Trainor, 1992). Lehman and Jukes (1988) view the development of the genetic code from the perspective of simultaneously evolving codons, anticodons and amino acids. In this model, codons become permanently associated with amino acids only when a codon–anticodon pairing is strong enough to permit rapid translation. Thus, codons specifying newer amino acids are not derived from codons encoding older amino acids (the classical view), but act as chain-termination or 'stop' codons until the development of tRNA adaptors capable of binding tightly to them. Another model (Yan *et al.*, 1991) shows a one-to-one correspondence between prime numbers and the genetic code by considering their combinatorial specificities.

In the near future we should be able to obtain information on the functional role of certain subsequences by making comparisons between different species. The method proposed by Goad and Kanehisa (1982) has allowed the identification of a deletion unit with a length of only one to three bases. Will this turn out to be a new functional unit — a 'deleton'? Another interesting application lies in the search for relationships between products that may seem unrelated *a priori*, for example the functional identification of oncogenic proteins among which a certain class has been found to be homologous to the variable regions of immunoglobulins and to Class I histocompatibility antigens. The availability of such powerful tools for investigation at the gene level may be expected to contribute to a better mathematical formulation of problems and thus to the construction of more satisfactory models to explain biological phenomena.

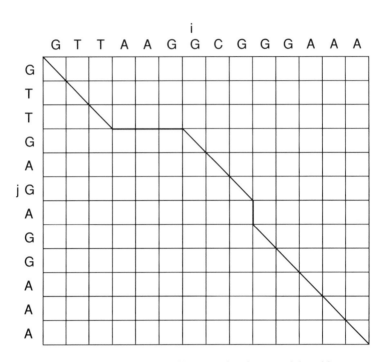

Fig. 5.10a. Example of a path constructed in a matrix of two nucleic acid sequences taking into account some possible coincidences/substitutions and insertions/deletions of bases.

Fig. 5.10b. Results obtained using the Kanehisa method for the local alignment of two nucleic acid sequences: AALVMC29MY (MC29 virus of chicken leukaemia, gene V-MYc) and AAV2LEFT (part of the genome of the adenoma-associated virus). For the values of the parameters and the weightings indicated, there are five locally homologous regions ($p < 0.66 \times 10^{-4}$).
(i) Values of the parameters;
(ii) description of the sequences used;
(iii) local homologies; and (iv) sequences visualised together with the homologies.

(i)

```
PARAMETERS:
 PMAX = 1. OE–04   DMAT = –1.0   DREP = 2.0   DEL = 3.0

PONDERATION FACTORS (Agreements and replacements):
      A    C    G    T    X
 A  –1.
 C   2.  –1.
 G   2.   2.  –1.
 T   2.   2.   2.  –1.
 X   0.   0.   0.   0.   0.

PONDERATION FACTORS (Deletions):
 #   1
     3.
```

(ii)

```
SEQUENCE1  AALVMC29MY         DE      1    A      100 TOTAL      100

        10              20              30              40              50              60
CCGCCCGGCG    CCAACCCCGC    GCTCTGCTGG    GGGTCGACAC    GCCGCCCACG    ACCAGCAGCG
        70              80              90             100
ACTCGGAAGA    AGAACAAGAA    GAAGATGAGG    AAATCGATGT

SEQUENCE2  AAV2LEFT          DE      1    A      100 TOTAL      100

        10              20              30              40              50              60
TTGGCCACTC    CCTCTCTGCG    CGCTCGCTCG    CTCACTGAGG    CCGGGCGACC    AAAGGTCGCC
        70              80              90             100
CGACGCCCGG    GCTTTGCCCG    GGCGGCCTCA    GTGAGCGAGC
```

(iii)

```
LOCALLY HOMOLOGOUS REGIONS BETWEEN AALVMC29MY AND AAV2LEFT

#1 (13) 0. 32E–04     #2 (11) 0.56E–04     #3 (10) 0.22E–04
   36          46         7                   22
GGTCGACACGCCGCCC      GCCCGGCG CCAA        CGCGCTCTGCT
: : : : : : : :  : : : : : :   : : : : : : : :   : : : : : : : : :   : : :
GGTCGCC CGACGCCC      GCCGGGCGACCAA        CGCGCTC GCT
   58          68         44                  23

#4 (10) 0 66E–04      #5 (10) 0. 18E–04
   20                     6
CCCGCGCTCTGC          CGCCCGGCGCC
: : : : : : : : : : : :      : : : : : : : : : :
CCCGGGCTTTGC          CGCCCGA CGCC
   70                    61
```

(iv)

```
SEQUENCE1  AALVMC29MY

        10              20              30              40              50              60
CCGCCCGGCGCCAACCCCGCGCTCTGCTGGGGGTCGACACGCCGCCCACGACCAGCAGCG
* * * * * * * * * *    * * * * * * * * * *    * * * * * * * * * * * * *
#5                     #4          * * * * * * * * * *    #1
* * * * * * * * * *                * * * * * * * * * *
#2                                 #3

        70              80              90             100
ACTCGGAAGAAGAACAAGAAGAAGAAGATGAGGAAATCGATGT

SEQUENCE2  AAV2LEFT

        10              20              30              40              50              60
TTGGCCACTCCCTCTCTGCGCGCTCGCTCGCTCACTGAGGCCGGGCGACCAAAGGTCGCC
                * * * * * * * * * *              * * * * * * * * * * *   * * * * * * *
                #3                               #2          #1
                                                             * * * *
                                                             #5

        70              80              90             100
CGACGCCCGGGCTTTGCCCGGGCGGCCTCAGTGAGCGAGC
* * * * * * * *
#1
* * * * * * *
#5
          * * * * * * * * * * * *
          #4
```

II.   Information theory and the genetic code

We have already seen how the infinite diversity of proteinic arrangements is based on the existence of an alphabet of 20 symbols, the amino acids, the arrangements of which in turn result from combinations of the elements of an alphabet of four symbols, the nucleotides, that constitute the nucleic acids, DNA and RNA. The principle of the invariability of nucleic acids is classically termed the 'central dogma' of molecular biology. The passage from one alphabet to the other corresponds to protein biosynthesis. Before tackling the dynamics of this phenomenon, let us see how the quantity of information contained by DNA (or RNA) may be measured using Shannon's H function as defined in Chapter 4.

1.   *Measurement of the quantity of information in the genetic code*

We have two alphabets available, the four nucleotide bases of the nucleic acids (D or R) and the 20 amino acids (AA). Let $i_v^{(\beta)}$ be the series of symbols corresponding to molecular species i (the information carriers) of the type: $(\alpha_{i_1}, \alpha_{i_2}, ..., \alpha_{i_v})$ where α is the symbol of the alphabet (α). Similarly:

$$i_v^{(\beta)} = (\beta_{i_1}, \beta_{i_2}, ..., \beta_{i_v}) \tag{5.1}$$

is a series of symbols of the alphabet (β). These information carriers have a 'length' v so that it is convenient to place them in the same 'class' (v). Each carrier is then characterised by:

- the *number* v of its symbols; and
- the *order* in which the symbols appear in the series. As repetition is possible, two elements of $i_v^{(\beta)}$ such as:

$$i_1, i_2, ..., i_v \quad \text{and} \quad i_2, i_1, ..., i_v \text{ will differ.}$$

Thus, if N_α is the number of symbols of the alphabet (α), there will be $(N_\alpha)^v$ series as above. Finally, information carriers will be noted $i_{v,k}^{(\alpha)}$, the index k specifying an element of class v for the alphabet (α). The set of such carriers is thus:

$$\{i_{v,k}^{(\alpha)}; k = 1, 2, ..., N_\alpha^v\}.$$

To illustrate this very practical notation, let us consider DNA, which has an alphabet (α) = {G, C, A, T}. The number of carriers of length $v = 2$, for example, will be $N_\alpha = 4^2$, i.e. GG, GC, GA, GT; CG, CC, CA, CT; AG, ..., TG, TC, TA, TT. One of these carriers, CT for instance, will thus be written: $i_{2,8}^{(\alpha)}$.

2.   *The genetic code*

Let us return to the problem of protein biosynthesis, briefly discussed above. This process may be represented by the communication diagram in Fig. 5.11 which is analogous to that in Fig. 4.2. The number of different triplet sequences that can be obtained from an alphabet (α) containing four symbols is $(N_\alpha)^3 = 4^3 = 64$, which is

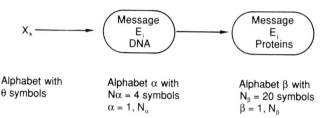

Fig. 5.11.   Protein biosynthesis from the point of view of information theory.

more than sufficient to code for the 20 amino acids. This is why the genetic code is said to be *degenerate*.

If we assume the symbols to be equiprobable, a molecule with N bases will have $N \log_2 4 = 2N$ bits of information. Thus, for DNA in a double helix, the quantity of information should be $2N/2 = N$ bits, while for a single-stranded RNA it should be $2N$ bits. This reasoning is obviously much too simple, and Gatlin (1968) offers a better approach to the problem, taking into consideration the following points:

(1) the genetic code is *degenerate* and the symbols are not equiprobable; and
(2) the genetic message is *redundant*.

The *redundance of the message* means that the use of additional symbols does not increase the quantity of information in the message but, as we shall see below, actually decreases it.

Let us first define a few terms:

- H_R is the quantity of information in the message taking into account the conditional probabilities of the appearance of the symbols and their correlations (the symbols are not independent);
- H_M is the quantity of information in the message with the hypothesis that the symbols (α) are equiprobable: $H_M = \log_2 N_\alpha$ (a completely random state); and
- R is the redundance which is zero when $H_M = H_R$.

Now we may naturally write:

$$R = 1 - \frac{H_R}{H_M} \qquad (5.2)$$

which implies that redundance decreases the quantity of information per symbol.

We have already said that the theory of information may be considered as a theory of communication. The recent theory of molecular machines is based on the application of one of Shannon's theorems for communication channels to the energy dissipation formula in the presence of thermal noise (Schneider, 1991a, b). Operations such as those of replication, control and translation of genetic material, sensing the outside environment, are performed by single molecules that make tiny decisions. For a molecular machine, making a decision means dissipating energy. Schneider has shown that there exists a precise upper bound on the number of choices a molecular machine can make for a given energy loss. This is a fundamental contribution towards the construction of molecular computers.

The theory of information has also been used to analyse the efficiency of information transduction for molecular recognition (Sarai, 1989). Information gain, which can be interpreted as the amount of information extracted from the sequence or structure by a molecule, represents the intrinsic ability of a molecule to recognise a specific sequence or structure. Unlike thermodynamic quantities, information gain is a normalised quantity that can measure specificity, and which can be experimentally determined. Gatlin (1968) indicates a method for the measurement of the total quantity of *information* I *stored by each symbol* and distinguishes this from the quantity of information transmitted. Thus:

$$I = RH_M \tag{5.3}$$

which means that the *information stored* is the product of Shannon's redundance term and the maximum quantity of information characteristic of the number of symbols of the alphabet. This quantity may be decomposed into two parts: $I = I_1 + I_2$, with:

(1) $$I_1 = H_M - H_{exp} \quad \text{with} \quad H_{exp} = -\sum_i f_i \log_2 f_i$$

where f_i is the observed frequency of the appearance of the bases. Thus, I_1 measures the difference with respect to the hypothetical case of the message composed of equiprobable symbols. This is, in a way, the asymmetry of the message, or the difference from the equiprobable state. For instance, I_1 measures the quantity of information stored in a message when certain bases are repeated while others are excluded.

(2) $I_2 = H_{exp} - H_R$ measures the difference with respect to a state of total independence because of the definition of H_R.

Gatlin introduces another function:

$$B = \frac{I_2}{R} \tag{5.4}$$

which is the fraction of redundance for information of type I_2.

Gatlin has calculated I_1, I_2, R and B from the composition of DNA for 36 species. He supposes that the probabilities of the appearance of the bases in the DNA correspond to first-order Markov chains, with known probabilities of transition Pr (j/i) allowing the determination of H_R. Figure 5.12 represents the result obtained. We may observe that:

(1) The approximation of the probability of appearance of the DNA bases by first-order Markov chains appears to be satisfactory.

(2) The graph of T_2 in terms of R shows vertebrates, bacteria and phages grouped distinctively. Vertebrates have high values of T_2, from 60 to 80% of the total value of T, which means that the information in vertebrate DNA is further from total independence than is the case for the other groups.

(3) Vertebrates may be said to optimise R and B as there appears to be a greater variation in the composition of DNA bases in bacteria and viruses than in vertebrates.

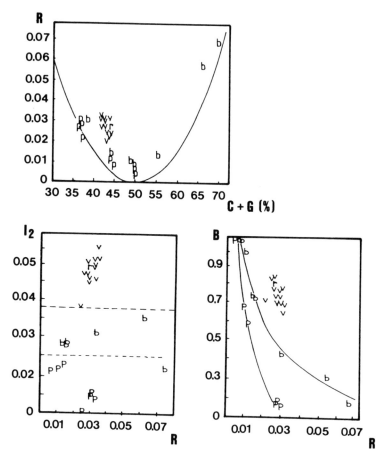

Fig. 5.12. Results obtained by Gatlin (1972) for DNA from 36 species. Bacteria (b), vertebrates (v), phages (p) and rheoviruses (r) appear to be 'classified' into distinctive groups (see text for comments).

(4)   The virulence of phages against *Escherichia coli*, for example, may be explained by competition between the DNA information of the two organisms.

In conclusion, the quantity of information I_2 reflects the increasing 'complexity' of organisms going from viruses to bacteria to invertebrates to vertebrates, while I_1 characterises the 'control' functions of DNA. High values of I_1, corresponding to the asymmetry caused by repeated base sequences, reduce the risk of errors during DNA transcription. According to Gatlin (1968), the living system introduces a redundance in its DNA (or RNA), either through I_1 or through I_2, and there are fundamental differences between the higher and the lower organisms in the relative use of these two processes. These results have been recently re-examined (Sibbald *et al.*, 1989). In Chapter 9, we shall see how this aspect of information, which is at the origin of some of the concepts of gene regulation, may be further explored. In particular, it

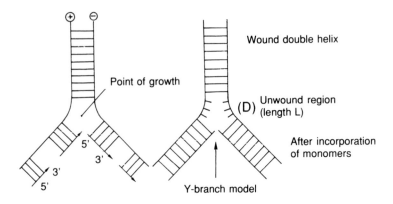

Fig. 5.13.   Diagram of DNA replication proceeding from a growth point.

would appear to be of interest to express the 'complexity' of an organism in terms of the information I_2, although the ideas used here, based on the theory of information, may seem too poor to lead to any highly significant deduction. This is why the larger notion of potential of *functional organization* will be introduced in Chapter 12.

III.   Chemical dynamics of heredity

Let us now tackle some of the problems involved in the kinetics of biopolymerisation, such as:

● the influence of the distribution of bases on the form of the denaturation curve;
● the sequence of the nucleotides;
● the relationship between the rate of DNA replication and the distribution of nucleotides; and
● the effect of the initiation and termination rates.

Only a description based on a model will be sufficiently accurate to take into account a certain number of experimental phenomena. As we have just seen, the comparison between DNA replication and protein biosynthesis shows some similarity and some difference. The similarity is that in both cases the copying process uses a nucleic acid template. The difference is that while replication serves either for DNA synthesis or for RNA synthesis (transcription), protein biosynthesis corresponds to translation only from mRNA.

How is the DNA molecule replicated? Does the synthesis begin at one end or at some point within the molecule? Direct observation has shown that the process always begins at a variable number of points and then extends in both directions. Only one enzyme, polymerase III, appears to be involved in the replication. The Y-conformation, at which there is simultaneous uncoiling of the parent double helix and the formation of the two daughter helices, provides a convenient model of the replication process (Fig. 5.13). It is believed that the formation of the daughter

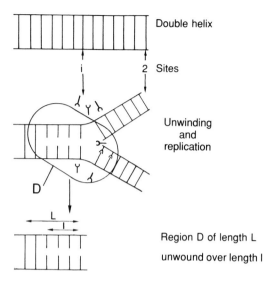

Double helix

2 Sites

Unwinding
and
replication

D

L

l

Region D of length L

unwound over length l

Fig. 5.14. DNA replication: D is the region of DNA uncoiling. The length of the region is equivalent to L sites of which l are unpaired.

chains takes place by the successive addition of smaller fragments. In spite of this complexity, our problem reduces to the description of the phenomenon at the growth point where the uncoiling of the parent helix is coupled with replication. Eigen *et al.* (1981) give a good account of this subject.

1. *Generalities*

The replication of the DNA double helix may be considered, following Zimmerman (1966), as the resultant of two processes (Fig. 5.14): first the 'liberation' of the two strands of the parent chain by local enzymic action in the region D causing the double helix to uncoil; and then the replication of the liberated strand over a part of D, where the process is identical to that of the duplication of a single strand of nucleic acid.

The overall result is that the region D advances one site further along the double helix. Let us define the 'length' of the chain by the number of sites N (each site on the chain being occupied by one of the nucleotides, each site of the double helix to be occupied by a base pair), the length of the region D by L sites and the length of the chain undergoing replication by $l(t)$, evidently with $l < L$ (Fig. 5.14).

To simplify, let us suppose that:

$$l(t) = L - 1$$

If $p(t)$ is the probability that an empty site be occupied by a monomer and k_{poly} the probability that this monomer binds by polymerisation to a monomer already fixed on the preceding site, then the overall result of the advance over one site in the region D will occur with the probability $p(t) \, k_{poly}$. This of course assumes the existence of the double chain, i.e. the prior existence of a pair of bases bound at this site, of which the probability of dissociation is $\alpha(t)$. *This term induces a coupling between the uncoiling of the double helix and the replication.* The probability of advancing over one site in the region D is thus:

$$\psi(t) = k_{poly} \cdot \alpha(t) \cdot p(t). \tag{5.5}$$

Here, by putting $\alpha = 1$, i.e. in the absence of DNA uncoiling, we evidently have the particular case of the replication of a single strand. As we shall see, protein biosynthesis can in fact be reduced to the study of this particular case.

Finally, the process may be described by a distribution of chain lengths $(n_i(t))$. In fact the number of chains of given length i for each template is $n_i(t)$, the sum of the length $n_{ij}^{(0)}$ weighted by the probabilities $P_j(t)$ that the template be in a given state E_j at time t, the state being defined by the number of growth points, the number of chains being formed, and so on. In what follows, we shall consider only the $n_i(t)$ values.

If we write:

$$\lambda_i(t) = \frac{n_i^{(t)}}{\displaystyle\sum_{i=2}^{N} n_i(t)} \qquad i = 2, 3, \ldots, N$$

then $\lambda_i(t)$ is the probability that the length of the growing chain in a given state be i at time t. The index i starts with the value 2 as the *dimer* produced by polymerisation is considered to be transported simultaneously with the enzyme. Growth presupposes that a monomer be initially fixed so that the smallest initial length will be equal to 2. Growth occurs simply by the addition of a monomer at the extremity of a growing chain. The probability of the transition of a chain of length i to a chain of length $i + 1$ is $\psi(t)\Delta t$ during the time interval Δt. Thus, as $n_2(t)$ is the number of elements of class 2, i.e. the number of chains of length 2 at time t, at the instant $t + \Delta t$ this number will be *diminished* by the number of elements leaving class 2 during Δt. Similarly, for all values of $i \in [3, N - 1]$, the number $n_i(t)$ will be *diminished* by the number of elements leaving class i but will be *augmented* by the number of elements leaving class $i - 1$ during the same time. Thus, the class N will be augmented only by the number of elements leaving class $N - 1$. All this may be mathematically expressed by the $N - 1$ equations below where $\psi(t)$ represents the flux of elements between the classes.

$$\left.\begin{array}{l} n_2(t + \Delta t) = n_2(t) - \psi(t) \, n_2(t) \, \Delta t \\[4pt] n_i(t + \Delta t) = n_i(t) - \psi(t) \, n_i(t) \, \Delta t + \psi(t) \, n_{i-1}(t) \, \Delta t \\[4pt] 3 \le i \le N - 1 \\[4pt] n_N(t + \Delta t) = n_N(t) + \psi(t) \, n_{N-1}(t) \, \Delta t \end{array}\right\} . \tag{5.6}$$

These are in fact identical to the equations obtained for chemical reactions in which $\psi(t)$ was defined as the rate constant. This is easily seen by dividing Eq. (5.6) by Δt and then letting Δt tend to 0, which gives a differential equation. In the present case,

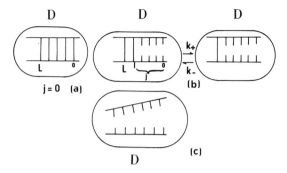

Fig. 5.15. Transformations in the region D: (a) all bindings are intact, $j = 0$; (b) chemical reaction leading from j dissociations to $j + 1$ dissociations; (c) the two strands are completely dissociated.

ψ is a function of time since it is expressed in the form of a product of probabilities which in general depend on time. The probability of the event: 'transition from i to $i + 1$ during δt' will thus be expressed as the product of a rate constant and the interval of time Δt.

The kinetics of biopolymerisation will have to estimate the following quantities, which it may be possible to determine experimentally:

$$c(t) = \text{conversion factor} = \frac{\text{replicated length}}{\text{length of matrix}} = \frac{\sum_{i=2}^{N} i n_i(t)}{N}$$

$$P_n(t) = \text{mean extent of polymerisation} = \sum_{i=2}^{N} i \lambda_i(t).$$

This is the mathematical expectation (Appendix D) of the discrete random variable i or the average replicated length at time t:

$$P_w(t) = \text{mean weighted length of chain} = \sum_{i=2}^{N} i^2 \lambda_i(t)/P_n(t).$$

Thus, the theoretical problem is to determine the quantities: $c(t)$, $P_n(t)$ and $P_w(t)$ from the distribution of the values $\lambda_i(t)$.

We have seen that the process of replication depends on the probability $\alpha(t)$ that at least one initial base pair in the region D be dissociated, i.e. on the probability of the double helix being *uncoiled*. If $\phi_0(t)$ is the probability that none of the base pairs is dissociated, then $\alpha(t) = 1 - \phi_0(t)$ is the probability that at least one free site is available. More generally, the replication process will start in D as soon as j successive pairs of sites will be dissociated. Let $\phi_j(t)$ be the probability of the state noted $|j\rangle$ and let $l(t)$ be the 'length' of the free sites in the region D, i.e. the number of sites with the property of being unoccupied after the separation of the two strands. This is the case shown in Fig. 5.15. The distribution $\{\phi_j(t)\}_{j=0,L}$ is necessary for the determination of the distribution $\{\lambda_i(t)\}_{i=2,L}$ since $\alpha(t) = 1 - \phi_0(t)$. Physically, this means that the double helix must be uncoiled *before* the replication.

Strictly speaking, DNA replication takes place in the region D of length L and is defined by the distribution $\{\phi_j(t)\}_{j=0,L}$ with: $l(t) = \sum_{j=1}^{L} j\phi_j(t)$, the mean length of the free sites in D. This corresponds to phase (I).

Phase (II), linked to phase (I), corresponds to the advance of the growing chain on the template. It is defined by the distribution of the values: $\{\lambda_i(t)\}_{i=2,N}$. Protein synthesis may thus be expressed in a manner analogous to phase (II).

2. DNA replication

Let us now consider the hypotheses and the concepts used in the study of phase (I) and phase (II) in the simple case where $l(t) = L - 1$, i.e. for the variation of the distribution $\{\phi_j\}$ with time when the enzyme acts on L sites, with dissociation possible for the first $L - 1$ sites but not for the Lth site.

Zimmerman and Simha (1965) make the following hypotheses:

(*H*1): Only one DNA strand is considered, i.e. it is assumed that the kinetics are identical on the two newly formed strands from the 5′ end towards the 3′ end. In reality the two strands are oriented in opposite directions, but the structural constraints of the molecules will be neglected here. DNA growth will be supposed to proceed in the direction of the arrow on the Y-model (Fig. 5.13).

(*H*2): From the point of view of the progression of the polymerisation, a particle may be considered to move over a *network* occupying first one site, then the next, and so on. The final result of the chemical reaction is to increase the number of free sites by one (Fig. 5.15). The rate 'constants' of the reaction, which modifies the number of free sites, are noted k_+ and k_- in the usual way.

(*H*3): The details of the reactions may be described as follows:

(i)   The uncoiling of the double helix in the region D is caused by an enzyme with the rate constant k_{unc}. Coupling of uncoiling and replication then occurs at the separated nucleic acid bases.

(ii)   The reaction proceeding in the opposite direction, with the rate constant k_{win}, causes rewinding of the two strands.

(iii) The monomer is incorporated into the DNA strand by diffusion with the rate constant k_{diff}, which is the same for all the bases. And as we have already seen, polymerisation occurs with the rate constant k_{poly}.

(iv)   It is assumed that there is no depolymerisation and that initiation is due to an enzyme that creates the *initial dimer* (the initial condition).

These three hypotheses correspond only approximately to the experimental description given above. On the one hand, we have neglected the notion of the reaction of the DNA fragments which serve as precursors to longer chains at the growth point. On the other, the topological reality has not been taken into account. This will

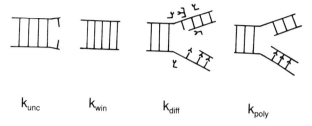

$$k_{unc} \qquad k_{win} \qquad k_{diff} \qquad k_{poly}$$

Fig. 5.16. Definitions of the chemical reaction constants of uncoiling (k_{unc}), of winding (k_{win}), of diffusion (k_{diff}) and of polymerisation (k_{poly}).

be considered in the next section. Here, it is the methodology used that is of special interest. Steps (i), (ii) and (iii) of (H3) are represented in Fig. 5.16. The systems of equations describing the uncoiling of the double helix and the replication of the separated strands are deduced from (5.6).

The occupation of the free sites may be described more simply using the hypotheses (H2) and (H3), supposing that:

$$k_{-} = k_{win} + k_{poly}\, p(t) \quad \text{and} \quad p(t) = 1 - e^{-k_{diff}\, t} \qquad (5.7)$$

with k_{-} *constant*, equal to its average value over a fairly large interval. This rate constant expresses the eventuality of the occupation of a free site with the probability $p(t)$, followed by polymerisation (probability k_{poly}) or else by rewinding (probability k_{win}) of the strands. It will then suffice to count the classes in which the process produces a unit increase or decrease of elements according to the system (5.6).

The rate constant for the creation of free sites is $k_{+} = k_{unc}$. Thus, initially, only one free site (class (1) defined by ϕ_1) may pass into class (0) (since $l = L - 1$) with the rate constant k_{-} (or the transition probability $k_{-}\, \Delta t$); the same class (0) may lose elements at the rate k_{+}. This corresponds to Eq. (5.8) (1)). Similarly, a chain with L free sites may lose them at the rate k_{-} and gain them from a class $(L - 1)$ at the rate k_{+}. In the general case ($1 \le j \le L - 1$), the transitions take place in both directions, $j \to j + 1$ and $j \to j - 1$. It is clear that the hypothesis $l = L - 1$ produces considerable simplification.

We have already seen that the probability of an event during time Δt is the product of the rate constant and the interval of time Δt. Then, according to (5.6) and the reasoning above, the system of equations satisfied by the functions $(\phi_j(t))$ may be written:

$$
\begin{aligned}
&(1) \quad \frac{d\phi_0}{dt} = -k_{+}\,\phi_0(t) + k_{-}\phi_1(t) \\[2mm]
&(2) \quad \frac{d\phi_L}{dt} = k_{+}\phi_{L-1}(t) - k_{-}\phi_L(t) \\[2mm]
&(3) \quad \frac{d\phi_j}{dt} = k_{+}\phi_{j-1}(t) - (k_{-} + k_{+})\phi_j(t) + k_{-}\phi_{j+1}(t) \quad j = 1, \ldots, L - 1
\end{aligned}
\qquad (5.8)
$$

with the initial conditions for the existence of a dimer:

$$\phi_2(0) = 1, \quad \phi_j(0) = 0 \quad \forall j \neq 2, \quad j \in [0, L].$$

Dividing Eqs (5.6) by Δt and with $\Delta t \to 0$, we obtain the system of equations (5.8), verified by the associated functions $(\lambda_i(t))$:

$$
\begin{array}{ll}
(4) \quad \dfrac{d\lambda_2}{dt} = -\psi(t)\,\lambda_2(t) & \\[2ex]
(5) \quad \dfrac{d\lambda_{i+1}}{dt} = \psi(t)\,(\lambda_i(t) - \lambda_{i+1}(t)) \quad 2 \leq i < N-1 & \quad (5.8) \\[2ex]
(6) \quad \dfrac{d\lambda_N}{dt} = \psi(t)\,\lambda_{N-1}(t) &
\end{array}
$$

where, according to (5.5) : $\psi(t) = k_{\text{poly}} P(t) \, (1 - \phi_0(t))$ in the case of replication.

The solution of these two differential equations coupled by ϕ_0 completely solves the problem posed. However, two questions still remain. First, using the hypothesis (H3), what do the terms k_+ and k_- signify? Secondly, recalling that we put $l = L - 1$ only for simplification, what happens to the system (5.6) when the region (D) extends far beyond the zone where the double helix is uncoiled? To put it simply, if we write: $l = L - c$, we would have to study the following cases:

$$j = 0, \quad 1 \leq j \leq c, \quad c + 1 \leq j \leq L - c, \quad l - c + 1 \leq j \leq L - 1, \quad j = L$$

and calculate the corresponding probabilities, which seriously complicates the resolution of the problem. The principal results of this calculation will be found in the original paper by Zimmerman and Simha (1965).

For a better understanding of the phenomena involved, it would be useful to study the variation of the mean value and the variance of the random variable $|j\rangle$, $j = 0$ to L, with $\phi_j(t)$ being the probability of the state $|j\rangle$.

$$m_p = \frac{d}{dt}\left\langle j^P \right\rangle = \frac{d}{dt}\sum_{j=0}^{L} j^P \phi_j(t)$$

is then time derivative of the moment (Appendix D) of order P, where $\langle j \rangle$ represents the *mean* of the values j, or the first-order moment. Thus, with:

$$\sum_{j=0}^{L} \phi_j(t) = 1 \quad \text{(by the definition of the probabilities } (\phi_j)_{j=0,L})$$

and:

$$\eta(t) = \sum_{j=0}^{L} j\phi_j(t)$$

we find

$$m_0 = \sum_{j=0}^{L} \frac{d\phi_j}{dt} = 0$$

by virtue of the system (5.6), which corresponds to the definition of the probabilities. The variation of the number of free sites is given by the equation:

$$m_1 = \frac{d\eta}{dt} = \sum_{j=1}^{L} j \frac{d\phi_j}{dt} = k_+\phi_0 + \sum_{j=1}^{L-1}(k_+ - k_-)\phi_j - k_-\phi_L$$

$$= k_+\phi_0 + (k_+ - k_-)\sum_{j=1}^{L-1}\phi_j - k_-\phi_L$$

$$= k_+\phi_0 + (k_+ - k_-)(1 - \phi_0 - \phi_L) - k_-\phi_L.$$

Finally:

$$m_1 = \frac{d}{dt}\langle j \rangle = (k_+ - k_-) + k_-\phi_0(t) - k_+\phi_L(t).$$

Similarly, the second-order moment is:

$$m_2 = \frac{d}{dt}\langle j^2 \rangle = \sum_{j=0}^{L} j^2 \dot{\phi}_j(t)$$

and the variance is:

$$\sigma^2 = \langle j^2 \rangle - \langle j \rangle^2.$$

Calculations identical to those above lead to:

$$\frac{d\sigma^2}{dt} = (k_+ + k_-) - k_-(2\langle j \rangle - 1)\phi_0(t) - k_+(2L + 1 - 2\langle j \rangle)\phi_L(t).$$

These equations give the rate of DNA replication in terms of the progression of the mean length of the free sites. Thus, at the beginning of the process: $\phi_0(0) = \phi_L(0) = 0$. The initial rate is $k_+ - k_-$ and the rate of change of the variance is $k_+ + k_-$. This is the rate of the process as long as ϕ_0 and ϕ_L are not too large. The time needed for the replication under this condition is:

$$t_0 = \frac{L}{k_+ - k_-}.$$

(5.9)

The rate of change of the variance σ^2 being $(k_+ + k_-)$, the value of σ^2 at the end of the replication will be:

$$\frac{k_+ + k_-}{k_+ - k_-} L.$$

The condition for weak dispersion, written: $\sigma^2 \ll L^2$, will be:

$$\frac{k_+ + k_-}{k_+ - k_-} \ll L.$$

Finally, the duration of a replication is of the order of $L/(k_+ - k_-)$ if the condition:

$$k_+ + k_- \ll (k_+ - k_-)L \tag{5.10}$$

holds good.

The values of k_+, k_{poly} and $c = L - l$ found by *in vitro* experiments on the synthesis of adenine 5-bromouracil (Wake and Baldwin, 1962) may be deduced by a simulation using the rate of conversion $dC/dt = 3.3 \times 10^{-4}$ sec^{-1}. Then, for various values of $c = L - l$, we obtain an estimation of the uncoiling rate k_+. For example, when $c = 1$, we have $k_+ = 0.34$ sec^{-1}, $k_{poly} = 3.4$ sec^{-1} and $t_0 = 2900$ sec at a conversion of 90%. But for $c = 10$ the rates diminish, falling to 0.036 and 0.36 respectively with $t_0 = 3400$. An optimum value for the time appears at $c = 2$.

The *in vivo* data obtained for T2-DNA indicate quite different rates: for $c = 1$ we have $k_+ = 1700$ sec^{-1} and $k_{poly} = 10\ 000$ sec^{-1} with $t_0 = 0.58$ sec. For $c = 10$, these values are respectively 170, 1700 and 0.71; and for $c = 2$ they are respectively 1100, 11 000 and 0.3. Thus, the *in vivo* replication time is of the order of 0.5 sec and k_+ of the order of 10^3. The advantage of such simulations is that they give an order of magnitude for certain parameters, for instance the parameter c and the process rates, which would otherwise be difficult to obtain. Obviously, this approach calls for further investigation.

3.   *Protein synthesis*

As we have seen, the growing polypeptide chain remains attached to the mRNA on the ribosome only by the tRNA coupled to the corresponding aminoacyl. The addition of a monomer displaces the fixation point on the template from one site to the next. The process is similar to the global phase (II) of DNA replication. The essential difference from this point on lies in the possibility of what is known as *multiple synthesis*, i.e. the possibility of the growth of several polypeptide chains at a given point of the template and at a given instant after repetitive initiation. This may be considered as a difusion of *several, non-overlapping segments* along a unidimensional network. This characteristic makes it necessary to formulate the problem in a way slightly different from that used above. However, we shall use the same method and in particular the system (5.6). We shall therefore express, as a function of time, the variation of the probability of occupation of a given site i of the network.

Let i be a site of the network ($1 \leq i \leq N$). The length of the segment L undergoing displacement on the template is the same as that of the ribosome. As before, let this region be called (D). The site i may be in several states since it may be occupied by one of the j sites of (D), ($1 \leq j \leq l = L - 1$), or by the site L, or remain unoccupied. Thus, there exist $L + 1$ states for i.

The segment of length L covering the site i of the network obviously occupies the following sites of the same network:

$$i - j + 1, \; i - j + 2, \; ..., \; i - j + L$$

for j varying from to 1 to L. Then the site i is either in the state $|0\rangle$ (empty) or in the state $|j\rangle$ ($1 \leq j \leq L$). These are the $L + 1$ states of the site i. Let $\phi_i^{(j)}$ be the probability that the site i be in one of the states $|j\rangle$. The set of states being *complete*, we have:

$$\sum_{j=0}^{L} \phi_i^{(j)}(t) = 1.$$

$\phi_i^{(j)}(t)$ and $\phi_{i-j+L}^{(L)}(t)$ are identical as both represent the probability of occupation of the site i in the state $|j\rangle$ or of the site $i - j + L$ in the state $|L\rangle$. In fact, a translation of $(j - L)$ sites of the network brings the site L of the ribosome in front of the site i of the network. This also means that each probability $\phi_i^{(L)}(t)$ may be expressed in terms of $\phi_i^{(L)}(t)$, and that in the dynamics of biopolymerisation it will suffice to consider the state $|L\rangle$.

The dynamics of the $\phi_i^{(j)}(t)$ values can be found in exactly the same way as for the system (5.6) by considering the statistical set of class contents. Let $q_i(t)$ be the flux of occupation in the state $|L\rangle$ between the sites i and $i + 1$, equal to the product of $\phi_i^{(L)}(t)$, and a transition probability to be calculated. Since a given segment may move in one direction, from left to right ($i \rightarrow i + 1$), or in the other, from right to left ($i + 1 \rightarrow i$), we may write:

$$q_i = q_i^+ - q_i^-$$

by noting $(+)$ and $(-)$ respectively the two contributions above. Then we clearly have:

$$q_i^+ = k_+ \phi_i^{(L)}$$

and:

$$q_i^- = k_- \phi_{i+1}^{(L)}.$$

But we also have:

$$\frac{d\phi_i^{(L)}}{dt} = q_{i-1}(t) - q_i(t)$$

since the number of elements in class (i) increases because of the elements entering from class $(i - 1)$ and decreases because of the elements leaving it. The polymerisation progresses in a given direction along the sites of the network but the displacement of the diffused segment may occur in either direction. Let us recall that protein synthesis is characterised with respect to DNA replication: (i) by the possibility of multiple synthesis; and (ii) by the diffusion of several, non-overlapping segments of length L over the one-dimensional network. The 'constants' k_+ and k_- depend on the conditional probability f_i that the site $i + 1$ be in state $|0\rangle$ when the site $|i\rangle$ is in state

$|L\rangle$ at the same instant. It can be shown (Pipkin and Gibbs, 1966; MacDonald *et al.*, 1968) that:

$$q_i(t) = k'_+ f_i^{(+)}(t)\, \phi_i^{(L)}(t) - k'_- f_i^{(-)}\phi_{i+1}^{(L)}(t)$$

where $f_i^{(+)}$ and $f_i^{(-)}$ are the conditional probabilities f_i depending on ϕ_i for each type of movement, respectively (+) and (−), with $2L - 1 \le i \le K - L$.

We then deduce:

$$\frac{d\phi_i^{(L)}}{dt} = k'_+ f_{i-1}^{(+)}\phi_{i-1}^{(L)} - k'_- f_{i-1}^{(-)}\phi_i^{(L)} - k'_+ f_i^{(+)}\phi_i^{(L)} + k'_- f_i^{(-)}\phi_{i+1}^{(L)}$$

or:

$$\frac{d\phi_i^{(L)}}{dt} = k'_+ f_{i-1}^{(+)}\phi_{i-1}^{(L)} - \left[k'_- f_{i-1}^{(-)} + k'_+ f_i^{(+)}\right]\phi_i^{(L)} + k'_- f_i^{(-)}\phi_{i+1}^{(L)} \quad 2L - 1 \le i \le K - L$$

$$(5.11)$$

This equation is analogous to (5.8(3)). The equations obtained by putting $i = 2L - 1$ and $i = K - L$ depend on the limit conditions. This method has the advantage of giving explicitly the rates of the diffusing segments, k_+ and k_-. In the case where the diffusing segments cannot interfere, the conditional probabilities are equal to 1, and we retrieve equation (5.8(3)). If we wish to go further, we will need to use approximations. Thus we may seek '*uniform*' solutions (with respect to space which is here simulated *discretely* by the index i), such that:

$$q_i = q(t)$$

which leads to:

$$\phi_i^{(L)} = \phi(t) = \phi.$$

With the explicit expression for $(f_i(\phi_j))$ we have:

$$q = (k_+ - k_-)\,\frac{\phi(1 - L\phi)}{1 - (L - 1)\phi}. \qquad (5.12)$$

The solutions of uniform probability lead to the stationariness of the system. Indeed, it has been shown that all the stationary solutions are close to these solutions. Figure 5.17 shows the variation of the flux of occupation $Q = \dfrac{Lq}{k_+ - k_-}$ in terms of $\Phi = L\phi$. These new 'normalised' quantities are deduced from the above equation in the form:

$$\frac{Lq}{k_+ - k_-} = \frac{\Phi(1 - \Phi)}{1 - \Phi + \dfrac{\Phi}{L}}.$$

They show that the maximum flux is attained when $\Phi = \sqrt{L}[1 + \sqrt{L}]^{-1}$ and is given by: $Q_{max} = (Lq_{max})/(k_+ - k_-)$ whence $q_{max} = (k_+ - k_-)Ln_m^2$ with $n_m = 1/[\sqrt{L}(1 + \sqrt{L})]$.

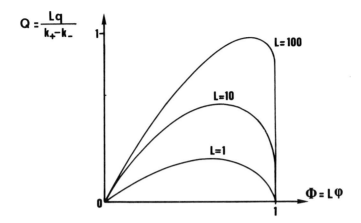

Fig. 5.17. Flux Q of the probability of occupation from one site to the next in terms of the reaction constants k_- and k_+ and of the density of probability.

As soon as the flux of occupation q is less than q_{max}, there are two solutions: the probability of one being low and that of the other very high. What does this result mean? The high density solution corresponds to a large flux that quickly approaches the maximum, and which is soon limited by the congestion caused by the diffusing segments, just as in a traffic jam. On the contrary, when the density is low (ϕ small), the flux may increase because of the rarity of the diffusing segments. We see that the increase in the length L of the segments leads to an increase not only of the traffic but also of the maximum flux, so that congestion then appears rapidly. This model is probably a good simulation of reality and has the advantage of bringing out clearly *the differences between DNA replication which is represented as the movement of a particle, and protein synthesis which is considered in terms of competitive traffic involving several diffusing segments.* The calculation also shows that in regions of high density: (i) the probability Φ may *oscillate* according to the site i of the network when L is great; and (ii) the oscillation is damped when i increases. Clearly the predictions of this model give considerable insight into the dynamics of protein biosynthesis and suggest an approach to experimental verification. It is true that the model requires further elaboration to take into account duplication errors, deletions, the role of the exons and the introns, and so on, but the day will surely come when full knowledge and total control of the kinetics of molecular processes will provide a fundamental underpinning to biotechnology.

IV.   Topological, structural and functional implications of nucleic acid chains

1.   *Topological concepts in DNA replication and structural consequences*

Following the work of Vinograd (1965), who first suggested that circular DNA may owe its properties to the supercoiling of the double helix caused by the reduced

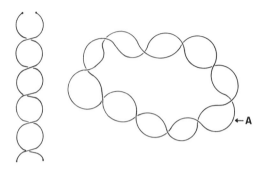

Fig. 5.18. A linear helicoidal chain is represented on the left and a circular chain on the right of the figure. The uncoiling of the chains at a point A by pulling the two strands apart will lead to a modification of the winding on either side of A.

number of *enlacements*, White (1969), Brock Fuller (1971), Bauer (1978) and Pohl and Roberts (1978) have studied the topological aspects of DNA replication. The mathematical arguments should be considered together with the experimental results reported by Rodley *et al.* (1976) and Stettler *et al.* (1979) which tend to demonstrate that the DNA structure is not a double helix. Even Crick (1976) admits that *there is no one-to-one correspondence between a double helix structure and its X-ray diagram, i.e. different structures may produce identical diagrams.* However, Wang (1979a) seems to have confirmed the double helix structure. The question thus remains open. The theoretical interest lies in the remarkable result, altogether novel in molecular biology, due to Pohl (1968): the passage from local properties to global properties imposes the *structure* and consequently the *function*. So how are we to apprehend a global situation knowing that the basic physical mechanism is essentially local? We shall see that this topological constraint leads to an alternative: either the chromosomal DNA is not a double helix, or else the supercoiling has to be suppressed before replication. It has recently been found that such suppression does not necessarily involve enzymic activity.

Here we shall present the problem in simple terms. A detailed description will be found in Pohl (1968), Pohl and Roberts (1978) and Bremermann (1979). The reader may consult Bauer *et al.* (1980) for a general review, and Wang (1982) for experimental results in favour of the double-helix structure of DNA.

a.   Topological findings

Let us consider two coiled strands, C_1 and C_2, fixed at each extremity, which may be visualised in the form of a chain. This would also be the case if the strands were attached to form a circular chain (Fig. 5.18). It is convenient to imagine the chain to be formed by the two edges of a twisted ribbon. The best known simple example is that of the Möbius strip which has its ends pasted together after being given a half twist. Such a ribbon is then characterised by the number of twists, called *enlacements*,

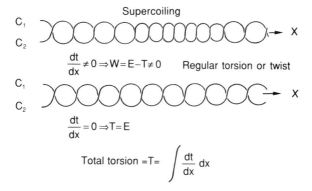

Supercoiling

C_1

C_2

$$\frac{dt}{dx} \neq 0 \Rightarrow W = E - T \neq 0 \qquad \text{Regular torsion or twist}$$

C_1

C_2

$$\frac{dt}{dx} = 0 \Rightarrow T = E$$

$$\text{Total torsion} = T = \int \frac{dt}{dx}\, dx$$

Fig. 5.19. Mathematical significance of supercoiling: there is a relation between the writhing number W, the total torsion T and the coiling E.

of one edge about the other. If the two strands C_1 and C_2 are oriented, the number of enlacements E will be an algebraic number $E(C_1, C_2)$, which reflects the topological property that no deformation will change the way in which the closed curves are enlaced, i.e. the number of enlacements cannot be modified without a rupture of one of the strands. Thus, E is a topological invariant corresponding to a global property.

Obviously, since the number E must be conserved, the uncoiling of the two strands at any point A will lead to a modification of the helical winding on either side of A. In simple terms this means a *local* increase in the number of spires. What is the mathematical significance of this property? Figure 5.19 shows how the invariance of E produces a local property: a torsion or twist about the axis of the ribbon. This local geometrical property is expressed by what is called the *writhing*, i.e. the number of rotations $dt(x)$ of a point M moving along an edge around the axis of the ribbon and advancing over a distance dx in the direction of the axis. Clearly, when the axis of the ribbon lies in a plane, the total writhing, or *tortility* T, will be equal to the enlacement E. But, in the general case, the problem is more complex and requires the introduction of a new number, the *writhing number* W, which depends uniquely on the curve described by the *axis* of the ribbon. This number W also expresses a *local* geometrical property associated with the local deformation of the axis. Consequently, W and T are not topological invariants.

To sum up, three concepts are needed to understand the topological constraints acting on DNA structure (Fig. 5.20):

- The number of *enlacements* E which corresponds to the number of times one of the curves C_1 or C_2 crosses the loop of the other. The sign of E depends on the direction of the curves. DNA may be considered to be a ribbon wound about itself, with the edges C_1 and C_2 oriented in opposite directions. The base sequences of DNA are in fact known to be organised in this way.
- The *tortility* T measures the torsion of the ribbon formed by C_1 and C_2 over a *given distance*. Mathematically, if we consider a segment perpendicular to the ribbon axis and directed towards its edge, the tortility is the integral of the

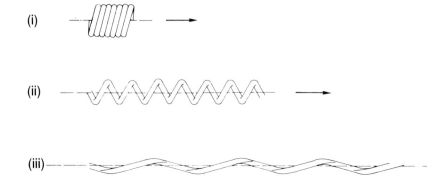

Fig. 5.20. Steps (i) to (iii) illustrate the phenomenon of supercoiling in a telephone cord the mechanical properties of which have been modified by torsion (after Bauer *et al.* (1980)).

variation of the angle of rotation of the segment in terms of the curvilinear abscissae on the ribbon axis. If the axis is rectilinear, T is the number of turns, i.e. the number of enlacements E, otherwise the calculation is delicate since the deformation of the axis will have to be considered. Consequently, circular DNA involves a mathematical relationship between E and T.

- The *writing* W describes the relationship between E and T. This depends exclusively on the curve followed by the ribbon axis. The mathematical problem is to determine W knowing E and T. White (1969) has shown that W may be calculated by a double integral, the Gauss integral, over the *entire chain* (C_1, C_2). However, it is easier to first calculate E and T and then deduce W from the simple relationship:

$$W = E - T. \tag{5.13}$$

Thus, if the ribbon axis lies in a plane, we have $W = 0$. Figure 5.20, which may be considered to represent a coiled telephone cord, is a good illustration of the mechanical properties of the cord: as we go from diagram (i) to diagram (iii), the 'ribbon' twists on itself while lengthening (T increases), and the 'spiral' reduces its rotation (W decreases) per *unit length*. This local geometrical characteristic is essentially related on the one hand to mechanical factors of minimum energy and on the other to the factor of topological invariance, E. And it is this characteristic which leads to the phenomenon of *supercoiling*.

The general problem which consists in finding the linking number for two closed, linked curves from local information is called the problem of *linear alignment*. Mathematically, the local structure determines the global structure, represented by the number of enlacements E, by a set of highly complex differential equations. The alignment problem in the case of a double helix produces a paradox that has been aptly described by Watson (1978). Thus, the number of enlacements for two closed, linked curves being fixed, the relationship $W + T = E = $ constant, necessarily implies

Fig. 5.21.   Configuration of supercoiled DNA (after Bauer *et al.* (1980)).

that any *local* modification of W leads to a modification of T and vice versa. This means that if we have two ribbons coiled about one another, with one ribbon having a smaller number of enlacements than the other, then the coiling and the writhing of the two ribbons will be different. The result will be a new structure caused by supercoiling, or the coiling of the ribbon about itself. In particular, two enlaced curves ($E \neq 0$), one of which is cut open, are said to be in the *relaxed* state. Their axis then lies in a plane so that $W = 0$ and $E = T$. If the cut is repaired and E reduced, then W and T will be modified and supercoiling will appear (Fig. 5.21).

The configuration of DNA as shown in Fig. 1.3 represents a minimum energy configuration: any tension in the molecule will lead to an increase in energy. If the diameter of circular DNA is great, the radius of curvature is large so that the energy is little different from that of a linear configuration.

Circular DNA with a number of enlacements smaller than in the relaxed state should uncoil. But then the energy of the configuration due to the deformation would increase. This is avoided by the supercoiling produced by increased writhing since there is a decrease in tortility which, as we have seen, represents the mechanical deformation inherent to matter. *A DNA molecule of which the double helix has lost a certain number of turns will adopt a compromise configuration which produces a maximum reduction of tortility while introducing a minimum amount of writhing.*

Thus, we are now in a position to explain the supercoiling of circular DNA.

b.   Experimental data

Bicatenary, circular forms as well as other forms of DNA have often been observed in certain plasmids. Forms I and II are bicatenary, but Form V is considered to be the juxtaposition of two, monocatenary, circular chains. The number of enlacements in the former is positive whereas in the latter $E = 0$.

It is clearly tempting to suppose that the bicatenary DNA of the chromosomes is not enlaced, which would eliminate the paradox discovered by Pohl and Roberts. This conjecture has in fact been used by Bremermann (1979).

Let us transpose the topological results seen above to the case of DNA. We may briefly recall the description of Chapter 1: the skeleton of the DNA molecule is made up of two sugar–phosphate chains coiled about an axis (Fig. 1.3) These correspond to the two curves C_1 and C_2 considered above. DNA in the linear state has a rectilinear axis, and in the relaxed circular state the axis lies in a plane.

It has been experimentally found that in the relaxed circular state the average number E is about 500 whereas in the unrelaxed circular state E is about 475. Thus, there is a deficit leading to supercoiling which can be measured not only by the mathematical formula $E = W + T$, but also by the physical constraints of minimum energy in the resulting conformation. The supercoiling is therefore the result of a topological constraint and a compromise between the energies of conformation in the stable state for a given writhing and a given tortility.

With this mechanism in operation, there are two ways in which the supercoiling of the double helix may be eliminated:

(i)   either one of the strands has to be cut so that the molecule attains the relaxed state (an enzyme — *helicase* — producing this result has actually been identified);

(ii)   or T has to be reduced so that W (negative) increases, i.e. tends to zero. A molecule — ethidium bromide — has indeed been found to be capable of reducing tortility (Fig. 5.22) by inserting itself between the base pairs of the double helix, but the process is not a physiological one.

Two interpretations may then be given, but these are controversial: the tortility T and the writhing W which characterise the supercoiling of the double helix describe a global property $E = W + T$. In case (i), the enzyme that cuts and then uncoils the two strands during the replication must therefore 'carry out' complicated calculations to determine, for a given E, the two numbers W and T, representing the local properties, while taking into account the variations of the geometrical deformation of the chromosome. The final aim, of course, is the reduction of the number of enlacements to zero for the daughter strands. How can this type of control be exercised during the replication of a bicatenary double helix of DNA? Pohl and Roberts have found five biological possibilities, all of which are thoroughly discussed in their original communication. They conclude that DNA structure cannot be a double helix, and that a possible configuration may be such as described by Rodley *et al.* (1976) (see Fig. 5.23).

Bremermann's conjecture may then be admitted. Of course, if it should prove true,

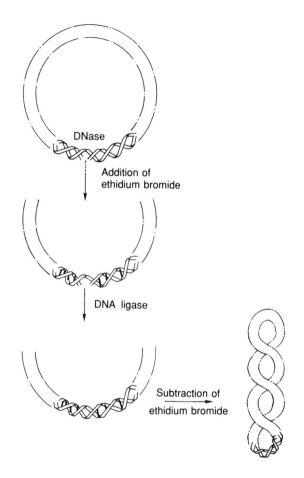

Fig. 5.22. Effect of ethidium bromide on the double helix: the molecule inserts itself between the base pairs of the DNA, reducing its writhing.

a certain number of consequences will apply to genetics. In particular, *heriditary zero enlacement* by replication without the cutting of the strand will have to be ensured. Similarly, the 'manipulations' of viruses and plasmids, the crossing-over, should preserve this property.

In case (ii), a molecule such as ethidium bromide may suppress the supercoiling. The experiments carried out by Bauer (1978) and Wang (1979b) seem to confirm the structure of the DNA double helix. So we have to suppose that the supercoiling existing in the initial state before replication is suppressed by the insertion of appropriate molecules during the replication.

This interesting aspect of replication clearly deserves further investigation. It is certainly remarkable that an abstract topological structure should lead to fundamental biological questions, perhaps as important to cell regulation as the allosteric effect discovered by Monod and Jacob. Indeed, we may wonder *how* and *why* DNA

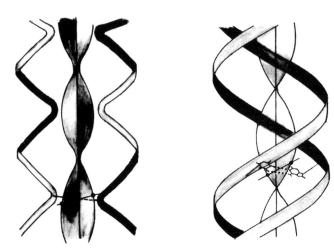

Fig. 5.23. Structure proposed by Rodley *et al.* (1976) to explain the properties of circular DNA, which the authors consider not to be in the form of a double helix (an interpretation which has been refuted, see text).

supercoiling originates. Do the dynamics of replication intervene in the process? As the double helix is coiled about a nucleosome to form an elementary chromosomal unit, the separation of DNA from the nucleosome will be bound to produce supercoiling. Thus, the solenoidal structure of the chromosome is an essential cause of the supercoiling. Several enzymic operations on the genome require local uncoiling of the DNA, so what are the functional implications of the supercoiling?

Wang (1982) has found enzymes that carry out 'mathematical' operations. Thus *type I topoisomerases*, discovered in 1971, may break a single strand to relax DNA supercoiling. They may also cause another part of the same strand to pass through the break, thus making a knot, or they may even interlock two monocatenary loops, i.e. create a supercoiled double helix. *Type II topoisomerases* may break both strands of the DNA helix. For example, DNA *gyrase* produces negative supercoiling in relaxed bicatenary loops, modifying the number of enlacements by multiples of two. In fact, gyrase makes a double-strand break, causing the DNA chain to pass through the break, and then joins the two ends.

Topoisomerases thus play an important role in DNA replication and transcription by controlling the *degree of supercoiling* (generally negative, as seen above), and modifying the separation of the DNA strands so as to adjust the rate of transcription to cell metabolism. We therefore have a new variable of topological origin: *the degree of supercoiling*, which intervenes in biological systems at the molecular level in a manner comparable to other variables such as the temperature or the ionic force. This variable clearly needs to be introduced into formalisms, such as that of Zimmerman, which describe the dynamics of DNA replication and transcription. *Current hypotheses concerning DNA structure and the phenomenon of supercoiling should then enable us to specify new parameters determining the rate of replication*

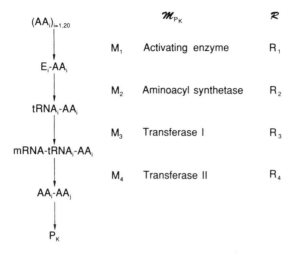

Fig. 5.24.   Diagram of protein biosynthesis in the form of an ($\mathcal{M}$–$\mathcal{R}$) system.

in function of the environment, i.e. allow us to act directly on cell regulatory mechanisms.

2.   *Topology and protein biosynthesis*

The formalism presented here is based not on structural observation but on Rosen's relational theory. In fact, the results obtained are a good illustration of the theory. The advantage here lies in the global treatment of the phenomena of protein biosynthesis and in the investigation of the consequences of the very existence of the elements of the system. As we have already seen in Chapter 3, Rosen's relational theory considers the structural variations of the components M of a system ($\mathcal{M}$) under environmental constraints, and its replication by associating a system ($\mathcal{R}$). This is called the ($\mathcal{M}$–$\mathcal{R}$) system.

Comorosan and Platica (1968) have applied this theory to protein biosynthesis. It must be admitted that Rosen's terminology not only provides a clear and elegant symbolism but also gives results of great interest.

The system of chemical reactions in Fig. 5.7 may be transcribed in the ($\mathcal{M}$–$\mathcal{R}$) form by the diagram in Fig. 5.24. The set of metabolic pathways leading to the synthesis of a protein P_k is identified to a system $\mathcal{M}_{P_k}$ which functions over the set of the 20 amino acids AA_i; the set of complexes E_i–AA_i, where E_i is the activator or Hoagland's pH 5 fraction; the set of complexes $tRNA_i$–AA_i, where $tRNA_i$ is the part of tRNA specific to AA_i; the set of complexes $mRNA$–$tRNA_i$–AA_i; and the set of AA_i–AA_j, which are the two amino acids with indices i and j in the protein P_K.

The application M_1 (in the mathematical sense) represents the effect of the activating enzyme. If $\mathcal{F}$ is the set of applications, then:

$$M_1 \in \mathcal{F}\left(\prod_{i=1}^{20} AA_i, \prod_{i=1}^{20} E_i - AA_i \right)$$

is a component of $\mathcal{M}$. The same is true for M_2, M_3 and M_4.

The input and output sets are the product spaces written:

$$\prod_{i=1}^{20} AA_i = AA_1 \times AA_2 \times \ldots \times AA_{20} \quad \text{and}$$

$$\prod_{i=1}^{20} E_i - AA_i = E_1 - AA_1 \times \ldots \times E_{20} - AA_{20}.$$

This may be done for each step M_1, M_2, M_3 and M_4.

a.   The concept of contractibility and its consequences

Let us recall the significance of the concept of contractibility: all the inputs received by the system are necessary for it to function. It is obvious that if the protein P_k made up of a number N_{P_k} amino acids, less than or equal to 20, the problem is to know if *all* the 20 amino acids are necessary. We know that all the parts of a protein are not equally essential to its function since the loss of some parts does not lead to the suppression of biological activity, for instance the enzymic activity. In fact, only certain amino acid residues constitute the active site of the enzyme. Let S_{P_k} be the set of amino acids *available*. We may define the set of *essential* amino acids, $\bar{S}_{P_k}$, i.e. the amino acids necessary to the biological function and constituting the active site of the enzyme. Then $S_{P_k} - \bar{S}_{P_k}$ represents the set of amino acids which may be replaced or remain absent, i.e. which may be considered *useless* (in terms of biological activity, though perhaps not in terms of molecular structure). The experimental observation that only the few amino acids belonging to the active site of the enzyme play a biochemical role is represented by the *factorisation* of the set S_{P_k} into $\bar{S}_{P_k}$ which is necessary, and $S_{P_k} - \bar{S}_{P_k}$ which is useless. The set $\bar{S}_{P_k}$ cannot be factorised further. Since each enzyme has a specific set of essential amino acids (constituting the active site), the factorisation of a metabolic system could correspond to the decomposition of the system into sets of essential elements.

One consequence of this construction is that, if $AA_i \in S_{P_k}$ with $AA_i \notin \bar{S}_{P_k}$, then the protein P_k may be synthesised even in the absence of AA_i in the metabolic pool. Another consequence is that the small number of the elements of $\bar{S}_{P_k}$ leads to the possibility of advantageous cellular adaptation. In molecular genetics, the *ambiguous coding* is the illustration of the possible substitution of one amino acid by another, since a specific tRNA may recognise two or more different amino acids. Indeed, ambiguous coding appears to be a biochemical mechanism necessary to adaptive processes (Comorosan and Platica, 1967).

We see that Rosen's formalism allows a protein to be attached to a functionally specific subset $\bar{S}_{P_k}$. The small number of elements in $\bar{S}_{P_k}$ provides considerable selective advantages in the phenomena of cellular adaptation: the existence of ambiguous codons thus appears to be an adaptive necessity.

b. The concept of restorability and its consequences

In the definition of Rosen's ($\mathcal{M}$–$\mathcal{R}$) system, $\mathcal{R}$ is the subsystem constituted by the components R_i, $i = 1$ to 4, which ensure the replication of the corresponding components M_i of $\mathcal{M}$. For example, R_1 is a biological subsystem which ensures the replication of M_1, i.e. the synthesis of the activating enzyme M_1. It is thus a particular system $\mathcal{M}_{P_K}$. Here we find a well known algorithmic property, called *recursion*, which poses several problems in computer science.

There are then two possibilities:

(i) P_K is one of the enzymes associated with the activation of $AA_i \in \bar{S}_{P_K}$: the amino acid is essential; it is a constituent of $P_K = E_i$ and P_K participates in the activity of AA_i, whence the recursion. Formally:

$$P_K = E_i \quad AA_i \in \bar{S}_{P_K}.$$

Let R_i be the component of $\mathcal{R}$ which replicates M_i, the process of the synthesis of E_i. Then:

$$\Phi E_i \in \mathcal{F}(\Pi E_k, E_i)$$

where E_k is associated with the activation of the $AA_i \in \bar{S}_{P_K}$. ΦE_i is an output from the component R_i. The protein P_K stimulates its own synthesis. At the molecular level, this is the ancient problem of the chicken and the egg, which we shall consider in detail in the next chapter. Obviously, in this case the restoration of $\mathcal{M}_{P_K}$ is not possible.

(ii) P_K is not an *essential* enzyme in the above sense. Formally:

$$P_K = E_K, \forall j, AA_j \in \bar{S}_{P_K} \Rightarrow E_K \neq E_j$$

or:

$$P_K = E_K; \quad AA_K \notin \bar{S}_{P_K}.$$

It is assumed that for the synthesis of P_K the property of non-contractibility holds good for n amino acids, all of which belong to $\bar{S}_{P_K}$. This means that E_K activates $AA_K \in \bar{S}_{P_K}$ but its active site is constituted by n amino acids AA_j different from the amino acids AA_K. For example, for $n = 3$, E_1 activates AA_1 and is constituted by AA_2, AA_3 and AA_4:

$$
\begin{array}{cc}
E_1 & \rightarrow \quad E \\
AA_2 - AA_3 - AA_4 & AA_1 - AA_3 - AA_4.
\end{array}
$$

It may then act on all the enzymes containing AA_1.

Then let AA_K be the amino acid activated by E_K, and S_{E_K} the set of enzymes associated with the activation of the amino acids of the set $S_{E_K} - AA_K$. Then the replication function is given by:

$$\Phi E_K \in \mathcal{F}(S_{E_K}^n, E_K)$$

where:

$$S_{E_K}^n = \underbrace{S_{E_K} \times S_{E_K} \times \ldots \times S_{E_K}}_{n \text{ times}}$$

which expresses the fact that n activating enzymes are necessary to the production of E_K which is made up of n essential amino acids. If $\overline{S}_{E_K}$ is the completed dependent set of E_K, then the relationship:

$$S_{E_K}^n \cap \overline{S}_{E_K} = \varnothing$$

is the condition of restoration in Rosen's theory (Theorem II). In fact, $\overline{S}_{E_K}$ represents the set of enzymes the synthesis of which depends on E_K. Clearly, if the number of amino acids in $\overline{S}_{E_K}$ increases, the possibility of restoration will decrease. It is believed that the active site of an enzyme is constituted by at least two amino acids (one for the recognition of the amino acid, the other for the selection of the tRNA) so that $2 \leq n \leq 19$. The upper limit is due to the fact that if AA_K is activated by E_K, the 19 amino acids remaining may be associated in the synthesis of E_K. When $n = 19$, each of the enzymes (activating) is involved in the synthesis of all of the others. In other terms:

$$S_{E_K}^{19} = \overline{S}_{E_K}$$

and the condition of restoration does not hold good. The example given for $n = 3$ (Chapter 3) is a good illustration of these concepts:

Each enzyme E_i ($i = 1$ to 4) activates the amino acid AA_i and possesses on its active site $n = 3$ amino acids, AA_k, $k \neq i$. To condition (ii) $\forall k = 1$ to 4, $AA_k \notin \overline{S}_{P_i}$, we have added the condition: $\forall k = 1$ to 4, $\forall j = 1$ to 4, $j \neq k \Rightarrow AA_j \in \overline{S}_{P_i}$. Thus, E_i is not necessary for its own biosynthesis. However, the other three enzymes are necessary for the synthesis of E_i.

If the enzymes are grouped in subsystems containing four elements, such that each enzyme is bound to all the other enzymes of the set (as shown above), we obtain the minimum completed dependent sets and the restorability is maximum. This result, which holds good when n, the number of amino acids on each active site, is equal to three, can be generalised for $2 < n < 19$. Comorosan and Platica (1967) also propose other models leading to minimum completed dependent sets using methods based on ambiguity and degeneracy of the genetic code.

We thus have:

$$S_{E_1} = \{E_2, E_3, E_4\}$$

$$\Phi E_1 \in \mathscr{F}(S_{E_1}^3, E_1)$$

$$\overline{S}_{E_1} = \{E_2, E_3, E_4\} = S_{E_1}$$

$$\overline{S}_{E_1} \cap S_{E_1} \neq \varnothing$$

The restoration of E_i is not possible because each E_i is involved in the synthesis of the other three enzymes.

These considerations have brought out an important concept concerning the *essential amino acid*, and a phenomenon necessary to the correct operation of the system, i.e. the formation of a $tRNA_i - AA_i$ complex within an appropriate completed dependent set. Rosen's theory allows us to see how certain functional relationships are necessary for the restoration of the system in the case of non-functioning. We may also see how the existence of degeneracy (an amino acid, for instance AA_1, may bind to $tRNA_1$ or to $tRNA_3$) or ambiguity of the genetic code ($tRNA_1$ may recognise either AA_1 or AA_3) appears necessary for the *stability* of the system, i.e. the system should be restorable in spite of the inhibition of enzyme E_1 constituted by AA_1 and AA_3.

The highly abstract formalism used here may appear difficult and esoteric. However, it only attempts to represent a biological reality: the multitude of functional relationships between the components of a system must be capable of undergoing development, operating substitutions, and executing self-reparation so as to ensure the correct functioning of the whole system. The notion of functional organisation, which we shall consider in Chapter 12, will also be used to represent and perhaps explain the phenomenon of self-reparation which is unique to the living world.

V. The hierarchical organisation of the replication–translation apparatus

We may now examine the more complex case of metabolic systems in which the metabolite synthesized is the end product of a metabolic chain. This chain involves several enzymes acting sequentially, in a definite order, on a substrate produced during the preceding step (see Fig. 2.9, Chapter 2). As we have seen above, each step in the series of reactions involves an enzyme produced by the replication–translation apparatus. An important issue is to represent the gene system by a simple model which, however, includes its most prominent properties. In Goodwin's well-known model (1978), enzymes result from clusters of structure genes G_i and polysomes X_i. Enzymes are a part of a metabolic pathway such as those given by schemas 4.47. The successive substrates, which together lead to the formation of the end product, define a level of organization that we shall call level 1. Each structural unit at this level

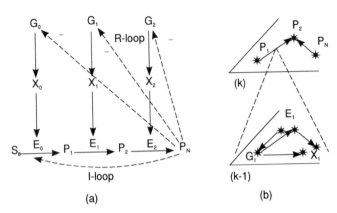

Fig. 5.25. Goodwin model of a regulated enzymatic pathway. (a) Enzymes E_i, $i = 1$, n result from clusters of structure genes such as G_i and polysomes X_i (top). They enter in a metabolic pathway which leads from substrate S_0 to the terminal product P_n. The synthesis of this product is regulated by an allosteric inhibitory feedback at G_i, and at enzyme E_0. (b) The same 2-level system where collective functions E_i and P_n are indicated at each level with their time scales (after Chauvet (1993a)).

contains the fundamental level, or level 0. We thus obtain what may be called a two-level *hierarchical system* (Fig. 5.25).

At each level, the dynamic development is associated with a specific time scale. Thus, protein biosynthesis works on a time scale which is much longer than that of the enzymatic reactions. Each global process, being the result of a set of partial processes, has its own time scale. The importance of this is well-illustrated by the rapidly advancing discipline of chronobiology, i.e. the systematic investigation of biological rhythms. It would thus appear reasonable to define the level of organisation by the time scale of the process such that a hierarchy of functions corresponds to each time scale.

The two lowest levels (noted 0 and 1) constitute what we have called the 'Goodwin system', i.e. a hierarchical system of regulated enzymes (second level) and genetical biochemical reactions (first level), both defining the metabolical unit (M) (Fig. 5.25). We generalize the metabolic system as an epigenetic system (Fig. 5.25a) into a metabolic pathway with an allosteric inhibitory control, and the same kind of feedback interaction with the structural gene. Specifically, it includes two control loops, one being an inhibition with a feedback at a point of the metabolic pathway (I-loop), and the other a repression of structural genes (R-loop).

According to the previous section, because of the very different time scales of these processes in the metabolic pathway and in the epigenetic system, this u_1-unit is a hierarchical system with two levels (Fig. 5.25b). Each enzyme E_i in the metabolic pathway, which transforms a given product P_i into another P_{i+1} at a higher level in the time scale {T2}, results from the collective behaviour, i.e. the dynamics, of the epigenetic system at the lower level in the time scale {T1}. The control between the two levels is given by the feedback loop (R) from the end-product P_n that acts on $X_i \equiv [mRNA]$, the concentration of messenger RNA. The allosteric inhibitory interaction is described by the term (see Eq. (1.12)):

$$f(P_n\,;\,\tilde{\omega},\,\kappa,\,\alpha_0) = \frac{\alpha}{\beta + \gamma P_n^{\tilde{\omega}}} = \frac{\alpha_0}{1 + \kappa P_n^{\tilde{\omega}}} \tag{5.14}$$

with $\alpha_0 = \alpha/\beta$, and $\kappa = \gamma/\beta$. In this equation, $\tilde{\omega}$ is the stoichiometry of the interaction, i.e. $\tilde{\omega}$ molecules of the end-product P_n bind with the aporepressor.

The u_1-units function according to the following two dynamical systems with their own time scales:

(1) *The epigenetic system with an R-loop* for the allosteric feedback repression is given by:

$$\frac{dX_i}{dt} = -\gamma_{X_i} X_i + f_{R,i}(P_n; \tilde{\omega}_R^i, \kappa_R^i, \alpha_{R,0}^i)$$

$$\frac{dE_i}{dt} = -\gamma_i E_i + \gamma'_{X_i} X_i \tag{5.15}$$

$$t \in \{T_1\}.$$

It is assumed that the catabolism of E_i is in direct relation with E_i, whether E_i is bound to P_i or not.

(2) *The metabolic system with an I-loop* for the allosteric feedback inhibition is given by Eq. (4.44(2)):

$$\frac{dP_1}{dt} = -\alpha_1 P_1 + f_I(P_n; \tilde{\omega}; \kappa; \alpha_0)$$

$$\cdots$$

$$\frac{dP_i}{dt} = -\alpha_i P_i + \alpha_{i-1} P_{i-1} \tag{5.16}$$

$$\frac{dP_{i+1}}{dt} = -\alpha_{i+1} P_{i+1} + \alpha_i P_i$$

$$t \in \{T_2\}$$

where γ_{X_i}, γ'_{X_i}, γ_i, α_i, $i = 1, 2, ...$, are the rate constants of the chemical reactions. The allosteric inhibition feedback term f_I is similar to that in equations (5.15), with different values of the parameters (see also Eq. (4.44)). This metabolic system is composed of enzymatic reactions such as $P_i \rightarrow P_{i+1}$ with rate $v_i = k_{3i} E_i P_i/(K_{M_i} + P_i)$, where $K_{M,i}$, k_{3i} are the Michaelis constants of enzyme E_i, and the rate constant of the reaction: $E_i P_i \rightarrow E_i + P_{i+1}$. If $P_i \ll K_{M,i}$ then $v_i = (k_{3i}/K_{M,i}) E_i P_i = \alpha_i P_i$.

The two systems (5.15) and (5.16), which correspond to two distinct levels of organisation, are decoupled in time. This means that the value of the concentration of the enzyme E_i is a constant during the dynamics of the metabolic pathway that leads to the end-product P_n. Therefore, *because of the functional hierarchy*, α_i is a

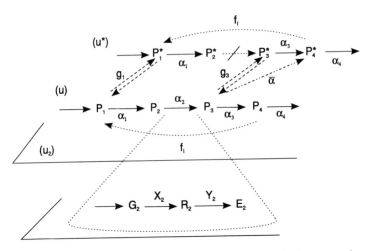

Fig. 5.26. Association of two structural units: the hierarchical system described in Fig. 5.25b, for which one link could be broken before P_3 in the biochemical pathway represented in Fig. 5.25a, for a unit u^*. The product P_3 in u is assumed to be carried in u^*, and act (functionally) on P_4^*, in order to maintain the biochemical reactions. As in Fig. 5.25a, P_4^* has a feedback on the first enzyme (after Chauvet (1993a)).

constant in the system (5.16). This natural simplification is of great mathematical importance in the deduction of the properties of the system (5.16).

Finally, the replication–translation apparatus may be represented as in Figure 5.26. The results we have shown in Chapter 4 concerning the channelling, which leads to the association of two metabolic pathways by increasing the stability of the system, may now be viewed as depending on the gene level because of the hierarchy: *a micromutation can break a biochemical pathway.* This mechanism will be described in Chapter 7 and will be shown to be fundamental to the relation between evolution and physiology.

6

Molecular Evolution and Organisation

Can the organisation, or rather the self-organisation, of macromolecular systems be explained at the molecular level? Eigen (1971) has attempted to characterise the degree of functional self-organisation using the concept of *value*, i.e. by an assessment of the properties of living organisms. In fact, while a random assembly of material elements is generally incapable of producing a functional structure (self-assembly excepted), any constraint applied to the system, by a process of selection for example, allows control of its evolution. This is what we usually call the *selective advantage* which, in the case of nucleic acids and enzymes, is related to the properties of control and of catalysis. The originality of Eigen's work lies in the introduction of the concept of selection in molecular dynamics and the correlation with known molecular parameters. For this, we need to determine *how the system makes use of its structural advantages and how the mechanism of valuation results from the dynamic behaviour of the system.* Recent work in the field of molecular evolution has led to theoretical insights concerning specific parameters (Iwasa, 1988) for nucleic acids (Luo *et al.*, 1988), proteins (Curnow, 1988), and RNA (Hickey *et al.*, 1989) as well as DNA organisation (Bodnar, 1988; Bodnar *et al.*, 1989).

While the Atlan–Yockey theory considers information *transfer* between messages E_i and E_j according to the quantity of information contained in their respective alphabets, Eigen's theory deals with the *dynamics of information carriers*, i.e. the dynamics of the symbols of the message, or rather that of the sequence of the symbols.

Following the copying of a message, there exists a certain number of carriers $X_{v,k}^{(\alpha)}$ of type $i_{v,k}^{(\alpha)}$ as defined in Chapter 4. The total number of copies in class (v) is then:

$$n_v^\alpha = \sum_{k=1}^{N_\alpha^v} X_{v,k}^{(\alpha)}. \tag{6.1}$$

We may therefore consider that, in a volume V composed of monomers, macromolecules and so on, it is possible to separate the symbols representing unit elementary information, or digits, into two subsets, according to whether they are organised or not, i.e. according to their potentiality in directing their own synthesis. Indeed, some symbols may have the capacity of transmitting the potential quantity of information they contain whereas others may not. The total number of organised symbols is evidently:

$$\sum_v v n_v^{(\alpha)}.$$

Let m_α be the concentration of the unorganised symbols. Then the total number of symbols of the alphabet (α) in the volume V is:

$$\sum_v v n_v^{(\alpha)} + \sum_\alpha m_\alpha.$$

It appears natural to suppose that the *degree of organisation* is given by the ratio:

$$D = \frac{\displaystyle\sum_v v n_v^{(\alpha)}}{\displaystyle\sum_v v n_v^{(\alpha)} + \sum_\alpha m_\alpha}. \tag{6.2}$$

This degree of organisation depends on the number of copies $n_v^{(\alpha)}$. We will therefore need to determine the evolution of the number $X_{v,k}^{(\alpha)}$ of information carriers as a function of time.

I.   Evolution of self-instructing information carriers

1.   *Phenomenological description of the evolution of chemical species*

To simplify the notation, let us write: $X_k \equiv X_{v,k}^{(\alpha)}$. The set of macromolecules composed of a sequence of v symbols of the alphabet (α), in an arrangement of the type k, thus forms a class of functional equivalence in terms of Delattre's formalism of transformation systems. According to Eigen, four hypotheses are necessary for selection to occur:

(i)   the system is open and not in thermodynamic equilibrium;
(ii)   the rate of the formation of carriers is greater than the rate of decomposition;
(iii)   in volume V, the number of carriers of length v is much smaller than the total number possible of carriers N_α^v, so that uninstructed formation is entirely negligible; and

(iv)   the environmental constraint is such that $\sum_k X_k$ = constant.

With the idea of rational internal coherence in mind, we should choose a theoretical language, such as Delattre's formalism of transformation systems (1980), which clearly reveals the underlying process often overlooked by macroscopic analyses. We shall therefore define the transformations between classes E_k for Eigen's theory. In addition to the hypotheses above, we shall suppose:

— that an element of the class E_k may either:
  • be synthesised (autocatalysed) by an underlying process that remains to be explained; or
  • be transformed, following copying errors, into an element of any class, say E_l with $l \neq k$; or
  • be destroyed in an element of class E_0, the class of unorganised molecules, i.e. that of the materials necessary for the synthesis of carriers; and
— that, following a copying error, an element of any class E_l, $l \neq k$, may be transformed into an element of class E_k.

Because of the hypothesis of selection (iv):

$$\Sigma X_k = \text{constant} \tag{6.3}$$

which leads to a reduction in dimension, Eigen's phenomenological equation does not bring out certain variables involved in a hidden process. Here, it is evident that particles can appear only from a class which is a pool of unorganised symbols, say E_0, and can disappear only by dissociation in the same pool. Similarly, hypothesis (iii) implies a second relationship between the variables X_k: for all the classes, the total number of mutants from any class E_k to a class E_i is equal to the total number of mutants from a class E_i to the class E_k (total balance).

With these conditions of linearity, Eqs (3.8) and (3.9) lead to:

$$\frac{dX_k}{dt} = \sum_l K_{kl} X_l, \quad k = 1, 2, \ldots, n$$

$$K_{kk} \leq 0, \quad K_{kl} \geq 0 \quad k \neq l. \tag{6.4}$$

The matrix $\mathbf{K}$ thus has a well-defined sign structure, and its elements may be determined by a graph corresponding to the postulates above.

In Eigen's theory, the behaviour of the carrier X_k is described by the phenomenological equation:

$$\frac{dX_k}{dt} = (A_k Q_k - D_k) X_k(t) + \sum_{l \neq k} \phi_{kl} X_l(t) - \Omega X_k(t) \quad k = 1, 2, \ldots, n \tag{6.5}$$

if there are n distinct types of carrier in volume V, where:

$A_k Q_k$ is the rate of formation of X_k with the *quality factor* Q_k. Q_k expresses the proportion of mutants from the X_k to other types X_l in a transition $E_k \to E_l$. If $Q_k =$

1, there is no copying error, and if $Q_k = 0$, then all the copies contain errors. Obviously, the reality corresponds to some intermediate value of Q_k;

D_k is the rate of destruction of carriers;

ϕ_{kl} is the rate of formation of X_k from transitions of the inverse type compared with the preceding, i.e. from E_l to E_k; and

Ω is the factor of dilution in function of time, the need for which is imposed by the hypotheses (iii) and (iv). Indeed, following (6.5) we have:

$$\frac{dX_k}{dt} = A_{kk}X_k + \sum_{l \neq k}\phi_{kl}X_l \quad k = 1, 2, \dots, n$$

with $A_{kk} = A_k Q_k - D_k - \Omega$, a term which is not always negative but, as we shall see, rather the contrary since it ensures the predominance of a molecular species.

In fact, the process is typically one of autocatalysis as described by Delattre (1980). The restriction of the number of variables is due to the constraints expressed by hypotheses (iii) and (iv) which are equivalent to the hypotheses of stationariness:

— constraint (iii) or the balance of errors:

$$\sum_k A_k(1 - Q_k)X_k = \sum_k\sum_{l \neq k}\phi_{kl}X_l$$

— constraint (iv) or the hypothesis of selection:

$$\frac{d}{dt}\sum_k X_k = 0 = \sum_k(A_k Q_k - D_k - \Omega)X_k + \sum_{\substack{k,l \\ l \neq k}}\phi_{kl}X_l$$

which imply:

$$\sum_k(A_k - D_k)X_k = \Omega\sum_k X_k = \Omega_0$$

so that the term Ω is determined by:

$$\frac{\sum_k(A_k - D_k)X_k}{\sum_k X_k}.$$

A finer description of the process is given by the formalism of transformation systems. The criterion of rationality obliges us to introduce a supplementary class E containing a number of elements supposed to be equal to X. Let us now consider the graph in Fig. 6.1. By the definition of autocatalysis (Delattre, 1980), the multiplicative balance of the process which starts from and returns to E_k must be superior to 1. Thus, an element leaving E_k must simultaneously 'combine' with an element of E to produce two elements of E_k, i.e. with $\beta = 2$. The process is such that the number of elements of the class E, i.e. X, is constant and the state of E_0, which is stationary, may be described by the system of equations, for $k = 1$ to n:

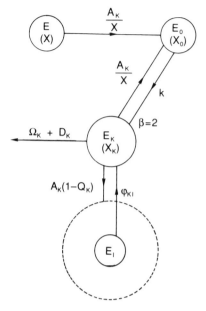

Fig. 6.1. Graph of autocatalysis according to the formalism of transformation systems.

$$\left. \begin{aligned}
\frac{dX_0}{dt} &= \left(\frac{A_k}{X}\right)XX_k - kX_0 = 0 \\
\frac{dX_k}{dt} &= -\left(\frac{A_k}{X}\right)XX_k + 2kX_0 - A_k(1 - Q_k)\,X_k - (\Omega_k + D_k)\,X_k + \sum_{l \ne k}\phi_{kl}X_l
\end{aligned} \right\}. \quad (6.6)$$

We deduce:

$$\frac{dX_k}{dt} = (A_kQ_k - \Omega_k - D_k)\,X_k + \sum_{l \ne k}\phi_{kl}X_l \qquad (6.7)$$

where the coefficient of X_k may be positive. The formalism of the transformations thus clearly shows that Eq. (6.7) is only, as Delattre (1980) puts it, an approximation of a system of a higher dimension respecting the criteria of rationality. In particular, it becomes evident that the appearance of the term $(A_k/X)XX_k$ in the first equation of (6.6) implies the existence of the term $-(A_k/X)XX_k$ in the second.

Having demonstrated that the factor $(A_kQ_k - \Omega_k - D_k)$ (6.7) can be positive, it appears legitimate to discuss this equation by posing, following Eigen, the physically significant expressions below, for $k = 1$ to n:

$$W_k = A_kQ_k - D_k$$
$$E_k = A_k - D_k = \text{excess production}$$

$$\overline{E} = \frac{\displaystyle\sum_{k=1}^{n} E_kX_k}{\displaystyle\sum_{k=1}^{n} X_k} = \frac{\Omega_0}{\displaystyle\sum_{k=1}^{n} X_k} = \text{mean production in function of time.}$$

Then:

$$\frac{dX_k}{dt} = W_k X_k + \sum_{l \neq k} \phi_{kl} X_l - \frac{\Omega_0 X_k}{\sum X_k} \quad k = 1, 2, ..., n.$$

or:

$$\frac{dX_k}{dt} = (W_k - \bar{E}(X_1, ..., X_n)) X_k + \sum_{l \neq k} \phi_{kl} X_l \quad k = 1, 2, ..., n \qquad (6.8)$$

This system of non-linear differential equations in X_k shows that:

(i) *If the error term* $\sum_{l \neq k} \phi_{kl} X_l$ *is negligible, then:*

$$\frac{dX_k}{dt} = (W_k - \bar{E})X_k \quad k = 1, 2, ..., n.$$

The discussion is straightforward: if there exists a subscript i such that for $k = i$ we have $W_i - \bar{E} > 0$, then X_i increases. Consequently, following (6.3), some X_j decrease. If we assume that $Q_k = 1 \ \forall k$, then $\dot{X}_k = (E_k - \bar{E})X_k, k = 1, 2, ...$. Since $\bar{E}$ is an average value, for some values of k we will have $E_k < \bar{E}$ and X_k will decrease, while for other values of k we will have $E_k > \bar{E}$ and X_k will increase. $\bar{E}$ varies so that:

$$\dot{\bar{E}} = \frac{\Sigma E_k \dot{X}_k}{\Sigma X_k} = \frac{\Sigma E_k (E_k - \bar{E}) X_k}{\Sigma X_k}$$

$$= \frac{\Sigma E_k^2 X_k}{\Sigma X_k} - \bar{E} \frac{\Sigma E_k X_k}{\Sigma X_k}$$

$$= \bar{E^2} - \bar{E}^2 = \text{Var}(E_k) \geq 0.$$

$\bar{E}$ tends asymptotically towards E_c, the maximum value of E_c corresponding to the privileged species X_c: $\bar{E} \to E_c$, $E_c - \bar{E} > 0$. Thus, when $t \to \infty$ the system X_k tends to a stationary state characterised by a privileged subscript c for which $\bar{E}(X_1^{st}, X_2^{st}, ..., X_n^{st}) \to W_c$. X_c^{st} is then the species selected under the influence of the environment by the condition (6.3). The number of carriers X_c^{st} of this chemical species increases whereas that of the others decreases. We see that W_c which characterises X_c^{st} plays the role of a *selective value*, since $W_c - \bar{E}(X_1^{st}, ..., X_c^{st}, ..., X_n^{st}) > 0$. It depends on the factor of quality Q_c. So we finally have an equilibrium of selection.

(ii) *If the error term* $\Sigma \phi_{kl} X_l$ *is not negligible, Eq. (6.8) will be conserved:*

$$\frac{dX_k}{dt} = (W_k - \bar{E}(X_1, ..., X_n)) X_k + \sum_{l \neq k} \phi_{kl} X_l.$$

The discussion above remains valid here, but the asymptotic stationary state will now be defined by:

$$(W_c - \overline{E}(X_1^{st}, \ldots, X_n^{st})) \, X_c^{st} + \sum_{l \neq c} \phi_{cl} X_l^{st} = 0.$$

Thus:

$$\overline{E}(X_1^{st}, \ldots, X_n^{st}) \rightarrow W_c + \sum_{l \neq c} \phi_{cl} \frac{X_l^{st}}{X_c^{st}}.$$

In short, the evolution of the chemical species X_c is determined by the selective advantage W_c, function of A_c, Q_c and D_c. A small variation in any one of these three parameters will lead to a variation of the selective value W, which in turn will immediately modify the behaviour of the system by a new distribution of the species. *The mechanism of this selection finally leads to the optimum state described by $\overline{E}$ for which the selective value W is maximum.*

Let us recall that this discussion is based on the environmental constraint (6.3), i.e. that the sum of the concentrations of the species remains constant in volume V. The same results will hold good for another type of 'ecological' constraint:

$$\sum_k D_k X_k = \sum_k A_k X_k = \text{constant} \tag{6.9}$$

which expresses the condition of constant flux across volume V.

2. Solution of the system of equations

To solve the system (6.8), let us use the method proposed by Jones *et al.* (1976) and write the equations in the form:

$$\dot{X}_i(t) = \Sigma A_{ik} X_k(t) - \overline{E}(t) X_i(t) \quad \begin{array}{l} i = 1, \ldots, n \\ k = 1, \ldots, n \end{array} \tag{6.10}$$

with:

$$A_{ik} = W_k \delta_{ik} + \phi_{ik}$$

where δ_{ik} is the Kronecker delta and $\phi_{ii} = 0$ $\forall i = 1, \ldots, n$. From the condition $\sum_{i=1}^{n} X_i(t) = c = \text{constant}$, it follows that:

$$\overline{E}(t) = \frac{1}{c} \sum_{i,k} A_{ik} X_k(t).$$

Changing to variables $Y_i(t)$ related to $X_i(t)$ by the expression

$$X_i(t) = \rho(t) \, Y_i(t) \quad i = 1, 2, \ldots, n$$

with

$$\rho(t) = \exp \left(-\int_0^t \overline{E}(\zeta) \, d\zeta \right)$$

leads to the new system in Y_i, which is formally simpler:

$$\dot{Y}_i = \sum_{k=1}^{n} A_{ik} Y_k \tag{6.11}$$

with

$$\bar{E} = \frac{\sum_{i,k=1}^{n} A_{ik} Y_k}{\sum_{k=1}^{n} Y_k}.$$

The condition of conservation

$$\sum_{i=1}^{n} X_i(t) = c$$

becomes

$$\sum_{i} Y_i(t) = \frac{c}{\rho}.$$

The differential equation in $\left(\sum_{i} Y_i \right)$:

$$\sum_{i} \dot{Y}_i(t) = \sum_{i,k} A_{ik} Y_k = \bar{E} \sum_{k} Y_k(t)$$

is easily integrated to give:

$$\sum_{i} Y_i(t) = c_1 \exp\left(\int_{0}^{t} \bar{E}(\zeta') d\zeta' \right)$$

where c_1 is the integration constant which can be calculated by $c = c_1$ and $c = \rho(t) \sum_{i} Y_i(t)$.

The solutions of (6.10) are then given by:

$$X_i(t) = \frac{c Y_i(t)}{\Sigma Y_i(t)}$$

where $\dot{Y}_i(t)$ is a solution of:

$$\dot{Y}_i(t) = \sum_{k} A_{ik} Y_k \quad i = 1, 2, ..., n. \tag{6.12}$$

3. Explicit solutions

If the matrix **A** is supposed to be independent of time, the systems (6.12) are linear and the solutions are given by:

$$X_i(t) = c \frac{\sum_k q_{ik} c_k e^{\lambda_k t}}{\sum_{p,k} q_{pk} c_k e^{\lambda_k t}} \tag{6.13}$$

where q_{ik}, the ith component of the kth eigenvalue, is an element of the matrix $\mathbf{Q}$ of eigenvectors defined by:

$$\sum_k A_{ik} q_{kl} = \lambda_l q_{il}$$

where $\mathbf{Q}_l$ is the eigenvector of elements $q_{1l}, q_{2l}, \ldots, q_{nl}$ associated with the lth eigenvalue λ_l of $\mathbf{A}$ and where c is defined by:

$$c = \sum_k X_k(t) = \text{constant}.$$

The c_k values are integration constants which depend on the initial conditions. For example:

$$c_k = \sum_i (Q^{-1})_{ki} X_i(0)$$

where $\mathbf{Q}^{-1}$ is the inverse matrix of $\mathbf{Q}$ of which the elements are noted q_{kj}. In matrix form (Eq. (6.10)), $\mathbf{A}$ may be written:

$$\mathbf{A} = \mathbf{W} + \mathbf{\Phi}$$

$\mathbf{W}$ being a diagonal matrix. The nature of the solutions related to the eigenvalues λ is specified by the matrix $\mathbf{\Phi}$. Thus, the $X_i(t)$ will have an oscillatory behaviour if $\mathbf{\Phi}$ is asymmetrical; this corresponds to the case where the transitions $E_i \to E_l$ do not occur at the same rate as the transitions $E_l \to E_i$. But if there exists an element ϕ_{ik}, strictly positive, in the matrix $\mathbf{\Phi}$, then there must exist at least one real eigenvalue, non-degenerate and positive, $\lambda_j^{\max}$ corresponding to a subscript j, such that:

$$\lambda_j^{\max} > |\lambda_k| \quad \forall k \neq j.$$

The components of the vector $\mathbf{Q}_j$ associated with the eigenvalue $\lambda_j^{\max}$ are real and positive. This is a very important result since from the expression (6.13) we may deduce that when $t \to \infty$, or at least as soon as $t \gg \dfrac{1}{\lambda_j^{\max}}$, the stationary state of any species i is given by:

$$X_i^{\text{st}} = c \frac{q_{ij}}{\sum_{k=1}^{n} q_{kj}} \quad i = 1, 2, \ldots, n \tag{6.14}$$

These values of X_i^{st}, corresponding to the greatest eigenvalue $\lambda_j^{\max}$, characterise the distribution of the molecular species in the stationary state. In particular, if the errors are few, $\mathbf{Q}$ will be a diagonal matrix, $q_{ij} \approx q_{ii}\delta_{ij}$, and only X_j will be the dominant species:

$$X_j^{st} \simeq \frac{cq_{jj}}{q_{jj}} = c.$$

4. Consequence: selection in molecular systems

It may be shown that the selection consists in a process of optimisation, or an extremum process, by changing the variables $X_i(t) \rightarrow Z_i(t)$ by the matrix of eigenvectors $\mathbf{Q}$:

$$X_i(t) = \sum_k q_{ik} b_k Z_k(t) \text{ with } b_k = \frac{1}{\sum_l q_{lk}}.$$
(6.15)

Then we readily obtain the equation in terms of the variables Z_i:

$$\dot{Z}_i(t) = (\lambda_i - \bar{E}) Z_i(t).$$
(6.16)

If suffices to calculate:

$$\sum_k q_{ik} b_k \dot{Z}_k(t) = \dot{X}_i(t) = \sum_k A_{ik}\left(\sum_l q_{kl} b_l Z_l(t)\right) - \bar{E} \sum_k q_{ik} b_k Z_k$$

$$= \sum_l \left(\sum_k A_{ik} q_{kl}\right) b_l Z_l - \bar{E} \sum_l q_{il} b_l Z_l$$

$$= \sum_l (\lambda_l q_{il}) b_l Z_l - \bar{E} \sum_l q_{il} b_l Z_l.$$

With this transformation we find that Z_i represents the evolution of a linear combination of species X_i. Z_i is called *a chemical 'quasi-species'*.

This equation makes it obvious that a 'quasi-species' Z_i, with an eigenvalue λ_i below the threshold value $\bar{E}$, will die out. A certain number of 'quasi-species' will thus survive. Using the variables Z_k, the mean production $\bar{E}$ may be written:

$$\bar{E} = \frac{1}{c}\sum_{i,k} A_{ik} X_k = \frac{1}{c}\sum_l \lambda_l Z_l$$
(6.17)

c is an invariant:

$$c = \sum_i X_i = \sum_i \sum_k q_{ik} b_k Z_k = \sum_k \left(\sum_i q_{ik}\right) b_k Z_k = \sum_k Z_k.$$
(6.18)

Above, with variables $X_i(t)$, we had $\bar{E}$ in terms of $X_1, X_2, ..., X_n$. The transformation (6.16) gives $\bar{E}$ as a linear combination of the $Z_i(t)$. From (6.17) we may deduce that:

$$\frac{\dot{Z}_i(t)}{Z_i(t)} = \lambda_i - \frac{1}{c}\sum_k \lambda_k Z_k \tag{6.19}$$

or:

$$\frac{d}{dt}\ln Z_i(t) = \lambda_i - \bar{E} \quad i = 1, 2, \ldots, n$$

which is a system of differential equations coupled by the second term $\bar{E}$. It is clear that if λ_i is greater than the mean productivity $\bar{E}$, then $Z_i(t)$ will increase, while if λ_i is smaller than $\bar{E}$, then $Z_i(t)$ will decrease. By integrating, we find:

$$\ln Z_i(t) = \lambda_i t - \int_0^t \bar{E}(\zeta)d\zeta + \ln Z_i(0)$$

where:

$$Z_i(t) = Z_i(0)\rho(t)e^{\lambda_i t}. \tag{6.20}$$

According to (6.19), a variation of Z_i implies a concomitant variation of Z_j and thus of $\bar{E}$. By inversing the relation of definition (6.15), we obtain:

$$Z_i(t) = \frac{1}{b_i}\sum_p (Q^{-1})_{ip} X_p(t)$$

with $\mathbf{Q}^{-1}$ representing the inverse matrix of $\mathbf{Q}$ with the elements q_{ik}, $i = 1$ to n, $k = 1$ to n.

With X_p obtained according to (6.13), we may write:

$$Z_i(t) = \frac{c}{b_i}\sum_p \left[(Q^{-1})_{ip}\sum_j q_{pj}c_j e^{\lambda_j t}\right] \bigg/ \sum_{p,l}q_{pl}c_l e^{\lambda_l t}$$

$$= \frac{c}{b_i}\sum_j \delta_{ij}c_j e^{\lambda_j t}\bigg/ \sum_{p,l}q_{pl}c_l e^{\lambda_l t}$$

$$= \frac{c}{b_i}\frac{c_i e^{\lambda_i t}}{\sum_{p,l}q_{pl}c_l e^{\lambda_l t}}$$

and using (6.17) we have:

$$\bar{E}(t) = \frac{1}{c}\sum_i \lambda_i Z_i = \frac{\sum_i \lambda_i \frac{c_i}{b_i}e^{\lambda_i t}}{\sum_{p,l}q_{pl}c_l e^{\lambda_l t}}. \tag{6.21}$$

Suppose there exists a j such that $\lambda_j t \gg 1$. Then, at the limit, we would have:

$$\bar{E}(t) \simeq \frac{\lambda_j \frac{c_j}{b_j}e^{\lambda_j t}}{\sum_l \left(\sum_p q_{pl}\right)c_l e^{\lambda_l t}} \simeq \frac{c_j}{b_j}\frac{\lambda_j e^{\lambda_j t}}{\left(\sum_p q_{pj}\right)c_j e^{\lambda_j t}} = \lambda_j$$

so that:

$$\lim_{t \to \infty} \overline{E}(t) = \lambda_j. \tag{6.22}$$

The process of selection is thus expressed by an optimisation of $\overline{E}$(t) with time until the mean productivity reaches the eigenvalue λ_j, i.e. it compensates for the mutations in the transitions i → j. This characterises the evolution of the system. Only one quasi-species will thus emerge from the primitive set of species. This quasi-species represents *an organised and selected combination of species* which may be called the 'wild type' of the population.

II.   Evolution with complementary instruction. The case of DNA or RNA replication

We have just considered the case of instructor systems, based on the copying of a sequence of symbols with the possibility of copying errors, from which we have deduced the *obligatory* phenomenon of the selection of a species corresponding to the greatest eigenvalue of the matrix **A**. This essentially mathematical result expresses the fact that the selected species has the greatest selective value, which appears evident. However, from the formal point of view, it is interesting to demonstrate a principle of optimisation, that of the mean productivity $\overline{E}$, in relation to the existence of a chemical *quasi-species*. The question which then naturally arises is whether this result holds good in the more complex case of *complementary instruction*. This refers to the possibility of self-instruction of the systems through reciprocal copying. Such is the case for the nucleic acids DNA and RNA. The rate equations should then describe a cyclic graph as in Fig. 6.2a where the chemical species k is represented by its two complementary forms (+k) and (−k), one form being obtained from the other by copying. In reality, the chemical species k may exist in one of the three forms α: as a single strand, called (+) for $\alpha = 1$, in the complementary form (−) for $\alpha = -1$, and in the complex form with two strands (+, −) for $\alpha = 0$, the division of which again produces the strands (+) and (−). The modified cyclic graph is shown in Fig. 6.2b. As the global result corresponds to the chemical reaction $S \xrightarrow{I} I$ starting from a substrate S, let us, for the sake of simplification, represent the self-replicating unit in Fig. 6.2b by the symbol I.

We may now write the following equations:

(i) The equation for the production of a (+) or a (−) strand by the division of a complex form (+, −) with, as in Eq. (6.5), the possibility of destruction described by D_k^α and the possibility of dilution described by $\overline{E}(t)$. The copying errors are contained in the term $\sum_\sigma \sum_{l \neq k} \phi_{kl}^{\alpha\sigma} X_l^\sigma$, where $\phi_{kl}^{\alpha\sigma}$ is the rate of production in a transition $E_l \to E_k$, in the form $\sigma \to \alpha$. The Greek superscripts represent the form (+), (−) or (+, −) of the chemical species. So we have:

$$\frac{dX_k^\alpha}{dt} = A_k^0 X_k^0 - D_k^\alpha X_k^\alpha - \overline{E}(t) X_k^\alpha + \sum_\sigma \sum_{l \neq k} \phi_{kl}^{\alpha\sigma} X_l^\sigma \tag{6.23(1)}$$

$$\alpha = \pm 1.$$

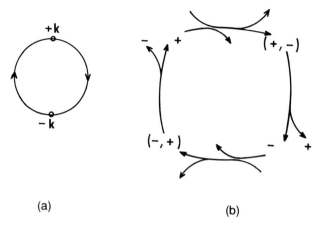

(a)   (b)

Fig. 6.2. Replication of nucleic acids (after Eigen, 1971): (a) complementary instruction between the two forms $+k$ and $-k$; (b) cyclic graph corresponding to the biosynthesis: $(+, -)$ is the two-strand complex, which on 'dividing' gives the strands $(+)$ and $(-)$.

(ii) The equation for the production of a complex $(+, -)$ starting from a single strand $(+)$ or $(-)$, with the rate of destruction D_k^0 and the same rate of dilution $\bar{E}(t)$ as above. With the same notation as in (6.5), we obtain:

$$\frac{dX_k^0}{dt} = A_k^{-1}Q_k^{-1}X_k^{-1} + A_k^{+1}Q_k^{+1}X_k^{+1} - (\bar{E}(t) + D_k^0)X_k^0$$

$$+ \sum_\sigma \sum_{l \neq k} \phi_{kl}^{\alpha\sigma} X_l^\sigma. \tag{6.23(2)}$$

Similarly, the environmental constant may be written:

$$\sum_{\sigma=0, +1, -1} \sum_k X_k^\sigma = \text{constant} = c$$

and the initial hypothesis (iii), of non-instructed formation being completely negligible, now becomes:

$$\sum_{\sigma=\pm1} \sum_k A_k^\sigma (1 - Q_k^\sigma) X_k^\sigma = \sum_{\substack{\sigma, \alpha, l \\ l \neq k}} \phi_{kl}^{\alpha\sigma} X_l^\sigma.$$

We then deduce an analogous condition for the adjustment of the dilution fluxes:

$$\frac{d}{dt}\left(\sum_\sigma \sum_k X_k^\sigma\right) = \sum_k \left(\sum_\sigma \frac{d}{dt} X_k^\sigma\right)$$

$$= \sum_k A_k^0 X_k^0 + \sum_k \sum_{\sigma=\pm1} A_k^\sigma Q_k^\sigma X_k^\sigma - \sum_k \sum_{\sigma=0, \pm1} (\bar{E}(t) + D_k^\sigma) X_k^\sigma + \sum_{\sigma, \alpha} \sum_{k, l \neq k} \phi_{kl}^{\sigma\alpha} X_l^\sigma$$

$$= 0$$

so that:

$$\bar{E}(t)\sum_{k,\sigma} X_k^{\sigma}(t) = \sum_{k,\sigma}(A_k^{\sigma} - D_k^{\sigma})\, X_k^{\sigma}(t) = \sum_{k,\sigma} E_k^{\sigma} X_k^{\sigma}(t)$$

and:

$$\bar{E}(t) = \frac{1}{c}\sum_{k,\sigma} E_k^{\sigma} X_k^{\sigma}(t). \tag{6.24}$$

The rate equations (6.23(1)) and (6.23(2)) may be written in matrix form, taking into account the chemical species k, in one of the three states α, represented by the three components of the vector $\mathbf{X}_k$:

$$\frac{d}{dt}X_k^{\alpha} = \sum_{\sigma,\,l}\mathcal{A}_{kl}^{\alpha\sigma}X_l^{\sigma} - \bar{E}(t)X_k^{\alpha}. \tag{6.25}$$

Thus, with notations analogous to those in (6.10), we find:

$$\mathcal{A}_{kl}^{\alpha\sigma} = W_{\alpha\sigma}^{k}\delta_{kl} + \phi_{kl}^{\alpha\sigma} \tag{6.26}$$

where the matrix $\mathbf{W}^k$ is defined by:

$$\mathbf{W}^k = \begin{bmatrix} -D_k^{-1} & A_k^0 & 0 \\ A_k^{-1}Q_k^{-1} & -D_k^0 & A_k^{+1}Q_k^{+1} \\ 0 & A_k^0 & -D_k^{+1} \end{bmatrix}$$

and $\phi_{kk}^{\alpha\sigma} = 0$.

By systematically using the superscripts to characterise the submatrices of $\mathcal{A}$ and the subscripts to identify the rank of the submatrices in $\mathcal{A}$, we see that the equation has a remarkable structure:

$$\frac{d}{dt}\begin{bmatrix} X_1^{-1} \\ X_1^{0} \\ X_1^{+1} \\ \vdots \\ X_k^{-1} \\ X_k^{0} \\ X_k^{+1} \\ \vdots \end{bmatrix} = \left(\begin{bmatrix} \mathbf{W}^1 & & \\ & \ddots & \\ & & \mathbf{W}^k \\ & & & \ddots \end{bmatrix} - \bar{E}\right)\begin{bmatrix} X_1^{-1} \\ X_1^{0} \\ X_1^{+1} \\ \vdots \\ X_k^{-1} \\ X_k^{0} \\ X_k^{+1} \\ \vdots \end{bmatrix}. \tag{6.27}$$

The structure shows that the right-hand member may be written in the form of a matrix of order $3N$, which itself is composed of submatrices $\mathbf{W}$ of order 3 along the diagonal, all the other terms being zero if the error term of the rate ϕ is neglected. Thus, even in the case of complementary instruction, it is possible to obtain an equation formally identical to (6.10), although explicitly more complex since it represents a matrix equation of order 3 for the chemical species k. The principal results already obtained may be immediately deduced from this equation. Thus, the result (6.12) may be expressed as follows:

The solutions of (6.27) are:

$$X_k^\alpha(t) = c \frac{Y_k^\alpha(t)}{\sum_\alpha \sum_k Y_k^\alpha(t)}$$

where Y_k^α is the solution of the equation analogous to (6.11):

$$\frac{d}{dt}Y_k^\alpha = \sum_{\sigma,l} A_{kl}^{\alpha\sigma} Y_l^\sigma.$$

From this we deduce the solution in X (Eq. (6.13)) where the λ_l^α are the eigenvalues of the matrix A and the Q_l^σ are the eigenvectors associated with the elements $q_{kl}^{\alpha\sigma}$:

$$X_k^\alpha(t) = c \frac{\sum_l \left(\sum_\sigma q_{kl}^{\alpha\sigma} c_l^\sigma e^{\lambda_l^\sigma t} \right)}{\sum_{\alpha,\sigma k,l} q_{kl}^{\alpha\sigma} c_l^\sigma e^{\lambda_l^\sigma t}}. \tag{6.28}$$

We obtain the expected result, i.e. the solution at a given t for the set of the species is a combination of solutions, each being obtained in the subspace restricted to three dimensions. In accordance with the results previously obtained, we see that there exists a selection among the various molecular triplets.

More precisely, it can easily be shown that if the error terms ϕ are negligible, there always exists for a given *collective k* (submatrix $\mathbf{W}^k$) one eigenvalue with a positive real part and two eigenvalues with a negative real part. When the error terms become large, certain eigenvalues may become complex, thereby indicating an oscillatory behaviour.

Then, for a *given chemical species k*, (i) the positive eigenvalue corresponds to an autocatalytic growth process of macromolecular triplets, this being the normal mode leading to the selection of the collective k; and (ii) the negative eigenvalues describe a relaxation process, with equilibrium between the forms $(+)$, $(-)$ and $(+, -)$. And for the *set of the species*, if $\lambda_j^{v,max}$ is the eigenvalue with the greater positive real part, at the end of a fairly long time: $t \gg \dfrac{1}{\lambda_j^{v,max}}$ then the solution (6.28) tends to:

$$\boxed{X_k^{v,st} = c \frac{q_{kj}^{v\sigma}}{\sum_{\sigma,k} q_{kj}^{\sigma v}} \qquad \begin{array}{l} \sigma = -1, 0, +1 \\ k = 1, 2, ..., n. \end{array}} \tag{6.29}$$

It follows that the distribution of the selected macromolecules in the stationary state corresponds to the maximum value $\lambda_j^{v,max}$. This parameter represents the selective value, the selection being characterised by the property of optimisation (6.30):

$$\boxed{\lim_{t \to \infty} \overline{E}(t) = \lambda_j^{v,\max}} \tag{6.30}$$

i.e. the selective value is the limit value of the mean productivity in time.

Let us sum up the results obtained. *If in a given population of information carriers (nucleotides in the present case) there is a great heterogeneity of composition, for example with the abundance of a single chemical type, there must always exist two different carriers so that, in the stationary state, i.e. in the long term, the species will be selected in practically equal quantities.* The possibility of mutation due to recopying errors will lead to macromolecular sequences composed of various chemical types (see Eq. (6.11)). This is what Eigen considers a prerequisite to the generation of any code, since such sequences contain information of much greater value. Thus, as Atlan has shown, order is created from noise. However, we may wonder whether nucleic acids are capable of molecular evolution without catalytic aid, i.e. whether they can organise themselves for reproduction. In fact, even the selected sequences have a low information content. As we shall see later, when such sequences participate in a self-instructed catalytic cycle (a catalytic hypercycle), they are capable of generating a function.

III.   Protein biosynthesis: self-organising enzymic cycles

1.   *Catalytic protein cycles*

We know that a set of biochemical reactions can be formally represented by a network (see Katchalsky's thermodynamic network theory in Oster *et al.*, 1973). Thus, certain enzymes, themselves the result of enzymic reactions, lead to the formation of polypeptide chains which in turn act as enzymes. We also know that their action is selective, multifunctional and catalytically controlled or capable of exercising catalytic control (Chapter 2). The problem that now arises concerns the conditions under which such a catalytic network will be *self-reproducing*, since the reproduction of a given enzyme of a network requires a certain number of other enzymes, and *selective*, which is a necessary condition because of the competition between species. Eigen (1971) and Eigen and Schuster (1979) have shown that two conditions will need to be fulfilled:

(i)   the network must contain one closed loop of reactions; and
(ii)   the coupling of the reactions must be catalytic.

Figure 6.3 represents a catalytic cycle satisfying these conditions. We may identify the functional classes E_i, $i = 1$ to m, with the enzymes noted in the same way. The enzyme E_i formed during the $(i-1)$th reaction serves as catalyser for the ith reaction (condition (ii)), and E_m retroacts on E_1 (condition (i)). Having posed these conditions, we may now use the same method as before to determine whether the catalytic cycles are self-reproducing and selective.

If X_k^i ($1 \le k \le n$) is the concentration of the enzyme E_i ($1 \le i \le m$) in the m-reaction cycle, we may write the system of phenomenological equations, neglecting the error terms, as follows:

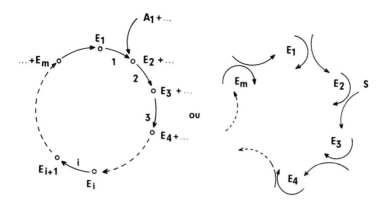

Fig. 6.3. Self-reproducing hypercycle or catalytic cycle with self-reproducing units: E_i catalysis of the reaction ending in E_{i+1}. The set for $i = 1$ to m is itself cyclic since it is closed (after Eigen, 1971).

$$\frac{dX_k^1}{dt} = A_1 Q_1 X_k^m - D_1 X_k^1$$

$$\frac{dX_k^2}{dt} = A_2 Q_2 X_k^1 - D_2 X_k^2$$

$$\vdots$$

$$\frac{dX_k^m}{dt} = A_m Q_m X_k^{m-1} - D_m X_k^m$$

(6.31)

or in matrix form as:

$$\frac{dX_k}{dt} = W^k X_k$$

with:

$$W^k = \begin{bmatrix} -D_1 & 0 & 0 & \dots & A_1 Q_1 \\ A_2 Q_2 & -D_2 & 0 & \dots & 0 \\ 0 & A_3 Q_3 & -D_3 & \dots & 0 \\ \vdots & \vdots & \vdots & & \vdots \\ 0 & 0 & 0 & \dots & A_m Q_m - D_m \end{bmatrix} \quad X_k = \begin{bmatrix} X_k^1 \\ X_k^2 \\ X_k^3 \\ \vdots \\ X_k^m \end{bmatrix}.$$

The terms A, Q and D depend on the subscript k as they are characteristic of the cycle.

Once again, we find one positive eigenvalue and $(m - 1)$ negative eigenvalues, so that the same conclusions may be drawn as for the system with complementary instruction:

- *The normal mode of the positive eigenvalue corresponds to an autocatalytic growth of the collective, and thus of the cycle as a whole;*

- *The other* (m − 1) *normal modes characterise the relaxation process within the cycle;*
- *For a set of* n *competing catalytic cycles, that with the greatest eigenvalue is selected: this then is the selective value of the cycle. If* $\lambda_j^{\mu,\,max}$ *is this eigenvalue corresponding to the normal mode* μ *in the cycle* j, *we have:*

$$\lim_{t \to \infty} \overline{E}(t) = \lambda_j^{\mu,\,max} \qquad (6.32)$$

which is characteristic of the optimisation process.

From the mathematical point of view, the evolution of the system is given by the following matrix equation of the order $N = mn$ with m being the number of reactions per cycle and n the number of cycles:

$$\frac{d\mathbf{X}}{dt} = \mathcal{W}\mathbf{X} \qquad (6.33)$$

with:

$$\mathcal{W} = \begin{bmatrix} \mathbf{W}_{11} & & & \mathbf{W}_{ik} & \\ & \mathbf{W}_{22} & & & \\ \hline & & \mathbf{W}_{kk} & & \\ & & & \mathbf{W}_{NN} & \end{bmatrix}$$

where the general term $\mathbf{W}_{ik}$ indicates the dependence or the independence of the cycles according as its value is zero or non-zero. Thus, Eigen has demonstrated a characteristic property of these systems: *if the selected cycle is coupled to another cycle by a lateral branch, then it will carry the set of these couplings all through its evolution.* This 'apparatus' could be a drawback for the selected cycle as it may reduce its selective value, i.e. decrease $\lambda_j^{\mu,\,max}$ in favour of some other cycle. A lateral branch of this type is called a 'parasitic' branch.

We first considered *self-instruction* and then *complementary instruction*, which is an inherent property of nucleic acids; we have seen that *catalytic protein cycles* also possess the same property. This therefore appears to be an essential characteristic of molecular behaviour in living organisms. The question that now arises is whether such cycles are self-organising.

Thus, let us suppose that in a given network each point has the same probability p of being the target of the catalytic activity of another point. The probability for a continuous loop of m reactions is then p^m, and that of a closed cycle of m reactions among n is thus: $\dfrac{n!}{m(n-m)!}p^m$. However, in fact the probabilities p are necessarily different since the probability of coupling depends essentially on the selective values.

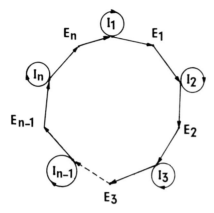

Fig. 6.4. Schematic representation of the cycle in Fig. 6.3. The cycle is made up of I_i self-instructing units noted I, and is itself cyclic and noted I (after Eigen, 1971).

Nevertheless, with this approximation we can show that, in large populations, the probability of the formation of a closed loop is almost certain. But, as we have just seen, parasitic branches are a handicap in the selection of self-reproducing loops. Moreover, the noise, i.e. the copying of errors by mutation, which generates selective advantages, does not allow the formation of a 'mutant cycle' capable of better performance. Indeed, this would imply the *simultaneous* mutation of the *m* enzymes forming the system. The probability of this event is of course very small, and the possibility of utilising such selective advantages is thus greatly reduced.

Finally, it appears that proteins do not automatically reproduce their mutants, in spite of the advantages these may offer, such as strong selective recognition due to tertiary structure or greater information capacity because of the length of the chain. Consequently, in a given population of macromolecular species, there is nothing to ensure an automatically favourable selection of species capable of forming a self-reproducing cycle. Eigen has demonstrated that *the basis of functional order is the association, or rather the cooperation, between the complementary instruction of nucleic acids and the self-reproduction of catalytic cycles.* The corresponding theoretical model, called the *self-reproducing hypercycle*, is discussed below.

2.   *Self-reproducing hypercycle*

The self-reproducing hypercycle is defined as a catalytic cycle of self-reproducing units. With the usual notation, such a cycle may be represented as in Fig. 6.3. A catalytic hypercycle is thus a cyclic network of reactions, which may themselves be cyclic or autocatalytic, i.e. self-replicating. The phenomenon of translation may for example be represented by a hypercycle of this kind. Here, each self-replicating unit, say I_i (Fig. 6.4), is capable of intrinsic self-instruction (level 1) and acts on another unit I_{i+1} by means of an enzyme E_i obtained by translation (level 2). Thus, we have here a second-order hypercycle. If other couplings, such as those due to metabolic

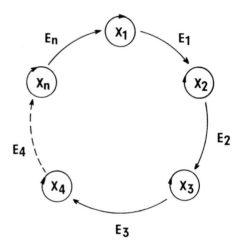

Fig. 6.5. Schematic representation of a hypercycle composed of species I_i in concentrations X_i and enzymes E_i in concentrations Y_i (after Eigen, 1971).

processes, are taken into account, a hypercycle of a higher order will be obtained. The first-order cycle is represented by the symbol ⓘ (Fig. 6.2) and the second-order hypercycle by ⓘ (Fig. 6.4).

The mathematical study of a hypercycle is evidently complicated by the fact it introduces chemical species simultaneously coupled in first-order cycles as well as in second-order hypercycles. Without going into detail, let us consider the hypercycle composed of species I_i in concentrations X_i, and enzymes E_i in concentrations Y_i, as represented in Fig. 6.5.

The rate of growth of the carriers I_i (the genetic code) is a function of the concentrations of E_{i-1} and I_i. But in this case we cannot consider that this rate is simply proportional to $Y_{i-1} X_i$, since the concentration of the compounds $E_{i-1} - I_i$, noted c_i, is not negligible. The rate may be calculated in function of Y_{i-1} and X_i. Eigen has shown that the evolution of the hypercycle may, in a simplified way, be represented by the system of coupled equations for $i = 1, 2, \ldots n$:

$$\left.\begin{aligned}
\frac{dX_i}{dt} &= (A_i Q_i - D_i) X_i + A_i' Q_i' c_i \\
\frac{dY_i}{dt} &= A_i'' Q_i'' X_i - D_i'' Y_i
\end{aligned}\right\} . \tag{6.34}$$

Finally, using the same method as before, we obtain the behaviour of the system of order $2nN$ in a population of N hypercycles, each being represented by the evolution of the preceding system of $2n$ equations, in the form:

$$\frac{d\mathbf{X}}{dt} = \mathbf{\mathcal{V}} \mathbf{X} \tag{6.35}$$

where $\mathbf{\mathcal{V}}$ is a matrix of order $2nN$.

Thus, the diagonalisation of the matrix $\mathcal{V}$ leads to the maximum selective value in function of the parameters of the system. The constraints of selection being given, we find that competition between hypercycles always results in the survival of only one hypercycle and the decay of all the others. This phenomenon of extremely severe selection may explain the two properties of the genetic code that are usually considered essential: *unicity* and *optimisation*:

- Let us suppose that in a population of hypercycles, with a chosen code for the reproduction of its functional properties, there exist imposed environmental conditions: then *only one cycle will emerge from the set* in the stationary state so that the code 'selected' at the beginning will remain unique. Of course, this property does not reveal the identity of the first code chosen.

- From the moment a given configuration is selected, *the evolution of the system will continue towards an optimisation of its functions.* Thus, contrary to first-order cycles, a hypercycle may make use of selective advantages so as to reproduce them by imitation. It will then undergo very rapid evolution. This means that a hypercyclic system will be selected to the detriment of independent competitors.

All the above properties are in favour of the hypercycle being the *principle of natural self-organisation.* Jejedor *et al.* (1988) and Galar (1989) have used other mathematical approaches to study the influence of Eigen's hypercycle. We shall try to show how these different theories, the thermodynamics of irreversible processes, the principle of order arising from noise, and the catalytic self-reproducing hypercycle, all contribute more to an interpretation than to a possible explanation of the evolution of the species. In this chapter, the reader will have recognised the methods of population dynamics: the introduction of different selective values leading mathematically to the existence of solutions that might have been predicted from the outset, although perhaps less easily in the case of complementary instruction or in that of the catalytic self-reproducing hypercycle. *However, we must admit that the explanation of the evolution of the species still seems a long way off. In particular, as we have already mentioned, we do not yet know how a regulated and controlled system could possibly undergo a process of evolution within its environment.*

7

Evolution and Physiology

I. Evolution and self-organisation of molecular systems

1. *Introduction*

The problem of evolution may be summed up in a single phrase: *how are we to reconcile the variety and the diversity of the living world with the molecular unity and the universality of the genetic code observed in all forms of life?*

Ever since the work of Darwin, Wallace and others, evolution has been considered to be the result of an enormous set of convergent phenomena, the essential elements being differential reproduction and successive mutations in a given environment. The basic principle of evolution, called Darwin's *principle of natural selection*, which corresponds to the survival of the fittest, was established long before the molecular mechanisms of heredity came to be known. Thus, without taking into account the events occurring at the subcellular level, the principle was postulated as being *the* connecting strand all through the immense variety of living organisms. Now we may adopt one of two positions: either we admit the definitive metaphysical explanation involving the occurrence of some 'extraordinary' random event breaking the cycle (nucleic acids $\rightarrow$ proteins $\rightarrow$ nucleic acids), which may be summed up by the expression 'origin of life', or we look for a possible explanation at the lowest level of the existence of the organism so as to deduce an essential principle valid for all forms of life at all the different levels, i.e. *a principle of optimisation of the evolutionary process*, analogous to physical laws such as the principle of least action, the conservation of energy, the conservation of momentum, and so on.

Does such a unique principle exist and, if so, on what is it based?

We first have to discuss *the* biological principle known as the principle of natural selection. If the principle is true, it would have to be a first principle, i.e. an axiom from which experimentally verifiable laws may be deduced. Now, it is well known from population genetics that natural selection is in fact a consequence of certain properties of species confronted with a given environment. The properties in question are none other than those previously described: metabolism, self-reproduction and mutagenesis. Indeed, it can be mathematically demonstrated, by the laws of population genetics, that these properties lead to a behaviour of selection and of evolution in a Mendelian population (Lewontin, 1967). However, as we saw in the concluding section of Chapter 6, the coefficients of selection (or the selective values) necessarily lead to a selective process.

But an even more suprising finding has come to light. In a remarkable synthesis from the standpoint of a physicist, Haken has shown that this behaviour is not limited to the living world but may be a characteristic of the material world as well. Thus, a theoretical study has demonstrated that the different modes of laser emission obey a 'principle of natural selection' (Haken, 1978). With his unifying approach, Haken has constructed the theory of synergetics.

2. *The three phases of evolution*

The evolution of the species is classically considered to consist of three phases which, at the 'beginning', are difficult to separate: (i) a prebiotic chemical phase; (ii) a phase of self-organisation and replication of existing entities; and (iii) the evolution of formed species.

The *first phase* is generally believed to have occurred in a prebiotic 'soup', a molecular chaos, i.e. a totally unorganised state. Let us recall the famous experiment in which Miller (1955) succeeded in synthesising amino acids under conditions thought to be identical to those prevailing in the early years of the Earth's existence. This first phase, which corresponds to the formation of chemical species necessary to living organisms, raises questions which are partly answered by the thermodynamics of irreversible processes. However, it is evident that this condition is necessary but not sufficient. We may wonder how matter managed to organise itself, even at the most elementary level, with characteristics fundamentally similar to those of living organisms:

- *metabolism* or the synthesis and degradation of molecular species;
- *self-reproduction* or the property of 'invariant' reproduction; and
- *mutagenesis* or the possibility of the creation of errors — and even the regulation of errors — during self-reproduction.

The *second phase* of evolution is an obligatory phenomenon. Indeed, a macro-molecular assembly thus organised could certainly not have been produced by chance. Although a small number of random events must have occurred at the very beginning, such as the formation of favourable chemical species or the diffusion and the local mixing of the species, and so on, we shall see that here also irreversible thermodynamic phenomena may have been involved. Self-organisation of molecular species must necessarily have occurred. But how? Here Eigen's theory provides an answer to the well-known problem: which came first, the chicken or the egg? The nucleic acids necessary for the formation of enzymes or the enzymes necessary for the synthesis of nucleic acids?

The random assembly of chemical species could only have become organised through the self-instructed formation of some of the species followed by their selection on account of certain specific properties or advantages (Chapter 6). But how are we to describe the 'degree' of evolution? From the quantitative point of view, we need a value characterising the degree of self-organisation of the selected system in terms of functional order, i.e. an order capable of realising a useful function having a certain 'value'. In this sense, the problem is to know if it is possible to describe the evolution of the species by certain molecular parameters representing the self-organisation, to determine the dynamics of the self-organisation and to define the conditions of favourable evolution, i.e. the conditions under which the selected system increases its functional capacity and thus its functional value.

The *third phase* of evolution is certainly the best understood thanks to the work of Lamarck and Darwin. Table 7.1 shows the main differences between Lamarckism and Darwinism with, in the latter case, the four constituent theories, in particular those of gradualism and natural selection, according to Mayr (1974). Let us recall that Darwinism is based on a certain number of postulates which have encountered some criticism. Putting it simply, Darwin supposes: (i) the existence of a variation in the properties of species to be necessary for the emergence of new species; (ii) the reproduction of some of these properties; (iii) competitive conditions in which certain species are selected; and (iv) constant environmental resources. Darwin then deduces the variation of the characteristic properties of species and their selection, leading to the evolution of the species. *Evolution is thus a result of two sequential phenomena: first variation, and then selection.* With these hypotheses, could two different populations possibly emerge? To avoid the competition, Darwin was led to suppose the isolation of certain species. Here we again find the notion of quasispecies described in phenomenological theory by Jones *et al.* (1976) (Chapter 6).

To go into further details, it will be necessary to specify what is meant by variation. In fact, variation is produced by two types of mutation:

(1)   The *micromutations*, which are frequent, analogous to the 'noise' we have mentioned, acting randomly on individuals of the population in given environmental conditions. This is called the *pressure of selection*: the selection operating until the individuals are totally adapted to their environment. A stationary state will be reached and this will remain unchanged as long as the environment is not modified. This may be summed up by saying that the environment is the conducting force of evolution. *There is total adaptation to the environment and no extinction.*

(2)   The *macromutations* which occur in a random manner: here also the influence of the environment is fundamental since each individual undergoing a

Table 7.1.   Differences between Lamarckism and Darwinism and their implications

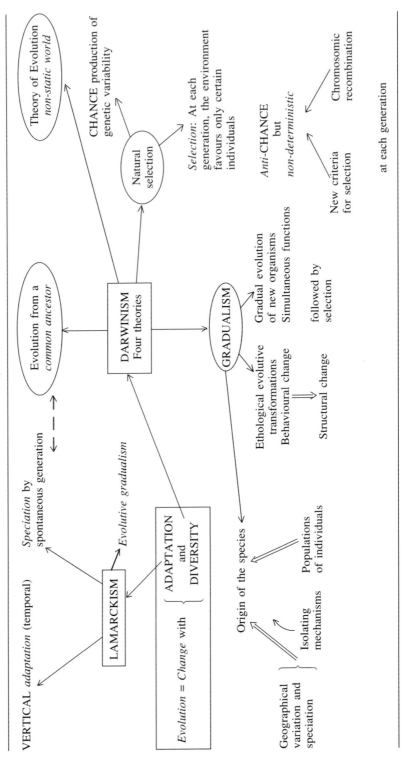

macromutation will have to adapt to the environment or else become extinct. There are thus two possibilities:

- either the macromutation produces an advantage which does not allow isolation so that the individual will compete with its predecessors, whence a high rate of selection;
- or the macromutation allows the individual to live isolated in a geographical niche.

In these two cases, the selection is a consequence of mutations but not of the environment. This is called the *pressure of mutation*: at each stage of evolution, the least adapted species becomes extinct. Finally, *with respect to the environment there is compatibility but not adaptation so that there is progressive direction of evolution with a great number of extinctions.*

These two schools of thought are completely opposed, both in terms of hypotheses and conclusions. Neo-Darwinism is a theory based on the Mendelian laws of heredity: supporters of micromutations claim that the conservation of information acquired by preceding generations explains the fact that an event leading to environmental adaptation appears at exactly the right moment. Mendel's laws thus support the theory of micromutations, since the major defect of this theory lies in the fact that a micromutation must appear simultaneously with a modification of environment: it is difficult to admit implicitly that all the micromutations are permanently available. Indeed, this is one of the essential criticisms of Darwinian theories. *Thus, the problem is to find out whether it is the mutation or the selection that is the creative agent of evolution.*

Experimental proof is likely to be difficult or perhaps even impossible to obtain. In any case, direct observation of the phenomena of evolution would require thousands of years. However, progress in developmental biology, in morphogenesis as well as in cell differentiation, may be expected to provide further information. In the present state of knowledge, the basic postulates of the theories remain controversial, making an objective choice of the creative agent in evolution rather difficult. In particular, we may mention the criticism levelled by Lewontin (1967), a population geneticist, against the axioms of the mathematical theory, and that of Grassé (1973), which we may sum up as follows. (i) How can we explain the *regressive evolution* of certain organs, compatible with mutations that are in general subtractive, while evolution is progressive? (ii) How do Darwinians reconcile the idea of the *survival of the fittest* with the tendency towards gigantism that led to the extinction of species, now fossilised, such as dinosaurs? (iii) Why do certain characteristics change *discontinuously* (successive doublings), in a manner which is incompatible with the neo-Darwinian theory of micromutations? (iv) How do we explain *orthogenesis* and *speciation*? Can we reason merely in terms of the barriers of reproduction due to geographical niches and to the accumulation of numerous small hereditary variations while knowing that physiological equilibrium will require to be maintained in spite of evolutionary jumps? Such then are the questions that arise, some remaining unanswered, and to which we shall return in the following sections, particularly with regard to the special relations between evolution and physiology.

The brief outline above is intended to present an approach to the current problems concerning the bases of the theory of evolution. In fact, at the molecular level, Eigen's axioms are derived from the theory of micromutations.

3. *Darwinian systems*

A system that obeys Darwin's principle of natural selection is called a *Darwinian* system. The conditions for the existence of such a system, at the macroscopic level, are fairly well known. At the molecular level, Eigen has indicated the properties of matter necessary for a system to be capable of selection and of evolution. Eigen describes the generating mechanism of evolution as follows.

In a biological system, certain structures must be stabilised at the expense of others according to the distribution of species present at a given instant. This means that stability is essentially a function of the other structures in competition and depends on the selective advantage, i.e. on the *value* acquired by a mutant during evolution. Moreover, the system must be capable of amplifying the advantage in its own favour. This type of procedure is characteristic of self-organisation as defined by Haken: the organisation that develops in the absence of forces external to the system.

Let us now state the properties of matter which appear to be necessary to a Darwinian system.

(a) *Metabolism* can only be carried out far from equilibrium. Consequently, according to the laws of thermodynamics, there must be regular compensation for the production of entropy. Living matter must escape from disorder, i.e. it must create or maintain order. To do this, it must consume energy drawn from the environment. While this is easy to understand, it is much more difficult to characterise the type of order mentioned previously. From the thermodynamic point of view, it is a structural order created far from equilibrium in a stationary state. This type of order is surely necessary *but quite insufficient, since the living world is essentially based on functional order.* As we have seen, the thermodynamics of irreversible processes show that chemical reactions of the autocatalytic type with mass diffusion could lead to spatial or spatiotemporal structures (dissipative structures). We shall see how important these are during organismal growth, particularly during morphogenesis.

(b) *Self-reproduction* is an essential characteristic since the growing molecular structure must conserve the information previously stored: this may be a selective advantage acquired by mutation or else a parasitic branch which will turn out to be a handicap during the evolution to come. We have seen the advantages of a structure capable of instructing its own synthesis.

(c) *Mutations* result from errors of replication and constitute the essential source of adaptation to the environment, a prerequisite to evolution. Indeed, copying errors constitute the principal source of evolution, which is a molecular way of expressing the principle of order arising from noise as described by information theory. It has been shown that the rate of mutation must satisfy a threshold condition so that the information accumulated during evolution be conserved.

A chemical species will not be able to participate in evolution unless it satisfies these three properties, which will have to be carried over from one generation to another. But how can we explain the existence, in the primitive molecular chaos, of what is generally known as the central dogma of molecular biology, i.e. the universal genetic code? The origin of protein synthesis clearly poses a very difficult problem. We have to explain how the primitive synthesis of proteins directed by a substrate of nucleic acids was able to evolve step by step towards a genetic code, the expression of which requires complex processes involving activator enzymes, transfer RNAs, ribosomes and so on.

Recently, Fukuchi and Otsuka (1992) have suggested that metabolic pathways may have been developed by the chance assembly of enzyme proteins generated from the sense and anti-sense strands of pre-existing genes. The authors cite the fermentation pathways and the pentose phosphate cycle driven by proteins synthesised by the genes of enzymes in the glycolytic pathway.

II.   A coherent interpretation of evolution

We have just seen that evolution may be studied from different points of view according to the phase we may be interested in. Moreover, information theory, which makes no hypothesis concerning the mode of functioning of the system described, appears to be clearly opposed to Eigen's theory of hypercycles. It must be admitted that the two theories apply to quite different fields: information theory is very general and may therefore be criticised for lack of precision, while Eigen's theory is phenomenological and thus essentially restricted to molecular systems or, at best, to very primitive organisms. Another aspect of these theories is that they deal mainly with functional order, so it would be interesting to find out how a stable structure could give rise to the development of functional order.

This is why we shall now consider successively, in as logical a manner as possible, first the creation of thermodynamically stable spatiotemporal structures in the prebiotic soup, then the molecular organisation determined by natural hypercycles, and, finally, the evolution of the species from the point of view of information theory.

1.   *Creation of thermodynamically stable spatiotemporal structures*

The results of the thermodynamic theory of irreversible processes may be summed up as follows: in the neighbourhood of thermodynamic equilibrium, order is destroyed whereas, far from equilibrium, order may be created under certain conditions.

For a critical value of the parameters, beyond a critical distance with respect to equilibrium, an environmental condition leads to instability and to a new organisation, in an environment in which there exists a set of non-linear autocatalytic chemical reactions. This structure is stabilised at a minimum level of dissipation, being 'fed' with entropy by the environment. Such dissipative structures therefore obey the *principle of order through fluctuation.*

Because of the non-linearity of the system and the conditions of non-equilibrium for certain values of the parameters (concentrations, diffusion constants, and so on), a small perturbation will lead to instabilities which in turn will generate the new spatio-temporal structure. A variety of situations may appear according to the parameters, and lead to a succession of instabilities and thus to different states and behaviour. We have already seen the thermodynamic interpretation of these phenomena.

The principle of order through fluctuation corresponds to an amplification of the fluctuations followed by a stabilisation of the new state by a continuous input of matter and energy from the environment.

The usefulness of such stable and spatially organised structures becomes evident when we consider the different phases of evolution, i.e. first, the existence of monomeric compounds necessary for the creation of living organisms in the prebiotic soup, then the molecular organisation of these chemical species, and, finally, the evolution of the organisms created towards 'increasing complexity'. Babloyantz and Nicolis (1972) have shown how the synthesis of biopolymers is facilitated and, in particular, how the cooperative effects allow the fixation of monomers on to the 'initial' template, the effects being analogous to the all-or-nothing transitions between stationary states.

Thus, the creation of dissipative structures may be considered to have led to the accumulation of polymers in the prebiotic conditions, which is a *prerequisite for a protein–nucleic acid interaction*. This supports Eigen's idea of catalytic hypercycles.

Following dissipative instabilities a small variation in the flux of monomers could produce an increase in the concentration of polymers. The localisation of chemical species, which would otherwise be greatly dispersed, within a very limited region, is evidently highly favourable to an eventual self-organisation; in any case, it increases the probability of the creation of catalytic cycles.

However, the niches may have been created in other ways. For example, Wong (1990) has used the recent techniques of neural networks to construct a model of prebiotic evolution based on the phenomenon of spin-glass, demonstrating the possibility of creating an environment that favours a linearly independent set of evolutionary niches. A surviving polymer acts on its environment according to a Hebbian learning algorithm (see Volume III). The method is based on the strong analogy between spin-glass formation and evolutionary systems composed of polymer ensembles.

2.   *Self-organisation and the evolution of molecular biosystems*

How could the stable structures existing, as demonstrated, in prebiotic conditions, become organised? Eigen, as we saw in Chapter 6, answers this question. *The existence of catalytic cycles followed by self-reproducing hypercycles became inevitable from the moment that the conditions of their appearance on Earth were satisfied.* Thus, the genetic code, the sole universal reality in the diversity of the species, could not but exist.

This determinism, in a population of interacting biopolymers (albeit structured as

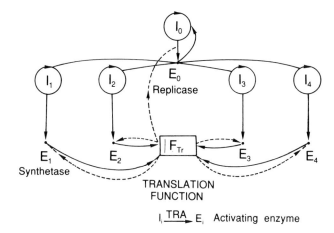

Fig. 7.1. Example of a system with all the functions necessary for efficacious replication–translation. However, calculation shows that such a system will not survive (after Eigen, 1971).

we have just seen), is the consequence of the properties necessary for a system to be Darwinian: metabolism (autocatalytic reactions), self-reproduction and mutation. The problem now is to find out how the replication–translation mechanism came to be organised: in other words, what was the starting point of the process?

Eigen has demonstrated that:

(1)   even a very 'primitive' system, in terms of weak efficacy in the finality of replication–translation, has greater advantages than any other system provided it is of the hypercyclic type; and
(2)   hypercyclic systems are realistic precursors to the apparatuses of replication–translation observed in procaryotic cells.

Does a system having all the properties of a Darwinian system, possessing all the material necessary for its metabolism, and being capable of replicating itself with errors, conserve its properties, i.e. the information already stored? Let us consider the example shown in Fig. 7.1 where I_0 is a messenger coding for a replicase E_0. This enzyme is also capable of specifically recognising the messengers I_1, I_2, I_3 and I_4. The strands (+), I_i^+ with $i = 1$ to 4, code for four synthetases E_i with $i = 1$ to 4. The strands (−), I_i^-, are the tRNAs for four amino acids. This system has all the functions necessary for efficacious replication–translation and works in a closed circuit. However, it can be mathematically demonstrated that a system of this type cannot survive: the existing couplings will not suffice to stabilise the reproducing components I_i. Numerical resolution shows that all the messengers I will eventually disappear.

Since the transfer RNA molecule establishes a relation between two languages — the language of nucleic acids and the language of proteins — it is one of the most important molecules ever to have appeared. However, the problem of its origin has

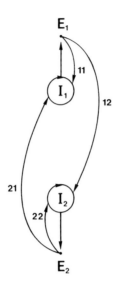

Fig. 7.2. System consisting of two mutant genes I_1 and I_2 (quasi-species), and behaving like a replication–translation apparatus. The functioning can be mathematically demonstrated.

not been completely elucidated. The molecule may have originated from a direct duplication event involving one of its two halves (Di Giulio, 1992). According to some authors (Lahav, 1991), RNA-like molecules must have played an exclusive role in early prebiotic self-replication and catalytic activity.

Whereas a simple self-reproducing unit such as an RNA is capable of multiplying itself in satisfactory environmental conditions, a hypercycle will not be able to reproduce itself as easily because of the complexity inherent to the number of catalysers involved. Indeed, only an organisation of the type described in the previous chapter will allow the evolution of a primitive replication–translation apparatus towards 'high complexity'. The theoretical existence of a hypercyclic precursor capable of transforming itself into a hypercycle with any number of elements has been demonstrated according to the Darwinian 'logic' of mutation–selection. What is missing in the above example that keeps it from evolving? We know (see Chapter 6, III.1) that there must be a reactive cyclic bond between the elements capable of reproduction. Let us therefore seek the simple 'primitive' hypercycle capable of evolution, but above all one which was formed initially by self-organisation. To do this, let us consider the very simple system composed of two mutant genes I_1 and I_2 which, as described in Chapter 6, are quasi-species, i.e. combinations of selected chemical species (Fig. 7.2).

Such a system will function like an apparatus of replication (through I_1 and I_2) and of translation. Let us suppose that I_i^+ (resp. I_i^-) acts in the same way as the tRNA of a pair of amino acids AA_1 (resp. AA_2). The same being true of I_2, a protein may be composed of the four types of amino acid. It follows that the products of the translation are capable of synthesis, complexification and duplication, and the system itself can function as a closed circuit. The two mutants I_1 and I_2 have thus generated a two-element hypercycle, the mathematical study of which allows us to examine their behaviour with respect to the couplings (11), (22), (21) and (12) (see Fig. 7.2). We find that the stable hypercycle is one which possesses a certain symmetry in the intensity of the bonds K_{ij} (K being the rate constant with i, j = 1, 2). The couplings must satisfy the condition of cross-influence, i.e. (12) and (21). The pre-domination of any other coupling will lead to the disappearance of the hypercycle. In addition, this type of two-element hypercycle may easily be transformed into a three-element hypercycle by the introduction of a new mutant I_1' or I_2'. The resulting hypercycle will either die out or be selected.

In conclusion, hypercyclic organisation can emerge as soon as the conditions of a Darwinian system are fulfilled and the quasi-species are selected; this organisa-tion may then undergo evolution following very small fluctuations and develop through the introduction of mutants to form a highly complex replication–translation mechanism.

Various types of autocatalytic reactions have been studied in the last decade. In particular, Stadler (1991; 1992) has considered autocatalytic reaction networks with two interacting mechanisms: one being an autocatalytic instruction: $I_k \to 2I_k$, and the other an additional catalytic activity of molecules I_j: $I_k + I_j \to 2I_k + I_j$. For example, a minimal prebiotic scenario with the potentiality to develop cooperation is derived from the inhomogeneous replicator equation. Cooperation emerges when the total concentration of replication material exceeds a certain threshold. A single species is selected below this value. Above this value, the second mechanism becomes impor-tant and may lead to cooperative behavior such as that involved in hypercycles.

3. Evolution of the species in terms of information theory

Let us recall the main results obtained in Chapter 4. The temporal evolution of a system (S), composed of substructures (E_i), is defined by its complexity $H(t)$ related to the redundance R by $H = (1 - R) H_{max}$, $R \in [0, 1]$. Intuitively, an organism is all the more 'complex' the more total information $H(t)$ it has, and $H(t)$ in turn depends on R and H_{max}. In information theory, the terms of complexity and self-organisation do have a formal significance, but this unfortunately lacks precision as the theory can neither take into account the privileged and specific functional relationships between the different parts of an organism, nor describe the dynamics of the underlying phenomena. Here again we observe that information theory which, because of its very general nature, is clearly well adapted to the communication of messages, tells us nothing about the evolution of processes. However although $R = 0$ defines total disorder, *a decrease in R characteristic of the structural disorder is accompanied by an increase in functional organisation* — at least as functional organisation is here defined — as long as H_{max} has not suffered too great a decrease because of the destructive ambiguity on the system–observer channel, according to Eq. (4.27):

$$\frac{\mathrm{d}H}{\mathrm{d}t} = H_{\max}\left(-\frac{\mathrm{d}R}{\mathrm{d}t}\right) + (1 - R)\frac{\mathrm{d}H_{\max}}{\mathrm{d}t} > 0 .\tag{7.1}$$

This functional order is generated from the noise which increases the total quantity of information in the system. We therefore suppose the system to be complex enough to organise itself, i.e. capable of functioning under other constraints. The evolution of H with time, i.e. of the functional complexity, is a function of two values, $R(t)$ and $H_{\max}(t)$ which in turn depend on the initial conditions $R(0)$ and $H_{\max}(0)$. These conditions define the type of behaviour of $H(t)$. We may try to characterise the evolution of *a* living organism or *an* individual of a given species using concepts based on information theory.

A living organism, considered as a self-organising system, may take on any value of R(0) *provided that the reliability be sufficient and that the condition of self-organisation* dH/dt > 0 *be satisfied.*

In general we observe that high reliability requires a long period of growth, training and adaptation. This leads to an interpretation of $R(0)$ as a measure of *evolutionary potential*, i.e. the capacity of increasing information during this period which itself is a measure of the possibility of development. Thus we may consider that dH/dt measures the *rate of training*. Let us consider a weak initial redundance $R(0)$ associated with a long period of growth characteristic of the higher organisms such as primates and *Homo sapiens*. Even if the rate of training is lower, the greater reliability in the 'more complex' organisms leads to better performance.

Superimposed on the evolution of a species is the evolution of *all of life* during millions of years. How can this evolution be described? According to information theory, the oldest species were 'simple' with a high evolutionary potential. They may thus be characterised by a high value of $R(0)$. *If we identify the global process of evolution since the origin with a self-organising system,* we see that the process may be defined by a very great $R(0)$, that of the primitive species, and a very great reliability.

This was followed by a decrease in redundance corresponding to a slow but not null variation in the quantity of information $H(t)$, i.e. in functional complexity (Fig. 7.3). This interpretation is justified by the following observations:

(i)   first, the more 'complicated' species were the last to appear and their evolutionary potential is much lower than is the case for primitive species; and

(ii)   secondly, the process of evolution has now slowed down to such an extent that it appears to have ceased (on our time scale).

From the concept above we may deduce two important consequences from the point of view of evolution (Atlan, 1972):

(1)   *The current species are all in their quasi-terminal period and cannot increase their functional complexity since the initial redundance* R(0) *is weak,* so that:
   - the simplest species do not evolve; and
   - the most complicated species do not evolve any further.

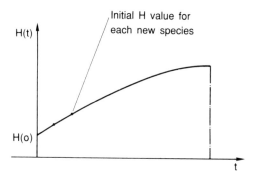

Fig. 7.3. Evolution of the functional complexity $H(t)$ in the sense of information theory (after Atlan, 1972): the system of living species undergoing evolution is identified with a self-organising system.

(2) *The structural and functional organisations of the oldest species were necessarily different from those of extant species.* In particular, this explains the difficulty of interpretation of the relationship of the species at the functional level. According to Atlan, the principle of order arising from noise avoids the pitfalls of finalism and takes into account the increase of complexity with time which has always intrigued biologists, as much in the case of individual development as in that of the evolution of the species. Finally, it would appear that, after millions of years, the organisation of the species is exhausted and will no longer admit any variation that would further decrease its redundance in order to attain greater complexity.

4. *Recapitulation: the scenario of evolution*

From the neo-Darwinian point of view, which considers only the occurrence of micromutations, evolution appears to be a continuous process. We may then, on the bases of the thermodynamic, hypercyclic and information theories developed above, *imagine* the evolution of the most primitive organisms from the first polynucleotides to protocells, and then to higher organisms.

(1) The very localised appearance in the prebiotic soup of macromolecules with spatial structures stabilised according to the thermodynamics of irreversible processes, in particular polymers of the RNA type, i.e. polynucleotides with the inherent property of self-reproduction.

(2) The predominance of G–C compounds, rich in guanine and cytosine, capable of forming very long reproducible sequences. In fact, the quality of the copy depends on the length of the sequences and *vice versa*. Certain substitutions A–U favour structural flexibility and facilitate rapid reproduction. This produces a first-order cycle, i.e. a quasi-species with Darwinian behaviour, noted $\mathcal{I}_1$.

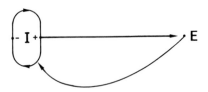

Fig. 7.4. The simplest possible autocatalytic system (after Eigen, 1971).

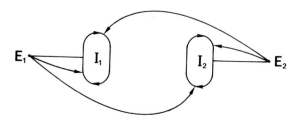

Fig. 7.5. The simplest possible hypercycle (after Eigen, 1971).

(3) The complementary strands of the reproducing unit formed play the part of a messenger and characterise codons. The set represented by $-I+ \rightarrow E$ is capable of generating translation products. At the 'origin', these were glycine and alanine (see Chapter 1 on the composition of proteins). However, such a set does not catalyse its messenger and cannot undergo evolution.

(4) Let us suppose that a translation product E catalyses the replication of its messenger I, as represented in Fig. 7.4. *Then this quasi-species will predominate and be selected.* Following copying errors, two mutants I_1 and I_2 will be selected and, conditions being appropriate, will form a hypercycle (Fig. 7.5) based on the code G–C. This will be capable of attributing not only the codons GLY and ALA but also, because of the appearance of the mutants, two other amino acids with a similar structure, such as ASP and VAL, and finally of generating enzymes possessing the functions of a replicase, a synthetase, and so on.

(5) In favourable conditions, the two-element hypercycle grows, increases in abundance and following a mutation generates a three-element hypercycle and then a four-element hypercycle, and so on. The primitive hypercycle which integrates its mutants is then capable of more complex functions: a more sophisticated code, longer chains, ribosomal precursors, and so on. Simultaneously with the increase in complexity appears a *spatial compartmentation*, i.e. a separation of functional molecular systems in space: there is enrichment of the messenger by localisation in the environment. The hypercyclic organisation of a compartment allows increased efficacy against external influences (Fig. 7.6). This problem is related to the existence of ecological niches.

(6) The hypercycle evolves into a highly sophisticated structure. It is completely isolated in space and forms a compartment. In addition to replication–translation,

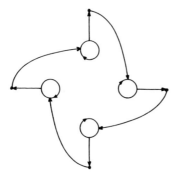

Fig. 7.6. A four-element hypercycle obtained by hypercyclic organisation of mutants (after Eigen, 1971).

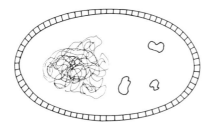

Fig. 7.7. A protocell obtained by successive spatial compartmentation (after Eigen, 1971).

the system develops control mechanisms and metabolic functions. In all likelihood, it forms a genome, i.e. a macromolecule possessing an enormous quantity of information that can be stored and transmitted.

(7) The genome is formed, constituting the protocell. The DNA directs the synthesis of a variety of enzymes with complex control systems. Spatial localisations lead to definitive functioning. The total quantity of information increases by restructuration (Fig. 7.7).

(8) Diversification occurs when initial redundance is very high. Copying errors lead to new functions. The noise gives rise to new functional structures and an increase in the total quantity of information: complexity increases while simultaneously the initial redundance for a given individual decreases. The evolutionary potential of the successive species decreases.

(9) The most recent species are also the most complicated but there is practically no further evolution as there is no increase in functional complexity.

Though this cannot be verified, certain authors believe that we have now reached the terminal period of evolution; *Homo sapiens*, having acquired the greatest complexity, will not be able to undergo any further evolution.

III.   Functional biology and evolutionary biology

1.   *Darwinism and physiology: the principle of vital coherence*

We have attempted to present a synthetic view of the theoretical methods of biological dynamics in general and of macromolecular species in particular, considering the properties of reproduction and the capacity for generating functional proteins as well the phenomena of selection and evolution since the origin of life.

It is now obvious that *the representations used operate at different levels* and that as yet there is no unifying theory of which these several hypotheses constitute particular cases at the levels of description considered, nor is there any explicit relationship between the concepts involved. What seems to us to be of importance is the definition of self-organisation of biological systems tending towards increasing complexity, taking into account structural as well as functional aspects. The theory of bifurcations clearly shows the progressive structural adaptation according to random fluctuations in the neighbourhood of bifurcation points when confronted by environmental constraints. Eigen's phenomenological theory is especially concerned with the functional aspect of the adaptive and progressive evolution of privileged molecular systems such as the self-reproducing catalytic hypercycle. But nowhere does the physiological finality, which is of course essential, find a place. Many biologists tend to oppose *functional* biology and *evolutionary* biology. For example, according to Mayr (1974), embryology, physiology and other branches of functional biology are quite distinct from evolutionary biology which specifically addresses the history of genetic programs. However as Dobzhansky and Boesigner (1968) put it: 'Nothing in biology makes sense except in the light of evolution', and this of course is precisely what led to our earlier discussion on the nature of biological complexity (Chapter 3). Consequently, it would appear difficult to oppose the two fundamental aspects of biology, the functional and the evolutionary. On the contrary, one should explain the other. Jacob's aphorism (1970), 'Creation is recombination', is a very suggestive way of saying that evolution is the product of natural selection. However, we may wonder how the entire regulatory mechanism, which tolerates certain harmful mutations leaving them dormant while allowing others to be expressed harmoniously through active genes, manages to ensure the functional stability of the organism. The two aspects of biology thus appear to be fundamentally interconnected, to such an extent that one cannot be studied without a knowledge of the other. Grassé (1973) goes even further by supposing the creation of new genes to be due to an 'internal motor' and by emphasising the distinction between gene *mutation* and gene *novation*:

> We thus have to admit that the determinism and the mechanism of evolution imply the intervention of internal factors of which we have given some idea when considering the acquisition of new genes . . . We believe that new information which is concretised and permanently incorporated into the genetic code in the form of nucleotide sequences can only result from preliminary intracellular operations.

This is a clear statement concerning the incursion of the functional aspect of biology in evolution.

In some ways the scenario of evolution constructed phase by phase on the bases of existing theories may appear attractive enough since it shows the filiation of the species, at least at the molecular level. But here, as in any other theory, it is obvious that the result depends directly on the initial postulates and, in this particular case, on neo-Darwinian postulates. We know how much biologists differ on this subject. Indeed, as we have briefly indicated in Chapter 1, the problem is real enough. The essential point is the relationship between molecular evolution and physiology, or rather the evolution of structures in favour of physiological functions, which we term 'vital coherence'. *But how can this problem be formalised?*

We already have empirical laws expressing, in particular, the taxonomy of the living world in space and time. All the biological disciplines, palaeontology, zoology, botany, genetics and, more recently, molecular biology, converge to an explanation of the filiation of the species within the neo-Darwinian framework. Eigen's theory, with its postulates of the existence of metabolism, self-reproduction and micromutations, is merely an extension. Moreover, the Darwinian system considered by Eigen is an exact copy of ecological systems, studied and formalised intensively: Axiom (iv) (Chapter 6), i.e. the condition of conservation of the number of informational macromolecules in a given volume, evidently leads to the 'natural' selection of molecular species. This condition, classical in ecology, seems clearly insufficient and, in any case, inapplicable from Phase 5 onwards to a description of the mechanism of the evolution of the species since a good many other causes may be involved. Indeed, the question is not the 'Why?' of evolution but the 'How?' This is the well-known position of the physicists Kirchoff and Mach who, in their time, refused to attempt any explanation of events *without a knowledge of the underlying mechanism*. In the current situation, this philosophical standpoint appears highly relevant, since — unfortunately — no coherent framework seems available for the theoretical laws of evolution. The laws, being merely empirical and not unanimously admitted by biologists, seem quite insufficient to allow any definite conclusions, and may even turn out to be very dangerous in terms of human and social consequences.

Molecular biology and the theories concerning permanent control and regulation, as presented in the following chapters, may be expected to make an essential contribution to the theory of evolution. The immediate problem is not to know *why* one species is transformed into another nor *why* the random existence of a micromutation or a set of micromutations occurs at just the right time and the right place at a given level of an organism or in a given spatial group on the planet, but to know *how* a transformation, whatever its nature, becomes compatible with the survival of the species, i.e. with the physiological functions of the species. As we shall see, from the biologist's point of view, physiological function results from the biosynthesis of a certain number of proteins at the molecular level.

The fundamental constraint of the environment is not the conservation of the total number of molecular species but, on the one hand, the conservation of the physiological functions essential to life and, on the other, that of the regulation of these functions according to given environmental variations. There must be adaptation to the environment through diversification of the functions: only those species that have 'learned' to acquire a harmonious behaviour will subsist. Moreover, we may readily conceive a mutagenesis compatible with life, with no variation of the environment.

This leads us to another question: if the passage from the structural to the functional occurs without ambiguity, how does a species react to the environment, i.e. what is the basic biofeedback involved? Let us consider two examples. The first is well known: did the existence of trees disseminated in the savannah lead to the lengthening of the necks of herbivores, of which the only survivor is the giraffe since it was the best at feeding itself? We believe this problem is not satisfactorily stated as the Lamarckian question inquires into the 'Why?' of the evolution. It would appear far more legitimate to ask how the giraffe managed, in the given environmental conditions, to develop its organism coherently at the level of the various cells, tissues and organs (neck vertebrae, muscles, circulatory system, locomotor apparatus, and so on). How does an external constraint modify the internal motor without causing it to stop functioning? The second example may be taken from the field of economics. During the worldwide crises of the 1980s, industrial society has undergone considerable transformation. For example, in the automobile industry, environmental conditions have been sharply modified in terms of the cost of raw materials, galloping inflation, foreign competition, and so on. In short, the automobile industry finds itself in pronounced Malthusian conditions, so that adaptation to the new situation is necessary for survival. What is essential here is the conservation of the physiological function, i.e. the transport of individuals. But it is not the conservation of the number of pieces constituting the automobile that is important but the harmonious modification of the whole structure which needs to remain compatible with transport.

The difference between the two examples is evident: in the first case, the organism is its own engineer and the phenomenon is one of self-organisation. In the second case, it is the external individuals, elements of the environment, which modify the system, causing it to evolve, so that here the phenomenon is one of organisation. We thus come to a fundamental point, which leads us to pose the principle that *a system may organise itself during evolution by successive adaptations while conserving its physiological functions: this is the principle of vital coherence.* We have seen how Eigen has dealt with the problem of the self-organisation of molecular species. In reality, this type of self-organisation, by the conservation of the number or the flux of macromolecules, appears to be structural rather than functional. This will be apparent below in the attempt at formalisation *taking into account the processes of regulation and control at the molecular level.* Let us once again emphasise the preferential relationship between evolution and physiology which we believe corresponds to the position of Grassé (1973), who rightly remarks that Darwinism is more interested in the variations and transformations within a species than in those between the different species. He concludes that there must exist an Evolutionary Project, a kind of internal motor at the molecular level based on the existence of a *novagenesis*, finally leading to transformations of species along the lines observed. Indeed, as we shall see in the next section, the points of view of Monod (1970) and

Jacob (1970), who believe in the transformation of species by random mutations related to the environment and *compatible with the life of the organism*, do not appear to be so very different.

We may wonder if evolution is oriented. Grassé, in opposition to molecular biologists, believes this to be the case. A solution to this problem would require that the direction of time be determined. But do we refer to classical time, which is a dynamic concept closely related to movement, or do we refer to time, yet to be defined, which may be directly deduced from a theory of the self-organisation of living species? Although information theory provides results that are more or less satisfactory according to the initial postulates used, it has the advantage of being based on concepts which are independent of dynamics as presently understood. We must also emphasise the importance of the remarkable property of the irreversibility of time, directly deduced by Prigogine from the theory of chemical kinetics interpreted on the bases of the second principle of thermodynamics. This aspect of time is evidently not revealed by the physical equations of classical or quantum dynamics and electromagnetism, which remain invariant when t is changed to $-t$. With the irreversibility of time, a phenomenon evolves in a deterministic, *oriented* and irreversible manner from one bifurcation point to another under the effect of *random* fluctuations of the parameters. This very new approach is directly applicable to the problem that interests us here. In fact, for a dynamic system, irreversibility could only occur in the neighbourhood of a bifurcation point.

Will such a scheme of self-organisation, based on the postulates (P_1) *of metabolism,* (P_2) *of self-reproduction,* (P_3) *of mutagenesis (or novagenesis?), and on the principle* (P_4) *of vital coherence, induce a phenomenon of directed evolution?* Does the *vital coherence*, which expresses the fundamental physiological relationship between the source (mutations) and the product (results of metabolism), and which describes the indispensable functional regulatory processes, lead to 'natural' selection? In the present context, the selection is not the direct cause of the evolution but rather a consequence of the vital coherence. This distinction is evidently essential for it is the organism that reacts to the environment in an adaptive and coordinated way according to external variations, and it is again the organism that reacts by self-regulation when internal variations appear.

We now need to demonstrate the possibility of directed evolution using the four basic postulates enunciated, i.e. to discover a theoretical law that will account for the experimental evolutionary laws observed. It is therefore necessary to formalise the dialectic above (Chauvet and Girou, 1983; Chauvet, 1993a).

2.   *An elementary model of evolution*

The explanation of the evolution of regulatory processes calls for formalisation. We have extensively discussed the problems related to the evolution of the species for three reasons: (i) functional biology cannot be independent of evolutionary biology and vice versa; (ii) there is no unanimity concerning the immanent mechanisms of evolution, which in any case are not perfectly known; and (iii) more importantly, we do not see how any certitude could be acquired without the knowledge of the consequences induced by the basic mechanism, or set of mechanisms, at the different hierarchical levels of the organism. We may mention, for example, the stability attained by a new biological system under the influence of the basic mechanism, the perturbation of the regulatory processes and finally of the behaviour of the system and its capacity to generate a new population, i.e. the passage from one level of organisation to another.

Is all this just pure speculation on the part of theoreticians eager for mathematical problems to solve? Considering the numerous critics of Darwin's theory, we certainly do not think so. Even a Darwinian as convinced as Mayr (1974) raises two fundamental questions at the gene level: first, that of the *enormous variability* of the enzyme genes and their extraordinary redundance in the higher organisms; and, secondly, that of the existence of *regulatory gene systems* (which allows Grassé (1973) to pose his hypothesis of novagenesis). A good account of the criticism of the theories of evolution is given by Lints (1981).

We have emphasised the fact that Eigen's theory demonstrates that *only some types of macromolecular species, functioning in a certain way (hypercycle), could evolve according to neo-Darwinian postulates.* This is already a considerable finding, even though incomplete because of the absence of physicochemical data.

The major defect of the theory is that it does not take into account the internal regulatory processes of the genome. To do so, we would have to use a level of organisation different from that of Eigen's model and integrate phenomena that are not yet understood. The internal coherence of representation in biology imposes the choice of several levels of organisation and a simultaneous consideration of the different levels, which adds to the inherent difficulties of the phenomena under study (Volume III). Formalisation is bound to become necessary as the increasing sophistication of measuring instruments pushes back the frontiers of the observable! We have a good example in the data banks of nucleic acid and protein sequences and the corresponding data-processing tools.

The model presented here does not, of course, pretend to solve the problem of evolution. However, it does attempt to pose the problem, hoping to initiate a promising line of enquiry within the framework of a spatiotemporal theory of biological organisation. The problem we would like to tackle is that of the *passage from one level of organisation to another, higher level,* i.e. in less mathematical terms, the influence of a regulatory process on the self-reproduction of the units. We shall consider a simple example, that of an $(\mathcal{M}-\mathcal{R})$ system consisting of a Goodwin-type model at the metabolic level and an Eigen-type model at the population level. A generalisation as well as the space effect will be discussed in Chapter 6 (Volume III). For the moment, let us recall the definitions given in Chapter 4, Section III and used in Chapter 5, Section V.

Definition 1: A biological system is a set of structural units that may be classified hierarchically according to more or less independent functional levels.

Definition 2: A level of organisation is defined by at least one physiological function. It is constructed on the subset of structural units of which the collective

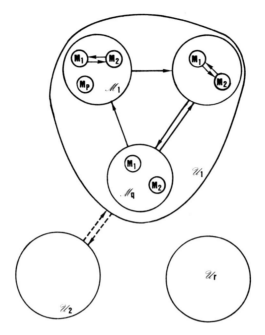

Fig. 7.8. Hierarchical set of structural units. $\mathcal{M}$ is a set of metabolic networks, $\mathcal{U}$ is a set of structural units.

behaviour defines the function. In this form, the identification of a level of organisation results from a process of construction of classes of units associated to realise a common physiological function which, in the widest sense, is considered as the physiological expression of a molecule.

a. Formalisation of the principle of vital coherence in terms of the levels of organisation

Let $\mathbf{O}^{(i)}$ be a population of elements $O^{(i)}$ considered according to their functioning as belonging to a level of organisation i. For a population of units satisfying Darwinian postulates, we will need to consider three levels within the framework of a hierarchical system, $i - 1$, i and $i + 1$, which may be identified with the set $\mathcal{C}$ of chemical reactions constructed on the biomolecules, the set $\mathcal{M}$ of metabolic and epigenetic networks constructed on the sets of reactions, and the set $\mathcal{U}$ of the self-replicating units constructed on the set of networks (Fig. 7.8). This process of construction evidently depends on the problem to be solved. Here, the levels i and $i + 1$ are of interest, and these are the levels we shall describe. Other hierarchical systems, with levels of organisation corresponding to a collective function, exist and may be associated because of their functional dependence. This is the case when physiological functions are closely linked, for example muscle cell respiration and movement due to the contraction of myofibrils.

Let us use the paradigm of self-association presented in Chapter 4, Section III.2. The hypotheses basic to the evolution of populations of units, considering the physiological constraints, may be stated explicitly according to the level of organisation:

Expression of neo-Darwinian postulates (P_1) and (P_2)
(level i, population $\mathcal{M}$)

(P_1)   A metabolic network $\mathcal{M}_j^{(i)}$ synthesises a protein P responsible for a physiological function necessary for the functioning of a self-replicating unit u. Henceforth we shall identify the protein with its physiological function.

(P_2)   Such a network is subject to genetic micromutations which could stop the synthesis of P.

Expression of the principle of vital coherence (P_3)

There are three possible outcomes for the units affected by a disadvantageous micromutation:

(1)   the unit dies out if the level of synthesis of the product P falls below a certain threshold value; or
(2)   the unit becomes associated with other units, not necessarily with identical properties but with the same physiological function P; or
(3)   the pathway of the metabolic network that is destroyed is replaced by a parallel pathway.

In reality, the third possibility is formally analogous to the second. Indeed, it suffices to diminish by unity the level of reference (here i) to find the same conclusions: the individual (or the unit) is then a metabolic pathway, the second level being the network. The knowledge of the behaviour of the system necessitates the following hypothesis:

Hypothesis 1: *The system organises itself according to a process analogous to that of a bi-molecular chemical reaction by the association of two units at most.*

This hypothesis is justified by the usual observation that the process of random encounter, or the expression of a physiological function in common, is highly improbable for more than two units at the same instant. The process depends on a rate constant which couples the two levels of organisation $\mathcal{M}$ and $\mathcal{U}$. In order to simplify, we shall not take into account the mechanism of transport: diffusion, chemotactism, etc. Later, within the context of a multiple field theory, the constant of association of the two units u_1 and u_1^* (obtained by random micromutation from u_1), will be introduced quite naturally.

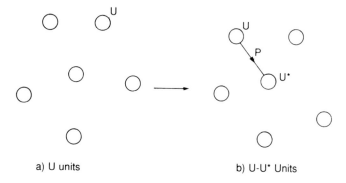

Fig. 7.9. Creation of a hierarchical system by the association of structural units: u^* is a unit which, after alteration, can only survive by association with u.

Thus, according to the principle of vital coherence, either u_1^* will die out or enter into association with another unit u_1 to form $u_2 \equiv (u_1, u_1^*)$ which will be the origin of a new population $\mathscr{U}_2$. In general, u_j will give rise to a population $\mathscr{U}_{j+1}$, with each element possessing a supplementary unit. This process, composed of successive associations, creates a hierarchical system (Fig. 7.9). It is closely analogous to the process of tissue specialisation and even to the biological concept of organogenesis, in which the micromutation is replaced by a controlled alteration of gene expression. Without going into the question of the underlying mechanism, such as that of cell differentiation which may bring a cell u_1 to a state u_1^* at a given time and in a given place according to a pre-established programme, we can perfectly simulate the evolution of a population of units. As seen in Chapter 10, the analogy is striking.

Organismal evolution may occur through an analogous mechanism with the micromutations, replications and recombinations subject to regulatory processes leading to the formation of new genes to satisfy the principle of vital coherence.

b. Description of an elementary model of evolution

This model should allow us to test the validity of the fundamental hypothesis basic to biological organisation as described above: the substitution of a unit by association is the consequence of a micromutation which breaks a chemical pathway and stops the synthesis of a given product. This hypothesis is deduced from the principle of vital coherence, i.e. the preservation of the physiological function necessary for survival. This is the mathematical translation of Jacob's statement (1970): 'Creation is recombination'.

It is important to determine the conditions of application. Let a population of self-replicating metabolic units be defined as follows (Postulate (P_2)):

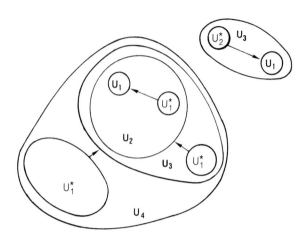

Fig. 7.10. Goodwin's metabolic chain: the regulation is carried out by a terminal product P_n(loop (R)) which represses one of the structural genes G_i (i = 1 to 3) or the initial enzyme E_1(loop (I)). R_i (i = 1 to 3) is a polysome representing the structures, illustrated in Fig. 5.8, on which mRNA, tRNA–AA and enzymes converge.

(i) *Each unit* u_1 *is a general metabolic system at the level of organisation* $\mathcal{M}$ *of the type described by Goodwin* (1978). In reality it consists of two coupled networks, one being metabolic and the other epigenetic. The metabolic network is a control loop based on the retroactive inhibition of a step of the metabolic pathway (loop I); the epigenetic network is a repression loop at the level of the structural genes (loop R). Figure 7.10 illustrates this very simple network. Let us refer to the hierarchical description of the replication–translation apparatus previously given in Chapter 5, Section V (see Fig. 5.26). To facilitate the mathematical treatment, let us consider the regulation of the metabolic system alone. A unit is described by a metabolic pathway (4.47) in which the number of steps is n = 4. This gives rise to an association between two such metabolic pathways (4.48):

$$\rightarrow P_1 \xrightarrow{\alpha_1} P_2 \xrightarrow{\alpha_2} P_3 \xrightarrow{\alpha_3} P_4 \xrightarrow{\alpha_4} \dots \tag{7.2}$$

The terminal product P_4 retroacts on the concentration of P_1 according to an inhibitory interaction of a cooperative nature. In other words, P_4 limits the rate of upake of P_1 according to the formula given by Eq. (5.14) derived from (1.12):

$$f(P_n; \tilde{\omega}, \kappa, \alpha_0) = \frac{\alpha}{\beta + \gamma P_n^{\tilde{\omega}}} = \frac{\alpha_0}{1 + \kappa P_n^{\tilde{\omega}}}$$

with $\alpha_0 = \alpha/\beta$, and $\kappa = \gamma/\beta$. In this equation, $\tilde{\omega}$ is the stoichiometry of the interaction, i.e. $\tilde{\omega}$ molecules of the end-product P_n bind with the aporepressor. We thus obtain the following system of equations:

$$\frac{dP_1}{dt} = -\alpha_1 P_1 + f_{1,4}(P_4; \tilde{\omega}, \kappa, \alpha_0)$$

$$\frac{dP_2}{dt} = -\alpha_2 P_2 + \alpha_1 P_1$$

(7.3)

$$\frac{dP_3}{dt} = -\alpha_3 P_3 + \alpha_2 P_2$$

$$\frac{dP_4}{dt} = -\alpha_4 P_4 + \alpha_3 P_3.$$

Let us suppose that a micromutation leads to the rupture of a link in the metabolic chain, for example the enzymic synthesis of P_3, i.e. because α_2 becomes null in unit u_1. We then have a different ('pathological') unit u_1^*. The principle of vital coherence implies that, for survival, u_1^* must be associated with other units synthesising the product P_3. Without, for the moment, making any further hypotheses concerning the transport process, we have the following metabolic schema 4.56 (also shown in Fig. 5.26):

$$
\begin{array}{c}
\downarrow \rule{0pt}{0pt} \\[2pt]
P_1^* \xrightarrow{\alpha_1} P_2^* \xrightarrow{\alpha_2} P_3^* \xrightarrow{\alpha_3} P_4^* \xrightarrow{\alpha_4} \\[2pt]
g_1 \Updownarrow g_3 \Updownarrow \\[2pt]
P_1 \xrightarrow[\alpha_1]{} P_2 \xrightarrow[\alpha_2]{} P_3 \xrightarrow[\alpha_3]{} P_4 \xrightarrow[\alpha_4]{} \\[2pt]
\uparrow \rule{0pt}{0pt}
\end{array}
$$

(7.4)

where the case of simple passive diffusion for P_1 and P_3 is obtained by putting:

$$g_1(P_1, P_1^*) = \beta_1(P_1 - P_1^*)$$

$$g_3(P_3, P_3^*) = \beta_3(P_3 - P_3^*).$$

It is easy to imagine such a molecular substitution for a function P_3 coming from u_1, for example by imposing the continuity of units and transmembrane diffusion. However, this is not indispensable for the theoretical principle exposed.

We can simplify the corresponding kinetic system by using the schema (4.53) rather than (4.52), i.e. by introducing the constant $\bar{\alpha}$ which simply describes the non-local contribution of product $P_3 \in u_1$ to the production of $P_4^* \in u_1^*$ (as in Chapter 4, it is assumed here that P_4^* may modify the synthesis of P_1 in an additive manner, in the same way as P_4, and that the coefficients for the degradation of P_4 and P_4^* are the same). P_1^* may also diffuse towards u_1, and the feedback of P_4^* on P_1 is assumed to be indirect. The final result is the functional association of u_1 and u_1^* (Fig. 5.26) to form a unit $u_2 \equiv (u_1 - u_1^*)$ satisfying the dynamic system (4.59):

$$\frac{dP_1}{dt} = -\alpha_1 P_1 + f_{1,4}(P_4; \tilde{\omega}, \kappa, \alpha_0) + f_{1,4}(P_4^*; \tilde{\omega}, \kappa, \alpha_0)$$

$$\frac{dP_2}{dt} = -\alpha_2 P_2 + \alpha_1 P_1$$

$$\frac{dP_3}{dt} = -(\alpha_3 + \overline{\alpha})P_3 + \alpha_2 P_2 \tag{7.5}$$

$$\frac{dP_4}{dt} = -\alpha_4 P_4 + \alpha_3 P_3$$

$$\frac{dP_4^*}{dt} = -\alpha_4 P_4^* + \overline{\alpha} P_3.$$

The functioning of each unit at the level of organisation $\mathcal{M}$ may then be studied.

(ii) *The evolution of populations of units* $u_j (1 \leq j \leq r)$, *if* r *is the degree of maximum organisation, follows Eigen's model (1971) which uses the phenomenological formalism of chemical reactions.* To take the environmental conditions into account, a constraint of conservation is imposed on the total number of 'elementary' species (of the type u_1).

Hypothesis 2: *Evolution of the population* $\mathcal{U}$. u_n *being the unit obtained by the bi-unitary association of* i + j *units for* $1 \leq i \leq n - 1$ *and* $1 \leq j \leq n - 1$, *the equation for the evolution of the population:*

$$\mathcal{U} = \bigcup_{n=1}^{r} \mathcal{U}_n$$

may be written:

$$\frac{du_n}{dt} = \left[a_n - \frac{1}{c}\lambda(u_1, u_2, \dots, u_r) \right] u_n + \sum_{i+j=n} k_{ij} u_i u_j$$

$$n = 1, 2, \dots, r \tag{7.6}$$

$$\sum_{n=1}^{r} n u_n = c: \quad \text{which is the condition for the conservation of the total number of elementary species } u_1$$

where λ is an explicit function of the condition of conservation, analogous to a dilution factor (Eigen's Ω factor, Eq. (6.5)).

By writing $\sum_{n=1}^{r} n \dfrac{d}{dt} u_n = 0$ we find:

$$\lambda(u_1, u_2, \dots, u_r) = \sum_{n=1}^{r} n a_n u_n + \sum_{n}\sum_{i+j=n} n k_{ij} u_i u_j. \tag{7.7}$$

In these equations the coefficient a_n corresponds to a phenomenon of autocatalysis, and k_{ij} is a *coupling parameter* between levels of organisation. In fact, according to Hypothesis 1, k is a function of the concentration of product P synthesised at an immediately lower level. Finally, these functions k_{ij}, generally asymmetric, express the process of functional self-organisation between, on the one hand, the levels of organization 𝓜 and 𝓤 and, on the other, the degrees of organisation i and j. They are therefore called the *parameters of self-organisation*.

This model represents at least a logical possibility for the functional self-organisation of a biological system. Among the questions that we might ask and which the theory should answer are the following. When an additional functional interaction turns out to be necessary because of the principle of vital coherence, are the rate of functioning and the stability of the new system modified? In this case, is there a new functional order created in the population 𝓤 by an increase in its degree of self-organisation, i.e. by a selection of units having increased their degree of organisation by association? Finally, does the combination of a selective value (the coefficient of u_i according to Eigen) and the parameter of self-organisation k_{ij} lead to a higher selective value for units of degree $i + j = n$?

The answers to these questions require rather lengthy calculations (see Chauvet and Girou, 1983), so only the essential elements will be presented below.

c. *Discussion and results: can this model be generalised?*

The first step in answering this question involves an investigation of the stability of the new system in $(P_1, P_2, P_3, P_4, P_4^*)$ obtained from that in (P_1, P_2, P_3, P_4). A transformation given by Walter (see Rapp, 1976) makes the system (7.5) dimensionless by using (4.60):

$$\xi = \left(\alpha_0 \alpha_X \alpha_Y \alpha_1 \, \kappa^{1/\bar{\omega}}\right)^{1/4} \quad t^* = \xi t \quad b_1 = \frac{\alpha_X}{\xi} \quad b_2 = \frac{\alpha_Y}{\xi} \quad b_{i+2} = \frac{\alpha_i}{\xi}, \quad i = 1, 2 \qquad (7.8)$$

which leads to a new system of equations (4.61):

$$(1) \quad \frac{dx_1}{dt} = -b_1 x_1 + \frac{1}{1 + x_4^{\bar{\omega}}} + \frac{1}{1 + \left(\dfrac{b_5}{b_3} x_5\right)^{\bar{\omega}}}$$

$$(2) \quad \frac{dx_2}{dt} = -b_2 x_2 + x_1$$

$$(3) \quad \frac{dx_3}{dt} = -(b_3 + b_5)x_3 + x_2 \qquad (7.9)$$

$$(4) \quad \frac{dx_4}{dt} = -b_4 x_4 + x_3$$

$$(5) \quad \frac{dx_5}{dt} = -b_4 x_5 + x_3$$

where:

$$t^* = a_0 t \qquad x_1(t^*) = a_1 X(t) \qquad x_2(t^*) = a_2 Y(t).$$

$$x_3(t^*) = a_3 P_1(t) \qquad x_4(t^*) = a_4 P_2(t) \qquad x_5(t^*) = a_5 P_2^*(t). \tag{7.10}$$

The linearisation of the system about the stationary states allows an analysis of the stability. Here again, as in Chapter 4, we find an important result: the region of stability increases when going from a system in (P_1, P_2, P_3, P_4) to a system in $(P_1, P_2, P_3, P_4, P_2^*)$ (Fig. 4.9). In other words, the functioning of the unit $u_2 \equiv (u_1, u_1^*)$ is metabolically more stable than that of unit u_1 since the instability of a linear system corresponds to stable periodic oscillations of a non-linear system. Thus, the system tends to diminish its periodically oscillating behaviour through association, and may be considered to lose its independence with respect to its own time scale by the creation of new functional behaviour. When a micromutation occurs, the tendency to associate is a characteristic at level $\mathcal{M}$. This answers the first question above. The answer to the two other questions involves a study of the system (7.6) at the level of organisation $\mathcal{U}$.

The study of the stability of the second- and third-order systems:

$$\left. \begin{array}{l} \dot{u}_1 = (\alpha - f) u_1 \\[4pt] \dot{u}_2 = (\beta - f) u_2 + k u_1^2 \\[4pt] u_1 + 2 u_2 = c \\[4pt] f(u_1, u_2) = \dfrac{\alpha}{c} u_1 + \dfrac{2\beta}{c} u_2 + \dfrac{2k}{c} u_1^2 \end{array} \right\} \tag{7.11}$$

and:

$$\left. \begin{array}{l} \dot{u}_3 = (\gamma - f) u_3 + k' u_1 u_2 + k'' u_2 u_1 \\[4pt] u_1 + 2 u_2 + 3 u_3 = c \\[4pt] f(u_1, u_2, u_3) = \dfrac{\alpha}{c} u_1 + \dfrac{2\beta}{c} u_2 + \dfrac{3\beta}{c} u_3 + 2k u_1^2 + 3k' u_1 u_2 + 3k'' u_2 u_1 \end{array} \right\} \tag{7.12}$$

shows the existence of two stable states, at least for the first system. Moreover, the evolution of the units u_1, u_2 and u_3 depends on the coupling parameters $k(P)$, $k'(P)$ and $k''(P)$, as well as on the coefficients α, β and γ. It is interesting to observe by simulation that if $k = k' = k''$, i.e. if all the parameters of self-organisation are identical for the associations $u_2 \equiv (u_1, u_1^*)$ and $u_3 \equiv (u_2, u_1^*)$ or (u_2^*, u_1), then it is the population $\mathcal{U}_3$ that becomes preponderant (Fig. 7.11). The species u_3 is selected *so that, in addition to Eigen's selective value γ, we have the intervention of the parameter of self-organisation* $k = k' = k''$, *the coupling between two levels of organisation.* Finally, we find a new selective value which leads to the selection of units of higher degree.

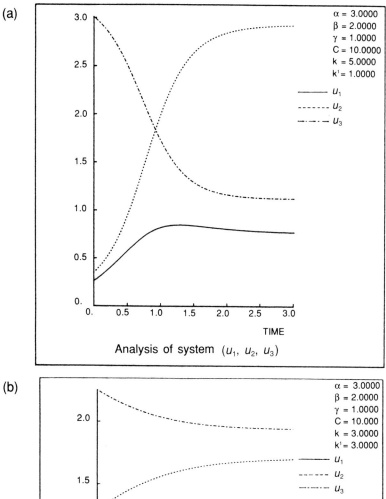

(a)

α = 3.0000
β = 2.0000
γ = 1.0000
C = 10.0000
k = 5.0000
k' = 1.0000

—— u_1
----- u_2
—·—·— u_3

TIME

Analysis of system (u_1, u_2, u_3)

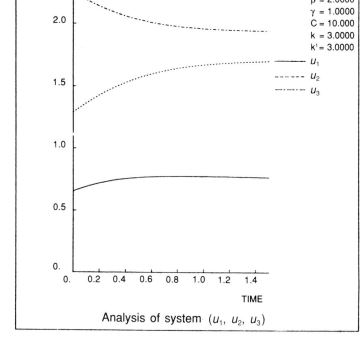

(b)

α = 3.0000
β = 2.0000
γ = 1.0000
C = 10.000
k = 3.000
k' = 3.000

—— u_1
----- u_2
—·—·— u_3

TIME

Analysis of system (u_1, u_2, u_3)

Fig. 7.11. Evolution of populations of units u_1, u_2 and u_3. The parameter of self-organisation k' which couples the two levels of organisation and acts as a selective value: (a) with $k' \neq k$, the population of u_2 is preponderant; (b) with $k' = k = 3$, the population of u_3 becomes preponderant.

The situation is more complicated when the spatial distribution of the units is taken into account. To solve this problem, we have studied the same evolutionary model in the light of the field theory (Chauvet, 1993c). Recently, the analysis of the spatial stability of Eigen's quasi-species model (presented in Chapter 6) has been reconsidered by Weinberger (see Vidybida, 1991) who has generalised Eigen's 'constant overall organisation' constraint into one of 'global population regulation'. With this constraint, the spatial generalisation of these two models is shown to be mathematically tractable.

Conclusion

The study of the evolution of populations of units satisfying the neo-Darwinian postulates (P_1), (P_2) and (P_3) as well as the principle of vital coherence (P_4) calls for the use of the conceptual framework of a functional biological organisation. It is certain that the difference between the life sciences and the physical sciences lies in the nature of their organisation, mainly functional for the former and structural for the latter. Moreover, it has been shown that the link may be found in the relationship between functional biology and evolutionary biology.

Thus, physical concepts cannot be directly applied to the problem of evolution, and it is necessary to define a new theoretical framework. This is certainly the most important result, and one of the advantages is that it allows the construction of what we have called the *multiple field theory, i.e. the description of non-local biological processes* by a field theory of hierarchical systems (Volume III). In fact, the phenomena of geographical speciation or ecological niches require that the equations of evolution take into account a non-local spatial effect.

Summing up this chapter

(i)   *Studies in evolutionary biology have to use the results acquired in functional biology and vice versa.* In other words, the evolution of a population of species must involve regulatory processes at all the levels of organisation.

(ii)   *An elementary model of evolution consists of the association of self-replicating units at a given level.* We have studied the metabolic level with regulatory processes of the Goodwin type and the population level with self-reproduction processes of the Eigen type. We have found that the association of units produces an 'increase of stability' and that the coupling parameter between different levels is involved in the selective value. This is a significant mathematical representation of the influence of regulatory processes.

(iii) *These results being valid for any of the levels of the functional organisation envisaged, the same reasoning may be extended to associations of units analogous to genes or portions of genes.* We may be sure that regulatory processes very similar to those described above operate between the different parts of the genome. We believe that the association of the self-replicating units leads to the creation of populations of new species based on the *new genes obtained by recombination at the most elementary level following the occurrence of micromutations.* Will the principle of vital coherence reconcile the points of view of Jacob (micromutagenesis) and Grassé (novagenesis)?

Summary of Part II

The description of biological organisation leads to the formalisation of the evolution of relationships between structural units rather than that of the evolution of the units themselves. Haken's method has the advantage of taking into account the hierarchy of systems through their time scales, i.e. the hierarchy of phenomena, some of which control the others. Von Foerster, followed by Yockey and Atlan, have described biological organisation using the theory of information. The formalisation of the celebrated principle of '*order arising from noise*' leads to the principle of the evolution of a self-organising system, the behaviour of which is based on the reciprocal influences of *redundancy* and *maximum information*. Since these theories do not take into account the *functional, physiological* aspect of biological systems, we have proposed a theory of functional organisation based on a minimum number of specific biological concepts, i.e. *the non-locality and the non-symmetry* of a *functional interaction*, which will be developed in the course of this book.

The process of *replication, transcription* and *translation* have been studied from a dynamic point of view, first by means of information theory, which allows measurement of the quantity of information contained in the genetic message, and secondly by a probabilistic approach to the double polymerisation at the growth point of DNA. A general method, adapted to the replication of nucleic acids as well as to the biosynthesis of proteins, allows determination of the order of magnitude of certain fundamental parameters otherwise inaccessible. The topological constraints imposed by the circular structure of DNA in replication have stimulated intensive research leading to the discovery of specific enzymes, the *topoisomerases*. Although this problem, remarkably posed in mathematical as well as in biochemical terms, has now been satisfactorily solved, new questions arise: How do the molecules recognise the global, topological structure of the DNA? For example, how do they determine the number of turns necessary for supercoiling? Lastly, Rosen's theory of $(\mathcal{M}\text{--}\mathcal{R})$

279

systems gives a definition of the concepts of *contractibility* and of *re-establishability* involved in the biosynthesis of proteins.

The notion of a neo-Darwinian molecular system has been presented in the framework of Eigen's theory which gives a relatively simple description of the evolution of a population of macromolecular species under constraints of the ecological type. The results obtained may appear trivial inasmuch as they are highly analogous to the well-known results of theoretical ecology. However, the method used introduces the concept of the *autocatalytic self-reproducing hypercycle*, and allows integration of the evolution of molecular systems within the general framework of the evolution of the species. The method also clearly shows how the 'noise', i.e. the sum of the errors of production, leads to the creation of a richer functional order.

Biological organisation cannot be considered without reference to the evolution of the species, which necessarily implies that every biological system studied at present is the end result of a slow process. Several theoretical problems then arise. In particular, how did the *physiological functions* manage to evolve coherently with respect to each other and with respect to the environment? Are the current concepts of evolutionary biology sufficient for the integration of functional biology? Can we describe an orthogenesis compatible with the results of physiology? In the framework of our theory of functional organisation, we show that a self-organising process, based on the *association* (with constraints) *of structural units at a lower level of organisation* (principle of vital coherence), increases the stability of the system and could be the basis of an elementary model of evolution. Moreover, such a process would allow unification of the concepts of *micromutagenesis* and *novagenesis*.

Part III: Cellular Organisation of Living Matter

But Nature creates infinity: for infinity is imperfect and Nature always seeks an end.

Aristotle, *On the Generation of Animals*

Introduction to Part III

The cell is the fundamental individual unit of all living organisms. This *cell theory*, first enunciated by Schleiden in 1838 and Schwann in 1839, led to a new approach to biology through studies on cell chemistry (metabolism, Chapter 8) and cell physiology (regulatory mechanisms, Chapter 9). But, even more importantly, it laid the foundations of developmental biology. When and how does cell division occur (Chapter 11)? How does a multicellular organism arise from a single cell, the egg (Chapter 10)? What is the relationship between growth, division and differentiation (Chapter 12)?

Cell biology is the basis of embryology. And here the unsolved problem concerning the development of a fertilised egg into an independent, autonomous organism remains the most intriguing mystery in the history of biology. Ever since Aristotle's treatise on the generation of animals and up to the work of Hamm and Leeuwenhoek in the seventeenth century, only the simplest ideas prevailed. The microscopic techniques of the latter led to the observation of the spermatozoon, the discovery of which immediately sparked the preformationist theory: the spermatozoon was held to contain the homunculus, a miniature human being. The opposition between epigenism (championed by Harvey in the seventeenth century and by Wolff in the eighteenth) and preformationism continued until the advent of experimental embryology, which seeks to investigate the determinism of the developmental process and the causal link directing the chronological succession of the different phases. This is indeed a most delicate task and, as we shall see, involves a molecular approach to embryology since the mechanisms of differentiation, division and growth are of molecular origin, in all likelihood at the gene level.

8

Cellular Organisation

How do the extremely complicated and highly organised molecular substructures finally give rise to an autonomous, functional, biological structure? This major biological problem remains to be solved. And the fundamental question of developmental biology, which concerns the process by which differentiated structures result from a totipotent egg, has not yet been satisfactorily answered.

We have seen above that, during the course of evolution, certain structures, hypercyclic at first and far more complex later, must have emerged as discrete entities in space for functional efficacy with respect to the environment. This is certainly best exemplified by the elementary individuality of the living cell, separated by a membrane from the environment. Other structures within the cell are also surrounded by membranes forming intracellular compartments. Coherent cell function, without which no survival is possible, obviously calls for highly complex regulatory processes controlling cell growth, division, differentiation, and movement with respect to the environment.

I. Cell description

Without going into a detailed description and classification of living cells, we shall here merely describe the typical cellular elements, necessary to what follows, as well as some examples useful for the understanding of the phenomena involved.

All cells can be classified in two categories: *procaryotes* and *eucaryotes*. The former lack a discrete nucleus and exist only in the form of unicellular organisms. The latter, with a well-defined nucleus surrounded by a nuclear membrane, consitute unicellular or multicellular organisms which, by extension, are called eucaryotic. The

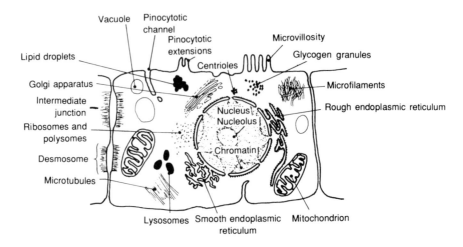

Fig. 8.1.   Essential organelles of a typical animal cell.

cells of such organisms possess an apparatus that allows cellular associations in the form of tissues and organs.

Figure 8.1 shows the various structures of a typical cell: nucleus, mitochondria, ribosomes, intracellular membranes, lysosomes, microtubules and microfilaments. Let us briefly review their functions.

The *nucleus* contains chromatin which condenses into chromosomes during cell division. Chromosomes carry the heriditary material of the eucaryotes and are made up of DNA as well as acid proteins, basic proteins (histones) and some RNA. The process of cell division will be considered in detail in Chapter 11. The nucleus is at the heart of the phenomena of self-reproduction characteristic of the living world. It is surrounded by a double membrane, the nuclear envelope, which has the usual structure of biological membranes, and disappears during cell division. The transformations that occur during this phase are extremely complicated and several of them remain obscure.

The *nucleolus* lies surrounded by nucleoplasm within the nucleus. Rich in RNA, the nucleolus is involved in the synthesis of ribosomal RNA. It also contains several particles that may be pre-ribosomal structures. The nucleoplasm contains several granules of unknown function.

The *mitochondria* are the biochemical power plants of the cell in which ATP, the fuel essential for cellular metabolism, is synthesised. ATP is formed following the oxidation of ingested food products. Mitochondrial structure is very complex. The pleated inner membrane of the two-membrane organelle forms crests studded with granules involved in membrane-coupled metabolic reactions. Mitochondria contain DNA and are capable of self-reproduction.

The *lysosome* contains several hydrolytic enzymes capable of digesting cellular material. The lysosomal membrane protects the cell from potential self-destruction.

In addition, the cell contains numerous structures of unidentified function. These may be surrounded by membranes or lie free within the cytoplasm in crystalline or granular form. Considering the extraordinary complexity of cell mechanisms, it appears unlikely that a precise understanding of the processes involved will be arrived at in the near future.

Apart from the organelles, membrane systems play a fundamental role by forming aqueous cell compartments. The plasma membrane isolates the cell from the environment and the elaborate network of endoplasmic reticulum which, in the form of rough endoplasmic reticulum, carries *ribosomes*, the sites of protein synthesis. Smooth endoplasmic reticulum, devoid of ribosomes, plays a part in lipid synthesis. The endoplasmic reticulum, which may be considered to be in functional continuity with the plasma membrane, leads to the development of the *Golgi apparatus*, a series of tightly packed smooth vesicles in parallel or semicircular arrays, involved in cell transport and metabolism. The *nuclear envelope* is a double membrane, equipped with pores, separating the nucleoplasm from the surrounding cytoplasm. The general organisation of the membrane systems is characterised by a continuity between the nuclear membrane and the plasma membrane via the reticulum endoplasmic and the Golgi membranes. The 'remainder' of the cell consists of an aqueous substance, the *hyaloplasm* or fundamental substance, and the cell organelles. The hyaloplasm contains a variety of microfilaments, similar to those involved in muscle cell contraction, as well as microtubules which contribute to the phenomena of cell motility.

II. Cellular organisation and regulation

1. *Formation of structures at the cellular level*

The organisation of a biological system, whether in the resting or the active state, as well as during growth, is subject to several processes of control. In simple terms we may distinguish three types of structure formation:

(i)   *self-assembly*, which, as described in the previous chapter, is the elaboration of a structure from its component parts and corresponds to the self-reproducing catalytic hypercycle;

(ii) *assisted assembly*, where molecules other than those involved in the final structure supply the information necessary; and

(iii) *directed assembly*, of which a simple example, taken from the material world, would be a crystal growing from a seed.

An example of self-assembly at the molecular level has been amply studied above. A structure of a slightly higher degree of complexity—in so far as a scale for the complexity of structures may be defined, a point we shall return to later—is that of the virus. The virus is perhaps the most elementary 'cell' and may be considered to be a living organism because of its properties of organisation and self-reproduction. The determinist behaviour of such a structure is most extraordinary. For example, the tobacco mosaic virus is cylindrical in shape, 16 nm in diameter and 300 nm in length, with a molecular weight of 40×10^6 Daltons. This

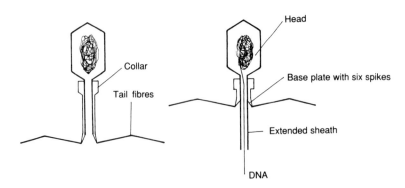

Fig. 8.2. A T_2 bacteriophage in the contracted state. The figure on the right shows the mechanism of penetration of viral DNA into a bacterium (after Loewy and Siekevitz (1974)).

is obviously a highly complex structure. With appropriate techniques the virus can be broken down into its component parts and then reconstructed, the sequence of operations being monitored through the electron microscope. By this means crown-shaped substructures have been visualised. In fact, the cylindrical structure is made up of a set of 2130 protein subunits arranged in helicoidal form around the viral RNA.

At the other end of the scale of viral complexity we have another well-known virus, the bacteriophage. In particular, the $T2$ variety may be found in one of the two states, normal or contracted (Fig. 8.2). The bacteriophage injects its DNA together with a small quantity of enzymes into a bacterium. This material takes control of the bacterial activity and enables the virus to reproduce itself using the substructures of its host.

We still do not know how the change from the molecular level to the multimolecular level of the biological structure is carried out in complex systems such as viruses. Advances are being made in this field by decomposing the system under study into its component parts and then reconstructing it *in vitro*. For example, it has been possible to break down ribosomes partially and then reassemble the fragments.

Moreover, some complex enzyme systems, such as those involved in the formation of fibrin, have been identified. This example is of great importance from the practical point of view in haematology and in cardiovascular disease since the formation of fibrin leads to the coagulation of blood. Certain reactions are now known to produce a fibrin polymer (Loewy and Siekewitz, 1974) and another enzyme transforms soluble fibrin into its insoluble form. The system of reactions may be represented by a dynamic system showing that the very precise regulation of the phenomenon is due to factors external to the final structure. Here we clearly have a case of *assisted assembly*.

Examples of this type are numerous. Before examining how the regulatory phenomena of biological systems may be described, let us consider some examples of *directed assembly*. This process requires the existence of a 'pre-structure' meant to pilot the subsequent polymerisation. Bacterial flagella, covering the surface of

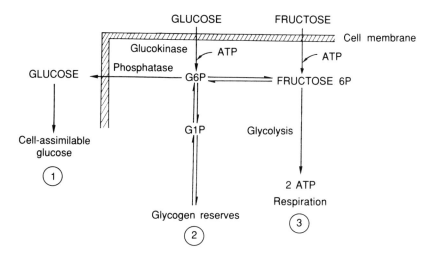

Fig. 8.3. The key compound in the energy metabolism of a liver cell, G6P, is at the origin of three functions leading to: (1) assimilable circulating glucose; (2) storage of glucose in the form of glycogen; and (3) anaerobic net production of 2 molecules of ATP or aerobic production of 36 molecules of ATP per molecule of glucose (see Volume II).

salmonellae (the cause of food-poisoning as well as diseases such as typhoid fever) are made up of molecules of flagellin. Flagella cannot be assembled without the prior availability of flagellin in the environment.

2. Regulation and metabolic pathways

Cell metabolism, which will be considered in greater detail with respect to physiological functions in Volume II, Chapter 6, involves complex mechanisms of degradation and synthesis which vary according to the function. Metabolic pathways specific to a given function may be considered to be superimposed on the biochemical processes regulating the life of the cell. Thus, for example, liver cells supply a good many proteins used in other parts of the human body; the alpha cells of the islets of Langerhans in the pancreas secrete hormones ensuring the regulation of energy metabolism; and thyroid cells produce thyroxine which controls cell metabolism in general.

Let us briefly describe two of the essential metabolic pathways: the glucose-6-phosphate (G6P) pathway and the Krebs cycle.

a. The glucose-6-phosphate pathway

G6P is an important compound in energy metabolism. It can be built up from free glucose, glycogen of fructose-6-phosphate (Fig. 8.3). Free glucose, assimilable by cells, is transported by the bloodstream and may be obtained by three pathways in the liver cell: (1) glycogen, a form of glucose storage; (2) glycerol, arising from the decomposition of triglycerides in fatty tissue; and (3) proteins broken down into amino acids and transformed into ketone bodies. This first set of metabolic reactions is triggered in the case of hypoglycaemia or fasting when free glucose is no longer available. The phenomena corresponding to reactions (2) and (3) are called *gluconeogenesis*. Conversely, glucose may be used by the cells in *four* forms: (1) free glucose used directly; (2) glycogen; (3) glycerol; and (4) energy release.

The second set of reactions corresponds to glucose saving in the postprandial state. Energy is essentially obtained from circulating glucose, and excess calories are put into reserve, mainly in the form of fats. In this example the variable regulated by the action of several hormones is the glycaemia. It should be borne in mind that *one single molecule may be at the origin of different metabolic pathways*. Thus, G6P is also at the origin of G5P and ribose-5P, the precursor of ribose, an essential component of ribonucleic acid. In all, there are at least three pathways leading from G6P to G5P, including the *pentose phosphate pathway* which itself occurs in two forms.

It is clear that a whole metabolic network is centred around a key component, in this case G6P, the regulation of which globally depends on several structures. This explains two points: (i) the prodigality of metabolic pathways in biological systems which allow adjustments according to input; and (ii) the ease of regulation which is based on only a few components.

b.   The Krebs cycle (the citric acid cycle)

The Krebs cycle is of prime importance in the conversion of the various molecules absorbed by an organism. This takes place along the following lines:

$$\text{Carbohydrates} \rightarrow \text{Lipids}$$

$$\text{Proteins} \rightarrow \text{Lipids} \rightarrow \text{Carbohydrates}$$

The organism is thus able to control its metabolism according to the environment by means of extremely precise regulation. Indeed, *the Krebs cycle is the central mechanism of cell metabolism*. The metabolic network is organised so as to use the pathways in all possible directions to meet changing external conditions.

In glucose metabolism, the glucose penetrates the cell in the presence of *insulin*, a hormone secreted by the islets of Langerhans in the pancreas. A series of enzymes belonging to the glycolytic pathway breaks down the glucose into pyruvic acid. Mitochondria take up the pyruvic acid and transform it into acetyl coenzyme A which then enters the Krebs cycle. This is followed by the formation of lipids.

What are the mechanisms that control the working of the cycle in one direction, i.e. production of energy, rather than in the other, i.e. production of lipids? Why is acetyl coenzyme A oxidised sometimes to produce energy and at other times to produce lipids? These processes of adjustment are as yet little understood.

c. Interpretation of regulatory phenomena

The role of theoretical biology should of course be to reduce the complexity of experimental results by appropriate methods of synthesis and interpretation. But the synthesis of such complex phenomena as appear to be involved in the two examples above may seem like attempting the impossible. However, the methods of network thermodynamics, described in Chapter III, may be expected to prove useful through automatic analysis of metabolic networks such as the Krebs cycle. More classical methods have been used to study glycolysis, considering the phenomenon as a dynamic system (Cramp and Carson, 1979); the results will be briefly described in Chapter 6, Volume II.

As the notions of stability and regulation are closely linked, the theory of dynamic systems generally offers an effective approach to the study of regulatory mechanisms. Indeed, the process of regulation consists in the stabilisation of the behaviour of a dynamic system working under the effect of internal or external constraints. Since such a system may be described by a set of differential equations as defined by Haken (Chapter 3), the regulation will allow a certain function of the state variables to be maintained constant. Let us suppose it possible to represent a metabolic network by a set of such equations:

$$\left. \begin{aligned} \frac{\mathrm{d}x_1}{\mathrm{d}t} &= f_1(x_1, \ldots, x_n) = v_{x_1}(t) \\[6pt] \frac{\mathrm{d}x_2}{\mathrm{d}t} &= f_2(x_1, \ldots, x_n) = v_{x_2}(t) \\[6pt] &\cdots\cdots\cdots\cdots\cdots\cdots\cdots \\[6pt] \frac{\mathrm{d}x_n}{\mathrm{d}t} &= f_n(x_1, \ldots, x_n) = v_{x_n}(t) \end{aligned} \right\} \tag{8.1}$$

where the variables x_i, $i = 1$ to n, represent the concentrations of the chemical species i involved in the system. The functions f_i, $i = 1$ to n, which are generally non-linear, contain the rate constants $k_{\pm}^{ij}$ and are obtained according to the laws of chemical kinetics. The study of this non-linear system is delicate but may be approached by the method devised by Higgins (1965). The influence of one reactant j at concentration x_j on another reactant i at concentration x_i is given by the variation term $\partial v_{x_i}/\partial x_j$, written for simplification as ∂_{ij}. Thus, the reactant j will be said to be:

an activator of i if $\partial_{ij} > 0$;

a self-activator if $i = j$ and $\partial_{ii} \geq 0$;

an inhibitor of i if $\partial_{ij} < 0$; and

a self-inhibitor if $i = j$ and $\partial_{ii} < 0$.

Consequently, the partial derivative ∂_{ij} indicates the coupling between the flux of reactant i and that of reactant j. Higgins' diagram (Fig. 8.4) gives a convenient representation of the qualitative behaviour of chemical reactions coupled in a given region in space. Mathematical analysis thus allows the definition of the signs of the

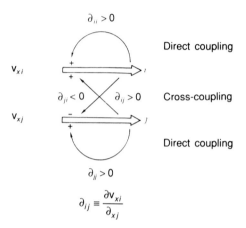

Fig. 8.4.   Direct coupling and cross-coupling (after Higgins (1965)).

partial derivatives in all the regions of space so that the corresponding diagrams may be deduced.

For example, Volterra's dynamic system:

$$\begin{cases} \dfrac{\mathrm{d}x_1}{\mathrm{d}t} = x_1(a - bx_2) \equiv v_{x_1} \\[2mm] \dfrac{\mathrm{d}x_2}{\mathrm{d}t} = (cx_1 - d) \equiv v_{x_2} \end{cases}$$

represents the interaction of two populations where x_1 is the density of the prey and x_2 the density of the predators. This system may be studied by Higgins' method, the partial derivatives being written:

$$\frac{\partial v_{x_1}}{\partial x_2} = -bx_1, \quad \frac{\partial v_{x_2}}{\partial x_1} = cx_2, \quad \frac{\partial v_{x_1}}{\partial x_1} = a - bx_2, \quad \frac{\partial v_{x_2}}{\partial x_2} = cx_1 - d.$$

This may be simplified by putting $a = b = c = d = 1$. Then, for each point (x_1, x_2) of the quadrant $(x_1 > 0, x_2 > 0)$, we have:

$$\partial_{12} < 0 \qquad \partial_{21} > 0.$$

The study of ∂_{11} and ∂_{22} is somewhat delicate as it is necessary to divide the quadrant $(x_1 \geq 0, x_2 \geq 0)$ into four parts: $0 \leq x_1 < 1$, $x_1 \geq 1$, $0 \leq x_2 < 1$ and $x_2 \geq 1$, for example, if $x_1 > 1$ and $x_2 > 1$, then $\partial_{11} < 0$ and $\partial_{22} > 0$. These relations characterise the regulation of the system at each point: x_2 is an inhibitor of x_1 and x_1 an activator of x_2 in the whole quadrant $(x_1 \geq 0, x_2 \geq 0)$; x_1 is a self-inhibitor and x_2 a self-activator in the subset $(x_1 > 1, x_2 > 1)$ of the quadrant.

Another well-known example of regulation is that of feedback to one of the links of a chain of enzyme reactions, as seen in the case of the catalytic hypercycle (Chapter 6). In a chain of reactions involving successive substrates S_i and enzymes E_i:

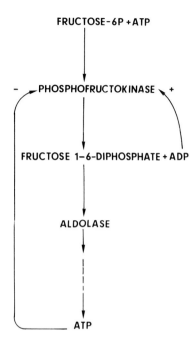

Fig. 8.5.   A glycolytic chain of reactions showing two feedback loops, one positive and the other negative.

$$S_1 \xrightarrow{E_1} S_2 \xrightarrow{E_2} \ldots \xrightarrow{E_{n-1}} S_n$$

the last reactant formed S_n, may be considered to have a positive or, as is more often the case, a negative feedback on the enzyme E_1. Complex metabolic networks would of course involve several feedback loops. Figure 8.5 shows a chain of reactions observed during glycolysis in which the two feedback loops lead to an unstable solution of the system. This phenomenon has been simulated by Higgins (1965). Thron (1991) has investigated a necessary condition for instability at a critical point with a constant concentration of S_1. The condition is expressed in terms of the orders of reaction, the order with respect to S_i being defined as $\partial \log V / \partial \log S_i$, where V is the rate of reaction.

The formalism proposed by Higgins (1967) has lately been developed into a metabolic control theory (Cornish-Bowden, 1989) quite distinct from the biochemical systems theory (Sorribas and Savageau, 1989a, b, c). The results obtained may be generalised as follows. The fluxes or the metabolic concentrations represent the variables of the system, the partial derivatives of the variables with respect to enzyme concentration being the control coefficients. These coefficients describe the influence of a particular enzyme E_i on a flux J, and are defined by:

$$C_i^J = \frac{\partial \ln J}{\partial \ln E_i}$$

The coefficients of elasticity express the influence of the concentrations of substrates, products and inhibitors on the properties of individual enzymes. These coefficients are closely related to the kinetic parameters and may be written:

$$\varepsilon^i_{x_j} = \frac{\partial \ln v_i}{\partial \ln x_j}.$$

However, these coefficients differ from kinetic parameters since they express the sensitivity of the rates of reaction to infinitesimal variations of the conditions prevailing within the cell.

The theory of metabolic control has been developed on the basis of two theorems, the theorem of summation and the theorem of connectivity, which give an elegant expression of the relations of conservation through a single matrix equation:

$$\varepsilon C = M$$

so that the product of the matrix of the coefficients of elasticity and the matrix of the coefficients of control is the unit matrix from 1 to −1 along the diagonal. For example, in the case of a metabolic pathway involving three enzymes:

$$X_0 \xrightarrow{e_1} X_1 \xrightarrow{e_2} X_2 \xrightarrow{e_3} X_3$$

where X_0 is a source and X_3 a sink. This gives:

$$\begin{bmatrix} 1 & 1 & 1 \\ \varepsilon^1_1 & \varepsilon^2_1 & 0 \\ 0 & \varepsilon^2_2 & \varepsilon^3_2 \end{bmatrix} \begin{bmatrix} C^J_1 & C^{x_1}_1 & C^{x_2}_1 \\ C^J_1 & C^{x_1}_2 & C^{x_2}_2 \\ C^J_1 & C^{x_1}_3 & C^{x_2}_3 \end{bmatrix} = \begin{bmatrix} 1 & 0 & 0 \\ 0 & -1 & 0 \\ 0 & 0 & -1 \end{bmatrix}.$$

This relationship includes:

(i)   the theorem of summation for the control of fluxes:

$$\sum C^J_i = 1;$$

(ii)   the theorem of summation for the control of concentrations:

$$\sum C^{x_j}_i = 0;$$

(iii)   the theorem of connectivity for the control of fluxes:

$$\sum \varepsilon^i_{x_j} C^J_i = 0; \text{ and}$$

(iv)   two theorems of connectivity for the control of concentrations:

$$\sum \varepsilon^i_{x_j} C^{x_j}_i = -1$$

$$\sum \varepsilon^i_{x_{k \neq j}} C^{x_j}_i = 0$$

where the summation applies to index i (enzyme).

This formalism is a set of rules leading to the solution C. It has been used for complex topological structures (Giersch, 1988; Small and Fell, 1989), and for time-dependent metabolic systems (Acerenza *et al.*, 1989). Reder (1988) has introduced a structural approach to describe the properties of systems depending only on the structure of the metabolic network and not on the kinetics, which are invariant for the system.

It is highly significant that coupling between reactions determines oscillating systems, i.e. a temporal order, a result already encountered frequently above, for example in the case of the Brusselator (Chapter 2) and of the hypercycle (Chapter 6). This is not surprising since the regulatory process is expressed through the stability of a system of coupled non-linear differential equations. The results just outlined here at the cellular level will again be found at the higher level of the whole organism, for example during hormonal control. In this case, a hormone molecule acts by feedback on one of the organs involved in the synthesis of the hormone itself.

This overall interpretation of regulation requires a complementary explanation in molecular terms. How does one molecule act on another to the point of inhibition or activation, or in other terms, as we saw in Chapter 1, how does the probability of reaction of one molecule on another decrease or increase so as to modify the rate of the chemical reaction? The solution to this important problem was first proposed by Monod and Jacob (1961) and verified a few years later. We have discussed at length, in Chapter 1, the regulation of enzyme reactions through molecular conformational changes due to the allosteric effect. The rate of the chemical reactions is modified by the enormous variety of protein states and molecular interactions. Thus, the allosteric inhibitor, usually an end product of a metabolic pathway, has a molecular structure which differs from that of the enzyme substrate. It cannot therefore react in the same way as the substrate at the same active site on the enzyme. It has been experimentally shown that the active sites are in fact situated on distinct subunits of the enzyme (see Chapter 1). Let us here merely recall that according to the Monod–Wyman–Changeux symmetrical model, the binding of the substrate to a subunit leads to a conformational change of all the subunits, thus increasing the reactivity of all the enzyme sites. In the more general model due to Koshland–Nemethy–Filmer, each subunit is considered to be more or less independent of the others, the overall result being expressed by a positive or negative cooperative effect.

Thus, at the molecular level, the regulatory process consists of a total or partial modification of the active sites through conformational changes. However, we must admit that, in the absence of more information concerning protein–ligand interactions, a certain number of questions remain unanswered.

Finally, let us mention the approach of Schuster and Heinrich (1991) who discuss the general properties of efficient solutions for networks of arbitrary complexity in terms of temporal hierarchy and demonstrate the relationships to the standard method of deriving steady-state models of enzyme kinetics. Their method, based on a principle of optimality which consists of minimizing all intermediate concentrations, shows that for all reaction systems there is a distinct time hierarchy since some reactions are infinitely fast and subsist in quasi-equilibrium. This principle of optimality reinforces our own approach to functional organization from the standpoint of a temporal hierarchy (see Chapter 4).

d.   The phenomenon of inverse regulation

Delattre (1977) devised the formalism of transformation systems (Chapter 3) which revealed the theoretical possibility of inverse regulation. As we shall see in Volume III, Chapter 1, this phenomenon had actually been observed by Bernard-Weil *et al.* (1975) in relation to the post-hypophysio-adrenal axis.

But what exactly is inverse regulation? Let X be a variable regulated by means of an input E. If the variations of E and X are of the same sign, the regulation is said to be direct. If not, we have a case of inverse regulation. Direct regulation occurs more frequently than inverse regulation, which is only found in non-linear systems. This property is readily demonstrated with the axioms of transformation systems. Indeed, the method is inductive and may be used to investigate the behaviour of any set of classes of elements defined by their functional equivalence.

The equation governing the cardinals N_j of the classes (j) in physically linear systems is given by Eq. (3.9):

$$\dot{N} = KN + \varepsilon$$

where K is the matrix of coefficients k_{ij} describing the transfer of elements between classes, with:

$$k_{ij} \geq 0 \text{ for all } i \neq j$$
$$k_{jj} \leq 0.$$

It can then be shown that:

$$|k_{jj}| \geq \sum_{\substack{i=1 \\ i \neq j}}^{n} k_{ij} \quad j = 1, \ldots, n. \tag{8.2}$$

In other words, K is a matrix with a dominant principal diagonal. Thus, if there exists a j such that the above inequality can be considered strict, it may be deduced that the eigenvalues λ_i of K have a real negative part so that the system is stable in the Lyapunov sense (Chapter 2). Delattre (1977) has given a detailed demonstration of this result, which may be summed up as follows:

- Any perturbation of N_i, $i = 1$ to n, will cancel itself.
- Any perturbation of (k_{ij}), i and j varying from 1 to n, due to internal or external field effects, will do the same, thus conserving stability. Consequently, if we consider that the regulation consists in the action of external 'forces' ε_i, $i = 1$ to n, on the function of state of the system $(f(N_j))_{j=1,n}$ about the equilibrium N_0 defined by:

$$KN_0 + \varepsilon = 0 \quad \text{or} \quad N_0 = -K^{-1}\varepsilon \tag{8.3}$$

we see that the elements b of the matrix $-K^{-1}$ have positive or zero values when $\forall j \neq k$, $k_{jk} \geq 0$ and $\mathrm{Re}(\lambda_i(K)) < 0$. Finally, a variation δN of the output about the state of equilibrium is linked to a variation $\delta\varepsilon$ of the input by:

$$(\delta N)_0 = (-K^{-1})_0 \, \delta\varepsilon \quad \text{with} \quad b_{kj} \geq 0 \; \forall k,j = 1 \text{ to } n \tag{8.4}$$

the matrix $-\mathbf{K}^{-1}$ being calculated in the neighbourhood of equilibrium. This corresponds to the direct regulation defined above. It may also be considered as 'regulation by compensation' as a perturbation in a class (k) may be compensated by a variation in a class (j).

How can this reasoning be extended to non-linear systems? The general equation describing the behaviour of classes (j) is given by Eq. (3.6):

$$\dot{N}_j = -S_j + Q_j + \varepsilon_j$$

where S_j and Q_j depend on N_k. About the stable singularity defined by:

$$-S_j + Q_j + \varepsilon_j = 0$$

we may write the variation of $\dot{N}_j$, that is $\delta \dot{N}_j$, in the form:

$$\delta(-S_j + Q_j)_0 + \delta\varepsilon_j = 0$$

or in the matrix form:

$$\delta(-\mathbf{S}_j + \mathbf{Q})_0 + \delta\boldsymbol{\varepsilon} = \mathbf{0}.$$

Let $\mathbf{V}$ be the Jacobian matrix (Appendix B) of $(-\mathbf{S} + \mathbf{Q})$ with:

$$V_{jk} = \left(\frac{\partial}{\partial N_k} (-S_j + Q_j) \right)_0 .$$

Then:

$$\delta(-\mathbf{S} + \mathbf{Q})_0 = \mathbf{V}\delta\mathbf{N}_0 = -\delta\boldsymbol{\varepsilon}.$$

Thus we have:

$$\boxed{\delta\mathbf{N}_0 = -\mathbf{V}^{-1}\delta\boldsymbol{\varepsilon}} \tag{8.5}$$

an expression comparable to (8.4). However, the structure of matrix $\mathbf{V}$ is not as well defined as that of matrix $\mathbf{K}$ because of the presence of non-linear terms of the type AN_jN_k in the equation for N_j. The results (8.3) and (8.4) above are no longer applicable and *a negative output may well correspond to a positive input*, which is the definition of inverse regulation. These results will be used later when we consider the endocrine system (Volume III, Chapter 1) and the nervous system (Volume III, Chapter 2).

III. Cell growth: an introduction

Growth, development, differentiation and morphogenesis are the four key concepts of developmental biology. They are of considerable importance as they are a fundamental characteristic of the living world. Let us draw on Waddington (1970), who has put forward a rigorous formalisation of these notions, for the essential definitions of these concepts.

1. *Growth*

The notion of growth appears to be intuitively evident but it turns out to be less so the deeper we go into the subject. Should growth be defined in terms of increased size or volume, due to the entry of new molecules? Or should the concept be restricted to the increase in the quantity of matter capable of self-reproduction, in other words, the increase in the total quantity of information?

Growth has been defined in various ways. In the present state of knowledge, it would seem legitimate to distinguish between the information-carrying molecules capable of reproduction and those involved in maintaining biological activity. We may thus admit that:

- *At the molecular level*, growth is expressed on the one hand as DNA replication and on the other as the synthesis of specific molecules.
- *At the cellular level*, at least three concepts are necessary:
  (1)   the rate of cell division;
  (2)   the rate of cell death and disappearance; and
  (3)   the rate of increase of the volume of individual cells.
- *At the tissue level*, the definition is rather more delicate as it depends strongly on the notion of the cell. However, we may consider the differential growth of the various parts of the system, in other words, examine the transformations that bring about changes from one substructure to another in a given system.

2. *Development*

Development is global expression and thus of relatively poor value. The term may be used to cover the gradual or random variations that an individual goes through during a lifetime.

3. *Differentiation*

The process of differentiation concerns the passage from a given state to a different one. The problem is the definition of what is meant by a *different* state. It is obviously necessary to distinguish between the gradual cellular differences arising during embryogenesis and those occuring at specific points of the developing system, as for instance in the course of regeneration of a part of the system. Examples of this are the regeneration of a hydra or the development of wings in a chicken.

The first type of differentiation is known as *histogenesis*. This is the gradual development of a substance homogeneous at given instant, or the transformation of a developing system. It is characterised by a *temporal order*. A cell contains all the information necessary to the future development which will lead to its differentiated form. This type of differentiation is of major importance. It is indeed the central problem of embryology, as summed up by the question: how does a cell, or a group of cells, 'know' the precise state of development attained in space and in time, and how does structural and functional differentiation occur? In other words, is there a *pre-established* program?

Following Waddington (1962) we must stress this type of differentiation which expresses the development of a cell, or a group of cells, the finality of which is known from the beginning.

The second type of differentiation, which should be clearly distinguished from the first, has been described by Waddington as *regionalisation*. This process consists of the development of two different structures, separate in space, arising from a previously homogeneous structure.

How are we to interpret the two spatial and temporal processes, regionalisation and histogenesis, defined respectively as a partition of a unitary system into spatially distinct subsystems and as a continuous variation of the chemical composition with time? Some examples will be considered in the chapters that follow.

4. *Morphogenesis*

Morphogenesis, closely linked to the two processes described above, refers to the *development of geometrical form*. Morphogenesis, histogenesis and regionalisation are three essential aspects of epigenesis.

Waddington (1962) and Thom (1972) have defined morphogenesis as the *set of all the processes that create or destroy form*. This very general sense of the expression calls for some qualification. The processes involved are not isolated but accompanied by changes in the nature of the components of the system. These variations may be due to the mutually dependent processes of histogenesis and regionalisation. This type of morphogenetic phenomenon is called *individualisation*. During the creation of forms, the constituent cells may then develop in the course of time, or follow different courses in different regions. Moreover, when individualisation is accompanied by an increase in mass, we again find the notion of growth as described above.

Given these definitions, it now remains to be seen how the phenomena may actually be described and what dimensions should be used to define the state of the system. Before going on to the mathematical models leading to the first explanations, it is worth noting that, if the variables of time and chemical concentration define the phenomenon of histogenesis at a given instant and if, in addition, the space variation describes the phenomenon of regionalisation, then the process of morphogenesis will be interpreted in space–time and that of individualisation in the same space–time supplemented by the dimensions of chemical concentration.

9

Regulation of Cell Function through Enzyme Activity

I. Introduction to the regulation of enzyme synthesis

Enzyme activity is regulated at two levels: according to the *quantity* of the enzymes formed and the *activity* of the enzymes involved in the various processes of biosynthesis and catalysis. The latter, called allosteric regulation since it occurs when enzyme reactions are caused by conformational changes, has already been described in Chapter 1. We shall now consider enzyme regulation in terms of the quantity of enzymes formed. The process takes place at the 'elementary' level during the transcription and the translation of the genetic message. The quantity of enzymes synthesised depends on environmental demands, i.e. on the existence or the non-existence of the molecules necessary to cell metabolism. This mechanism is essentially economic in terms of energy when the cell is in the 'equilibrium state', with fairly constant molecular concentrations.

In this connection, let us recall the experimental results obtained by Jacob and Monod (1961):

(i)   *escherichia coli*, cultured in a medium made up only of lactose, contains large quantities of the enzyme β-galactosidase;

(ii)   when cultured in a glycerol medium, the bacteria contain no β-galactosidase;

(iii)   however, the introduction of lactose into the glycerol medium leads to synthesis of the enzyme. Thus, lactose induces the synthesis of β-galactosidase; and

(iv)   another enzyme, the permease, allows lactose to pass through the cell membrane.

The lactose (glucose and galactose bound by a β-galactosidase link) is used as an energy source by *E. coli* after the compound has been decomposed into glucose and galactose. Steps (i) and (ii) do not follow the same metabolic pathways, and enzyme synthesis occurs only in case of need. This corresponds to an *environmental adaptation*. Moreover, as soon as the medium contains lactose, the bacteria synthesise the permease which allows lactose uptake.

How are we to explain this interaction between environment and organism from the lower level, i.e. the genes that govern the synthesis of the enzyme? Between the maximum synthesis of the enzyme and no synthesis at all there is of course a wide range of intermediate values. How does the regulation take place? We shall see, in the following section, that the process may be represented by a system of differential equations and that certain conditions imposed on the solutions then correspond to mechanisms of regulation.

Let us consider a gene, an independent chromosomal unit, controlling the synthesis of an enzyme E. Let S be the gene and $E(S)$ the enzyme formed. We have seen that S is made up of a sequence of nucleotide bases. A mutation may therefore transform the gene S into a gene $\bar{S}$ coding for an inactive enzyme. S is called the *structural gene*. Let S_1 and S_2 be the structural genes controlling the synthesis of two enzymes involved in a single function (in the present case, permease and β-galactosidase). Similarly, let R be the inductor gene governing the production of a protein R_e necessary to the synthesis of E. The bacterium is said to be *inductive* as an extrinsic inductor is required to counter the action of R_e. When this protein is not necessary to the synthesis of E, the gene which governs the synthesis of a protein inactive in the formation of E is called $\bar{R}$. In this case, the bacterium is said to be *constitutive*.

It requires to be shown that the set $\{R, S_1, S_2, ..., S_n\}$ composed of the inductor gene R and the structural genes S_i, $i = 1$ to n, is a functional unit. It may be remarked that R also is a structural gene since it ensures the synthesis of a protein contributing to enzyme synthesis. Genetic recombination experiments have been designed to investigate the role played by R. Following mutation, the gene S, which is a sequence of nucleotides, may be considered as a new structure. For example, the mutation of S into $\bar{S}$ may lead to the replacement of amino acids A or B by amino acids $\bar{A}$ or $\bar{B}$ respectively. Consequently, we have the following configurations:

$$S = AB$$

and

$$\bar{S} = \bar{A}\bar{B} \text{ or } A\bar{B} \text{ or } \bar{A}B.$$

It is interesting to study the consequences of a mutation which leads to a change of an amino acid at locus A or at locus B. Intuition is here confirmed by experimental results: the functional role of S depends on the localisation of the mutation, i.e. $A\bar{B}$ and $\bar{A}B$ are incapable of producing β-galactosidase. It may therefore be concluded that S constitutes a functional unit. Such functional units are called *cistrons*.

Table 9.1. Results of experiments in genetic recombination in *Escherichia coli*. The bacteria, male $\overline{RS}$ and female RS, cannot synthesise enzyme E because, in the first case, the structural gene S has undergone a mutation and, in the second case, there exists a repressor R_c in the cytoplasm. However, recombinant bacteria can synthesise the enzyme (the first operand indicated corresponds to the male bacterium). We should distinguish between the inductive or constitutive cytoplasm and the inductive or constitutive gene. (See text for the notations used)

	Genes	Observation	Cause
Bacteria	$\overline{RS}$ constitutive	No synthesis of E	Mutation: $\overline{S}$
	RS inductive		The cytoplasm contains R_c
Recombinants	1. $\overline{RS} \times RS$	No synthesis of E	The cytoplasm of the female contains R_c
	2. $RS \times \overline{RS}$	Synthesis of E *during limited time Δt*	No repressor in cytoplasm of the female: RS constitutive gene of the male
		After Δt: no synthesis of E	R synthesises R_c whence repression

Similar experiments in genetic recombination have revealed the role of the inductor R. Let X be the operator of the recombination between bacteria of opposite sexes, the male being represented by the left operand and the female by the right, thus producing a zygote.

The results are summed up in Table 9.1 which clearly shows a property of 'non-commutativity' for the recombinants. However, during recombination, genetic material from the male is injected into the cytoplasm of the female so that it is the presence of R or $\overline{R}$, i.e. of R_e, in the cytoplasm that leads respectively to the non-synthesis or to the synthesis of E. With this hypothesis, the recombinant 2 (Table 9.1), which possesses R and S, should after a certain time, i.e. the time necessary for the synthesis of R_e in the cytoplasm of the zygote, inhibit the synthesis of E by S. Experiments have confirmed *the existence of a repressor R_e synthesised by the gene R so that the induction is in fact a de-repression*. From these results, Jacob and Monod have deduced the two models of regulation shown in Fig. 9.1. However, in certain mutants, the heterozygotes 1 and 2 synthesise E even in the presence of R_e. This is why a further hypothesis, later verified experimentally, was proposed concerning an operator gene O, not involved in the synthesis of any cytoplasmic substance, the set $\{ROS\}$ acting as a 'dominant inductive gene'.

Finally, although a certain number of points still remain to be cleared up, we are led to the representation of regulation in the biosynthesis of proteins as shown in Fig. 9.2.

Induction is a de-repression controlled by the operator gene. When the operator is bound to the repressor, no synthesis takes place. When the operator is not bound to

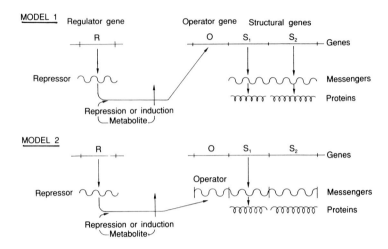

Fig. 9.1.   Gene regulation according to Jacob and Monod (1961).

the repressor — which occurs when the inductor is bound to the repressor — then the corresponding structural genes are expressed. The set composed of the operator gene and the structural gene is called the operon.

But which precisely are the variables controlled during the process of regulation, and how is this control exercised? Judging by the examples already given, it is obvious that the variables controlled are the products of metabolism: excess of a product must stop the synthesis of a key enzyme in its own metabolic pathway, and, conversely, lack of the product must accelerate the process. Then the product P will have to act on the repressor. But in what form? It appears that P acts on the repressor synthesised in an inactive form (called *aporepressor* for the regulator gene) by transforming it structurally into the active form.

In short, *the unemployed product of a metabolic pathway binds to the aporepressor, causing a change in structure and producing the active form of the repressor. This then binds to the operator which in turn blocks the synthesis of the key enzyme or enzymes of the metabolic pathway.* The structure of this system of control is very simple and appears to be valid at least in the case of bacteria.

As we shall see below, the system of control in the higher organisms is far more complex. For example, the system described by Revel *et al.* (1972) implies such precise regulation that a mathematical description becomes a very delicate affair.

The differences in the frequency and distribution of nucleotides and nucleotide substitutions in the DNA regions involved in the regulation of gene expression provide valuable information about the mechanisms of gene regulation. In some cases, regions from different loci of the same species, or from homologous loci of different species, are located at similar positions and possess similar sequences. It is thus likely that such regions perform similar functions (Shabalina *et al.*, 1991).

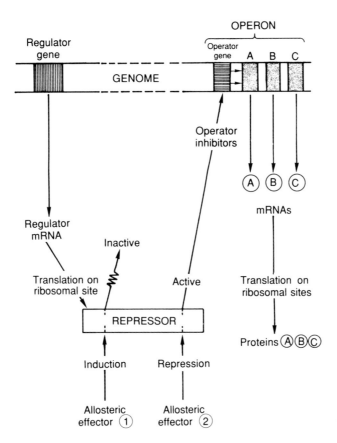

Fig. 9.2.   The regulation of protein biosynthesis in bacteria. The allosteric effect changes the structure of the aporepressor producing an active repressor that blocks enzyme synthesis.

II.   Theoretical model of regulation

According to the theoretical model of regulation proposed by Jacob and Monod, the level of concentration of a product acts on the concentration of the active repressor which then blocks the synthesis. We may therefore consider that regulation in a metabolic network takes place as indicated in Fig. 9.3 where the structural genes GS_1 of operon I lead to the synthesis of an enzyme E_1, which works through P_1 on operon II and activates the aporepressor R_2.

Using the formalism of chemistry, it can be shown that the Jacob–Monod regulation process is adequately represented by the set of reactions in Figs 9.3 and 9.4.

The state variables of the system are the concentrations of enzymes E_1 and E_2 synthesised from messenger RNA X_1 and X_2 copied off the structural genes GS_1 and GS_2. These variables, noted $[E_1]$, $[E_2]$, $[X_1]$ and $[X_2]$, satisfy the system of differential equations when appropriate units are used:

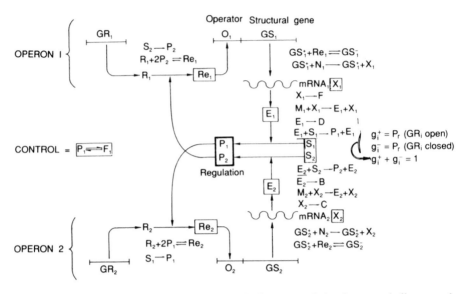

Fig. 9.3. Jacob and Monod's theoretical model of gene regulation in a metabolic network (after Babloyantz and Nicolis (1972)).

$$R_1 + 2P_2 \underset{k'_1}{\overset{k_1}{\rightleftharpoons}} Re_1 \qquad\qquad 2P_1 + R_2 \underset{k'_8}{\overset{k_8}{\rightleftharpoons}} Re_2$$

$$G_1^+ + Re_1 \underset{k'_2}{\overset{k_2}{\rightleftharpoons}} G_1^- \qquad\qquad G_2^+ + Re_2 \underset{k'_9}{\overset{k_9}{\rightleftharpoons}} G_2^-$$

$$G_1^+ + N_1 \xrightarrow{k_3} G_1^+ + X_1 \qquad\qquad G_2^+ + N_2 \xrightarrow{k_{10}} G_2^+ + X_2$$

$$X_1 \xrightarrow{k_4} F \qquad\qquad X_2 \xrightarrow{k_{11}} C$$

$$M_1 + X_1 \xrightarrow{k_5} E_1 + X_1 \qquad\qquad E_2 \xrightarrow{k_{12}} B$$

$$E_1 \xrightarrow{k_6} D \qquad\qquad M_2 + X_2 \xrightarrow{k_{13}} E_2 + X_2$$

$$E_1 + S_1 \xrightarrow{k_7} P_1 + E_1 \qquad\qquad E_2 + S_2 \xrightarrow{k_{14}} P_2 + E_2$$

$$P_1 \xrightarrow{k'_7} S_1 \qquad\qquad P_2 \xrightarrow{k'_{14}} S_2$$

$$P_1 \underset{k'_{15}}{\overset{k_{15}}{\rightleftharpoons}} F_1$$

Fig. 9.4. The set of chemical reactions corresponding to operons 1 and 2 in Fig. 9.3. (For simplification we have used the notation $k_1 = k_+^1$, $k'_1 = k'_-$ ....)

$$\begin{cases} \dfrac{d[E_1]}{dt} = k_5[X_1] - k_6[E_1] \\[2mm] \dfrac{d[E_2]}{dt} = k_{13}[X_2] - k_{12}[E_2] \\[2mm] \dfrac{d[X_1]}{dt} = -k_4[X_1] + \dfrac{A}{K + [E_2]^2} \\[2mm] \dfrac{d[X_2]}{dt} = -k_{11}[X_2] + \dfrac{A}{K + [E_1]^2 + F'^2 + 2F'[E_1]} \end{cases}$$

obtained from the chemical reactions in Fig. 9.4. The first two equations assume that the concentrations $[M_1]$ and $[M_2]$ of the precursors remain unchanged. To simplify we have written:

(i) $K = \dfrac{k_2}{k_2'} = \dfrac{k_9}{k_9'}$

where K is the equilibrium constant of the reactions which lead the structural gene GS_i ($i = 1$ or 2) from the 'open' to the 'closed' form (respectively noted G_i^+ and G_i^-), by binding to the repressor Re_i

(ii) $A = k_3[N_1]K = k_{10}[N_2]K$

where A is the quantity involving the irreversible reaction of mRNA X_i synthesis from its precursors N_i, the 'closed' form gene concentration G_i^-, and the repressor Re_i concentration;

(iii) $\dfrac{k_1}{k_1'}[R_1]\left(\dfrac{k_{14}[S_2]}{k_{14}'}\right)^2 = \dfrac{k_8[R_2]}{k_8'}\left(\dfrac{k_7[S_1]}{k_7' + k_{15}}\right)^2 = 1.$

(iv) $\dfrac{k_8[R_2]}{k_8'}\left(\dfrac{k_{15}'F_1}{k_7' + k_{15}}\right)^2 = F'^2.$

By equating the second members of the set of differential equations above to zero, we obtain a fifth-degree algebraic polynomial equation which, as a consequence of the non-linearity of the system, shows multiple stationary states. Babloyantz and Nicolis (1972) have shown that this equation in E_1 describing the stationary states of the system has three solutions (Fig. 9.5) in a given interval of the parameters when the product F_1 is considered as a *constraint* of the thermodynamical system. This result is of interest as the system can go from one stationary state to another in a small interval I_1 for a given variation of the other parameters of the system. In other words, in a two-way regulating system where the *command* is operated by the reaction:

$$P_1 \rightleftharpoons F_1$$

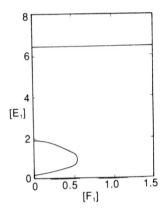

Fig. 9.5.   The Jacob and Monod model showing multiple stationary states for E with instability in the intermediate state leading to transitions of the 'all-or-nothing' type. The values of the parameters are: $k_5 = k_{13} = 0.5$, $k_4 = k_{11} = 5$, $A = 20$, $K = 1$ and $k_6 = k_{12} = 0.3$ (after Babloyantz and Nicolis (1972)).

which describes the system as *open*, in the thermodynamic sense, with respect to a reservoir containing the final product of degradation F_1, a small variation of the parameters will lead to a radical change in the state of the system. Moreover, it can be shown that this transition occurs far from thermodynamic equilibrium.

Finally, as the *regulation* is imposed by the equation:

$$\frac{k_1}{k_1'} [R_1] \left(\frac{k_{14}[S_2]}{k_{14}'} \right)^2 = 1 \qquad (9.1)$$

or, using appropriate units:

$$[Re_1] = [E_2]^2 \qquad (9.2)$$

and

$$[Re_2] = ([E_1] + F')^2$$

it can be seen that a decrease in $[E_1]$ leads to an increase in $[E_2]$.

Consequently, in this model of *cross-regulation*, of which a formal representation was given in Chapter 8, the stable stationary states correspond to one of the pathways, the others being inhibited. The command, as we have just seen, is operated by the variation of the concentration of the metabolic product F_1. Commutation towards a state of high concentration is extremely sensitive and takes place without any modification of the information carried by the genome. Only the values of the other parameters $[S_1]$ and $[S_2]$ play a role, $[F_1]$ being fixed. According to Babloyantz and Nicolis (1972), transition phenomena are related to the asymmetry of the system because of the non-equilibrium coupling with the environment. Thus, the phenomena of commutation between different pathways would seem to be of the same nature as the transitions leading to dissipative structures.

Although in the main the results obtained by Babloyantz and Nicolis appear to be justified, further investigation of the control system in more complex cases is still necessary. The thermodynamic interpretation of open systems brings to light an essential element, namely the surprising analogy between an irreversible transition occurring beyond a point of chemical instability and the sudden, all-or-nothing change that takes place following slight modifications of the parameters of the system. The thermodynamic theory of irreversible processes (see Chapter 3) provides an interesting physical explanation for biological processes which, from the mathematical point of view, are precisely situations of catastrophic transition related to the structural stability of the systems. Chevalet *et al.* (1981, 1983), using recent experimental data, have worked out an elegant mathematical model for the operon. The representation is highly complex as the dynamic systems used take account of time-lag terms. The numerical solutions obtained reveal the extraordinary diversity of possible states and the indetermination that consequently results.

Segel and Perelson (1992) have used the mechanism controlling the number of plasmids per cell, i.e. the plasmid copy number, to illustrate the quasi-steady-state assumption (Segel, 1988) in the regulation of DNA replication (see Chapter 2, Section I.2). We may recall that plasmid replication, unlike chromosome replication, can be experimentally perturbed without lethal effects to the cell. Thus, plasmids can be employed as vectors for the cloning of genes. An increase in the plasmid copy number results from the deletion of a specific gene, the Rom gene, Rom being a 63 amino acid protein that encodes downstream from the origin of replication. Rom increases the interaction between two RNA molecules, RNA I and RNA II, transcribed from opposite strands of DNA, thereby increasing the inhibition of replication (see Segel and Perelson (1992) for the molecular mechanisms involved, or Cesarini and Banner (1985) for a review of the topic). The study of the regulation of gene expression is currently based on a general, abstract approach using linguistic terms. Thus, Collado-Vides (1989) has proposed the use of a generative grammar to construct an integrative paradigm for genomic organisation.

III. Regulation of protein biosynthesis in higher organisms

The Jacob–Monod model described above was initially designed to explain the regulation of enzyme biosynthesis in bacteria, in particular that of β-galactosidase in *E. coli*. Not surprisingly, this model is not totally valid for higher organisms although the principle, *which supposes the existence of regulator, operator and structural genes*, still holds good. Indeed, because of the inherent complexity of higher organisms, a stimulus, produced for example by a hormone, will activate a great number of structural genes. This is a very different situation from the simple case of *E. coli* in which only two structural genes are involved in the enzymic response to the availability of lactose in the medium, those that control the synthesis of permease and β-galactosidase.

The difficulty arises from the fact that, in higher organisms, the model must take into account a multiple response originating from several sources widely dispersed over the genome. The fact that the quantity of DNA varies in different species according to the degree of organisation suggests that the size of the genome increases with complexity. Britten and Davidson (1969) have found evidence in favour of a system of gene regulation in higher organisms. However, as the fundamental pathways of biosynthesis are found at all levels of complexity of the organisms, it appears difficult to establish a direct relationship between the size of the genome and its 'complexity'.

The only explanation possible is given by the *hypothesis of redundance*, or the repetition of DNA sequences of at least several hundred nucleotides in length. This phenomenon has been observed in 15 to 80% of total DNA according to the species studied. Two other factors have been examined: the frequency of repetitions which is of the order of 10^2 to 10^6, and the accuracy of the copies which varies from 1/3 to 2/3. Britten and Davidson conclude that the main difference between a sponge and a mammal may lie in the higher degree of integration of cell activity, and thereby *in the greater complexity of regulation*, rather than in larger number of structural genes.

Thus, an increase in the number of controlling and regulating processes appears to be more likely to characterise the degree of organisation of a species than the number of biochemical pathways in enzyme biosynthesis. This, of course, is not a new idea, and has been discussed earlier, particularly in Chapters 4 and 5. The difficulty lies in constructing a model that takes into account these hypotheses since in fact it is the number of combinations of the structural genes, or rather the number of associations of operon-like units in the active state, which creates the *functional complexity*. Thus, there must exist batteries of genes which are inactive in their normal state but which, when activated, govern a definite function. Moreover, a given gene may belong to several different batteries of genes.

The process described by Britten and Davidson corresponds to an instruction loop as shown in Fig. 9.6.

The response to an external signal takes place in five steps, each corresponding to a distinct gene. Britten and Davidson give a precise significance to each of the five types of RNA. *Producer genes*, corresponding to the structural genes described by Jacob and Monod in bacteria, produce an RNA substrate, for example messenger RNA. *Receptor genes* are bound to producer genes. The producer genes ensure translation only when the specific complex: RNA-*activator*–receptor gene–producer gene is formed. The RNA–activator is synthesised from the *integrator gene*, which itself is activated by the *sensor gene*. The sensor gene is essentially the part of the genome that acts as the site of fixation for the product molecules. The sensor gene thus plays a specific role.

The different steps of this system are shown in Fig. 9.7. The two levels of interaction, the first at the integrator gene level and the second at the receptor gene level, are clearly indicated.

The sensor gene responds specifically to the action of a molecule by activating an integrator gene which, by transcription, produces an activator gene. The activator gene then stimulates the receptor gene. Consequently, the synthesis of the final

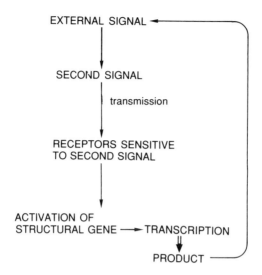

Fig. 9.6. A succession of steps in the Britten and Davidson model of regulation in higher organisms.

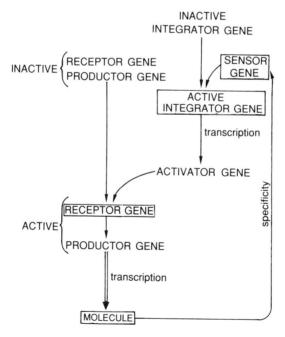

Fig. 9.7. Gene regulation in Britten and Davidson's model.

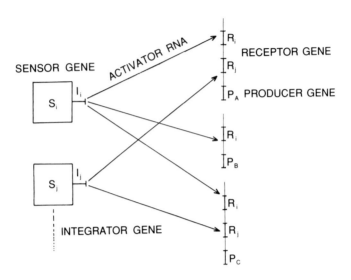

Fig. 9.8.   The various types of gene taken into account by Britten and Davidson.

product may involve the activation of either the battery of integrator genes or the battery of receptor genes (Fig. 9.8). However, these two variations of the same model lead to identical results. DNA repetition, or redundance, is thus explained in the context of the 'increasing complexity' of organisms, since the action of a specific molecule leads to the expression of several genes. This model will also be used when we examine the phenomena of cell differentiation (Chapter 12).

It is difficult to understand how the activity of transcription is controlled with respect to time. Zuckerkandl (1978a), commenting on the temporal control of chromosomes reported by Ashburner *et al.* (1973), considers that a mere *positive* specific control would be insufficient to explain the findings. He believes there must exist an additional transconformational mechanism involving non-specific histones. However, according to Britten and Davidson, it is the non-specific histones that satisfy the conditions of a *negative* control. The Britten and Davidson model is further criticised for not including genes of the *emitter* type although the gene regulation unit is made up of contiguous parts corresponding to the structural gene and the receptor gene. Then, using the analogy of the operon in procaryotes, Zuckerkandl defines the functional unit of gene action (*fuga*) as a continuous length of DNA containing functionally distinguishable parts which make up a unit of transcription and control. Although the various types of gene encountered (or likely to be encountered!) have hitherto been qualified as well as possible, it would now appear indispensable to provide precise definitions based on a sound taxonomic rule. Zuckerkandl's classification, based on the two well-established steps of gene transcription and translation, is compatible with the *fuga* concept. Thus, the polypeptide chains arising from transcribed and translated structural genes may have two functions: the regulatory function and the 'cellular' function, i.e. participation in cell metabolism. A structural gene may then be considered as a 'regulator'.

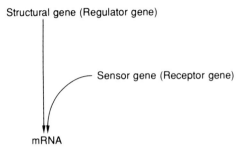

Structural gene (Regulator gene)

Sensor gene (Receptor gene)

mRNA

Fig. 9.9. A simplified diagram showing the role of the sensor gene in Britten and Davidson's model of protein synthesis.

A gene may be non-transcribed or be such that the products of its transcription do not act on the genome. If made up of a short sequence of nucleotides, it is a *receptor gene*. If the sequence is long, it is *conformational* DNA, controlling the activity of the *fuga*. If the gene is transcribed but not translated, the RNA formed may either have a cellular function (the gene in question is then analogous to a structural gene) or may interact with a regulator RNA.

Obviously the subtle distinction between structural genes and regulatory genes needs to be clarified. Let us therefore adopt the following definitions proposed by Zuckerkandl (1978b):

- a *regulator gene* belongs to a particular class of structural genes;
- a *regulator molecule* is either an RNA or a protein synthesised under the control of a regulator gene;
- a *receptor gene* is a target portion of DNA situated near a structural gene;
- a *control gene* is a regulator gene or a receptor gene; and
- an *emitter gene* is a part of the DNA that generates products of translation and/ or transcription. It is therefore also a structural gene or a regulator gene.

The simplified representation in Fig. 9.9 suggests that, in Britten and Davidson's model, the binding of a protein to a sensor gene activates, by an allosteric effect, an integrator gene leading to the synthesis of a messenger RNA under the control of a structural gene.

The system thus appears to follow only one direction in which a receptor gene bound to a regulator gene controls its transcription activity. Zuckerkandl's criticism is mainly aimed at the idea of a double level of interaction corresponding to a double transcription activity. He considers this incompatible with the protein–sensor gene specificity which varies during the development of the organism. The problem may be solved by negative control being exercised by an emitting sequence integrated into the *fuga*, the functional regulating unit composed of:

(i) structural genes and associated sequences of transcribed mRNA;
(ii) transcribed sequences of hnRNA (heterogeneous nuclear RNA); and
(iii) non-transcribed DNA sequences involved in transcription control.

It is not possible to go into further details here as this is a field of research in rapid development. The reader is invited to consult the work of Zuckerkandl for the implications of the *fuga* concept in molecular anthropology. In particular, the author shows how gene regulation is involved in tissue development and differentiation. Given the complexity of these phenomena, with direct and inverse feedback control, the interpretation of such a model would surely gain a great deal if it were expressed in mathematical terms. This would undoubtedly lead to better definitions of the fundamental concepts and to the construction of new propositions.

10

Cell Growth and Morphogenesis

I. General aspects

How do cells and tissues, so different from each other in complex organisms, arise from a single original cell, the fertilised egg? As we shall see, the answer to this question lies in *cell differentiation*, a process closely related to the phenomena of cell control and regulation. Another associated process, *morphogenesis*, which is fundamentally similar, determines the form of the organism through cell differentiation. The two processes correspond to different levels of description, the first being essentially molecular while the second deals with the structure of tissues and organisms. Bouligand (1980) gives a good overview of this topic.

The important problem of cell differentiation should be considered, on the one hand, with respect to *embryogenesis*, i.e. the transformation of the fertilised egg not into a multicellular mass of identical cells but into an organism made up of distinct cells generally organised in the form of separate tissues (*histogenesis*) and, on the other, with respect to the *regeneration* of tissues after they have been formed (*regionalisation*).

The initiation of complex genetic programs under the effect of the internal cellular environment is a mysterious process. It appears to depend on the activation of specific parts of some of the genes among the extraordinarily large variety of genes available to the cell. Indeed, cell differentiation remains one of the most fascinating and elusive problems in biology today. However, recent advances in genetic engineering offer some hope that in the coming years a beginning of an explanation may emerge concerning the execution of these genetic programs.

Let us now consider the basic concepts underlying the phenomena of cell differentiation and morphogenesis, bearing in mind that research in these fields is soon

315

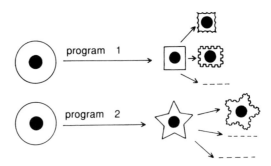

Fig. 10.1.   Program of cell differentiation: each step gives rise to a specific cell.

likely to produce considerable changes. Although the theoretical studies available at present are incomplete, they are already proving useful in the formalisation of the problems, and some solutions may be expected in the near future.

Since cells reproduce themselves by binary division, an essentially discrete process, *where, when* and *how* does cell differentiation actually take place?

As the quantity of information stored in the genes of a cell is known to be invariant, it appears quite likely that cell differentiation is based on differential gene activity, i.e. the initiation of activity in certain parts of the genes by specific molecules. But here two questions arise. First, is differential genetic activity the cause rather than the effect of differentiation? Secondly, how should genetic activity be defined? Does genetic activity refer to the transcription of a gene on to RNA or to its translation into a specific protein, or does it necessarily involve both these phenomena? Indeed, it is now known that gene transcription is not always accompanied by translation.

Above, we have used the term 'initiation of genetic programs' as it is now becoming increasingly clear that the process is first triggered in a definite *place* and at a definite *time*. Moreover, the process appears to resemble the working of a computer program, i.e. the automatic and sequential execution of a set of instructions. The observation that all the cells produced by a differentiated cell are cytologically very similar is in favour of this idea. Furthermore, even the most complex multicellular organisms are composed of only a few hundred types of cell. The number of types is very small if we consider the billions of cells that make up an individual. The only explanation currently held to be valid is that each of the fundamental cell types has its own program of differentiation and that all the other cells are produced by particular steps in this program (Fig. 10.1).

Differentiation has been defined as the process by which different cells arise from one initial cell. But in what ways do the cells differ? If we admit that different cells have different *functions*, then even cells with relatively little organisation, such as procaryote bacteria, undergo something like differentiation during their lifetime. The triggering of an operon engages the bacterium in a metabolic network with a highly specific function such as division, sporulation or aggregation. The differentiation in this case is a minor one but can be more extensive in some procaryotes,

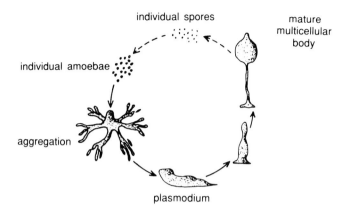

Fig. 10.2. Life cycle of *Dictyostelium discoideum*.

such as the Myxobacteria, in which cell aggregation is followed by the formation of a fruiting body containing multitudes of spherical cells (McLean, 1975). The Acrasiales (slime moulds) are another unicellular variety which go through a similar process of differentiation. Because of their deterministic social behaviour, a theoretical model of the Acrasiales would seem to be of interest.

II. Unicellular organisms

1. *Differentiation in the Acrasiales*

Experimental observations on *Dictyostelium discoideum*, which is a species of the Acrasiales (Myxomycetes), show long-range intercellular interactions of repulsion or attraction according to the stage of the cell cycle. Repulsion occurs after germination whereas attraction appears in the presence of food and is oriented towards the source. In the absence of food, the distribution of the organisms first tends to be uniform and then shows points of aggregation that are capable of differentiation.

This behaviour is closely analogous to a reaction of order obtained by fluctuations in a uniform distribution (Chapter 3). But can we, in the case of amoebae, really refer to the *structuring of order* or, as Prigogine (1977) puts it, to social behaviour of the type '*order through fluctuations*'?

The problem has been formulated by Keller and Segel (1970) and Segel (1980). It is known that the aggregation of the cells is caused by a chemical substance, acrasine (identified as cyclic AMP in *Dictyostelium discoideum*) which can be broken down by an enzyme, acrasinase. The life cycle of the organism is shown in Fig. 10.2 and the aggregation phase in Fig. 10.3. Two types of phenomena appear: (i) morphogenesis or the aggregation phase; and (ii) differentiation from the plasmodium ending up in mature fruiting body. Here, we shall consider only the phenomena of morphogenesis.

Let $a(\mathbf{r}, t)$ be the concentration of the amoebae, $\rho(\mathbf{r}, t)$ and $\eta(\mathbf{r}, t)$, the concentrations of acrasine and acrasinase respectively, at a point $(\mathbf{r}, t)$.

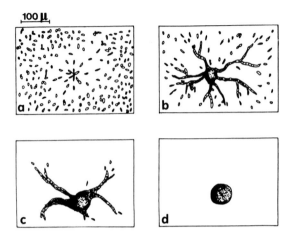

Fig. 10.3. Four steps in the aggregative phase of *Dictyostelium discoideum*. (a) Individual spores; (b) aggregation; (c) plasmodium; (d) multicellular body.

The model requires the following hypotheses:

(i)   acrasine is produced at the rate of $f(\rho)$ per amoeba;
(ii)   acrasinase is produced at the rate of $g(\rho, \eta)$ per amoeba;
(iii) acrasine and acrasinase interact according to the equation:

$$\rho + \eta \underset{k_{-1}}{\overset{k_1}{\rightleftharpoons}} C \overset{k_2}{\longrightarrow} \eta + P \qquad (10.1)$$

which is the fundamental equation of enzyme kinetics (Chapter 2) where P is the product;

(iv) acrasine, acrasinase and the product P diffuse according to Fick's Law; and

(v) the concentration of the amoebae varies *chemotactically* because of the acrasine, and *randomly* because of the diffusion. This double influence is represented by the equation:

$$\frac{\partial a}{\partial t} = -\mathbf{V} \cdot (D_1 \mathbf{V} \rho) + \mathbf{V} \cdot (D_2 \mathbf{V} a) \qquad (10.2)$$

where the second member is the sum of two terms, the first describing the chemotaxis and the second the diffusion effect. The parameter D_1 is the chemotactic coefficient, with the form $D_1 = \alpha \dfrac{a}{\rho}$ (it has been shown that the flux of amoebae is proportional to the density and that amoebae are sensitive to the gradient $\mathbf{V}\rho/\rho$), and D_2 is the coefficient of diffusion. The solution to the problem may be simplified by supposing that the complex C is in the stationary state and that the total enzyme concentration is a constant $\eta_0 = \eta + C$, so that, according to Eq. (10.1), we have:

$$\frac{\partial C}{\partial t} = k_1 \rho \eta - (k_{-1} + k_2) C = 0$$

thus:

$$\eta = \frac{\eta_0}{1 + \dfrac{k_1}{k_{-1} + k_2}\rho}.$$

Using hypotheses (i) and (ii) together with Eq. (10.1), the variation of the concentration of acrasine may be written:

$$\frac{\partial \rho}{\partial t} = -k_1 \rho \eta + k_{-1} C + af(\rho) + D_\rho \nabla^2 \rho$$

or:

$$\frac{\partial \rho}{\partial t} = -k(\rho)\rho + af(\rho) + D_\rho \nabla^2 \rho \tag{10.3}$$

with:

$$k(\rho) = k_1 \eta - \eta_0 k_{-1} \frac{K}{1 + K\rho}$$

and $K = k_1/(k_{-1} + k_2)$. Equations (10.2) and (10.3) constitute the simplified dynamic system describing the behaviour of the amoebae. The homogeneous solution of the system corresponds to the thermodynamic branch. At equilibrium, $a = a_0$ and $\rho = \rho_0$ are related by Eq. (10.3):

$$a_0 f(\rho_0) = k(\rho_0)\rho_0.$$

The condition for stability about the equilibrium may be sought by linearising the system. For small peturbations $|\delta a| \ll a_0$ and $|\delta \rho| \ll \rho_0$, by writing:

$$\bar{k} = k(\rho_0) + \delta \rho \left(\frac{\partial k}{\partial \rho} \right)_0$$

we find that the system becomes unstable if:

$$\frac{D_1 f(\rho_0)}{D_2 \bar{k}} + \frac{a_0 \left(\dfrac{\partial f}{\partial \rho} \right)_0}{\bar{k}} > 1$$

or in the neighbourhood of equilibrium:

$$\boxed{\frac{\alpha}{D_2} + \frac{a_0}{\bar{k}}\left(\frac{\partial f}{\partial \rho} \right)_0 > 1} . \tag{10.4}$$

The *first term* of the inequality, $\alpha/D_2 = \rho D_1/a D_2$, contains the chemotactic factor D_1 which becomes preponderant as soon as $f(\rho)$ is roughly constant, i.e. as soon as the production of acrasine is independent of the concentration of acrasine in the medium. When this rate of production is constant, the local concentration of amoebae and of acrasine, and thus the stability of the system, is affected to some extent, on the one hand, because of the diffusion by D_2 and the degradation by $\bar{k}$ and, on the other, because of the production of acrasine by $f(\rho)$ and the chemotactic effect by D_1. The *second term* of the inequality, characterised by $a_0 \left(\dfrac{\partial f}{\partial \rho} \right)_0 > \bar{k}$, shows an instability in the acrasine level ρ.

To sum up, the system will be unstable if:

α increases, i.e. if the acrasine gradient expressed by D_1 increases, or if D_2 decreases;

$a_0 \dfrac{\partial f}{\partial \rho}$ is large enough, i.e. if the acrasine production f increases, or if the acrasine concentration ρ increases.

Thus, it may be stated that amoebae aggregations will necessarily occur as soon as these conditions exist. The conditions are created by the properties of the microorganisms and may be expressed by an increased sensitivity to the acrasine gradient and to the rate of acrasine production. These results have in fact been substantiated by experimental observations.

We have here a remarkable application of the theory to a singular behaviour which is finally explained on the one hand by the sensitivity of amoebae to a chemical substance they secrete, cyclic AMP, i.e. by the double phenomenon of chemotaxis and of diffusion in the biochemically active medium, and on the other by a spatiotemporal order created by the fluctuations of initially uniform conditions. This spatiotemporal order constitutes a morphogenetic structure. Other examples will be considered later.

In conclusion, the study of some of the simplest living organisms allows us to establish some general rules:

- *The form of the organism is the consequence of cell differentiation.* This principle will be applied to morphogenesis.
- *The constituent parts of a differentiated organism become interdependent from the structural and the functional points of view. Plasticity is lost*, the process of differentiation having become irreversible. The loss of plasticity varies according to the 'complexity' of the organism. This principle governs the regeneration of the whole or part of the organism.
- *There is stability in the differentiated state*: the cell retains its original function.
- *It is always the most 'complex' part that becomes dominant during the lifetime of the organism.*

2. *Human red blood cells*

Human erythrocytes are peculiar in that they lose their nuclei during development and do not divide. However, as these cells are constantly renewed at the estimated

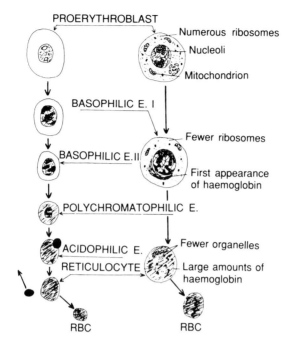

PROERYTHROBLAST

Numerous ribosomes

Nucleoli

Mitochondrion

BASOPHILIC E. I

Fewer ribosomes

BASOPHILIC E.II

First appearance
of haemoglobin

POLYCHROMATOPHILIC E.

ACIDOPHILIC E.

Fewer organelles

RETICULOCYTE

Large amounts of
haemoglobin

RBC   RBC

Fig. 10.4. The erythropoietic system: development of the erythrocyte series. E = erythrocyte.

rate of 2×10^{11} cells per day, there must necessarily exist stem cells giving rise to erythrocytes.

Figure 10.4 shows the different cell types arising from the stem cells of the erythropoietic system. There are three distinct ways in which a stem cell may divide, producing: two stem cells, a stem cell and a pro-eythroblast, or two pro-erythroblasts. Erythropoiesis is regulated by a hormone, *erythropoietin*, which controls the concentration of the different cell types.

Thus, stem cells may be considered to be poorly differentiated cells that allow the organism to compensate for the loss of plasticity of the fully differentiated cells incapable of regeneration. Many other examples of this type of development are now well known: the lymphoid cells of the immune system, the regenerating cells of the skin, and so on.

3. *Some examples of cell differentiation at the molecular level*

Cells are said to be differentiated if they have same genome and if certain parts of the genome are differently expressed, i.e. if the cells synthesise different molecules. With this definition, bacteria cannot be considered to differentiate in any durable fashion other than by a temporary adaptation to the environment as, for example, by the triggering of the operon when lactose becomes available in the medium. These transient variations are part of the normal cell cycle. What we are interested in are the permanent differences which, at the macroscopic level, reflect changes occurring at the molecular level. The richer the biochemical pathways of an organism, the greater will be its tendency to specialise in a given function. Indeed, this is precisely what is observed in higher organisms and has been described by Zuckerkandl (1976) as the functional unit of gene action.

In this respect, the molecular study of haemoglobin (Chapter 1) is of considerable interest. The comparison of the amino acid sequences of proteins in different tissues and in different organisms, and in particular the determination of the monomers synthesised in a given cell together with the combinations of monomers found in the tissues, should yield a useful indirect description of tissue differentiation.

It has been established that the different globins, which are finally assembled to form the haemoglobin molecule, are synthesised at different times during cell development. Moreover, the synthesis of the different human globins is independently controlled by specific genes. Thus, the process can in fact be considered as true cell differentiation.

III.   Higher organisms

1. *Embryogenesis*

A good understanding of the essential characteristics of living organisms can only be obtained through a study of their development. Indeed, it is during the developmental stage that the processes of cell division and differentiation occur most vigorously. The successive — often overlapping — steps of metazoan development transform the egg in the unicellular stage into the fully autonomous multicellular organism.

a.   *Principal steps of embryogenesis*

Let us briefly recall the terms used in human embryogenesis. Details will of course be found in any standard textbook on embryology (for example, Houillon, 1969), but here we shall merely summarise the main steps: fertilisation, segmentation, implantation, gastrulation, neurulation, and organogenesis.

Fertilisation

Fertilisation is the step during which the ovule is transformed into an egg, the *ovocyte*, with a diameter of about 0.1 mm, located within the follicular envelope. At first, the development of the ovocyte is similar to that of the other higher vertebrates, but soon it becomes more complex as the placenta is formed. The placenta is a protective envelope which isolates the ovocyte from the maternal body while allowing the passage of nourishing substances.

The rupture of a graafian follicle, a periodic occurrence in the ovary, releases an ovocyte which then migrates to the fimbriae at the upper end of a uterine tube. Fertilisation occurs as soon as a spermatozoid penetrates the cell membrane of the ovule which is then transformed into an *egg*. The fertilisation is followed by the steps below, which occur more slowly in mammals than in other vertebrates.

Segmentation

Segmentation begins with the egg being cleaved, probably along a meridian, into two *blastomeres*. The second cleavage takes place asynchronously on the two blastomeres, and the segmentation then continues until 16 or 32 blastomeres are formed. The end of the segmentation corresponds to the appearance of the *morula*. A reorganisation which occurs during the segmentation leads within 3–4 days to the formation of a *primary blastula* or *blastocyst* (Fig. 10.5). The blastocyst contains, in particular, a cavity, of which the peripheral wall gives rise to part of the embryonic appendages, and an inner cell mass or *embryoblast* which gives rise to the organic regions of the embryo.

The final step of segmentation leads to the formation of the *secondary blastula*, in which the peripheral layer, called the *trophectoderm*, extends to form the *hypoblast* covering the inner surface of the embryoblast, at this stage called the *epiblast*, before spreading around the blastocyst cavity.

Thus, the segmentation consists of a series of binary divisions leading to the production of blastomeres, organised in three formations: *trophectoderm*, *epiblast* and *hypoblast*, arranged around the blastocyst cavity (Fig. 10.6).

Implantation

By the second week of development, the free blastocyst reaches the uterus where it undergoes *implantation* in a mucosal crypt. As soon as this occurs, *amniogenesis* and *placentation*, i.e. the formation of the embryonic structures, begins.

Amniogenesis takes place by cavitation (Fig. 10.7) since the epiblast is still covered by the trophectoderm. The amniotic cavity is formed in the space surrounded by the *ectoamnios*, the *ectoblast* and the *mesoblast*. Simultaneously, the *primary yolk sac*, which is very small in the human embryo, appears within the endoblast, giving rise to the viteline circulation which carries the first blood cells. Then, the trophectoderm forms the *chorion*, which eventually gives rise to the fetal part of the placenta.

Placentation, the formation of the placenta, occurs at almost the same time as the nutritional demands of the early stages of embryogenesis practically exhaust the resources of the egg. The *allantois*, formed by a diverticulum of the endoblast, grows towards the chorion. At this stage there are two influences at work due to the blastocyst and the uterine mucosa. The *placenta*, strictly speaking, consists of the *fetal* placenta together with the *maternal* placenta.

At the end of the second week, the yolk sac of the human blastocyst is larger than the amniotic cavity.

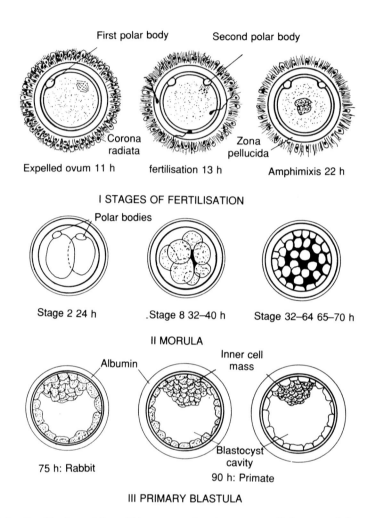

Fig. 10.5. Fertilisation of the rabbit ovum and various stages of cleavage of the early embryo. The initial time corresponds to the mating period (after Houillon (1969)).

Gastrulation

Gastrulation corresponds to the set of morphogenetic processes leading to the formation of the three *primary germ layers*: the *ectoderm*, the *mesoderm* and the *endoderm*. This follows the development of the hypoblast and the epiblast that occurs during segmentation, as seen above, but it would be difficult to distinguish the different steps chronologically as they are roughly synchronous. Just after the formation of the yolk sac and the amniotic cavity, the fundamental layers of the zone of separation between these structures give rise to the *primitive streak* (Fig. 10.8) which then lengthens to form the *primitive knot* and the *neural plate*. At day 19 the primitive streak is completed within the embryonic disc (Figs 10.9 and 10.10).

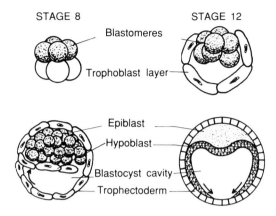

Fig. 10.6. Formation of the endophyll in mammals (after Houillon (1969)).

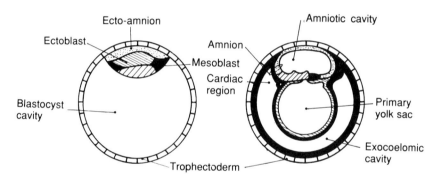

Fig. 10.7. Two steps in the formation of the amniotic cavity in mammals (after Houillon (1969)).

Neurulation

Neurulation starts at the beginning of the fourth week, corresponding to the appearance of the first *somite*, i.e. the first segmetal formation of the mesoderm at day 21 or 22. This is followed by the development of the neural plate and the neural tube, the fusion of the neural folds which occurs at the stage when seven or eight somites are present, and the closure of the neural tube. After neurulation, the development of the intestines, the lungs and the thyroid begins.

Organogenesis

In the fifth week the region containing the primitive streak becomes very small compared with the neural plate, and is finally transformed into the *end bud*. The embryo is then fairly well formed and this first phase of development is followed by *organogenesis*, during which the various tissues are assembled to form the different organs of the body.

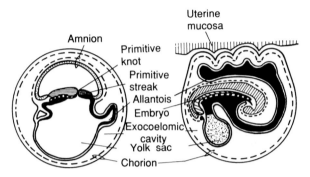

PRIMITIVE STREAK STAGE   DEVELOPMENT OF THE ALLANTOIS

Fig. 10.8. Development of embryonic appendages in mammals: left, the primitive streak which will later become the embryo at the base of the amniotic cavity; right, the embryo and the allantois develop while the yolk sac shrinks. The association of the allantois and the chorion leads to the formation of the fetal placenta (after Houillon (1969)).

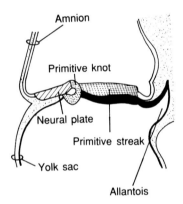

Fig. 10.9. The primitive streak at day 19 (see also Fig. 10.8; after Houillon (1969)).

These then, in brief, are the various steps occuring at the very beginning of life. The question that immediately arises is: how are we to explain the changes that take place in an organism during its development? Is the mature organ of the adult entirely 'contained' in the initial egg? Or does the organ undergo progressive change, with a permanent regulation of the process, until adulthood? These two approaches, respectively known as the *preformation theory* and the *epigenetic theory*, have been carefully explored. Indeed, since the development of organisms seems to follow a deterministic pattern, considerable effort has been devoted to finding the laws of development. This, of course, is the aim of experimental embryology.

Thus, over the past century there has been much controversy over the two supposedly opposed standpoints:

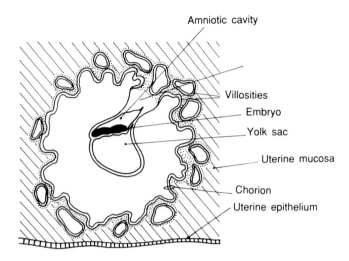

Fig. 10.10.   Implanted human blastocyst at day 19. Neurulation appears at the fourth week (after Houillon (1969)).

- a *mosaic* behaviour, in which the organogenesis *corresponds exactly* to the real potential of the egg, with the formation of distinct territories, the fate of each being precisely determined; and
- a *regulated* behaviour, in which the organogenesis induced is *superior* to the real potential of the egg, i.e. the egg is capable of regulation.

But it has now been demonstrated that the two ideas are in fact *successively valid in time: the egg, homogeneous at first, develops under a regulating process,* gradually breaking up into a partition of *independent territories.* This finding points to the existence of a certain number of *morphogenetic fields* (defined below), which progressively enter into action as the power of the regulation decreases. The final result is the appearance of a set of regions, the future of each being totally determined. Before going on to a mathematical interpretation, let us try to justify the concepts involved.

b.   *Concepts of cell differentiation*

(1)   *Regulation*

Regulation has been defined as the process by which a part of an embryonic system reacts so as to carry out the full set of its normal morphogenetic functions. The reason for the regulation is easily understood but how the regulation is actually achieved is a delicate question. One answer may lie in the existence of gradient fields such as those experimentally demonstrated in sea-urchin embryos: (i) an *animal gradient* decreasing from the animal pole to the vegetal pole; and (ii) a corresponding *vegetal gradient* in the opposite direction.

We have already seen that a two-variable system of the activator–inhibitor type allows a good regulation of local equilibrium. However, regulation over a region would require three gradients.

(2)   *Morphogenetic fields*

Morphogenetic fields are defined as *embryonic regions, with no visible morphological or histological differences, destined to develop eventually into specific organs.*

Clearly defined regions appear progressively as the embryo develops. Thus, for example, the neuroblast is determined as nervous tissue at the level of the neural plate as early as the neurulation stage.

(3)   *Induction*

The phenomenon of induction is defined by a double manifestation: (i) the *transfer of potentiality from one region to another*; and (ii) the *harmonisation of the resulting organism*.

The notion of induction has been defined on the basis of the experimental results of grafts on blastulae. The interesting point is that during normal development a strongly determined region may act by induction on a mass of poorly determined cells. This could be explained by the existence of an *organising centre* or a *differentiating centre* capable of the long-distance induction of specific regions through the emission of chemical substances, the inductors, acting on appropriate receptors.

More precisely, it may be considered that different inductors correspond to the various regions of the organising centre, and that the induction is all the better (in terms of the quality of the development of the organism) that the inductor and the receptor match perfectly. This phenomenon is called the *regionalisation of induction*, and clearly justifies the notion presented in Section I of this chapter.

We have dwelt at some length on the phenomena of embryogenesis for, while the fate of all organisms is of course birth, growth and death, it must be acknowledged that any solution to the riddle of life must first explain the processes by which a single cell, the egg, gives rise to an autonomous, functional living organism. These are, of course, the highly intricate and inconceivably complex processes of cell division, cell differentiation and intercellular communication. Developmental biology is surely one of the most fascinating aspects of the extraordinary phenomenon we call life.

2.   *Positional information and cell differentiation*

How does a cell actually 'know' the state of its own development and how does it react to a given situation? Many attempts have been made to determine the factors involved in the triggering of the differentiation process, which may be taken to reflect the 'knowledge' a cell has of its state of development. The first idea, which is also the simplest, is to consider that the cell acquires '*positional information*' with respect to a specified origin. Another idea is that of epigenetic selection (Sachs, 1988), the mechanism of which is quite different from that of pattern formation.

As indicated above, the results of embryological studies suggest the existence of physiological gradients during development. Such positional information must necessarily be related to a system of coordinates within which each cell can be located. Whatever the mechanism leading to the existence of *cell coordinates* in the cell mass, it must first produce an *invisible primary pattern* which then becomes visible as a *morphological pattern*.

Some interpretations of cell location have led to theoretical models based on general properties due to lateral inhibition with local autocatalysis (Oster, 1988) and even to some mathematical models (Newman *et al.*, 1988; Nagorcka, 1989). Wolpert (1969, 1981) has carried out extensive investigations involving planned cell permutation on hydra, an organism particularly suitable for experimental grafting and regeneration. He concludes that there exists a universal gradient in the organism, created either by a substance, the *morphogen*, or by a signal (the point remains to be cleared up) which is recognised by all the cells immersed in the gradient. Thus, the eventual development of the cell is determined by its specific position in the field.

Wolpert (1968) illustrates his idea with the 'problem of the French flag', an image that conveniently fits a one-dimensional field. Such fields are characterised by three properties:

(i)   The property of *polarity*, which indicates the orientation of the coordinate system with respect to an origin: the field is unipolar if cell position is defined with respect to only one end, and bipolar if cell position is defined with respect to both ends.

(ii) The property of *morphallaxis*, illustrated in Fig. 10.11. If the flag is cut along xx', a new gradient $M'M_2$ is established instead of M_1M_2 and the three regions, blue, white and red, are regenerated according to this gradient; and

(iii) The property of *epimorphosis*, an embryological term corresponding to *invariance of size*, which — as opposed to morphallaxis — leads to a new growth until the *initial* positional information is attained, i.e. the length is conserved. Thus, if part of the morphogenetic field of an organism is suppressed and if the two ends are joined at the site of the cut, the remaining part will be capable of regulation and the initial *pattern* will be restored. Therefore, *the respective proportions of the different parts of the pattern are independent of the total size of the organism.*

Mathematically, let L be the length of the system and M the concentration of the morphogen at x. After reduction, the length of the system decreases to L'. The size will remain invariant if the concentration of the morphogen M at the point x of the initial system is equal to the concentration at x' of the modified system, where x and x' satisfy the relationship:

$$\frac{x}{x'} = \frac{L}{L'} \quad \text{with} \quad M(x) = M(x'). \tag{10.5}$$

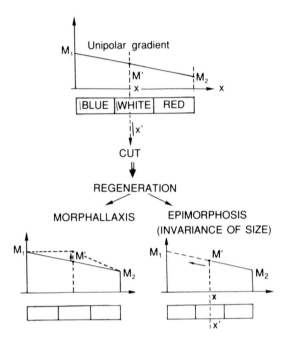

Fig. 10.11. Wolpert's 'problem of the French flag' illustrates positional information in a one-dimensional morphogenetic field. Two properties may appear: morphallaxis and epimorphosis (see text) (after Wolpert (1968)).

In a morphogenetic field in which the morphogen is transported by diffusion, it can be shown that the invariance of size can exist only if defined as above. Consider the simplest equation of diffusion:

$$\frac{\partial M}{\partial t} = D_M \frac{\partial^2 M}{\partial x^2}$$

with the stationary solution:

$$M = \frac{P - S}{L} x + S \tag{10.6}$$

where S and P are respectively the concentrations of the source and the sink. The condition (10.5) is satisfied since the coefficient D_M, the cell characteristic, is not involved. Now, with the reaction term $f(M)$, we have:

$$\frac{\partial M}{\partial t} = f(M) + D_M \frac{\partial^2 M}{\partial x^2}$$

and in the stationary state:

$$f(M) = -\frac{D_M}{L^2} \frac{\partial^2 M}{\partial \alpha^2} \quad \text{with} \quad \alpha = x/L.$$

The invariance of size is indicated by $\alpha = \alpha'$ and $f(M(x)) = f(M(x'))$, which implies that the diffusion coefficient changes from D_M to $D_{M'}$ to compensate for the change in length from L to L', that is:

$$D_M/L^2 = D_{M'}/L'^2 \ .$$

(10.7)

This important result depends on the sole hypothesis that morphogen transport occurs by diffusion and there exists a reaction term $f(M)$.

This idea is supported by several observations:

(1) *Cell differentiation may result from cell movement and, conversely, cell differentiation may lead to displacement of cells*, as in the phenomenon of the sorting out of cells. Of course, we do not here refer to the locomotive behaviour of certain cells, but rather to the interrelationship *between a field of spatial information in the mathematical sense and a cell subjected to such a field*. We have already seen an example of this behaviour in the case of *Dictyostelium discoideum*. Another example, rather surprising in this context, is provided by the extensive displacement certain cells undergo before becoming functional: *lymphoid cells*, for instance, are formed in the thymus but then migrate to the spleen and the ganglia before playing a part in the immune defence system.

Cohen and Robertson (1972a, b) have identified five types of morphogenetic movement:

(i)   local rearrangements during particular stages of embryogenesis, such as the cleavage and formation of the blastula;
(ii) independent movement of individual cells, as the in the migration of lymphoid cells and in the aggregation of microorganisms such as the Acrasiales;
(iii) coordinated movements of groups of cells in one dimension (forming a line of cells) or in two dimensions (forming a leaflet), as during neurulation;
(iv) extension of cell processes while the cell itself remains stationary; and
(v)   variations of form leading to morphogenetic movement, such as the movement resulting from a change in the sign of the surface curvature as the blastula gives rise to the gastrula, and again as the gastrula gives rise to the neurula.

Movements of type (i) and type (v) are difficult to analyse because of the co-operative phenomena involved. But movements of types (ii) (iii) and (iv) are quite well represented by the example of the amoebae *Dictyostelium discoideum* discussed above, in which the movements are totally deterministic, resulting from chemotaxis (positive or negative) and random diffusion.

(2) *Cell-to-cell contacts play a fundamental role in cell differentiation*. This aspect is compatible with the existence of a morphogenetic field created by a substance migrating from one cell to another. A discussion of the different cell-junctions will be found in Volume II, but for the present it will suffice to say that all junctions, whatever the type, allow the transfer of molecules. In particular, intercellular junctions

appear to be of crucial importance in embryogenesis during which cell differentiation is at its peak.

Thus, intercellular contacts appear necessary, not only for good cell adhesion, but also for the process of differentiation. McMahon (1973) suggests that intercellular contacts are the bases of the transmission of positional information, and hypothesises the existence of complementary 'contact-sensitive' molecules with the following properties:

(i)   these molecules are active as soon as they enter into contact with the specific complementary molecule on the membrane of the adjacent cell. Activation consists of the regulation of a morphogenetic substance A;

(ii)   more precisely, there exist two types of active molecule, one increasing the concentration of A and the other decreasing it;

(iii) the direction of the field — called the *polarity* of the field — is defined by the distribution of the sensitive molecules in the field; and

(iv)   the concentration of A is modulated by negative retroaction, i.e. a stimulation decreases the concentration and vice versa.

The experimental bases of this model will be found in McMahon's original communication. Here again, cyclic AMP is considered to be the 'signal' allowing the cells to locate their positions in the field. Cyclic AMP is known to act as a second messenger to several hormones through an intermediate membrane-bound molecule, adenyl cyclase, which plays the role of a regulator. Moreover, cyclic AMP is involved in the regulation of enzyme activity, in genetic transcription and in cell permeability, all of which are essential to cell differentiation.

Let us consider a discrete, one-dimensional network in which the ith cell is identified by the index i. F_i^* and R_i^* are molecules sensitive to cell contact, respectively of the 'front' and 'rear' types. When all the possible contacts between the molecules F_i^* and the R_{i-1}^* are established, contact activation appears, and the molecules are called F_i, R_{i-1} (Fig. 10.12a). Postulate (1) states that the complementary sensitive molecules in contact, F^* and R^*, are activated under the form F and R and that the activation is transmitted from one cell to another, where:

$$F_i = R_{i-1} = \min (F_i^*, R_{i-1}^*). \tag{10.8}$$

Moreover, if we consider Postulate (2), F decreases A, the concentration of cyclic AMP, whereas R increases this concentration. A inhibits the production of R^* and increases that of F^*. We may then, omitting the index i, write the kinetic equations with the rate constants l_1, m_1, l_2 and m_2:

$$\frac{\mathrm{d}F^*}{\mathrm{d}t} = l_1 A - m_1$$

$$\frac{\mathrm{d}R^*}{\mathrm{d}t} = -l_2 A - m_2 + \left(\frac{\mathrm{d}R^*}{\mathrm{d}t}\right)_0. \tag{10.9}$$

Postulate (2) may be expressed in the form:

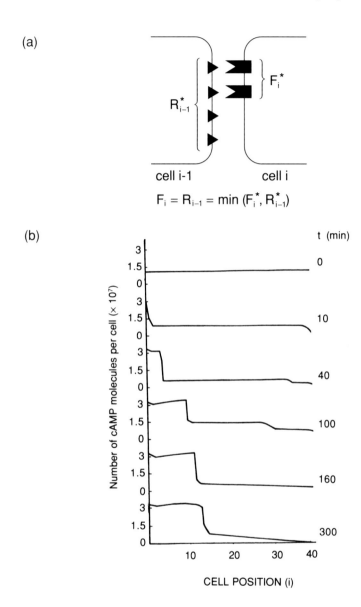

(a)

$$F_i = R_{i-1} = \min (F_i^*, R_{i-1}^*)$$

(b)

Fig. 10.12. Model of cell differentiation according to McMahon (1973). (a) Contact activation; (b) the number of cAMP molecules depends on cell position, whence a morphogenetic field.

$$\frac{dA}{dt} = k_2 R - k_1 FA \qquad (10.10)$$

where k_1 is the rate constant of destruction of A, and k_2 is the rate constant of formation.

This kinetic formulation calls for some remarks from the standpoint of usual chemical kinetics. First, it is supposed that R^* is produced even in the absence of A (since we have the term $(dR^*/dt)_0$). Secondly, the rates l_1 and l_2 are truly independent of F^* and R^* respectively. And finally, $(-k_1FA)$, the rate of destruction of A, depends on A. But, on the contrary, from a formal point of view, its rate of formation depends only on R. In effect, the ATP concentration is much greater than the adenyl cyclase constant K_M. We may therefore suppose that the process involves a pseudo-first-order reaction. These hypotheses have been substantiated by the cyclic AMP assays carried out on *Dictyostelium discoideum*.

The numerical solution of the system gives the concentration of A in terms of the position of the cell (Fig. 10.12b). A *regulatory morphogenetic field* is clearly at work. This model, together with other more recent models, will be discussed later.

(3) *The action of molecules on cell surfaces*. Certain hormones can intervene in the process of cell differentiation. We have seen an example of this with erythropoietin in the differentiation of stem cells leading to the formation of red blood cells. Other examples are the response of vertebrate leucocytes to the action of phytohaemag-glutinin, and that of lymphocytes to tuberculinin. The model above supplies a further example, as the quantity of cyclic AMP within a cell depends on the polypeptide hormones secreted at a distance. This may explain the interactions that occur between the distant, fully differentiated cells in higher organisms.

3.   Control and cell differentiation

While the mechanisms of control and regulation have been relatively well established in the case of bacteria, through the tested model of Jacob and Monod (Chapter 9), in higher organisms these mechanisms are still matters of speculation (Britten and Davidson, 1969; Georgiev, 1969; Zuckerkandl, 1978b). However, it is admitted that there are two types of control: *transcriptional* and *post-transcriptional*.

a.   Transcriptional control

It is currently believed that acid proteins act as specific gene regulators in the chromatin of eucaryote cells, and that non-histone proteins probably play a role in transcription. It may be considered, while awaiting further experimental proof, that acid proteins or RNA molecules hold the key to the regulatory function of specific gene transcription.

As for the mechanism, we have seen how Britten and Davidson (1969), intrigued by the enormous quantity of DNA in eucaryote cells, were led to suppose that the majority of genes are merely transcribed but not translated, so that it is likely that these play only a regulatory role. Another important observation is that the position of the gene may influence its expression and thus the timing of its transcription. In fact, it has been experimentally demonstrated that:

(i)   a specific enzyme appears at a particular moment of the cell cycle; and
(ii) there exists a bijection between the chromosome map and the chronological succession of enzyme syntheses.

It may be deduced that a sequential reading of the DNA would lead to progressive differentiation of the cells. Simple molecular mechanisms can account for sequential activation of appropriate groups of genes throughout development and for specific constraints on developmental pathways (Bodnar *et al.*, 1989). Indeed, the existence of a strong coupling between the regulation of the cell cycle and cell differentiation has been established. At the theoretical level, the analysis made by Babloyantz and Kaczmarek (1979) would appear to be fruitful. However, the abundance of models in this field clearly reflects our present state of ignorance.

b.   Post-transcriptional control

This type of control concerns all the processes that may alter or modify the transcribed message before or during its translation into a protein (McLean, 1975). However, it is not known whether control takes place at the translational level alone; it appears more likely that a control would occur at different levels in time and in space, particularly within the nucleus and the cytoplasm. Boncinelli (1978) has proposed a model of this type for eucaryote cells.

c.   Post-translational control

Mechanisms of post-translational control of cell differentiation are important as they act at the level of enzyme activity within the major metabolic networks. For example, the existence of systems controlling enzyme breakdown is well known. Here we may sum up three essential results:

(i)   the rate of protein breakdown is specific to the cell;
(ii)   the breakdown rate is also tissue specific; and
(iii) protein breakdown follows first order kinetics.

The main point is that the mechanisms of cell differentiation are situated at all possible levels and may operate wherever modulated, differential enzyme activity occurs.

d.   Mathematical models of control in cell differentiation

Edelstein (1972) and Babloyantz and Hiernaux (1975) have proposed a quantitative description of the relationship between genetic control and cell differentiation. The authors deduce the properties of a model consisting of:

- a morphogenetic field created by the diffusion of a morphogenetic substance between a source and a sink by passive or by active or assisted transport; and
- a system of genetic control or regulation of the Jacob–Monod type.

Considering here again, for simplification, a one-dimensional network containing N cells, $i = 1$ to N, the phenomena of the control and the transport of the morphogen I may be described as follows:

(i)   the operon in its active state leads to the synthesis of two proteins: protein E which represents the differentiated state of the cell, and protein M which allows the transport of the morphogen I from one cell to a neighbouring cell (permease, for instance); and

(ii)   the substance I can inactivate the repressor.

Satisfactory operation of the model should be reflected by a non-linear function $E(I)$ increasing sharply beyond a certain threshold. This corresponds to an 'all-or-nothing' response often considered a good description of a new cell function. As we shall see, MacWilliams and Papageorgiou (1978) have found a subtler response which is probably closer to the truth. The chemical equations ensuring genetic control of the Jacob–Monod type are as follows:

$$R' \rightleftharpoons R_i$$
$$R_i + O_i^+ \rightleftharpoons O_i^-$$
$$R_i + 2I_i \rightleftharpoons F_1 \qquad \text{Condition (ii)} \qquad (10.11(1))$$
$$\alpha + O_i^+ \rightarrow O_i^+ + E_i + M_i \qquad \text{Condition (i)}$$

R, a repressor molecule synthesised from its precursor, R', can repress the operator from O^+ to O^-. The substance I can combine with the repressor R to inactivate it. The proteins E and M can then be synthesised. M is a permease which ensures the entry of substance I into the cell. In a very general way, the transport equations may be written:

$$M_i + I_{i-1} \underset{k_-}{\overset{k_+}{\rightleftharpoons}} M_i + I_i$$
$$M_i + I_{i+1} \underset{k_-}{\overset{k_+}{\rightleftharpoons}} M_i + I_i$$
$$M_i \rightleftharpoons F \qquad (10.11(2))$$
$$E_i \rightleftharpoons G$$

With $k_+ = k_-$, and Eq. (10.11(1)) for certain values of the constants, we have *facilitated transport* across the membrane, i.e. a transport superior to that which would occur by diffusion (see Volume II).

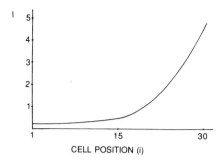

Fig. 10.13. Primary pattern $I(i)$ according to the model due to Babloyantz and Hiernaux (1975) with a Jacob–Monod type gene control.

With $k_+ > k_-$, we have *active transport*, i.e. the morphogen accumulates in the cell against a gradient of concentration. In the *quasi-stationary* state in which:

$$\frac{dR_i}{dt} = \frac{dO_i^+}{dt} = \frac{dO_i^-}{dt} = 0 \quad \text{for} \quad i = 2 \text{ to } N - 1 \qquad (10.12)$$

we obtain the following system of differential equations where f, g and h are functions deduced from kinetic theory (the details of which will be found elsewhere):

$$\left.\begin{aligned}
\frac{dE_i}{dt} &= f_i(E_i, I_i) \\[6pt]
\frac{dM_i}{dt} &= g_i(M_i, I_i) \\[6pt]
\frac{dI_i}{dt} &= h_i(M_{i-1}, M_i, M_{i+1}, I_{i-1}, I_i, I_{i+1})
\end{aligned}\right\} \qquad (10.13)$$

The limit concentration values are such that the function $E(I)$ shows a sudden increase. Thus, this model supposes the existence of a source and a sink. These two extreme values at $i = 1$ and $i = N$, by active transport of the morphogen, first determine a gradient, the *primary pattern* I(i) (Fig. 10.13), and then the *morphological pattern* E(i) (Fig. 10.14). *It has been found that when the pattern is established by active transport, only the value of the source is determinant in the number of cells undergoing induction. On the contrary, if we suppose the existence of diffusion, the length of the field and the limit concentration values play a part. This is because, in the latter case, we take no account of the phenomenon of epimorphosis.*

The connection between the theory of positional information, here established by active transport, and the theory of genetic control according to Jacob and Monod leads us, in spite of its simplicity, to known properties of cell differentiation: *stability* with respect to perturbations of enzyme concentrations, *polarity* of the pattern obtained, and *epimorphosis* under certain conditions in the number of cells undergoing induction (*invariance of size*); in other words, the number of cells undergoing induction is not modified when the number of cells in the field varies.

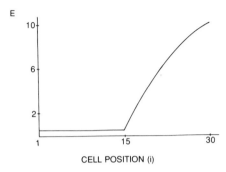

Fig. 10.14. Morphological pattern corresponding to the primary pattern in Fig. 10.13. E = enzyme.

Edelstein (1972) uses the phenomenological kinetics of the induction of the lac-operon of an enzyme E, the synthesis of which is induced by an enzyme M which diffuses from one cell to another and which itself is synthesised by the enzyme E. The phenomenological equation which describes the variation of E is:

$$\frac{dE}{dt} = 900k \frac{0.002\,(35 + M^2)}{35 + 0.002\,(35 + M^2)} - kE \qquad (10.14)$$

and the equation which describes the variation of M, due to synthesis and diffusion, is:

$$\frac{\partial M}{\partial t} = gE - k'M + D\nabla^2 M. \qquad (10.15)$$

Finally, for a zero flux of morphogen at the limit values, an identical behaviour of the system is determined. The difficulty that arises with this model is due to the existence of unstable solutions which obviously cannot be associated with morphogenetic processes. As we shall see, the equations of this model do not satisfy Turing's general condition.

IV. Morphogenesis: Turing's theory

Two views of pattern generation dominate the field of morphogenesis: the *Turing chemical appoach* (Turing, 1952) and the *Oster–Murray mechanical approach* (Odell et al., 1981; Murray et al., 1983; Murray and Oster, 1984a, b). In the former, the reaction and diffusion of chemicals (see Chapter 2) produces the spatial steady-state pattern of chemical concentration. Turing suggested the conditions under which heterogeneous spatial patterns would be obtained: if two substances A and B tend to a stable uniform steady state in the absence of diffusion (i.e. when their diffusion coefficients are equal), then heterogeneous spatial patterns will be developed by diffusion-driven instability *when their diffusion coefficients are unequal*. This is a very profound idea because diffusion is considered to be a stabilising process. We shall first study this chemical approach since it involves the spatiotemporal variation of the cell population. The mechanical approach will be presented in Chapter 12, Section III.

1. *General aspects*

Two essential concepts emerge from the analysis above:

(i)   *the existence of a morphogenetic field C* characterising a set of coupled inter-acting cells controlled by the same system; and

(ii)   *the spatiotemporal dependence of the field C(**r**, t)*.

The spatial dependence $C(\mathbf{r}, t_0)$ corresponds to the *positional information* of the cells in the field while the temporal dependence $C(\mathbf{r}_0, t)$ corresponds to the '*biological clock*'. Various theoretical interpretations have been given for this field, the general properties of which are the following (McMahon, 1973):

(i)   *active* cells cannot be distinguished from *reactive* cells;

(ii)   a large number of cells gives rise to a constant number of cell types after differentiation; and

(iii) the field is often *self-regulating*, i.e. it reacts to all perturbations.

In a general way, morphogenesis is admitted to be a two-step process:

(i)   First, a *primary pattern* is created through a mechanism generating a gradient established by one or more chemical substances, the morphogens. This mechanism introduces positional information thus creating the morphogenetic field. Two types of theory have been put forward to explain this:

   • the creation of a gradient by a mechanism diffusing morphogen in a set of cells containing a source and a sink, by active transport of morphogen (Babloyantz and Hiernaux, 1975) or by cell-to-cell contact (McMahon, 1973); and

   • the difference of phase between the periodic signals emitted by a pace-maker cell being propagated at different speeds in the cell network (Goodwin and Cohen, 1969).

(ii)   Secondly, a *morphological pattern*, deduced from the primary pattern, in which the positional information triggers the expression of certain proteins in the metabolic network, leading to the activation of specific functions. This, in fact, is the process of cell differentiation under genetic control.

In the preceding sections of this chapter, we have considered the main lines of current ideas concerning morphogenesis, the central problem of developmental biology. However, the fact that several theoretical models have been constructed clearly suggests that the true answer has yet to be found. The multiplicity of the solutions proposed, all equally plausible (Harrison and Kolar, 1988), constitutes a powerful stimulus to the search for the fundamental mechanism. The recent contributions, which we shall now discuss, considerably improve on the existing concepts.

2.   *Theoretical models of morphogenesis*

a.   Creation of the gradient

Gierer and Meinhardt (1972), on the basis of experimental results in embryology, have laid down three postulates necessary to the creation of a morphogenetic gradient:

- *Short-range activation* (induction of an organ by small transplants);
- *Long-range inhibition* (two identical neighbouring organs inhibit each other); and
- *Distribution of a density of sources* for the two substances, activator and inhibitor, supposed to be due to the distribution of cell types or cell substructures. However, the authors consider that there is an essential difference between the density of the sources and the concentration of the substances emitted by the sources: the density of the sources is maintained constant in spite of a transplantation while the source activation varies rapidly, i.e. the primary pattern does not result from the source distribution but rather from the concentrations of substances synthesised by the sources. With these conditions, the problem is formulated as follows. Let $\rho(x)$ be the density of the activator sources and $\bar{\rho}(x)$ that of the inhibitor sources. Gierer and Meinhardt suppose the concentration of the inhibitor sources to be a function of the concentration of the activator M and of the average density of the inhibitor sources. Putting this formally, we have:

$$\bar{\rho}(x) = f(M(x), \langle \bar{\rho} \rangle_x). \tag{10.16}$$

The concentration of the activator is supposed to vary according to the equation:

$$\frac{\partial M}{\partial t} = \gamma \rho \, \frac{M^k}{\langle M \rangle^l} \left(1 - \frac{\beta}{\rho} \frac{\langle M \rangle^n}{M^m} \right) \quad \text{with} \quad n > m > 0 \,. \tag{10.17}$$

Considering the hypotheses above, this is the most general equation capable of generating different models of cell differentiation. The first term, the *production* term, is a function of the density of sources, of the concentration M, and of the concentration of the inhibitor via the factor $\langle M \rangle$ by cross-catalysis according to the factor γ. The second term, the *destruction* term, independent of the density of the sources of activators, depends only on M and $\langle M \rangle$.

The characteristics of the morphogenetic field are determined from various hypotheses concerning ρ, M, $\bar{\rho}$ and $\bar{M}$. In particular, we have to suppose a gradient of density ρ of the sources of activators. With these conditions, Eq. (10.17) will generate various models. *We shall consider the 'activator–inhibitor' model with different sources,* described by the two-equation system:

$$
\left.
\begin{aligned}
\frac{\partial M}{\partial t} &= \rho_0 \rho + c\rho \frac{M^2}{\overline{M}} - \mu M + D_M \frac{\partial^2 M}{\partial x^2} \\
\frac{\partial \overline{M}}{\partial t} &= c' \overline{\rho} M^2 - v\overline{M} + D_{\overline{M}} \frac{\partial^2 \overline{M}}{\partial x^2}
\end{aligned}
\right\}
\tag{10.18}
$$

where it is supposed that M and $\overline{M}$ obey first-order breakdown kinetics, that $\overline{M}$ diffuses more rapidly than M, and, finally, that the base production of the activator is proportional to M. It can be shown that these equations may be deduced from (10.17) written in the form:

$$
\frac{\partial M}{\partial t} = \rho \frac{M^r}{\langle M \rangle^{\frac{st}{u+1}}} \left(1 - \frac{\beta \langle M \rangle^{\frac{st}{u+1}}}{\rho M^{r-1}} \right)
\tag{10.19}
$$

when there exists a spatial 'hierarchy' of the processes $D_{\overline{M}} \gg D_M$. The gradient will be initiated if $st/(u+1) > r - 1 > 0$, and, with different sources ($\rho \neq \overline{\rho}$), $r = s = 2$, $t = 1$ and $u = 0$, we obtain the system (10.18).

This theory explaining the creation of the gradient is interesting for it clearly shows how a primary pattern may be generated from *two* morphogens. However, it supposes the prior existence of a slight gradient in the density of the sources. So, while the model above gives a good description of the phenomenon of 'amplification' that generates the primary pattern, *it does not explain the heterogeneity that actually triggers the process in the initially uniform tissue.* Babloyantz and Hiernaux (1975) give an elegant demonstration of the feasibility of such a gradient in a biochemical system with three components. The authors demonstrate that dissipative processes (Chapter 3) must be at work, so that the instability of the initially uniform state finally leads to the appearance of the pattern. Thus, development appears to be spontaneously determined by fluctuations capable of generating a dissipative spatial structure and, thereby, morphogen sources and sinks.

As we have seen, a mechanism of gene control of the Jacob–Monod type, associated with the active transport, facilitated transport or diffusion of a substance I from one cell to another, can lead to a strong non-linear variation of enzyme E in function of I. This 'all-or-nothing' behaviour is characteristic of the process of cell differentiation. However, *as such a system contains only one morphogen, no stable pattern can be formed.*

Let us now consider a system with two morphogens, M and $\overline{M}$, and a substance I, described by the following reaction equations:

$$
\begin{aligned}
A &\xrightarrow{k_1} M \\
\overline{B} + 2M &\xrightarrow{k_2} \overline{M} + 2M \\
M &\xrightarrow{k_3} F \\
\overline{M} &\xrightarrow[k_5]{k_4} G \\
\overline{M} + I &\rightleftharpoons M \\
B + M + I &\xrightarrow[k_7]{k_6} I + 2M
\end{aligned}
\tag{10.20}
$$

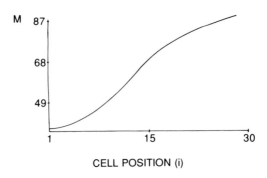

Fig. 10.15. The gradient of a substance M in the space field according to the model due to Babloyantz and Hiernaux (1975).

where B and $\overline{B}$ are the precursors of M and $\overline{M}$. The irreversibility of the system is ensured by maintaining the concentrations of A, B, $\overline{B}$, F and G constant, in time and in space.

Far from thermodynamic equilibrium, it can be deduced that this system can be represented by the two equations:

$$
\left.
\begin{aligned}
\frac{\partial M}{\partial t} &= k_1 A + k_7 \frac{k_6}{k_5} B \frac{M^2}{\overline{M}} - k_3 M + D_M \frac{\partial^2 M}{\partial x^2} \\
\frac{\partial \overline{M}}{\partial t} &= k_2 \overline{B} M^2 - k_4 \overline{M} + D_{\overline{M}} \frac{\partial^2 \overline{M}}{\partial x^2}
\end{aligned}
\right\}
\tag{10.21}
$$

where it is supposed that $\partial I/\partial t \approx 0$ in the quasi-stationary state. This is a remarkable result as these equations are formally identical to Eqs (10.18) obtained by Gierer and Meinhardt (1972).

Let L be the length of the one-dimensional field. The study of the stability of the system (10.21) with respect to perturbations of the type:

$$
M = M_0 + r \cos \frac{n\pi x}{L} \qquad \frac{r}{M_0} \ll 1
$$

$$
\overline{M} = \overline{M}_0 + \overline{r} \cos \frac{n\pi x}{L} \qquad \frac{\overline{r}}{\overline{M}_0} \ll 1
\tag{10.22}
$$

shows that, for L between two characteristic values L_n^1 and L_n^2 with $n = 1$, a gradient of the activator morphogen M is spontaneously established in the field (Fig. 10.15). When L increases, the behaviour becomes unstable for values of $n > 1$. In fact, the linearised dispersion equation is a quadratic equation of the type $X^2 + B_n X + C_n = 0$, with coefficients B_n and C_n depending on the integer n, for which there are two solutions for each value of n. If $L_2^1 < L < L_2^2$, the inhibitor morphogen $\overline{M}$ varies as in Fig. 10.16, with one peak value for $n = 1$ and two peak values for $n = 2$. *Thus, different forms of the pattern may appear during development.*

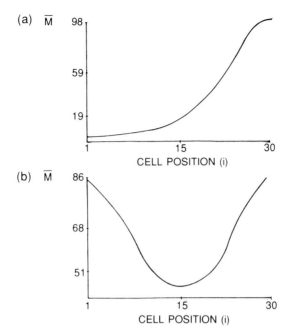

Fig. 10.16.   Different forms of the $\overline{M}$ pattern corresponding to particular values of the parameters L and n (see text), according to Babloyantz and Hiernaux (1975). (a) $n = 1$; (b) $n = 2$.

Babloyantz and Hiernaux (1975) have also described the spontaneous creation of a gradient by fluctuations in an initially homogeneous tissue, giving an interpretation of the gradient under genetic control. The treatment is similar to that presented in the preceding section of this chapter.

Tarumi and Mueller (1989) have shown that for large systems the Gierer–Meinhardt model (1972) can explain the observation that pattern-generating mechanisms are remarkably insensitive to a wide range of environmental and experimental conditions: a definite wavelength of the ultimate spatial pattern is selected when the unstable homogeneous steady state is locally disturbed. This result may also be valid for other reaction–diffusion systems.

b.   Interpretation of the gradient of positional information

In the existing models of the interpretation of the gradient (Edelstein, 1972; Babloyantz and Hiernaux, 1975), the morphogen inhibits the repressor. MacWilliams and Papageorgiou (1978) have suggested another possibility: the gradient may be created by the binding of an allosteric protein to one or more morphogens. The protein would then be in the active state, the concentration varying with the position according to the concentration of the morphogens. The approach used here is particularly simple: since the concentration of the complex PS, obtained by the binding of a protein P to a substance S, passes through a maximum for certain values of $[S]$, it may be supposed that this maximum defines a region of cell differentiation in a field of which the gradient is given by S.

This reasoning is especially fruitful with a finite number of morphogens and allows the construction of morphological patterns. It also explains the phenomenon of the amplification of structures. We shall first consider the case of a single morphogen S, and then that of two morphogens S and T.

The equation of reaction between P, the concentration of which is the same at all points of the field, and S, the concentration of which varies according to a function $f(x)$, may be written:

$$P + 2S \underset{k_-^1}{\overset{k_+^1}{\rightleftharpoons}} PS + S \underset{k_-^2}{\overset{k_+^2}{\rightleftharpoons}} PSS. \tag{10.23}$$

According to the values of $[S]$, the equilibrium will be displaced. We may thus obtain the concentrations of P, PS and PSS in function of $\log [S] = \log f(x)$, i.e. in function of the position, when a given function is chosen for f, for example $f(x) = e^x$. We have:

$$[PS] = \frac{[P] + [PSS]}{\dfrac{1}{K_1[S]} + K_2[S]}. \tag{10.24}$$

The values of $K_1 = k_+^1/k_-^1$ and $K_2 = k_+^2/k_-^2$ give the position and the width of the peak $[PS](x)$, and *the maximum value of $[PS]$ then defines a differentiated region.*

MacWilliams and Papageorgiou also deal with the problem of the binding of P with two molecules of S and two molecules of T, considering an active state to be that in which one molecule of P is bound to one molecule of S and to one molecule of T, i.e. the complex $[PST]$ (Fig. 10.17). If the concentrations of S and T, equal to $f(x)$ and $g(y)$, define two perpendicular directions, then $[PST]$, which varies with the concentrations $[S]$ and $[T]$, will vary with the position of the couple (x, y). Here, a similar but rather more complicated treatment than in the preceding case gives:

$$[PST] = \frac{[P] + [PS] + [PSS] + [PSST] + [PSSTT] + [PSTT] + [PTT] + [PT]}{\dfrac{K_1}{[T]} + \dfrac{K_2[S]}{[T]} + K_3[S] + K_4[S][T] + K_5[T] + K_6\dfrac{[T]}{[S]} + K_7\dfrac{1}{[S]} + K_8\dfrac{1}{[S][T]}} \tag{10.25}$$

taking into account the nine possible states of coordination for P. The triplet $([PST]$, x, y) defines the surface of concentration passing through a maximum in a position which is function of the K_i, $i = 1$ to 8. Here also it may be supposed that differentiation occurs when the concentration of the active state exceeds a certain threshold value.

Up to now we have considered only a single binding protein P. But several such proteins may be envisaged, the differentiation being ensured by the most 'active' protein, i.e. the protein with the highest concentration in the active state. Such a mechanism is particularly appropriate for spatial differentiation as it allows for a certain regional 'independence', each region being represented by a specialised

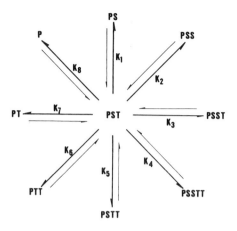

Fig. 10.17. Model due to MacWilliams and Papageorgiou (1978). The *PST* coupling is obtained from among eight possible compounds, so that there are nine states of coordination possible for *P*.

morphogen. Thus, the simulation of the primary pattern of a quadruped forelimb by MacWilliams and Papageorgiou, as shown in Fig. 10.18 and in Table 10.1, uses 13 proteins bound to the morphogen.

However, it is not clear how the system and, in particular, the hypothetical allosteric protein *P* are controlled. Thus, Meinhardt (1978) extends the notion of inhibition–activation, as used in the formation of the primary pattern, to the system of genetic control. As we have seen, one of the mechanisms used for the formation of the primary pattern (Gierer and Meinhardt, 1972) is based on an autocatalytic feedback of the activator $\overline{M}$ diffusing more rapidly than the inhibitor M (Eq. (10.18)), called *autocatalysis* and *lateral inhibition*. This model implies the mutual inhibition of two sources close together, the distinction between the 'active' and the 'inactive' cells being made by the equations of the morphogenetic field. The gradient may therefore be interpreted through a similar mechanism: *a gene undergoes selective activation by autocatalysis and lateral inhibition (or rather repression in this case).*

Then, if we consider a set of control genes $G_1, G_2, ..., G_i, ...$, the gene G_i will be active if there exists a high concentration g_i of the corresponding activator gene. This phenomenological theory must satisfy the following conditions:

(i)   The activated gene maintains its state of activity by autocatalysis according to the equation:

$$\frac{dg_i}{dt} = \frac{c_i}{r} g_i^2 - \alpha g_i \tag{10.26}$$

where $c_i g_i^2$ describes the autocatalytic feedback and $-\alpha g_i$ the destruction of the activator.

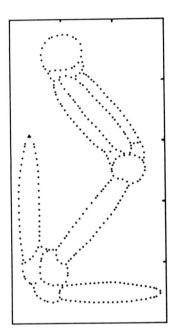

Fig. 10.18. Model due to MacWilliams and Papageorgiou (1978). Primary pattern of a quadruped forelimb obtained with 13 proteins. Table 10.1 shows the corresponding numerical data.

(ii)   The repressor r acts on all the genes and is produced by each activated gene:

$$\frac{dr}{dt} = \sum_i c_i g_i - \beta r \tag{10.27}$$

where $-\beta r$ describes the destruction of the repressor.

(iii) There is only one active gene in each cell. If there were two active genes, competition would lead to the suppression of one of them, leaving the other in stable equilibrium with the repressor.

(iv)   The active gene is controlled by the concentration of the local morphogen.

By introducing particular values into the equations above, we can see how the system produces the morphological pattern. Meinhardt gives an example which, under certain hypotheses (the overlapping of the activator genes, hierarchy in the efficacy of autocatalysis ($c_{i+1} < c_i$), maximum operating level of the activated genes, and the influence of the morphogen on the production of the activator genes) leads to the pattern shown in Fig. 10.19.

Table 10.1. The values $K_i (i = 1$ to $8)$ are those that would define an identical form centred at $[S] = [T] = 1$. X_{max} and Y_{max} are the coordinates of the centre of the territory, $0 < X < 50$, $0 < Y < 30$. C_{max} is the maximum concentration of the 'active' state at the centre of each territory, obtained for certain values of ΣP calculated for each protein. A low value of C leads to a significant variation of the form. ΣP is the sum of the concentrations of all the states of coordination of P.

Territory	K_1	K_2	K_3	K_4	K_5	K_6	K_7	K_8	C_{max}	X_{max}	Y_{max}
Primary centre of clavicle	40	0	1	0	40	0	1	0	0.985	17.5	4
Epiphysis of clavicle	0.1	0.1	0.1	0.1	0.1	0.1	0.1	0.1	0.983	8.5	5.5
Primary centre of coracoid	1	0	40	0	1	0	40	0	0.985	4	17.5
Epiphysis of coracoid	0.1	0.1	0.1	0.1	0.1	0.1	0.1	0.1	0.984	5.5	7.5
Proximal epiphysis of humerus	10	10	10	10	10	10	10	10	1	8	8
Primary centre of humerus	1	70	1	1	1	70	1	1	1.006	15	15
Distal epiphysis of humerus	10	10	10	10	10	10	10	10	1	24.5	23
Proximal epiphysis of ulna–radius	0.01	0.1	0.1	0.07	0.01	0.1	0.1	0.07	0.982	26.7	22.6
Primary centre of ulna–radius	0	0.005	0	0.05	0	0.005	0	0.05	0.982	33.2	17.2
Distal epiphysis of ulna–radius	0.01	0.1	0.1	0.09	0.01	0.1	0.1	0.09	0.982	40.6	10.9
Ulna–radial space	0	3.5	0	100	0	3.5	0	100	1.006	33.3	16.7
Hand	0.15	1	1	0.85	0.15	1	1	0.85	1.007	43	10
Threshold	0	0	0	0	0	0	0	0	0.97	—	—

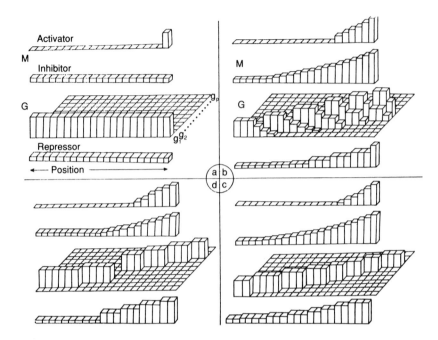

Fig. 10.19. Model due to Meinhardt (1978). Formation of a pattern in a linear network of cells: positional information is generated by an inhibitor–activator system of morphogens. Each figure represents the concentration of inhibitor and activator morphogens M, the concentration of the activator g_i of the p genes activated (each gene is situated perpendicular to the one-dimensional network of cells), and the concentration of the repressor r. Each block corresponds to a cell in a given position i, $i = 1$ to n.

3. *Morphogenesis: local and global theories*

a. *Turing's theory: a synthesis of theoretical models*

Let us examine the underlying unity of the various models of morphogenesis. The basis of the interpretation of the phenomenon is a morphogenetic field in which a two-step process — the creation of a primary pattern followed by that of a morphological pattern — leads to cell differentiation.

But what triggers this process which consists, particularly in the case of regionalisation, of the breaking up of a homogeneous state following an instability?

Prigogine *et al.* (1972) consider that the final state obtained is a dissipative structure. But Turing (1952) was the first to propose a mechanism basic to all the current theoretical explanations of morphogenetic processes. His important pioneering work provides a connecting link between all the recent theories we have discussed.

Turing's approach concerns the development of a dissipative structure (Prigogine and Lefever, 1968) arising from an instability of the initial homogeneous state. This is what Chernavskii and Ruijgrok (1978) have called the 'instructed' process. Let us consider the classical system already studied expressed in Turing's general form:

$$\frac{\partial X}{\partial t} = P(X, Y) + D_X \left. \frac{\partial^2 X}{\partial r^2} \right\}$$

$$\frac{\partial Y}{\partial t} = Q(X, Y) + D_Y \left. \frac{\partial^2 Y}{\partial r^2} \right\}$$

(10.28)

where r is the space coordinate. This system has the trivial solution (X_0, Y_0) corresponding to the structureless stationary state:

$$P(X_0, Y_0) = Q(X_0, Y_0) = 0.$$

According to Turing, the condition for the instructed process is that the homogeneous solution (X_0, Y_0) be unstable with respect to perturbations of the order of the wavelength $\lambda_0 = 2\pi/k_0$ and stable with respect to any other perturbation. Mathematically, this means that the limit conditions must be periodic on a ring of radius R, since the solution of the system (10.28) must be independent of conditions external to the system, particularly of the environment. This is expressed by:

$$L = 2\pi R = n\lambda_0, \; n \in \mathbb{N}$$

and the initial conditions must undergo small differences at (X_0, Y_0) if we put:

$$X - X_0 = X_0 e^{\omega_\pm t} \cos kr$$

$$Y - Y_0 = Y_0 e^{\omega_\pm t} \sin kr.$$

(10.29)

$\omega_\pm$ represents the positive frequency ω_+ and the negative frequency ω_- in the solution of the system (10.28) linearised about (X_0, Y_0) where the temporal frequencies $\omega_\pm$ depend on k, the space wave number. Taking into account the dispersion equation, Turing's condition may then be written:

for

$$\boxed{\begin{array}{c} \operatorname{Re}(\omega_+) > 0 \\ 0 < k_-^2 < k_0^2 = \dfrac{n^2}{R^2} < k_+^2 \end{array}}.$$

(10.30)

In other words, the system (10.28) must have a positive frequency ω_+ within a certain interval $[k_-, k_+]$ defined by:

$$k_+^2 < \left(\frac{n+1}{R}\right)^2$$

$$k_-^2 > \left(\frac{n-1}{R}\right)^2$$

(10.31)

function of the diffusion constants D_X and D_Y and other parameters of the system. When $k_-^2 = k_+^2$ for a certain set of values of the parameters, we obtain a *bifurcation of the Turing type*. When k_- and k_+ have complex values in a certain region of the space of the parameters, the initial homogeneous state is stable and no morphogenetic process will occur.

Finally, if Turing's condition is satisfied, any perturbation of wavelength $\lambda_0 = 2\pi/k_0$ will be amplified and a temporal order, independent of the initial conditions, will be created, starting up an 'instructed' process.

All the theoretical models based on the reaction–diffusion equations that we have discussed in various parts of this work are particular cases of Turing's model. Thus, the 'Brusselator' (Chapter 1) is described by the system (10.28) where:

$$P(X, Y) = A + X^2Y - (B + 1)X$$
$$Q(X, Y) = BX - X^2Y. \tag{10.32}$$

We have seen that when the concentrations A and B are fixed, the dissipative structure depends on A and B and that the structure is formed under the condition of being far from thermodynamic equilibrium:

$$|X - X_0| \gg X_0 \quad \text{and} \quad |Y - Y_0| \gg Y_0. \tag{10.33}$$

Chernavskii and Ruijgrok (1978) give a good discussion of morphogenetic models constructed along the lines of Turing's theory. The main results are:

(i)   *Turing's condition for the appearance of a final, dissipative structure is necessary but not sufficient.* Thus, Edelstein's model, which does not satisfy this condition, does not lead to a stable dissipative structure.

(ii)   *Dissipative structures of great amplitude, i.e. far from thermodynamic equilibrium, can only arise in systems with multiple stationary states.*

For example, the model due to Babloyantz and Hiernaux, as well as that due to Gierer and Meinhardt, which can be deduced from Turing's condition, have more than one homogeneous stationary state. With a Turing-type model, Martinez (1972) obtains stable patterns by the perturbation of the initial concentrations while varying the length of the linear network of cells. Here also the 'Brusselator' is used for the development of the two morphogens X and Y. It further supposes the existence of a third substance Z which, as soon as it exceeds the threshold value Z_c, triggers cell division. The variation of Z depends on X and Y through the equations:

$$\left. \begin{aligned} \frac{dZ}{dt} &= a(X - Y) - bZ \quad \text{if} \quad X - Y > 0 \\ \frac{dZ}{dt} &= -bZ \qquad\qquad\quad \text{if} \quad X - Y \le 0 \end{aligned} \right\}. \tag{10.34}$$

Unlike X and Z, Y may diffuse from one cell to another. The conditions at the limits are periodic as in Turing's description.

This original mechanism allows the development of a cell network of increasing length in function of time by binary cell division: *here cell differentiation is closely related to cell kinetics.* The results obtained thus seem to confirm that cell division during morphogenesis is under spatial control according to the distribution of stable de-repressors in the stationary state, and that this distribution is influenced, if not completely determined, by a Turing-type mechanism. Indeed, it is certain that gene control supports the process of cell differentiation and that a cell, before reaching a differentiated state, must divide a certain number of times.

To sum up, before going on to the more general theory proposed by Thom, we may say that Turing's essentially non-linear theory concerns a reaction–diffusion system which must satisfy two conditions:

(i)   *the stationary states, homogeneous or non-homogeneous, must be unstable;*
(ii) *a random perturbation takes the system from the homogeneous to the non-homogeneous state, with intermediate steps as described by Chernavskii and Ruijgrok.*

Further details lie beyond the scope of our work which is developed mainly from a physiological viewpoint. A full mathematical treatment of these problems has been given by Murray (1989), who discusses the important role of the dispersion relation in determining the growth factor as a function of the wavenumber or wavelength. The reader will also find many interesting applications, e.g. to animal coat patterns. Murray suggests that a single mechanism (a reaction–diffusion system that can be diffusively driven unstable) could be responsible for generating almost all the common patterns observed. The spatial pattern of morphogen concentration in the morphogen pre-patterns determines the subsequent differentiation of the melanin producing cells. The illustrations below show various examples demonstrating the prominent role of the scale factor γ, i.e. a measure of the domain size B.

Murray (1980; 1981) has chosen a reaction–diffusion system, first considered by Thomas (1975), given in a non-dimensional form by:

$$\frac{\partial u}{\partial t} = \gamma f(u, v) + \nabla^2 u \tag{10.35}$$

$$\frac{\partial v}{\partial t} = \gamma g(u, v) + d\nabla^2 v \tag{10.36}$$

$$f(u, v) = a - u - h(u, v), \quad g(u, v) = \alpha(b - v) - h(u, v), \quad h(u, v) = \frac{\rho u v}{1 + u + K u^2}$$

with the boundary conditions:

$$(\mathbf{n} \cdot \nabla)u = 0, \quad (\mathbf{n} \cdot \nabla)v = 0, \quad \mathbf{r} \text{ on } \partial B$$
$$u(\mathbf{r}, 0), \qquad v(\mathbf{r}, 0) \qquad \text{given}$$

where ∂B is the closed boundary of the domain B and $\mathbf{n}$ is the outward unit normal to ∂B. This non-dimensioned form of the reaction-diffusion system includes the reaction kinetics f and g, the coefficient γ which is proportional to the area (and thus defines a scale factor), and the diffusion coefficient d. The two latter parameters have an important biological meaning since the results can be viewed in (γ, d) space.

According to Murray, the colour pattern on the mammalian skin results from an underlying pre-pattern that is formed in the early stages of embryonic development. Melanoblasts, which are genetically determined cells, migrate and become melanocytes, the specialised pigment cells in the basal layer of the epidermis.

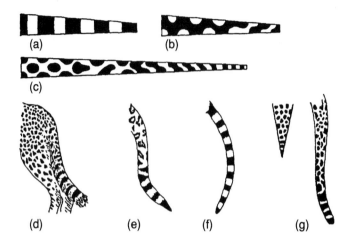

Fig. 10.20. Computed solutions of the non-linear reaction–diffusion system with zero flux boundary conditions and initial conditions taken as random perturbations about the steady-state. Dark regions have a morphogen concentration above the steady-state. (After J. D. Murray (1989)). (a) Scale factor = 9; (b) Scale factor = 15: the pattern bifurcates to more complex patterns as the scale factor increases; (c) Scale factor = 25: spot-to-stripe transition with a concentration of morphogen under the steady-state; (d)–(g): Typical tail patterns for various species.

Melanocytes generate melanin, the substance which determines hair colour. The actual production of melanin by the melanocytes is attributed to the action of an unidentified chemical substance constituting the pre-pattern of the observed coat colour pattern in the animal. Murray suggests that the pre-pattern is created by a reaction–diffusion system. There is some evidence in favour of this theory, such as the scale factor and the geometry in the tail patterns that are typical of many spotted animals (Fig. 10.20), or in the case of the zebra's stripes. This is consistent with the embryonic tail structure of these animals at the time at which the pattern formation mechanism is supposed to be operative. The domain *B* is a closed surface and appropriate conditions are random perturbations around a steady state. The process of pattern formation in this domain is activated at a specific time of development. For a given domain, the size and the geometry govern the precise spatial patterns of morphogen concentration. Some initiation switch might operate the bifurcation parameter of the reaction–diffusion system, thus activating the diffusion-driven instability. Numerical simulations have produced satisfactory results for the scapular stripes on the foreleg of zebra (Fig. 10.21); for the effect of the body surface scale (Fig. 10.22); for the butterfly (with a different reaction–diffusion mechanism) (Fig. 10.23); and for a giant unicellular organism, the green marine alga *Acetabularia* for which the morphogen could be free Ca^{++}. A full description of these examples has been given by Murray (1989) and Dowse and Ringo (1989).

Fig. 10.21.   (a) Examples of scapular stripes on the foreleg of zebra. (b) Predicted spatial pattern using the reaction–diffusion mechanism (after J. D. Murray (1989)).

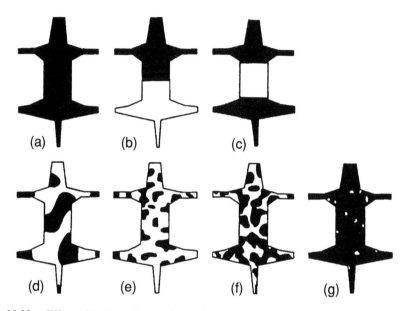

Fig. 10.22.   Effect of body surface scale on the spatial patterns using the reaction–diffusion mechanism. The value of the scale factor goes from less than 0.1 (a) to 5000 (g). The domain dimension is related directly to this scale factor (values: 0.5, 25, 250, 1250 and 3000 respectively in figures (b) to (f)). (After J. D. Murray (1989).)

b.   Thom's theory: towards a general theory of development

In the study of developmental mechanisms above, we have seen how the difficult passage from the invisible primary pattern to the morphological pattern may be explained by postulating the existence of biochemical substances — morphogens — giving each cell its specificity within the developing tissue. We now have to face *the delicate mathematical problem of going from the local to the global situation.*

Here we can do no better than quote from Thom's *Stability structurale et morphogénèse* (1972):

> ... *the biologist should therefore, from the start, postulate the existence of a local determinism to explain each partial micro-phenomenon within the living organism, and then try to integrate all the local determinism into a*

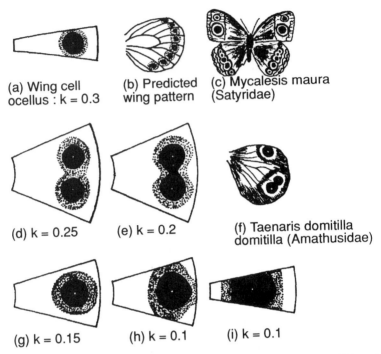

(a) Wing cell
ocellus : k = 0.3

(b) Predicted
wing pattern

(c) Mycalesis maura
(Satyridae)

(d) k = 0.25

(e) k = 0.2

(f) Taenaris domitilla
domitilla (Amathusidae)

(g) k = 0.15

(h) k = 0.1

(i) k = 0.1

Fig. 10.23.   (a) Two emissions of morphogen generate a patterned eyespot within a wing cell. The dark region had less morphogen injected than that with the shaded domain, for the same values of parameters. (b) Predicted wing if an eyespot was located in each wing cell. (c) Typical example. (d)–(e) Effect of the parameter k decreasing ((f) actual example). (g)–(i) Effect of different geometries. (After J. D. Murray (1989).)

coherent and stable global structure. From this point of view, the funda-
mental problem of biology is a problem of topology, since topology is pre-
cisely the mathematical discipline that allows the passage from the local to
the global ...

The task proposed in these lines, which we believe corresponds to the main pre-occupation of theoretical biology, is indeed an enormous one. It implies the final existence of a global structure supported by what Thom calls the *vital field*. Without even venturing to speculate on the ultimate nature of such a field, it must be admitted that the reductionist attitude of interpreting the set of metabolic networks of a living organism on the basis of a field in which such organisms may be considered as 'structurally stable particles or singularities' is a very challenging one indeed. From this standpoint, the global problem of morphogenesis could be tackled in terms of mathematical topology.

A discussion of the fundamental notions of morphogenesis within the context of the theory of catastrophes may be difficult for those unfamiliar with topology as it requires a thorough knowledge of the precise definitions and concepts involved. Thom's elementary theory of catastrophes was briefly presented in Chapter 3 using an intuitive approach. Let us now specify the space of the internal and external variables or parameters. We have seen that, according to Thom, morphogenesis is the study of the origin and the development of structures and that, in this sense, the stability and the reproduction of the spatiotemporal structure constitute a problem of structural stability along the lines of qualitative dynamics and differential topology. The creation of forms necessarily implies the existence of discontinuities in the properties of the medium. We have seen, at the beginning of this chapter, that this is particularly true of embryogenesis. But, of course, the morphology must remain stable for small variations of the initial conditions. Such a morphogenetic process is said to be the *support of a morphogenetic field*, otherwise called a *chreod*. Thom's definitions are drawn from the analyses of Waddington (1957), and it is interesting to see how difficult it was for these authors to agree on the meaning of the terms. The correspondence exchanged on the subject is to be found in Thom's *Modèles mathématiques de la morphogenèse* (1974). Finally, Thom defines a chreod as a *developmental pathway associated with a local system of biochemical kinetics*. A chreod is thus attached to an attractor of the biochemical kinetics. In a way, this corresponds to the emergence of a structurally stable process or, as Thom puts it, an 'islet of determinism', isolated from other structurally unstable zones in the epigenetic landscape. From a mathematical point of view, the system of differential equations (2.41):

$$\frac{\partial x_i}{\partial t} = f_i^\lambda(x_1, x_2, \ldots, x_n) + D_i \nabla^2 x_i \qquad i = 1 \text{ to } n$$

which represents classical chemical kinetics with variables of concentration $x_i(\mathbf{r}, t)$, defines in $\mathbb{R}^n$ a field of vectors f^λ of which the components are the f_i^λ in the space with n concentration-coordinates x_i. When t increases, the x_i vary and tend to a concentration $x_i^{(0)}$ of which the components are those of a limit state called the *attractor* of the system. In fact, the concentrations also depend on their localisation in space $\mathbf{r}$. If the diffusion term is small (which means a perturbation of the x_i), the system evolves according to *local dynamics*, in other words, in $\mathbf{r}$ is associated a field of vectors f_λ or a dynamic system. Then, the domain (D) is divided into several domains, each associated with a given attractor. The separation of these domains occurs at a set of points forming the morphology of the process. These are the catastrophic points or subset K_D of the substrate space (D) (see Chapter 3). This separation of the domains is also called a *shock wave*.

We have discussed the deformation of these local dynamics when the parameter λ describes the space Λ of the parameters of the system. If the field is derived from a potential F, the system moves towards the local minimum of F when λ varies. Therefore, there exists a subset of Λ containing all the possible deformations of the critical dynamics. It is supposed that the potential $F(x_1 \ldots x_n; \lambda_1 \ldots \lambda_n)$, called for convenience $F_\lambda(x)$, possesses a singularity at the origin. But there exists a family of functions $F(x, (U_i))$ possessing the properties necessary to a physical potential. This family of curves, called *universal unfolding* by Thom, is written (Eq. (3.49)):

$$F(x, (U)) = f(x) + U_1 f_1(x) + \ldots + U_k f_k(x)$$

about the singularity $f(x)$, such that $F(x, 0) = f(x)$. $F(x, u)$ is said to be the polynomial that unfolds f, and $f(0)$ is the singularity at the origin. This origin is the *organising centre* of the process, since the phenomenon of morphogenesis begins at the point where $f(0) = 0$ and the process then operates according to the equation:

$$\boxed{\frac{\mathrm{d}\,x_i}{\mathrm{d}\,t} = -\frac{\partial F_\lambda(x_i)}{\partial x_i}} \qquad i = 1 \text{ to } n \qquad (10.37)$$

depending on the values of the parameters λ. It is also known (Guckenheimer's theorem) that the system is structurally stable for a potential which is a realisation of $F(x, (U_i))$. And this is where Thom's theory reveals all its power: *morphology is merely the observable expression of a structurally stable system, developing according to successive chreods, when the 'external' parameters λ vary*. It is clear how the potential $F_\lambda(x)$ is deformed, but the relationship between the theoretical 'mathematical' parameters u and the physical parameters λ is less evident. To clarify this point, we simplify by identifying u with $(\mathbf{r}, t)$, which reduces the number of useful parameters to a maximum of four. This identification leads to an upper limit of *seven* elementary catastrophes (see Chapter 3). It follows that the application $G: D \to \Lambda$ physically represents the development of the morphological process. This is called the *growth wave* or the *developmental wave of morphology*.

The phenomenon of organogenesis may now be interpreted as the development of several metabolic fields corresponding to local potentials of a dynamic biochemical system. It would surely be most interesting to determine or at least to make certain of the existence of each of the chreods of the epigenetic landscape. We know that organogenesis requires the interaction of tissues of different types, each requiring to be controlled by a different attractor. Thus, there must exist several specialised chreods, each corresponding to competing attractors within the same morphogenetic field. The separating surfaces between the attractors present a limited number of stable singularities depending on the potential $F_\lambda(x)$. With this approach, the epigenetic landscape is a 'catastrophic set' associated with the singularity compatible with a local dynamic system; in other words, it is made up of catastrophes associated with the growth wave. *Finally, morphogenesis is viewed as the result of a dynamic system of which the growth wave progresses according to the successive catastrophes in the space–time* $\mathbb{R}^4$.

This representation is evidently highly abstract but it does have the advantage of revealing, qualitatively and without complicated calculations, the interactions between dynamic systems, the formation of basins, the continuous deformations of the substrate space, the competition between attractors in a given dynamic system, and so on. Further details are unfortunately beyond the scope of this work, but let us leave the last word to Thom:

> *In conclusion, we shall then admit that embryonic development is controlled by 'directing leaflets' which may be assimilated to wave fronts. These leaflets consequently present the stable singularities of wave fronts that are at the origin of catastrophes provoking later differentiation. It is only in exceptional*

cases such as that of a particularly thick and compact tissue, egg yolk for example, that we are likely to find shock wave surfaces with a qualitative aspect described by a rule like the Maxwell convention. We shall therefore try to identify the singularities of embryonic morphogenesis with elementary catastrophes.

At the close of this section we cannot but acknowledge the present state of ignorance concerning the exact mechanism of cell differentiation. While it is true that the theory gives coherence to the set of concepts, definitions and properties observed, what is still missing is of course the essential *experimental discovery of the morphogen or the morphogens involved in the differentiation.* In particular, Thom's theory of morphogenesis provides an unambiguous model of embryonic development by the successive identification of the singularities observed with the elementary catastrophes of the theory. If this identification turns out to be valid, it will clearly be easier to discover the spatiotemporal mechanisms at the basis of the morphogenetic process.

V. A description of growth for functional organisation

Recently, Goodwin and Kauffman (1990) have discussed the relationship between the dynamically-changing spatial patterns of gene activity and the emergence of phenotypic patterns in the early development of *Drosophila*. Similarly, Lacallili (1990) has modelled the pair-rule pattern of *Drosophila* by a reaction–diffusion system containing four morphogens. Indeed, a great deal is now known about gene expression in the *Drosophila*. Oster (1988) has discussed lateral inhibition models of the developmental processes. However, the current models based on the Turing chemical mechanisms, or even the Oster–Murray mechano-chemical mechanisms (see Chapter 12), describe the structural rather than the functional organisation.

Thus, from the functional viewpoint, we still have to find out how an organ, or any other structure of an organism, actually acts on another. Our theory (Chauvet, 1993a, b), the details of which will be found in Chapter 12, describes the development of a functional organisation, i.e. the emergence of a functional hierarchy. Let us here introduce the formal, qualitative aspect of this theory.

1. *A break in the functional interaction: consequences on the stability of biological systems*

a. *A break in the functional interaction: the choice between Life and Death*

What happens when an interaction in a functional organisation is suppressed because of internal constraints, such as mutations at the genetic level, and/or external constraints, such as the presence or absence of food at the metabolic level? If the functional interaction under consideration is vital for the system, signifying that the product (the elementary function) which is carried from the source to the sink is necessary to the life of the system, then there are two eventualities for the system: either this product comes from another structural unit in the system, or the system dies. The choice depends on what happens at the lower levels of organisation, i.e. in each source that makes the product. We already have presented this idea as a paradigm for the self-association of metabolic pathways (Chapter 5, Section V), and for a description of the relation between evolution and physiology (Chapter 7, Section III):

> *If, at a given time, a structural unit does not produce the elementary physiological function* P_j *that is necessary to life, then in order to survive it must receive this function from another structural unit that possesses it. In that case, a new elementary function is created.*

Let us recall that within the classical scheme of protein biosynthesis, the repressor is emitted by the regulator gene, and acts on structure genes. There is an elementary function created from the source, i.e. the regulator, to the sink, i.e. the structure genes, which may be identified with the messenger RNA. The same analysis can be made for the metabolic pathway:

$$S_1 \to S_2 \ldots \to S_{i-1} \to S_i \to \ldots S_n \qquad (10.38)$$

where an enzyme E_i, which can be identified as an elementary function, acts sequentially on a product S_{i-1} to create S_i. In this example, S_n would be the final product of a metabolic pathway. If one enzyme in the chain is suppressed, then the survival of the system implies the use of another pathway, i.e. another elementary function that originates in another structure gene, resulting in S_i. This kind of substitution is often observed at the metabolic level.

Such a paradigm gives a causal interpretation to the existence of a functional interaction, since if each structural unit could produce the set of enzymes necessary to the life of the system *there would be no functional interactions*. With a break in the functional interaction, a source is transformed into a sink for a given product, with an elementary function then being created from the source to the sink.

The hypothesis above constitutes the basis for generating the different levels of organisation of the biological system. Many examples can be found to justify this hypothesis, for instance the passage from one metabolic pathway to another when environmental conditions vary, or the grouping of cells when the environment changes (e.g. *Dictyostelium discoideum*, see Section II). The application of this hypothesis to physiological mechanisms leads to what we have called the *principle of vital coherence* (Chauvet, 1993a).

b.   Functional hierarchical organisation: the consequence of the choice

The organisation of the system into a hierarchical one is a consequence of the choice made by the system. Let us consider a set of v structural units which have the same μ individual physiological products P_α, $1 \le \alpha \le \mu$, (i.e. the same potentialities) these products being necessary for the 'life' (i.e. the functioning) of this set. If several units denoted u^*, have lost one or more such physiological products P_α, then u^* die unless P_α is provided by another unit u. With the present description, we may say that an elementary function has been created from u to u^*. This mechanism of

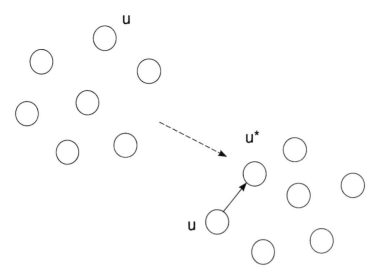

Fig. 10.24. Association between two units. A modified unit u^*, i.e. a unit u in the space of units which has lost a physiological function (or a product), associates with a 'normal' unit u, in order to retrieve this product (see also Fig. 7.9).

functional self-association explains why there are particular functional links in the system. Many such links could be realised to satisfy several combinations between a subset of u^*-structural units and a subset of u-structural units. This idea will be expressed through the concept of *functional complexity* (see also Eqs 2.1 and 3.2) or *potential of organisation*.

According to the paradigm enunciated above, and following the genetic program that imposes the 'choice' mentioned, either u_1^* will die out or enter into association with another unit u_1 to form $u_2 \equiv (u_1, u_1^*)$, which will be the origin of a new population $\mathcal{U}_2$. In general, u_j will give rise to a population $\mathcal{U}_{j+1}$, with each element possessing a supplementary unit. This process, composed of successive associations, creates a hierarchical system (Fig. 10.24) which is at the origin of *tissue specialisation* and leads to the biological concept of organogenesis, in which the alteration of gene expression is controlled by the genetic program.

Let $\mathcal{U}_j$ be the population of elements u_j, each containing j units. These elements can be obtained in different ways by asociations of the type:

$$(u_{j-1}, u_1^*), \ \ldots \ , (u_{j-P}, u_P^*), \ \ldots \ , (u_1, u_{j-1}^*)$$

For example, Fig. 10.25 contains the units $u_4 \equiv (u_1^*, u_3) \equiv (u_1^*, ((u_2, u_1^*))) \equiv (u_1^*, ((u_1, u_1^*)u_1^*))$, and $u_3 \equiv (u_2, u_1^*) \equiv ((u_1, u_1^*), u_1^*)$. If, in this description of populations of units, we take into account the physiological functions affected by non-permanent micromutations, we see in particular how tissue specialisation may occur (Fig. 10.26). In Fig. 10.26 the numbers in parentheses and the arrows of the hierarchical graph show which products have been lost by a unit. Let us suppose, for example, that u_1 'initially' possesses three physiological functions P_1, P_2 and P_3; that P_1, P_2 are

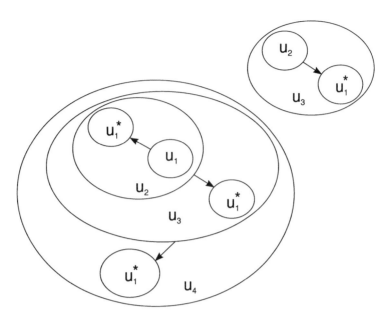

Fig. 10.25. Functional self-association hypothesis. Units u_2, u_3, u_4 are built by association between a pathological *unit u_1^** and a normal one u_1. For example, units $u_2 \equiv (u_1, u_1^*)$, u_3 ... are built by association (after Chauvet (1993a)).

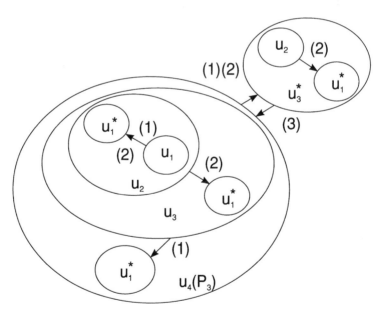

Fig. 10.26. Tissue specialisation. Numbers in parentheses represent the product transported from the source to the sink. Three such products are assumed to be necessary for the living cell. At a given time, the two products P_1 and P_2 are assumed to be missing. Thus, u_1^* associates with u_1 to create a unit u_2. If P_2 is eliminated from *another* unit u_1, this unit can associate with u_2 to create u_3, and so on. In the figure, such a set of transformations at given successive times, e.g. u_4 loses P_1 and P_2, leads to a unit u_4 that is finally specialised in the synthesis of P_3 (after Chauvet (1993a)).

eliminated from the unit u_1 (give u_1^*) leading to the creation of (u_1^*, u_1); that P_2 is then eliminated from the unit u_1 which associates with (u_1^*, u_1); and that finally P_1 is eliminated in a unit u_1 (giving u_1^*) which then associates with (u_1^*, u_1). If we now assume that u_4 loses P_1, P_2 at a given time, then unit u_4 will be specialised in the synthesis of P_3. The population $\mathcal{U}_4$ thus constructed is by definition a specialised tissue. On the contrary, u_3^* obtained from u_3, for example by the loss of P_3 in u_1, would be forced to associate with u_4. Finally, a unit of type u_7, and thus a population $\mathcal{U}_7$, obtained by self-reproduction, will be created. This population possesses an important property since $\mathcal{U}_7$ is made up of tissues, one being identical to u_4 specialised in the synthesis of P_3, and the other being identical to u_3 specialised in the synthesis of P_1 and P_2. Thus, an organ composed of differentiated tissues is formed.

As we shall see, this very simple, formal schema constitutes an understandable basis for a definition of a physiological system considered from the functional viewpoint. The example above shows a process that leads to units called u_4, specialised in the synthesis of a specific product P_3. It appears that the sequence of functional interactions is organised in order to carry out this specific elementary function. These functional interactions involve dynamical processes that vary on a common time scale. Thus, units u_1, u_2, u_3, which are associated within u_4 to produce P_3, constitute a level of organisation. For reasons that have already been suggested (Chapter 4, Section I), time scales have been chosen to specify a level of organisation in the hierarchy of the physiological system (see Chapter 4, Section IV). This idea of the *structuration of the functional hierarchical system* will be explored in Volume III (Chapter 6), and will be shown to be related to the regulation processes. It implies a certain 'order' in the system, tending to reduce the complexity of the system.

2. *Evidence for the existence of self-association: an increase in stability*

Let $\mathcal{U}_1$ be the population of units u_1. Let us suppose that a given unit $u^* \in \mathcal{U}_1$ is affected by a micromutation and/or some perturbation of a physiological mechanism. According to the principle of vital coherence, this unit will survive if, and only if, it can be associated with another normal unit in $\mathcal{U}_1$, which has the same physiological properties. The association between u_1 and u_1^* generates a new unit called $u_2 \equiv (u_1, u_1^*)$, and increases the complexity of the dynamics at the level of metabolism. Then, the level of organisation for $\mathcal{U}_2$, the new population of units such as u_2, is one unity higher than the level of organisation for $\mathcal{U}_1$. We may note that the self-association is bi-unitary, i.e. it can be realised with at most two units at the same time. One way of determining whether such a self-association could occur between two units, which are two hierarchical systems, is through the study of *the stability of the dynamics before and after association*, whatever the nature of the mechanism involved. An increase in the domain of stability of the new dynamical system obtained by association will favour the existence of that association between units. This hypothesis has been tested for a particular model with two levels of organisation: the level of metabolism inside the elements, and the level of replication of these elements (Chapter 7, Section III).

This process of self-association can be easily generalised. Let $u_{k+1} \equiv (u_k, u_1^*)$ be a unit in $\mathcal{U}_k$, which is created by an association between a unit in $\mathcal{U}_k$ and a perturbed unit u_1^* in $\mathcal{U}_1$, or an association between a perturbed unit u_k^* in $\mathcal{U}_k$ and a normal unit u_1 (see Fig. 10.26). All intermediates provide the possibility of self-association (Chauvet, 1993a). Such a process leads to the construction of recurrent models for $k = 1, 2, ..., j$. When the corresponding dynamical systems are transformed into non-dimensional systems, the condition of stability of the linearised system around equilibrium points can be determined, and the linear and the non-linear systems may be numerically investigated. *The stability of the system that corresponds to a higher level of organisation is shown to be greater, although its complexity, i.e. the number of elementary functions, increases.*

Thus, the paradigm proposed above leads to a mathematical interpretation of the developmental process. For the present, this description is merely schematic as it does not take into account the underlying physiological mechanisms. However, in Chapter 12, we shall introduce the notion of the *potential of functional organisation* which will enable us to describe the dynamics of the development. Let us now consider some models of cell division and the dynamics of cell populations.

11

Cell Division

The process of cell division, mitosis, an essential feature of the reproduction of living organisms, underlies all the phenomena hitherto discussed, whether at the molecular level with DNA replication, or at the cellular level with differentiation and embryogenesis.

The first question that comes to mind concerns the relationship between mitosis and cell differentiation. Of course, the two phenomena are intricately linked, but it would be interesting to determine whether the differentiation takes place during mitosis and, if so, at which phase of the process. Figure 11.1 sums up the situation. Differentiation may occur not only during the growth of an organism but also in the adult state since certain cells, such as blood cells and skin cells, are constantly renewed. For example, the existence of stem cells as undifferentiated precursor cells is characteristic of developing tissues. The various kinds of mitosis in relation to differentiation may be classified as follows:

(i)   *proliferative mitosis*, in which the cells *reproduce identically*;
(ii)   mitosis yielding two daughter cells *differentiated non-identically*;
(iii) mitosis yielding two daughter cells *differentiated identically*;
(iv)   mitosis yielding two daughter cells, *one identical to the mother and the other differentiated*, or the case of a line of stem cells giving rise to *two types of differentiated cell*.

We have already seen (Chapter 10) that cell differentiation is due to the action of *inductors* during embryonic development, and to that of *hormones* during adult life. It is now known that this action takes place only at certain moments of the cell cycle. Moreover, some cells divide *before* differentiation whereas others are capable of division *after* differentiation.

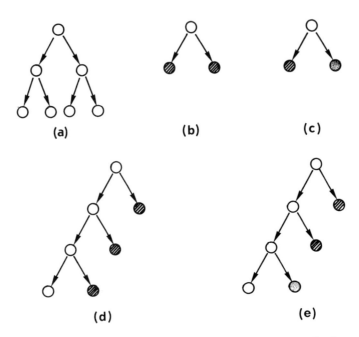

Fig. 11.1.   Various relationships between mitosis and differentiation: (a) simple proliferative mitosis; (b) mitosis leading to symmetrical differentiation; (c) mitosis leading to asymmetrical differentiation; (d) continuation of a cell line with a single type of differentiation; (e) continuation of a cell line with two types of differentiation.

How then do we account for the '*stability of the differentiated state*'? Many authors have been interested in the hereditary aspect of the differentiated state, or the 'determined' state as it is called in embryology. One explanation involves the interaction between cytoplasmic proteins and the chromosomes during part of the cell cycle, followed by a disappearance of the interaction during mitosis. This observation may imply that the genome restructures itself according to new molecular information, thereby developing a new program of genetic activity. If so, the quantity of information during successive divisions would require to be regulated in order to stabilise the differentiated state.

We are thus led to the conclusion that, in a system as complex as that of a cell, all the phenomena are chemically linked both in terms of time and of space: the precise moment of the cell cycle, the mitosis and the differentiation; and the substance involved, such as the cytoplasm, the nucleus and the cell membranes.

I.   The cell cycle

1.   *Description*

The duration of the cell cycle (incorrectly called the mitotic cycle) is defined as the time Δt between the middle of the mitosis of a given cell and the middle of the mitosis of one of the daughter cells. The concept of the cell cycle thus supposes that the state

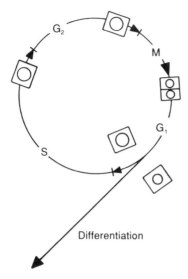

Fig. 11.2.   Cell cycle showing the four phases: $G1$, S, $G2$ and M. The differentiated state appears from the beginning of $G1$.

of the daughter cell is identical to that of the mother cell, since the notion of the cycle implies the periodic repetition of the event. Although, as we have just seen, this is not always the case, the concept is still useful for it allows a convenient represen-tation (Fig. 11.2) of the life of a cell.

Observations have shown that the cell cycle is closely related to changes in the nucleus, and particularly to the transformations of the chromosomes. If we consider only the modifications of the DNA, the cycle may be divided into four phases:

(1)   The $G1$ phase, 30 to 40% of the cycle, during which the quantity of DNA remains constant;

(2)   The S phase, 30 to 50% of the cycle, during which DNA and the associated basic proteins, the histones, are synthesised resulting in the duplication of the DNA;

(3)   The $G2$ phase, 10 to 20% of the cycle, which may be considered as the post-synthetic or pre-mitotic stage, during which the DNA again remains constant; and

(4)   The M phase, 5 to 10% of the cycle, in which the mitosis actually occurs.

These four phases are characteristic of the proliferative cell cycle. Cells that do not divide may be considered to be in one of the two non-proliferative states: (i) a resting period $R1$ (sometimes called $G0$) in the course of phase $G1$, or $R2$ in the course of phase $G2$; and (ii) a period of specific activity during which the cell plays its physio-logical role, for example the period of postprandial activity during which the liver cell stores glucose in the form of glycogen. Obviously, this state should not be mistaken for a differentiated state. Cooper (1988) discusses the cell cycle in relation to the continuum model, as well as the criteria of cell cycle regulation.

In short, *the cell has a double potentiality, cellular and functional, which it may*

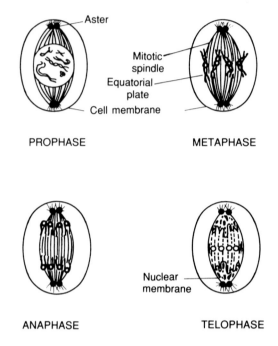

Fig. 11.3.   Morphological changes during the four phases of cell division.

conserve all through its life, either of these courses being triggered by a specific stimulation. This invariant property, which we shall call *idempotence,* will be used below to deduce the time-variation of the functional organisation (see Chapter 12, Section III). However, in reality, after a period depending on the type of the cell, it will no longer be able to re-enter the cell cycle but will follow the course to cell death. The discontinuity of the cell cycle in terms of DNA synthesis and the variability of the cellular state provide two important criteria for the study of cell populations, the *age* of the cell and its *maturity*.

The different steps of cell division in phase *M*, accompanied by striking changes in the nucleus (Fig. 11.3), are classified as follows:

Prophase: the chromosomes become shorter and thicker through coiling of the chromatids held together by the centromere. The chromosomes move towards the equatorial plane of the nucleus while the cell spindle develops between the centriole pairs migrating to the poles of the cell.

Metaphase: the nucleolus and the nuclear membrane disappear while the chromosomes acquire their characteristic shapes as they settle in the equatorial plane of the fully developed cell spindle.

Anaphase: each chromosome splits at the centromere into individual chromatids which then migrate along the cell spindle to the opposite poles of the cell.

Telophase: the chromatids at each pole of the cell begin to uncoil and lengthen while the cell spindle disappears and an active constriction splits the cell into two daughter cells.

Fig. 11.4. The coiling of the chromosome fibre producing a contraction during the metaphase (after Bak *et al.* (1979)).

Interphase: in each daughter cell, the structure of the nucleus gradually reappears, the nuclear membrane and the nucleus becoming visible, while the chromatids continue to uncoil and take on the aspect of chromatin.

This classical description of the cell cycle is generally valid, and has even been simulated by specific programs (Czihak and Linhart, 1992). However, it has not been possible to identify the *G*1 phase in some eucaryotes. Studies of cell cycles in several species have shown that certain enzymes undergo variations represented by different patterns of peaks and plateaux. *Thus, the classical concept of the phenomenon, based simply on the quantity of DNA, appears insufficient. The analysis of enzyme activity leads to a more satisfactory explanation of the cell cycle.*

The transition points between the different steps of the cell cycle have received much attention. Experiments using various techniques, such as thermal shock, the blocking action of actinomycin D or vincomycin, or irradiation, have shown certain points of transition: the beginning and the end of DNA synthesis, the different steps of DNA replication, and so on.

What are the modifications taking place at the chromosome level? The structure of the chromosome is fairly well known (Chapter 5): the chromatin is a DNA–histone complex in the form of a chain of nucleosomes (Kornberg and Klug, 1981). This basic organisation leads to an approximately sevenfold increase in DNA concentration.

During mitosis, the main changes are directed towards the eventual sharing of the genetic material between the two daughter cells so that each new cell has the same initial DNA concentration. Indeed, the length of the DNA is some 7000 times greater than that of the chromosomes during the metaphase, so that the DNA is far more tightly packed in the chromatid than it is during the interphase. It is currently believed (Latt *et al.*, 1974) that only one molecule of DNA passes through the centromere of each chromatid. Bak *et al.* (1979) have described three levels of DNA organisation in chromosomes during the metaphase: (i) 11 nm nucleosome fibres; (ii) 25 to 30 nm helicoidal fibres; and (iii) 400 nm unit fibres. This accounts for a DNA contraction by a factor of about 40. Therefore, during the process of shortening which occurs from the beginning of the prophase to the end of the metaphase, a further coiling of the unit fibre must take place (see Fig. 11.4). This probably corresponds to the helices frequently observed in the chromosome body.

During the interphase, the chromatin in the nucleus is mainly in the form of the 11 nm nucleosome fibres. However, higher order DNA structures have also been found, for example a loop-type structure (Igo-Kemenes and Zachau, 1977).

It is clear that *the organisation of the chromosomes varies with the different steps of cell division*, the chromosomes being in the most contracted form during the metaphase. Bak *et al.* (1979) have suggested that the DNA satellite sequences that have been observed could lead to the organisation of chromosomes in loops during the interphase, finally contributing to DNA condensation.

But how is the contraction actually triggered? And how do the helices relax to just the right degree without being totally disrupted? According to Bak *et al.* (1979), the structure of the unit fibre in metaphase chromosomes is mainly determined by the physical properties accompanying the coiling of the 25 to 30 nm fibre, due to the association with the $H1$ histones which suppresses the DNA loops. These physical properties evidently include the long- and short-range binding forces within the molecular complexes involved. A mechanistic hypothesis suggests that surface tension forces acting on a matrix could induce the contraction. But does such a matrix really exist?

Another theory favours the idea of a spontaneous contraction under appropriate environmental conditions. In any case, all the observations suggest that the morphological sequence of events seems to be predetermined and, once initiated, develops spontaneously. As we shall see below, Kaufmann (1977) gives an explanation based on the existence of a *biological cell clock* in which the phases corresponding to cell division *are not the elements of the clock itself.* This corresponds to the observed fact that *all cells entering a cell cycle will emerge from the cycle whatever the conditions of the mitosis may be.*

Various other modifications, other than those of the chromosomes, appear during cell division (see for example Tyson (1989)): the gradual disruption of the nuclear membrane and the sudden disappearance of the nucleolus at the end of the prophase, the splitting of the chromatids at the centromere, and so on. But the explanations of such phenomena still remain rather obscure.

Observers have long been fascinated by the cytoplasmic movements of a dividing cell seen through a microscope. This is why theories were very soon proposed to explain the phenomena of cytokinesis. Most of the theories were based on the idea of dynamic instability due to forces of surface tension acting within the cell. The binary splitting of an oil drop suspended in water, following changes in the surface tension, is an old observation which immediately suggested a similar mechanism for cell division. This has led to a model of cytokinesis which satisfies the laws of fluid mechanics and illustrates the forces acting during mitosis. The method used is a good introduction to the study of the movement of cells in their environment (Greenspan, 1978).

2.   *Models of the cell cycle*

The description of the cell cycle given above leads to a *causal loop model* in which a discontinuous series of events is related by a simple causal loop. Experimental support for the model has been provided by Hartwell (1974), but although this interpretation explains the periodicity of the cycle, with the repetition of 'identical' events and the maintenance of the causal relationship between the events, the nature of the relationship itself remains a major problem. For example, Lloyd *et al.* (1992) find chaotic solutions using a physiologically plausible model of the cell division cycle consisting of a mitotic oscillator with two time scales. Dispersed and quantised cell cycle times appear to be the consequences of a chaotic trajectory with a deterministic basis rather than a probabilistic one as assumed in many models.

According to Tauro *et al.* (1968), the execution of the cell cycle is an expression of the linear reading of the genes along the chromosomes. This explains the succession of the phases of the cycle, but not the periodicity, i.e. the enzyme synthesis from one cycle to the next. This difficulty is overcome by another type of model which, by analogy with cell differentiation, supposes the existence of a *mitogen*, a specific substance that triggers mitosis as soon as the accumulation of the substance exceeds a certain threshold value. Here we see how one cycle is followed by the next, but what maintains the causal relation between the different phases of the cycle still remains a mystery.

With either of these models, the sequential execution of the phases of the cell cycle, or even the cycles themselves, immediately call to mind a *biological cell clock* piloting the series of events exactly as the computer clock times the sequence of operating instructions. The problem is to determine whether mitosis itself *is* the biological clock or whether it is *conducted* by the clock.

In a general way we can say that a step *A* is an element of the clock in the causal loop model if it is an element of the closed causal loop producing the periodicity. Thus, blocking *A* leads to stopping the clock so that the cells accumulate in the blocked phase *A*, exactly as in the case of the product of a metabolic pathway. Conversely, if the other phases are executed in spite of *A* being blocked, then *A* will not be considered to be part of the clock. For instance, Hartwell's experiments suggest that DNA synthesis, which occurs in phase *S*, is not an element of the clock since the other cell phases can be executed without it.

Mutants have been used to block cell division at chosen steps of the cell cycle. This suggests a hybrid model consisting partly of a closed causal loop and partly of a separate central clock. Other experiments have shown that bi-nucleated cells obtained by fusion go through synchronous mitoses. In the case of *Physarum*, for example, the phase after fusion corresponds to the mean of the phases of the two cells weighted by the cytoplasms (Harris, 1970), exactly as if particular phases of the cycle were characterised by specific levels of chemical concentrations.

These observations seem to fit two complementary ideas: *(i) the existence of causal relationships between the discrete events constituting the phases of the cell cycle; and (ii) the existence of a phase characterised by a continuously varying concentration of certain substances.*

3. *The limit cycle model of biochemical oscillations*

All periodic events may of course be attributed to a clock in the primitive sense of the term. The experiments carried out by Mano (1970) and Hartwell (1974) show that if the cell clock does exist it may be independent of mitosis as well as of DNA and RNA synthesis. The results of the experiments suggest that such a clock, identified by the biochemical oscillations of several variables, may in fact intervene at the level of post-transcriptional control.

Following Kaufmann (1977), let us suppose that the cell clock consists of a protein X synthesised at a constant rate A, and a protein Y, an inactive form of X, produced at a rate proportional to X, say BX, with Y catalysing its own production from X at a rate proportional to X and to Y^2. This simple model of the clock may be represented by the non-linear system:

$$\left.\begin{aligned}\frac{dX}{dt} &= A - BX - XY^2 \\[2mm] \frac{dY}{dt} &= BX + XY^2 - Y\end{aligned}\right\}. \qquad (11.1)$$

According to Kaufmann, the biochemical bases of the model are not yet sound enough, even in the case of *Physarum*. However, the histone $H1$ and the protein in its activated state following phosphorylation have been reported to play a similar role in the control of mitosis (Bradbury *et al.*, 1973).

Our aim here is to show that this model implies a very general mechanism whatever the identity of the underlying substances. It is interesting that the same simple kinetic scheme has been successfully used to interpret periodic phenomena such as glycolytic oscillations (Ghosh *et al.*, 1979). Let the set of reactions corresponding to the system (11.1) be:

$$A \to X$$
$$B + X \to Y + D$$
$$X + 2Y \to 3Y$$
$$Y \to E.$$

The stationary state is obtained in $X^{(0)} = A/(B + A^2)$ and $Y^{(0)} = A$. It can be shown by the usual methods that there exists a limit cycle for:

$$\boxed{2A^2 > (B + A^2)^2 + A^2 + B} \qquad (11.2)$$

Figure 11.5 shows the limit cycle in the space of the states (X, Y), and Fig. 11.6 shows the isochronous lines for each hour of the cycle. The characteristics of the cycle may be summed up as follows:

(1)   Mitosis is triggered off as soon as $Y \geq Y_c$. This is the point of intersection of the straight line $Y = Y_c$ and the limit cycle. In other words, *mitosis is not an element of the cell clock*.

(2)   The system remains in the stationary state for the corresponding X and Y values. For the other values, we obtain a limit rotation about the stationary state which determines the phase of the cycle. For values other than that of the stationary

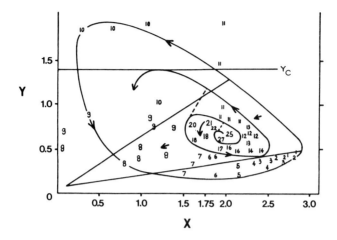

Fig. 11.5. Simulation of the cell cycle by a limit cycle in the space of states (X, Y) satisfying the system (11.1). Y_c is the critical concentration threshold for triggering mitosis. The numbers indicate the time required, in hours, for the path from the corresponding state on the diagram to intersect $Y = Y_c$ so that mitosis occurs (after Kaufmann, 1977).

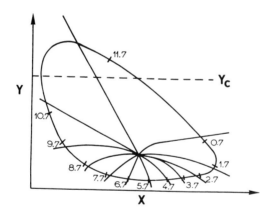

Fig. 11.6. The mitotic oscillator (Kaufmann, 1977). The curves starting from a point situated inside the limit cycle are isochronous lines, defined as the loci of points separating equal time intervals along the pathways. Consequently, the intersections of the isochronous lines with the limit cycle indicate the number of hours before mitosis.

state, the system tends to a limit cycle: the periodic phenomenon is stable for all perturbations.

(3) Several systems starting from an isochronous point (Fig. 11.6) will be synchronised after a certain lapse of time while continuing to 'turn' towards the limit cycle.

(4) The set of states (E) surrounding the stationary state has a property very different from that of the set of states (E') which are close to the limit cycle:

identical systems starting from (E) will be completely asynchronous whereas those starting from (E') will be quasi-synchronous.

(5)   The number of cycles required to reach the threshold Y_c depends on the initial state (Fig. 11.6). Within the limit cycle, this time is longer the nearer the system is to the stationary state.

This simple model has been validated by studies on *Physarum*, and some of its predictions have been substantiated. Can this be taken to mean that a system with two variables, such as the histone $H1$ and its phosphorylated form, corresponds to reality at the cellular level? The truth is likely to be found at a more complex level where the qualitative behaviour is identical. For example, we may envisage a complex network of biochemical reactions, some of which are discrete whereas others are continuous, leading to a global oscillation piloted by a set of parameters.

Kaufmann's vision of the nature of the phenomenon of cell division is highly potent and should eventually lead to a control of cell proliferation. The theory shows that the continuous variation of the parameters of the system (A, B) leads to a variation of the form, of the period and of the amplitude of the limit cycle, and thereby to oscillations. These oscillations and the stationary states may appear and disappear. When the value of A decreases, the amplitude of the limit cycle decreases and may even be reduced to a stationary stable point (the resting state G_0), which means that the cell cycle may be controlled by the simple variation of a parameter.

But how do these variations occur? It has long been experimentally observed that cells in close contact with each other stop proliferating. This property, which disappears in cancer cells, is called *contact inhibition*. It may be hypothesised that *a substance emitted by one cell diffuses into a neighbouring cell, acting on its cell cycle*. Kaufmann has investigated this problem from a theoretical point of view by considering the case of two cells with coupled biological clocks. In particular, he has shown that cell fusion, which allows a 'diffusive' coupling, can lead to temporary suppression of mitoses and that a gradual modification of the ratio of the coefficients of diffusion can suddenly lead either to synchronised mitoses or else to contact inhibition and the cessation of mitoses.

Thus, although a very general mechanism has been suggested for cell division, the problem at the cellular level is far from being solved. Because of possible applications in the field of cancer, one of the major lines of research over the last few decades has been oriented towards the understanding of the phenomena of cell division in the context of cell populations (McElwain and Pettet, 1993; Gyllenberg and Webb, 1990), as well as functional properties (Smolle and Stettner, 1993) and gene mutations (Bois and Compton-Quintana, 1992).

II.   Development of a cell population

Let us first define the characteristics of a population of cells:

-    The *age* of a cell in a given class, the time being measured from the instant at which the cell possesses all the features of the class. This variable very simply describes age-dependent cell properties.

● The *maturity* then represents the set of variations which are not part of the features of the class.

These variables, being simple, represent only very roughly the intrinsic properties of cells. It should be noted that a class differs from a compartment in the usual sense of the term and that the formalism based on compartments is not easily applied here. For we would have to write:

$$\frac{dN_i}{dt} = \mathcal{E}_i - S_i$$

for each number N_i of cells in the i-class, where $\mathcal{E}_i$ represents the flux of cells entering the class through birth or immigration, and S_i the flux of cells leaving the class through death or transfer to other classes, during the acquisition of new properties. At the limit, from this point of view, each cell would constitute a class by itself since the properties of the cells vary independently of time.

It is therefore necessary to take into account the *heterogeneity* of the classes. Since the age a and the *maturity* μ are linked, the age indicating the maturity of the cell, we may consider as independent variables either the couple (t, a) or the couple (t, μ). The former formalism is due to Von Foerster (1959) and the latter to Rubinow (1968).

Recently, Lasota *et al.* (1992) have introduced an *internal* or *physiological time* that is reset when an event occurs in the biological system. The rate of maturation is therefore defined by $d\tau/dt$ which depends on the amount of a certain activator. The lifespan of the organism is shorter when its rate of maturation is increased. The authors derive a recurrence relation for the values of the activator at the time of the occurrence of the events, and offer an interpretation of the cell division cycle in the context of this general model.

1. *Kinetics with variables* (t, a)

A class of cells $(\mathcal{C}_a)$ is defined by the number of cells $\Delta N_a(t)$ between the ages a and $a + \Delta a$. It is implicitly supposed that maturation acts by very short, discrete steps taking all the cells from the age class a to the age class $a + \Delta a$. At the limit, the density function of the ages is given at any instant t by:

$$n(t, a) = \lim_{\Delta a \to 0} \frac{\Delta N_a(t)}{\Delta a}. \tag{11.3}$$

Thus, at time t, the class $\mathcal{C}_{a_1}^{a_2}$ is composed of a number of cells equal to:

$$\int_{a_1}^{a_2} n(t, a)\, da.$$

Let us now establish the law for the development of the population.

During the time Δt, the age increases by Δt. In the age class $(\mathcal{C}_a)$, between a and $a + \Delta a$, the number of cells at instant t is $n(t, a)\Delta a$. Therefore, at time $t + dt$, the number of cells in class $(\mathcal{C}_a)$ is $n(t + dt, a + dt)\Delta a$. The difference in the number of

374	Cellular Organisation of Living Matter

cells between the times t and $t + dt$ evidently represents the loss of cells from the class $\mathcal{C}_a$ during the time dt. With the hypothesis that cell loss is proportional to the number of cells in class $(\mathcal{C}_a)$ and to time dt, in other words equal to:

$$-\lambda n(t, a) \, \Delta a \, dt$$

where the coefficient λ is a function of t, a and other parameters, we may write the equation for the development of the population:

$$n(t + dt, a + dt) \, \Delta a - n(t, a) \, \Delta a = -\lambda n(t, a) \, \Delta a \, dt.$$

Using Taylor's formula in the neighbourhood of (t, a) we have:

$$n(t + dt, a + dt) = n(t, a) + \frac{\partial n}{\partial t} dt + \frac{\partial n}{\partial a} dt + \varepsilon(dt)dt \quad \text{with} \quad \lim_{t \to 0} \varepsilon(t) = 0$$

whence the equation:

$$\frac{\partial n}{\partial t} dt + \frac{\partial n}{\partial a} dt = -\lambda n(t, a) \, dt.$$

Finally:

$$\boxed{\frac{\partial n}{\partial t} + \frac{\partial n}{\partial a} = -\lambda n(t, a)} \qquad (11.4)$$

This equation with partial derivatives t and a gives a local description of the number of cells conserved in class $(\mathcal{C}_a)$. If N is the total number of cells at time t, it is possible to determine the equation governing the development of the population. We have:

$$N(t) = \int_0^\infty n(t, a) \, da$$

and, integrating Eq. (11.4) from 0 to ∞, we have:

$$\int_0^\infty \frac{\partial n}{\partial t} da + \int_0^\infty \frac{\partial n}{\partial a} da = -\int_0^\infty \lambda(t, a...) \, n(t, a) \, da$$

or:

$$\frac{dN}{dt} + n(t, \infty) - n(t, 0) = -\int_0^\infty \lambda n(t, a) \, da.$$

The term $n(t, \infty)$, which represents the number of cells of infinite age, is null. The term $n(t, 0)$ gives the number of births at time t in the class $(\mathcal{C}_a)$, obtained either through maturation from the other classes or through mitoses. This is a function $\alpha(t)$, such that:

$$\boxed{\frac{dN}{dt} = \alpha(t) - \int_0^\infty \lambda n(t, a) \, da} \qquad (11.5)$$

2. Kinetics with variables (t, μ)

Maturity may be considered as an age-linked phenomenon consisting of variations which are external to the characteristics of a given class. However, it is more convenient to represent maturity as a variable, increasing in value from 0 to 1 from the beginning to the end of life. Thus a cell, born with a maturity $\mu = 0$, divides after a certain time, called the *generation time*, to form two daughter cells when its maturity $\mu = 1$. There exists a discontinuity that may be described by the condition of mitosis between the function of cell density $n(t, \mu)$ and that of the generation time of the cell population $\tau(t, \mu)$. This generation time is defined by:

$$a(t, \mu) = \tau(t, \mu)\, \mu.$$

Such a definition is phenomenological and, since the degree of maturation of a cell is not constant, we have to introduce the rate of maturation v_μ, function of time, of the degree of maturation, and characteristic cell parameters. For example, it is known that the rate of maturation depends on the cell density n. Moreover, a cell divides as soon as it reaches a maturity $\mu = 1$. The generation time τ, defined as the time necessary for the cell to go from $\mu = 0$ to $\mu = 1$, is therefore linked to the rate of maturation v_μ by a relationship of the type:

$$\tau = \frac{1}{v_\mu}. \tag{11.6}$$

If all the generation times were identical, the cell flux per unit time for maturity $\mu = 0$ and for maturity $\mu = 1$ should satisfy the condition:

$$n(t, 0) = 2n(t, 1^+)$$

since a cell of maturity $\mu = 1$, which has a density $n(t, 1^-) = n(t, 0)$, disappears giving birth to two daughter cells of maturity $\mu = 0$ with a density $n(t, 1^+)$. Here, 1^+ (resp. 1^-) is a value infinitely close to 1, approaching unity from higher (resp. lower) values. As the generation times are distinct, the condition for mitosis between the cell fluxes per unit time becomes:

$$\frac{n(t, 0)}{\tau(t, 0)} = 2\,\frac{n(t, 1)}{\tau(t, 1)}. \tag{11.7}$$

The reasoning behind the equation of development of $n(t, \mu)$ is identical to that used in the formalism (t, a), except that we take into account the variation of μ when going from t to $t + dt$. The variation of $\Delta\mu$ in terms of the rate of maturation v_μ may be written:

$$\text{Variation in } \Delta\mu = \frac{\partial v_\mu}{\partial \mu}\,\Delta\mu\,\Delta t.$$

Thus the equation of the conservation of the number of cells in the class $(\mathcal{C}_\mu)$ is:

$$n(t + \Delta t, \mu + \Delta\mu)\left(\Delta\mu + \frac{\partial v_\mu}{\partial \mu}\,\Delta\mu\,\Delta t\right) - n(t, \mu)\Delta\mu = -\lambda n(t, \mu)\,\Delta t\,\Delta\mu.$$

Applying Taylor's formula, we have:

$$\frac{\partial n}{\partial t} + \frac{\partial}{\partial \mu}(v_\mu n) = -\lambda n(t, \mu)$$ (11.8)

This equation is identical to Eq. (11.4) if we put age a equal to $\tau\mu$, with τ constant:

$$\frac{\partial n}{\partial t} + \frac{\partial}{\partial \mu}\left(\frac{n}{\tau}\right) = \frac{\partial n}{\partial t} + \frac{\partial}{\partial a}\left(\frac{n}{\tau}\right)\frac{\partial a}{\partial \mu}$$

$$= \frac{\partial n}{\partial t} + \tau\frac{\partial}{\partial a}\left(\frac{n}{\tau}\right) = \frac{\partial n}{\partial t} + \frac{\partial n}{\partial a}.$$

If, because of the heterogeneity of the cell population, τ is not constant, then the age a and the maturation μ are not equivalent variations. Following Rubinow (1968), let us consider μ as the 'cytological age' and a as the 'chronological age'. The same results apply and, in particular, the number of cells with a generation time τ at time t, and with a density $w(\tau)$ at point τ, can be written:

$$n(t, \mu) = \int_0^\infty w(\tau)\, n_\tau(t, \mu)\, d\tau$$ (11.9)

where $n_\tau(t, \mu)$ is the number of cells with a maturity μ and with a generation time τ at time t.

Finally, the total number of cells is deduced by double integration over the maturity time τ:

$$N(t) = \int_0^\infty w(\tau) \int_0^1 n_\tau(t, \mu)\, d\mu\, d\tau.$$

Rubinow's formalism may be applied to the investigation of the action of a cytotoxic drug on a cell population. It has been experimentally found that the development of the total number of cells $N(t)$ follows an exponential law such as:

$$N(t) = N_0 2^{t/\tau}.$$ (11.10)

We shall now show that the representation described above allows us to deduce this behaviour. The drug acts on a cell population defined by its generation time and, if we consider only this class, we should seek a solution for the equation:

$$\frac{\partial n}{\partial t} + \frac{1}{\tau}\frac{\partial n}{\partial \mu} = -\lambda n$$

such that:

$$n(t, 0) = 2n(t, 1)$$

where the function λ takes into account all cell losses.

Let us look for a solution of the form:

$$n(t, \mu) = N_0 e^{-L(t)} p(\mu) e^{\beta t} \text{ where } L(t) = \int_0^t \lambda(u)\, du.$$

This should satisfy the equation;

$$\beta p(\mu) + \frac{1}{\tau}\frac{dp}{d\mu} = 0$$

of which the solution is:

$$p(\mu) = p(0)e^{-\beta\tau\mu}.$$

Thus:

$$n(t, \mu) = N_0 e^{-L(t)}p(0)e^{-\beta\tau\mu}e^{\beta t}.$$

In particular, at the initial time:

$$n(0, \mu) = N_0 p(0)e^{-\beta\tau\mu}.$$

If we integrate over the domain $\mu \in [0, 1]$, we obtain the total population at the initial time, that is:

$$N_0 = \int_0^1 n(0, \mu)\,d\mu = N_0 p(0)\int_0^1 e^{-\beta\tau\mu}d\mu$$

$$= N_0 p(0)\frac{1 - e^{-\beta\tau}}{\beta\tau}$$

and finally:

$$p(\mu) = \frac{e^{-\beta\tau\mu}}{1 - e^{-\beta\tau}}\beta\tau.$$

β may now be determined by using the condition at the limits of mitotic division:

$$n(t, 0) = N_0 e^{-L(t)}p(0)e^{\beta t} = 2n(t, 1) = 2N_0 e^{-L(t)}p(0)e^{-\beta\tau}e^{\beta t}$$

or:

$$e^{-\beta\tau} = \frac{1}{2} \quad \text{or} \quad \beta = \frac{1}{\tau}\ln 2$$

whence the expression for $n(t, \mu)$:

$$n(t, \mu) = 2N_0 e^{-L(t)}\ln 2\,\exp\left[\left(\frac{t}{\tau} - \mu\right)\ln 2\right]$$

which represents the cell density in the class $\mathcal{C}_\mu$.

The total number of cells at time t is obtained by integration over $\mu \in [0, 1]$:

$$N(t) = \int_0^1 n(t, \mu)\,d\mu = 2N_0 \ln 2\, e^{-L(t)}\int_0^1 \exp\left[\left(\frac{t}{\tau} - \mu\right)\ln 2\right]d\mu.$$

The integral may be written:

$$\frac{1}{2\ln 2}e^{\frac{t}{\tau}\ln 2} = \frac{1}{2\ln 2}2^{t/\tau}$$

whence:

$$N(t) = N_0 2^{t/\tau} e^{-L(t)}.$$

We see here that the variation of the total number of cells in the population is in keeping with experimental findings and that the theory takes into account *cell loss* which is a function of time.

We may wonder which of the two formalisms discussed would be the more appropriate in a given situation. In particular, the variable 'maturity' is abstract and rather difficult to interpret. When the maturation rate v_μ is constant, μ can represent a certain number of time units and thus the age of the cell, which is the case described by Von Foerster. In the (t, μ) representation, mitosis always occurs at maximum maturity $\mu = 1$, while in the (t, a) representation, the daughter cells have no 'memory' of the maturity of the parent cell, i.e. of generation time. Consequently, the cell loss function λ, which depends on (t, μ), allows the distinction between mitosis occuring at $\mu = 1$ and cell death. For example, Trucco (1965), using the formalism (t, a), writes $\lambda = M + D + T$, with M, D and T representing respectively cell losses through mitosis, cell death and transformation.

The importance attached to work of this nature is justified by possible consequences in cancer treatment (see for example White (1990)). A cancerous tumour is caused by the uncontrolled, anarchic proliferation of abnormal cells deriving from cells of a particular type. A cell population of this kind is bound to have various generation times, so that a solution to the problem would lie in the selective control over the action of a drug with respect to the time τ and the histological type of the cells. Further on, we shall consider a mathematical representation of this problem.

III.   Analysis of the cell cycle: population theory

The analysis of the cell cycle is based on *fraction labelled mitosis* (FLM), in which a cell population is briefly exposed to a radioactive marker, tritiated thymidine. This molecule is selectively incorporated during the S-phase, and the different phases of the cell cycle as well as the homogeneity of the cell population can then be evaluated by autoradiography.

The first interpretation of the FLM curves was proposed by Quastler (1963) who constructed a model assuming that the phases $G1$, S, $G2$ and M were constant. The curve predicted by the theory is shown in Fig. 11.7 and the experimental observation is represented in Fig. 11.8b. The difference seen here is due to the deterministic interpretation of the phenomenon. In fact, the phases of the cell cycle vary according to probabilistic laws. Indeed, using a Monte Carlo method, Barrett (1966) has found the best results for a log-normal distribution of *transit times* in each phase. Models of this type, as well as others along the same principle, are rather difficult to use in practice since running them on a computer is long and expensive (see, for example, Takahashi (1968), Steel and Hanes (1971), Valleron and Frindel (1973), and Valleron

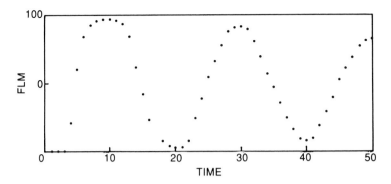

Fig. 11.7. Theoretical fraction labelled mitosis (FLM) in function of time, as predicted by Quastler's model (1963).

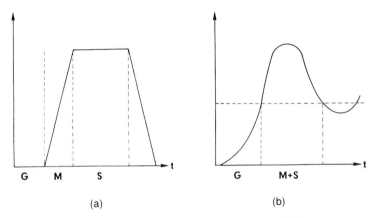

Fig. 11.8. Comparison of (a) theoretical, and (b) experimental FLM curves in function of time.

and MacDonald (1978); (White, 1989, 1991)). Other mathematical techniques have been recently introduced (Tyrcha, 1988; Arino and Kimmel, 1989) and applied to a study of the influence of a time-lag in a model of cell proliferation (Arino and Mortabit, 1992) and to cell cycle kinetics and unequal division (Kimmel and Arino, 1991). The mother–daughter correlation has been used to measure the variability of the duration of the cell cycle (Hejblum *et al.*, 1988) and has also been applied to the continuum model proposed by Cooper 1988). The formalism of the Leslie matrices, developed for population biology, has been successfully used by White (1978) to study the cell cycle. We shall now describe this powerful method.

1. *Leslie matrices applied to population studies*

In a general way, the study of populations involves the study of the individuals in different age-classes under the influence of three phenomena: *reproduction, ageing*

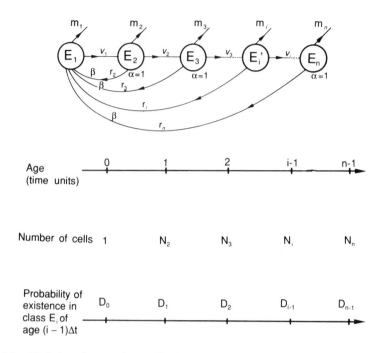

Fig. 11.9. Variations in age classes E_1, E_2, ... , E_n. as a result of cell death, ageing and reproduction.

and *death*. We then need to establish either differential equations, continuous with respect to time, or recurrent equations corresponding to the different steps of the evolution, discrete with respect to time. We can pass mathematically from the continuous differential formalism to the discrete recurrent formalism, but this may lead to contradictory results. Delattre (1981b), applying the theory of transformation systems (Chapter 3) to a population containing three classes, has shown that the equations lead to a stationary state in one case but not in the other. Hence precautions, of the physical or the epistemological order, will have to be taken. Let us follow Delattre's reasoning.

Let E_i be the class containing individuals of average age $i - 1$, and let δ_i be the 'width' of the class in terms of age. Figure 11.9 shows the inputs and outputs for each class E_i where each individual reproduces itself by binary division, as is the case of cell division. Here, *per unit time*:

m_i is the probability of the death of an element of E_i;

v_i is the probability that an element of E_i passes into E_{i+1} through ageing; and

r_i is the probability that an element of E_i disappears giving birth to β daughter cells by binary reproduction.

We may now write the equations for the variation of the number of elements N_i in the class i. Since E_1 is the class of age 0, it should be treated separately as indicated in Fig. 11.9. Then, for $i \neq 1$, we have:

$$\frac{\mathrm{d}N_i}{\mathrm{d}t} = v_{i-1}N_{i-1} - (m_i + v_i + r_i)N_i$$

$$i = 2 \text{ to } n$$

(11.11)

The functional identity of the elements belonging to a given class allows us to associate the same probabilities, m_i, v_i and r_i, to all the elements of E_i. This corresponds to Axiom 1 of the theory of transformation systems (Chapter 3). For this reason the variation of m_i and r_i in the age group δ_i should be negligible and the distribution of the individuals between the ages $i - \dfrac{\delta_i}{2}$ and $i + \dfrac{\delta_i}{2}$ should be uniform, i.e. δ_i should be sufficiently small. Moreover, internal coherence implies a relationship between δ_i and $(m_i + v_i + r_i)$: the total number of elements leaving E_i between t and $t + \delta_i$ is equal to the total content of E_i at time t since all elements leave by one of the pathways: death, ageing or reproduction. Thus, with $\delta_i = \mathrm{d}t$:

$$(m_i + v_i + r_i)N_i(t)\,\mathrm{d}t = N_i(t)$$

and:

$$m_i + v_i + r_i = \frac{1}{\delta_i}.$$

Equations (11.11) can be written:

$$\frac{\mathrm{d}N_i}{\mathrm{d}t} = v_{i-1}N_{i-1} - \frac{1}{\delta_i}N_i.$$

For $i = 1$, each class E_j ($j \neq 1$) contributes elements through reproduction:

$$\frac{\mathrm{d}N_1}{\mathrm{d}t} = \sum_{j=2}^{n} \beta r_j N_j - \frac{1}{\delta_1}N_1.$$

In matrix form, we have the differential system:

$$\frac{\mathrm{d}\mathbf{N}}{\mathrm{d}t} = \mathbf{AN}$$

(11.12)

where

$$\mathbf{A} = \begin{bmatrix}
-\dfrac{1}{\delta_1} & \beta r_2 & \beta r_3 & \cdots & \beta r_n \\
v_1 & -\dfrac{1}{\delta_2} & 0 & \cdots & 0 \\
0 & v_2 & -\dfrac{1}{\delta_3} & & 0 \\
\vdots & \vdots & v_3 & \ddots & \vdots \\
0 & 0 & 0 & v_{n-1} & -\dfrac{1}{\delta_n}
\end{bmatrix}.$$

The corresponding system of recurrent equations must satisfy:

$$N_i(t + dt) = N_i(t) = dN_i(t)$$

or in matrix form:

$$\mathbf{N}(t + dt) = \mathbf{IN}(t) + d\mathbf{N}(t)$$

$\mathbf{I}$ being the n-order unit matrix. Using Eq. (11.12) we may write:

$$\mathbf{N}(t + dt) = (\mathbf{I} + \mathbf{A}dt)\mathbf{N}(t)$$

$$\mathbf{N}(t + dt) = \mathscr{A}\mathbf{N}(t)$$

with

$$\mathscr{A} = \mathbf{I} + \mathbf{A}dt.$$

We can then obtain the successive states of the system from the initial time t_0 by putting:

$$dt = j\Delta t \text{ where } j = 1, 2, 3 \dot{\dot{\,}}.. :$$

$$\boxed{\mathbf{N}(t_0 + j\Delta t) = \mathscr{A}^{\,j}\mathbf{N}(t_0)} \,. \tag{11.13}$$

The matrix $\mathbf{A} = \mathbf{I} + \mathbf{A}\Delta t$ is written:

$$A + \begin{bmatrix} 1 - \dfrac{\Delta t}{\delta_1} & \beta r_2 \Delta t & \beta r_3 \Delta t & \cdots & & \beta r_n \Delta t \\[1.2em] v_1 \Delta t & 1 - \dfrac{\Delta t}{\delta_2} & 0 & \cdots & & 0 \\[1.2em] 0 & v_2 \Delta t & 1 - \dfrac{\Delta t}{\delta_3} & \cdots & & 0 \\[1.2em] \vdots & \vdots & & & & \vdots \\[1.2em] 0 & 0 & 0 & \cdots & v_{n-1}\Delta t & 1 - \dfrac{\Delta t}{\delta_n} \end{bmatrix}$$

and if $\delta_i = \Delta t = 1$ for all i, then we obtain the Leslie matrix $\mathbf{L}$:

$$L = \begin{bmatrix} 0 & \beta r_2 & \beta r_3 & \cdots & \beta r_{n-1} & \beta r_n \\ v_1 & 0 & 0 & \cdots & 0 & 0 \\ 0 & v_2 & 0 & \cdots & 0 & 0 \\ \vdots & \vdots & \vdots & & \vdots & \vdots \\ 0 & 0 & 0 & \cdots & v_{n-1} & 0 \end{bmatrix}.$$

Finally, the Leslie matrix corresponds to class widths all equal to Δt, the time-interval chosen for the study of the successive states of the system. However, it should be remarked that the generalised use of this formalism in population studies gives rise to difficulties inherent to the juxtaposition of competing formalisms.

Although the interpretation of the Leslie matrix is immediate, with the first line corresponding to newborn individuals and the main sub-diagonal corresponding to transfers from one class to the next through ageing, each case will require very careful consideration. It would be useful to apply the theorems proposed by Demetrius (1980) stating the conditions for the overall stability of multiplicative recurrent systems. According to Delattre (1971b), the mere internal mathematical coherence of a mathematical formalism is obviously not sufficient to justify its use, whence the interest of an axiomatic formalism, such as that of transformation systems, which emphasises the coherent procedure of the formalisation applied to the problems being studied.

2. Interpretation of the FLM curve by population theory

The FLM method involves the *distribution of transit times*. Therefore, the probabilities m_i, v_i and r_i of the Leslie matrix should be expressed in terms of the probabilities of transit times P_i in each class E_i. Thus let $P_i(t)$ be the probability that a cell entering the cycle at time $t = 0$ leaves between times $(i - 1)\Delta t$ and $i\Delta t$, where Δt is the width of the age-class, as in the case of the matrix $\mathcal{A}$ (Eq. (11.13)). It can be immediately deduced that the probability that a cell of age $i\Delta t$ has left the cycle at any time is the sum of the probabilities of such events, each excluding the others, occurring between $(0$ and $\Delta t)$ or $(\Delta t$ and $2\Delta t)$ or ... $(i - 1)\Delta t$ and $i\Delta t)$, that is:

$$\sum_{k=0}^{i-1} P_{k+1}(t).$$

Thus, the probability that a cell has an age $i\Delta t$, i.e. it has not left the cycle at this time, is:

$$D_i = 1 - \sum_{k=0}^{i-1} P_{k+1}(t).$$

To simplify, let us suppose that there is only one cell in E_i at time $t = 0$, so that $N(0) = (1, 0, \ldots, 0)$. Let us first consider the phenomenon of ageing. The probability v_i is given by the number of cells of age $i\Delta t$ (class E_{i+1}) divided by the number of cells of age $(i - 1)\Delta t$, that is:

$$v_i = \frac{N_{i+1}(t + \Delta t)}{N_i(t)}$$

with $t = (i - 1)\Delta t$. At time $t = (i - 1)\Delta t$, and with respect to the age-class 0, the number of cells of age $(i - 1)\Delta t$ is:

$$N_i(t) = 1 - \sum_{k=0}^{i-2} P_{k+1}(t) = D_{i-1}.$$

Similarly, at time $t + \Delta t = i\Delta t$, the number of cells of age $i\Delta t$ is:

$$N_{i+1}(t + \Delta t) = 1 - \sum_{k=0}^{i-1} P_{k+1}(t) = D_i.$$

Then:

$$\boxed{\begin{aligned} v_i &= \frac{D_i}{D_{i-1}} \quad i = 2, \dots, n \\ v_1 &= D_1 \end{aligned}}$$

(11.14)

Let us now consider the phenomenon of reproduction. For each class $i \neq 1$, the probability r_i is given by the number of daughter cells, appearing at time $t + \Delta t$ in the class E_1 of age 0, divided by the number of mother cells, i.e. cells that *do not age*. Since the number of cells subject to ageing in an age-class $(i-1)\Delta t$ is $(D_i/D_{i-1})N_i$, we deduce that the number of cells that will reproduce themselves is:

$$\left(1 - \frac{D_i}{D_{i-1}}\right)N_i = \frac{P_i}{D_{i-1}}N_i.$$

As the factor of reproduction is $\beta(1 \leq \beta \leq 2)$, we have:

$$N_1(t + \Delta t) = \beta \sum_{i=2}^{n} \frac{P_i}{D_{i-1}}N_i(t) + \beta N_1(t).$$

The last term indicates the presence of the unique cell at time $t = 0$, with $N_1(t) = 1$. By identification we find:

$$r_i = \frac{P_i}{D_{i-1}} \quad i = 2, \dots, n$$

$$r_1 = 1.$$

Thus, the Leslie matrix can be written:

$$\mathbf{L} = \begin{bmatrix} 0 & \beta\dfrac{P_2}{D_1} & \beta\dfrac{P_3}{D_2} & \cdots & \cdots & \beta\dfrac{P_n}{D_{n-1}} \\ D_1 & 0 & 0 & \cdots & \cdots & 0 \\ 0 & D_2/D_1 & 0 & \cdots & \cdots & 0 \\ \vdots & \vdots & \vdots & \cdots & \cdots & \cdots \\ 0 & 0 & 0 & \cdots & \dfrac{D_{n-1}}{D_{n-2}} & 0 \end{bmatrix}.$$

But how do we adapt the theory to the practical problem, i.e. the analysis of the FLM curves which, as we have seen, break down into several parts each corresponding to a step in cell growth, that is $G1$, S, $G2$ and M? Let us see how White (1978) has applied the theory to each of these steps. If $\hat{\mathbf{N}}$ is a vector with four components, each of which is related to a given step (j = 1, 2, 3, 4), then, by extension, we have the matrix equation:

$$\hat{N}(t + \Delta t) = \hat{T}\hat{N}(t). \tag{11.15}$$

Here again, $\hat{T}$ is an operator that projects the variations of the phases of the cell population according to the chronological age. This is a Leslie matrix of which the elements are functions of the parameters D_i^j and P_i^j at time $i\Delta t$ of the phase j. A *stable* distribution of the cell population is obtained by diagonalising $\hat{T}$. In fact this particular case corresponds to the frequent occurrence of an increase in the number of cells, which means a constant proportion with respect to the chronological age. A stable age-distribution should follow from the definition:

$$\hat{T}\hat{N}_i = \lambda_i\hat{N}_i$$

where $\hat{N}_i$ is an eigenvector associated with an eigenvalue λ_i of $\hat{T}$. Let $\hat{N} = (V_1 V_2 V_3 V_4)$ be a distribution of this kind. It can be explicitly calculated in function of D and P for each eigenvalue λ.

Let us suppose that, for a given type of cell, the form of the initial distribution is $\hat{N}(0) = (0\ V_2\ 0\ 0)$, as is the case of a population in stable growth, uniformly marked during the S phase. Following the matrix equation (11.15), we have:

$$\hat{N}(2\Delta t) = \hat{T}\hat{N}(\Delta t) = \hat{T}^2\hat{N}(0)$$

and, by recurrence:

$$\hat{N}(i\Delta t) = \hat{T}^i\hat{N}(0).$$

For the type of cell considered, the number of marked cells at time $i\Delta t$ is:

$$\hat{T}^i(0\ V_2\ 0\ 0)$$

for the set of cells undergoing mitosis:

$$\sum_{\text{Mitoses}=M} \hat{T}^i(0\ V_2\ 0\ 0).$$

Finally, the total number of cells at time $i\Delta t$ is:

$$\sum_M \hat{T}^i(V_1\ V_2\ V_3\ V_4).$$

From the definition of FLM we deduce that the value at the point ($i\Delta t$) is:

$$\boxed{\text{FLM}(i\Delta t) = \frac{\displaystyle\sum_M \hat{T}^i(0\ V_2\ 0\ 0)}{\displaystyle\sum_M \hat{T}^i(V_1\ V_2\ V_3\ V_4)} = \frac{\displaystyle\sum_M \hat{T}^i(0\ V_2\ 0\ 0)}{\lambda^i \displaystyle\sum_M (V_1\ V_2\ V_3\ V_4)}.} \tag{11.16}$$

This is the ratio of the number of marked cells to the total number of cells.

Thus, this procedure allows an interpretation of the FLM curve for a distribution

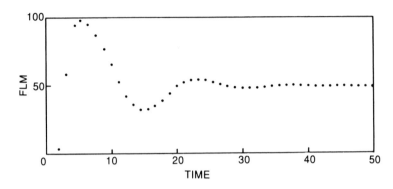

Fig. 11.10.   Theoretical FLM curves obtained by White (1978) with the assumption that the distribution of transit times follows a negative binomial law for a highly dispersed population.

(P_i) of transit times, given *a priori*, over the cell cycle. We may note that the probability distributions most often used are the negative binomial law, simulating a highly dispersed population, i.e. an asynchronous population, and Pearson's law simulating a synchronous population. The FLM curve thus obtained is shown in Fig. 11.10. The reader may consult White (1978) for further details.

12

Cell Growth, Division and Differentiation

For the sake of simplicity, cell growth, division and differentiation have been treated above as separate topics, but of course all these phenomena should be studied globally. As we have seen, the process of cell differentiation is fairly well understood when it occurs after mitosis, but some essential questions remain unanswered. *Does cell differentiation take place before or after mitosis, i.e. does the process affect the parent cell or only the daughter cells? Is cell differentiation symmetrical or antisymmetrical? And, finally, what precisely is the relationship between cell growth and cell division and differentiation?*

Let us see how Ortoleva and Ross (1973) have attempted to answer the first question on the basis of the chemical processes involved in cell differentiation.

I. Asymmetrical cell division

Daughter cells produced by cell division may be differentiated or undifferentiated. Here, we shall consider the case of *asymmetrical* differentiation as represented by the diagrams $A \rightarrow (A$ and $B)$ or $A \rightarrow (C$ and $B)$ in Fig. 12.1. It may be considered that symmetrical division or differentiation raises no special problem of interpretation since division and differentiation occur successively in time. The separation of the two phenomena is not as evident as in the case of asymmetry.

How does asymmetry occur? When is it initiated? If the process of division continues to be symmetrical during the mitosis, asymmetry may be produced before or after nuclear division. Obviously, any attempt at answering these questions will first require that the nature of the asymmetry be defined (Treinen and Feitelson, 1993).

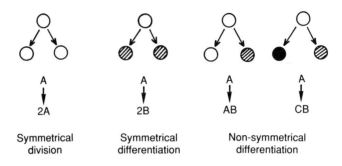

Symmetrical division	Symmetrical differentiation	Non-symmetrical differentiation

Fig. 12.1.   Symmetrical and asymmetrical cell differentiation.

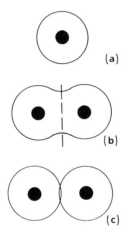

Fig. 12.2.   Cell division according to a model due to Ortoleva and Ross (1973).

Differentiation, as we have seen in Chapter 10, is described by a particular cellular state brought about by a gradient mechanism. The state is defined by a set of concentrations of chemical species c_i and a specific state of gene activation. Cell division starts with the division of the nucleus which transforms the parent cell into two potential daughter cells with a strong coupling between them. This is followed by the development of a membrane separating the two daughter cells which are then only weakly coupled (Fig. 12.2).

The model of asymmetrical differentiation, due to Ortoleva and Ross (1973), is based on the possibility of the *transfer of molecules across the membrane being formed during mitosis between the two daughter cells*. Random fluctuations of the chemical concentrations existing between the two cells tend to produce differences leading to physicochemical asymmetry. Thus, an instability produced by a high-amplitude fluctuation could in fact lead to an asymmetrical perturbation of the differentiated state.

The mechanism here is not new; indeed, it is often encountered (see Chapter 11).

This particular case involves the reaction–diffusion of chemical species. Let $\mathbf{C}^{(i)}$ be the vector characterising the chemical state of the cell concentrations of the potential daughter cell, number i, $i = 1$ or 2. Three hypotheses may now be made:

(1) the intercellular transport is described by the flux:

$$\mathbf{J} = \mathbf{H}(\mathbf{C}^{(1)} - \mathbf{C}^{(2)})$$

where $\mathbf{H}$ is the matrix of transmembrane permeabilities (see Volume II, Part IV);

(2) the transport between the daughter cell i and the environment is described by the flux:

$$\mathbf{J}_{\text{env}}^{(i)} = \mathbf{G}(\mathbf{C}_{\text{env}}^{(i)} - \mathbf{C}^{(i)}(t))$$

where $\mathbf{G}$ is the transmembrane permeability; and

(3) the molecular species are subject to chemical reactions satisfying a system of the type:

$$\frac{d\mathbf{C}^{(i)}}{dt} = \mathbf{F}(\mathbf{C}^{(i)}) \quad i = 1, 2.$$

Initially, the state is perfectly symmetrical, that is:

$$\mathbf{C}^{(1)}(0) = \mathbf{C}^{(2)}(0) = \mathbf{C}(0).$$

The problem is to determine the condition under which this symmetry is destroyed. As we have seen, Turing has already studied this condition and Prigogine has given a complete formulation: it is the equilibrium between the reaction and the diffusion which leads to the destruction of the symmetry. If the difference from the symmetrical state is represented by the perturbation $\delta\mathbf{C}^{(i)}(t)$ such that:

$$\mathbf{C}^{(i)}(t) = \mathbf{C}(t) + \delta\mathbf{C}^{(i)}(t)$$

and the equations satisfied by $\delta\mathbf{C}^{(i)}(t)$ are deduced from the hypotheses (1), (2) and (3), giving:

$$\frac{d\mathbf{C}^{(i)}}{dt} = \mathbf{G}[\mathbf{C}_{\text{env}}^{(i)} - \mathbf{C}^{(i)}] + \mathbf{H}[\mathbf{C}^{(j)} - \mathbf{C}^{(i)}] + \mathbf{F}(\mathbf{C}^{(i)}) \quad i, j = 1, 2 \quad i \neq j \quad (12.1)$$

then the linearisation of these equations in terms of the variables of perturbation $\delta\mathbf{C}^{(i)}(t)$ leads to a condition of instability (see Chapter 2). This condition may be satisfied by various mechanisms such as autocatalysis and cross-catalysis of chemical reactions between cytoplasmic components, the influence of membrane permeability on the diffusion of these molecules, reactions at the genome level, and so on, all these processes being capable of modifying the chemical concentrations.

Let us now sum up the results:

(1) *Instability leads to competitive events between the different types of differentiation, symmetrical and asymmetrical.*

(2)   *As the asymmetry develops, various situations may be observed:*
- *a strong coupling between the daughter cells will maintain a lasting asymmetrical state; and*
- *a weak coupling will lead to multiple stable stationary states.*

This is the same result as already obtained by Babloyantz *et al.* (Chapter 9) in the case of gene regulation. Thus it would appear possible to interpret many biological phenomena as multiple stationary states occurring far from thermodynamic equilibrium (see, for example, Turner, 1974).

(3)   *If this theory holds good, the consequences would obviously be of the greatest importance.* For example, it should be possible to stabilise the process of asymmetrical differentiation by the use of chemical substances acting on the gradient, *without any anti-mitotic effect on the target cells.* Melanocytogenesis, the transformation of a melanoblast into a melanocyte capable of melanin synthesis, could then be blocked by chemical agents without any anti-mitotic activity. Such qualitative results would substantiate the notion of a breakdown in the symmetry.

Other mechanisms have been suggested to explain asymmetrical cell division: for example, the existence of an asymmetry in the parent cell, the action of an external gradient, the initiation of asymmetry by hormone activation of random sites or by the incorrect replication of the genome. However, all these seem less relevant than the mechanism described above.

II. Cell growth

The interpretation of cell growth is a rather delicate matter, especially as the definition of the term itself lacks precision. As we have already attributed a molecular and cellular significance to the phenomenon (Chapter 10), let us now attempt to describe these aspects. In particular, increase in volume is characteristic of cell growth but, on closer examination, we are soon confronted with the notion of the growth of a cell population or even that of a cell culture, where the link between increased volume and proliferation due to cell division is far from clear.

1. *Analytical description: mass and volume*

A relationship is known to exist between the *mass*, the *length* and the *rate of growth* of a cell. The question that arises is: why does a cell have to attain a certain size before dividing?

Let τ be the generation time of a cell. The increase in cell mass follows a law analogous to that expressed by Eq. (11.10):

$$m(t; \tau) = m(0; \tau)2^{t/\tau}. \tag{12.2}$$

This is the mass of the cell t units of time after its birth. It has also been shown (Schaechter *et al.*, 1958) that the initial mass is an exponential function of the generation time of the form:

$$m(0;\ \tau) = m*2^{T/\tau} \tag{12.3}$$

where $1/T$ is called the characteristic time constant. But what does this constant represent? The growth of a cell going through the cell cycle suggests a relationship between an increase in the mass and the beginning of the phase S of DNA synthesis. In fact, it has been found (Donachie and Masters, 1968) that T is the characteristic time between the initiation of the phase S and the cell division that follows. More precisely, let n_i be the number of initiation sites for DNA replication, i.e. the number of sites at which replication begins all along the DNA molecule. Let $m_i(t;\ \tau)$ be the mass of the cell at time t defined by the integer i such that $t = (i + 1)\tau - T$.

This mass m_i may now be calculated from Eqs (12.2) and (12.3) as follows: At time $t = \tau - \theta$, i.e. θ minutes before a division, we have:

$$m(t;\ \tau) = m(\tau - \theta;\ \tau) = m(0;\ \tau)2^{(\tau-\theta)/\tau}$$

$$= 2m*2^{T/\tau}2^{-\theta/\tau} \tag{12.4}$$

$$= 2m*2^{(T-\theta)/\tau}$$

where $m*$ is an 'experimental' mass given by Eq. (12.3). Thus in the particular case of the mass T units of time *before* an $(i + 1)$-order division, such that:

$$\theta = T - i\tau,\ t = (i + 1)\tau - T$$

we obtain:

$$m_i(t;\ \tau) = 2m* 2^{\frac{T-(T-i\tau)}{\tau}} = 2m*2^i$$

$$\boxed{m_i(t;\ \tau) = 2^i(2\ m*)} \ . \tag{12.5}$$

Finally, we find that the masses m_i *are whole unit multiples of the quantity* $(2\ m*)$*, independent of the generation time* τ*.*

This is indeed a most curious result for which no satisfactory molecular explanation has yet been put forward. $(2\ m*)$ has been called the *magic mass* since:

$$m_0(\tau - T;\ \tau) = 2\ m*$$

$$m_1(2\tau - T;\ \tau) = 2(2\ m*) \text{ first division} \tag{12.6}$$

$$m_i((i + 1)\tau - T;\ \tau) = 2^i(2\ m*) \text{ } i\text{th division.}$$

One consequence is that the difference between the instant $(i + 1)\tau - T$ at which the initiation occurs and the eventual division $(i + 1)\tau$ represents the dimension T introduced in the phenomenological equation (12.3). Another consequence, following upon the recurrent equations (12.6), is that the number n_i is equal to 2^i, so that the ratio:

$$m_i(t; \tau)/n_i = 2\, m^* \qquad (12.7)$$

is a constant. This result corresponds to the observed finding that, at the beginning of DNA replication, the ratio of cell mass to the number of initiation sites is constant. However, it should be noted that when the rate of growth is slower than one division per hour, i.e. the generation time is greater than an hour, the above results do not fit quite so well.

We shall now try to solve the following problem: what is the relationship between the length of a cell at birth, i.e. with a maturity $\mu = 0$, and its length at $\mu = 1$?

Let $L(t; \tau)$ be the length of the cell at the instant t for a generation time τ. We now need to calculate $L(0; \tau)$ and $L(\tau; \tau)$. Observations show that, to a first approximation:

$$L(0; \tau) = k(m(0; \tau))^{1/3} \qquad (12.8)$$

where k is a constant of proportionality. We may therefore adopt this empirical law as a working hypothesis.

The results obtained above (Eqs (12.2) and the following) were valid only for an 'average' cell and in particular for a generation time somewhat shorter than the mean of the individual generation times. Here, the reasoning is based on observations carried out on cell cultures. The average value $\bar\tau$ will therefore be defined as the time after which the values of the mass and the length are twice the initial values. To avoid the difficulties that may arise from such a distribution of generation times about the mean value, let us, following Krasnow (1978), adopt the hypothesis that the dynamic laws governing the fluctuations about the mean are the same as those that describe the mean value itself. Let the ratio of the length of the cell at a given time to its length at time corresponding to the mean generation time be called simply the *length*. Then, following Eq. (12.8), the length at birth may be written:

$$l(0; \tau) = \frac{L(0; \tau)}{L(0; \bar\tau)} = \left(\frac{m(0; \tau)}{m(0; \bar\tau)}\right)^{1/3}$$

but since $m(0; \tau) = m^* 2^{T/\tau}$ (Eq. 12.3), we deduce:

$$l(0; \tau) = 2^{(T/3)(1/\tau - 1/\bar\tau)}. \qquad (12.9)$$

If we put $s = T\ln 2$, we may interpret s as the replication time so that s/τ is the ratio of the time for the doubling of the DNA to the time for the doubling of the cell, and thus the quantity causing the increase in mass is:

$$m(0; \tau) = m^* e^{s/\tau}. \qquad (12.10)$$

Similarly, at maturity, we have:

$$l(\tau; \tau) = \frac{L(\tau; \tau)}{L(0; \bar\tau)}. \qquad (12.11)$$

The process of cell division may therefore be analysed as follows: for a generation time τ_0, the length of a cell at birth is $l(0; \tau_0)$. At time τ_0, its length becomes $l(\tau_0; \tau_0)$ and it then gives birth to two daughter cells, each with a length of $\frac{1}{2}l(\tau_0; \tau_0)$.

Evidently, if the generation time for the daughter cells is the same, τ_1, then:

$$\frac{1}{2} l(\tau_0; \tau_0) = l(0; \tau_1)$$

and similarly:

$$\frac{1}{2} m(\tau_0; \tau_0) = m(0; \tau_1).$$

If the properties are invariant from one generation to another, we have:

$$l(0; \tau_1) = \left(\frac{m(0; \tau_1)}{m(0; \bar{\tau})}\right)^{1/3} = \left(\frac{\frac{1}{2}m(\tau_0; \tau_0)}{m(0; \bar{\tau})}\right)^{1/3}$$

(12.12)

$$l(0; \tau_1) = \frac{1}{2} l(\tau_0; \tau_0) = \frac{1}{2} l(0; \tau_0) f(\tau_0; \tau_0)$$

where we introduce the *elongation factor* $f(t; \tau)$ of the cell at time t for a generation time τ. Finally:

$$\left(\frac{\frac{1}{2} m(\tau_0; \tau_0)}{m(0; \bar{\tau})}\right)^{1/3} = \frac{1}{2} l(0; \tau_0) f(\tau_0; \tau_0)$$

and since:

$$l(0; \tau_0) = \left(\frac{m(0; \tau_0)}{m(0; \bar{\tau})}\right)^{1/3}$$

by simple substitution, we deduce:

$$\frac{m(\tau_0; \tau_0)}{m(0; \tau_0)} = \frac{1}{4} f(\tau_0; \tau_0)^3.$$

In more general terms:

$$\frac{m(\tau; \tau)}{m(0; \tau)} = \frac{1}{4} f(\tau; \tau)^3.$$

When $\tau = \bar{\tau}$ we have:

$$\frac{m(\bar{\tau}; \bar{\tau})}{m(0; \bar{\tau})} = \frac{1}{4} \cdot 2^3 = 2$$

since by definition the mass (or length) is doubled when $t = \bar{\tau}$. *The important result here is that the final mass (or length) of a cell is not twice the initial mass (or length).*

Let us consider the ratio:

$$\alpha(\tau) = \frac{L(\tau)}{R(\tau)} = \left[\frac{\pi(L(\tau))^3}{V} \right]^{1/2} \tag{12.13}$$

where V is the volume of a cylinder of length L and radius R. In terms of the mass m with a density ρ, we have:

$$\alpha(\tau) = \left[\frac{\pi \rho (L(\tau))^3}{m(\tau)} \right]^{1/2}.$$

For convenience, let us write $f(\tau) \equiv f(\tau; \tau)$. And since:

$$(L(\tau))^3 = (L(0))^3 f^3(\tau)$$

we obtain:

$$\alpha(\tau) = \left[\frac{\pi \rho (L(0))^3 f^3(\tau)}{m(0)f^3(\tau)/4} \right]^{1/2} = 2\left[\frac{\pi \rho (L(0))^3}{m(0)} \right]^{1/2} = 2\alpha(0). \tag{12.14}$$

In other words, the ratio $\alpha(t) = L/R$, called the 'aspect ratio', doubles its value during the generation time. Therefore, except when $\tau = \bar{\tau}$, i.e. when $f(\bar{\tau}) = 2$, it is the only dimension that is doubled during cell division. Consequently, the elongation factor $f(t; \tau)$ satisfies the equation $f(t; \tau) = 2^{t/\bar{\tau}}$. The *mass factor* $F(t; \tau)$ is a solution of the equation:

$$F(t; \tilde{\tau}) = 2^{t/\tilde{\tau}} \tag{12.15}$$

where $\tilde{\tau}$ is defined as the time required for the doubling of F. There is therefore a relationship between $\bar{\tau}$ and $\tilde{\tau}$:

$$\tilde{\tau} = \tau \Big/ \left(\frac{3\tau}{\bar{\tau}} - 2 \right). \tag{12.16}$$

The three quantities, α, f and F, defined above in Eqs (12.12), (12.13) and (12.15), are characteristic of cell growth when the generation time τ differs from the mean time $\bar{\tau}$. Figure 12.3 shows how these dimensions vary with time. Krasnow thus deduces the variation of cell length:

$$L(t; \tau) = L_\infty 2^{\frac{T}{3\tau}} 2^{\frac{t}{\bar{\tau}}} \tag{12.17}$$

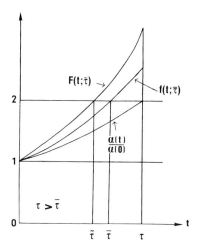

Fig. 12.3. The upper curve shows the mass factor $F(t; \bar{\tau}) = 2^{t/\bar{\tau}}$ as a function of time t for a cell with a generation time $\tau > \bar{\tau}$, where $\bar{\tau}$ is the mean value for the cell culture. The lower curves correspond to the elongation factor $f(t; \tau) = 2^{t/\bar{\tau}}$ and the aspect ratio $\alpha(t)/\alpha(0) = 2^{t/\tau}$. At $t = \tau$, the length of the cell doubles in value. Cell division occurs at $t = \tau > \bar{\tau}$, when $\alpha(t) = 2\alpha(0)$. (After Krasnow, 1978.)

where L_∞ is the minimum length of the cell at birth, experimentally found to be $k(m^*)^{1/3}$ where k is a constant. Similarly, we may calculate the variation of the cell mass:

$$\boxed{m(t; \tau) = m^* 2^{T/\tau} 2^{\left(\frac{3\tau}{\bar{\tau}} - 2\right)\frac{t}{\tau}}}.$$
(12.18)

All the results obtained above are of a geometrical nature, based on a description which is mathematically correct but which unfortunately *lacks an underlying physiological interpretation.* Starting from phenomenological laws, we have seen how the cell divides as soon as the aspect ratio α doubles in value, independently of any changes in the other dimensions, and that the generation time τ is completely determined by the initial cell mass, regardless of the mean generation time $\bar{\tau}$ of the cell culture. The importance of this result should not be underestimated, bearing in mind that its mysterious nature merely reveals our present state of ignorance. Further explanation would require determination of the relationship existing between cell growth, division and DNA replication. Krasnow's findings are of interest as they allow a good inference of the successive steps of cell growth from measurable parameters:

(*i*)   The initial mass determines the aspect ratio and thus the generation time (Eq. 12.14).

(*ii*) The generation time τ and the mean culture time $\bar{\tau}$ determine the effective generation time $\tilde{\tau}$ (Eq. 12.16) and thus the rate of growth:

$$\frac{d}{dt} \ln m(t;\tau) = \frac{2}{T}\left[\ln m\left(\frac{T}{2};\bar{\tau}\right) - \ln m(0;\tau)\right] \qquad (12.19)$$

where $m(T/2; \tau)$ is a unique mass characteristic since, at $t = T/2$, all the cells in a given culture have the same mass $m(T/2; \tau) = m*2^{3T/2\bar{\tau}}$ (Eq. 12.18).

(iii)   *The cell continues to grow at this rate until the aspect ratio doubles in value. At time $\bar{\tau}$, the mass is equal to $2m*2^{T/\tau}$. The replication then begins.*

(iv)   *The end of the replication does not trigger cell division. However, the growth, and thus the initial size, depend on the moment of the initiation of DNA replication.*

We can hardly go any further for the present as too little is known about the relationships between cell growth, DNA replication and cell division. The results above show a good experimental fit but we should bear in mind that they are based on the *phenomenological* equations (12.2) and (12.3). However, in spite of this, these results have encouraged several attempts at interpreting cell growth at the molecular level (see, for example, Reiner, 1973), but there is no general agreement about the underlying hypotheses.

The problem of the role of the environment on a cell has been the object of what is called the *cellular sociology* (Marcelpoil and Usson, 1992). Obviously, we must take into account the cellular interactions occurring during the growth of a cell. A diagram involving three parameters: the average roundness of the forms, the degree of the homogeneity of the roundness, and the degree of the heterogeneity of the area (area disorder) appears to be a powerful tool for the determination of the intrinsic disorder of a cellular population. It has been possible to simulate the spatial perturbations of theoretical populations such as the aggregation and randomisation of positions. The formation of spatial structures in bacterial colonies has been shown to depend on intercellular regulation (Budriene *et al.*, 1988). The process of growth of a population of cells suspended in a liquid medium has been studied by Fuxman (1993).

2.   Global description of the behaviour of a cell

Like the analytical description above, the global approach is of great interest. Davison (1975) has attempted to describe the growth of a cell by simulating its behaviour in a culture medium. The methodology used is in itself most instructive. Rather than trying to show that the global behaviour of a system as complex as that of a cell can be reduced to the solution of a large dynamic system, Davison uses certain standard variables to characterise cell behaviour. A dynamic system may be used to represent the variation of a large set of variables which are components of a vector of state X. Let us consider a system:

$$\dot{X} = F(X, U, K)$$

where X is the state vector, U is the vector of inputs and K is the vector of parameters. Following Heinmets (1966), Davison has chosen a system of biochemical reactions to represent the behaviour of the cell (Fig. 12.4). This is obviously the crucial

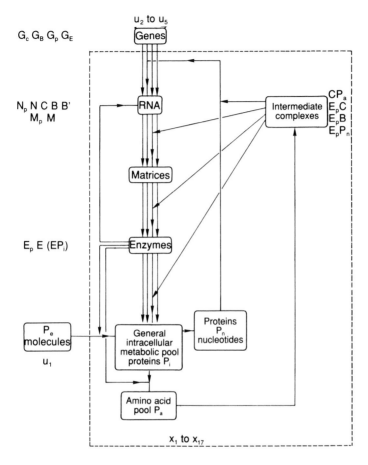

Fig. 12.4. Metabolic pathways of cell function according to a model due to Davison (1975). (See also Table 12.2 for a description of the variables.)

point of the model describing protein synthesis (Fig. 12.5 and Table 12.1) on the basis of the fundamental cell components, limited to the nucleus, i.e. genes, DNA, operons, and so on, details of which we have already considered.

Evidently, the prime difficulty lies in describing cell growth on the basis of the chemistry of the nucleus (considered, of course, in extremely simple terms). We have just seen that the relationship between cell growth and mitosis is far from being perfectly known. Davison introduces the mechanism of cell division as follows.

Cell division corresponds to the doubling of the concentration of nuclear substances $x_i(0)$ which are components of the state vector of the cell, present at the initial time. Thus, division will occur at time t such that the distance between state $\mathbf{X}(t)$ and $2\,\mathbf{X}(0)$ is:

(i)   minimum;
(ii) strictly smaller than $\|\mathbf{X}(0)\|$, where $\|\ \ \|$ represents the norm of the vector.

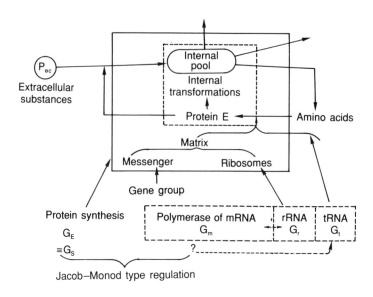

Fig. 12.5. Protein synthesis with regulation of the Jacob–Monod type in Davison's model.

This definition of mitosis is equivalent to minimising the functional:

$$J(t) = \| X(t) - 2\, X(0) \|.$$

This results in the definition of cell growth. Let $(X^i;\ t_i^*)$ be a cell of generation i dividing at time t_i^*. The *clone* is then defined by the succession of states $(X^i;\ t_i^*)$, $i = 1, 2, \ldots$, satisfying the recurrent dynamic system:

$$0 \le t \le t_1^* \quad \begin{cases} \mathbf{X}^1(0) = \mathbf{X}^0 \\ \dot{\mathbf{X}}^1(t) = \mathbf{F}(\mathbf{X}^1, \mathbf{U}, \mathbf{K}) \end{cases}$$

$$\vdots$$

$$t_i^* \le t \le t_{i+1}^* \quad \begin{cases} \mathbf{X}^{i+1}(0) = \dfrac{1}{2}\, \mathbf{X}^i(t_i^*) \\ \dot{\mathbf{X}}^{i+1}(t) = \mathbf{F}(\mathbf{X}^{i+1}, \mathbf{U}, \mathbf{K}) \end{cases}$$

where the values of t_i are obtained by using the criteria of minimisation:

$$\boxed{\begin{aligned} &\| \mathbf{X}^i(t_i) - 2\, \mathbf{X}^i(0) \| \text{ minimal in } t_i^* \text{ and} \\ &\| \mathbf{X}^i(t_i) - 2\, \mathbf{X}^i(0) \| < \| \mathbf{X}^i(0) \| \end{aligned}}.$$

With this condition of cell division, the problem may be solved by putting down the initial conditions $\mathbf{X}(0)$, the input parameters (vector $\mathbf{U}$) and the internal parameters (vector $\mathbf{K}$) which are the constants of the biochemical reactions.

Table 12.1. The set of equations representing the 'functioning' of the cell (see Table 12.2 for a description of the variables used) (after Davison, 1975).

RNA polymerase

$$\frac{dE_p}{dt} = k_{11}N_p[CP_a] - k_{17}E_p - k_{21}E_pC + k_{31}[E_pC] - k_{23}E_pB + k_{33}[E_pB] - k_1E_pP_n + k_6G_E[E_pP_n]$$

Nucleotide pool

$$\frac{dP_n}{dt} = k_{19}EP_i - k_1E_pP_n + k_{12}M + k_{13}M_p - k_4G_cP_n - k_5G_pP_n - k_3G_BP_n$$

Polymerase–nucleotide complex

$$\frac{d[E_pP_n]}{dt} = k_1E_pP_n - k_6G_E[E_pP_n]$$

mRNA (for E)

$$\frac{dM}{dt} = k_6G_E[E_pP_n] - k_7BM - k_{12}M + k_{10}N[CP_a]$$

(B–M) matrix for the synthesis of E

$$\frac{dN}{dt} = k_7BM - k_{10}N[CP_a] - k_{28}N$$

Amino acid pool

$$\frac{dP_a}{dt} = k_{20}P_iE - k_9CP_a$$

tRNA–Amino acid complex

$$\frac{d[CP_a]}{dt} = k_9CP_a - k_{10}N[CP_a] - k_{11}N_p[CP_a]$$

Total protein

$$\frac{dE}{dt} = k_{10}N[CP_a] - k_{16}E - k_{22}EP_i + k_{32}[EP_i] - k_2EB'$$

Intercellular metabolite–protein complex

$$\frac{d[EP_i]}{dt} = k_{22}EP_i - k_{32}[EP_i]$$

mRNA (for E_p)

$$\frac{dM_p}{dt} = k_5G_pP_n - k_8BM_p + k_{11}N_p[CP_a] - k_{13}M_p$$

Matrix

$$\frac{dN_p}{dt} = k_8BM_p - k_{11}N_p[CP_a] - k_{29}N_p$$

Ribosomal RNA fraction

$$\frac{dB'}{dt} = k_3G_BP_N - k_2B'E$$

Table 12.1.   (*cont.*)

Ribosome

$$\frac{dB}{dt} = k_2 B'E - k_7 BM - k_8 BM_p - k_{14} B + k_{10} N[CP_a] + k_{11} N_p[CP_a] - k_{23} E_p B + k_{33}[E_p B]$$

Transport RNA

$$\frac{dC}{dt} = k_4 G_c P_n - k_9 CP_a + k_{10} N[CP_a] + k_{11} N_p[CP_a] - k_{15} C - k_{21} E_p C + k_{31}[E_p C]$$

General intercellular metabolic pool

$$\frac{dP_i}{dt} = k_{18} EP_e - k_{19} EP_i - k_{20} EP_i + k_{32}[EP_i] + k_{14} B - k_{22} EP_i + k_{15} C + k_{16} E + k_{17} E_p$$

$$+ k_{28} N + k_{29} N_p - k_{30} P_i$$

Polymerase–tRNA complex

$$\frac{d[E_p C]}{dt} = k_{21} E_p C - k_{31}[E_p C]$$

rRNA–polymerase complex

$$\frac{d[E_p B]}{dt} = k_{23} E_p B - k_{33}[E_p B]$$

Using 26 chemical reactions involving 17 different substances (Table 12.2), Davison obtains the simulated behaviour of a normal cell. From the qualitative point of view the results are excellent and enable us to visualise the effect of the modification of any one of the variables of the system. For example, Fig. 12.6 shows the effect of a variation of the internal pool P_i, of the amino acid pool P_a, and of the nucleotide pool P_n for a normal cell. *This example illustrates the power of the simulation when an extremum condition of the solution of the dynamic system is known.* Of course, in the particular case considered here, the choice of the chemical system cannot be considered entirely representative. But the principle of the method, applied to the more comprehensive and accurate experimental results we may expect in the future, should lead to a far more refined simulation of cell behaviour.

We may mention another global approach based on the uptake, storage and utilisation of energy, during embryonic development (Zonneveld and Kooijman, 1993). The metabolism, without being analysed in detail, is presented by means of phenomenological curves.

III. Mechano-chemical approach to morphogenesis: Murray's mechanical model for mesenchymal morphogenesis

As mentioned in the introduction to Chapter 10, a new approach to the generation of pattern and form in development has been made in the past few years by Murray

Table 12.2. Description of the variables in Table 12.1. See also Fig. 12.4 for their significance in the metabolic pathways of cell function.

State of the cell:

$x_1 = E_p$ = RNA polymerase for the synthesis of mRNA
$x_2 = P_n$ = Nucleotide pool for RNA synthesis
$x_3 = (E_p P_n)$ = Intermediate complex
$x_4 = B'$ = Ribosomal RNA fraction
$x_5 = B$ = Ribosomal fraction
$x_6 = C$ = tRNA
$x_7 = M$ = mRNA (for E)
$x_8 = M_p$ = mRNA (for E_p)
$x_9 = N$ = (B–M) matrix for the synthesis of E
$x_{10} = N_p$ = (B–M_p) matrix for the synthesis of E_p
$x_{11} = P_a$ = Amino acid pool
$x_{12} = (C P_a)$ = Intermediate complex
$x_{13} = E$ = Total protein
$x_{14} = P_i$ = General intracellular metabolic pool
$x_{15} = (E_p C)$
$x_{16} = (E_p B)$ } Intermediate complexes
$x_{17} = (E P_i)$

External inputs to the cell:

$U_1 = P_e$ = Extracellular metabolic pool
$U_2 = G_E$ = Genes for mRNA synthesis (synthesis of M)
$U_3 = G_C$ = Genes for tRNA synthesis (synthesis of C)
$U_4 = G_B$ = Genes for rRNA synthesis
$U_5 = G_p$ = Genes for mRNA synthesis (synthesis of M_p)

Rate constants of reactions: (equations in Table 12.1)

$k = k_1, k_2, \ldots, k_{33}$

and his colleagues on the basis of mechanical models. These concern the morphogenetic processes involved in the coordinated movement, or patterning, of *populations* of cells. In the chemical pre-pattern approach (see for example Nagorcka (1989) for a review), the two processes of pattern formation and morphogenesis occur sequentially. In the mechano-chemical approach, these two processes are considered to be simultaneous, i.e. they constitute a single process. The two approaches are quite different, and a major argument in favour of the second is its capacity of being self-correcting, i.e. of reacting to disturbances as development proceeds through the continuous mechanical deformations of the tissues. For a simplified version of this model for morphogenesis, see Britton (1988).

Let us now consider the pattern formation of mesenchymal cells (fibroblasts) during early embryogenesis (Murray *et al.*, 1983; Oster *et al.*, 1983; Murray, 1989; Rao and Talwalker, 1989). Fibroblasts move independently in the body, secreting the fibrous material which forms the extracellular matrix (ECM) of tissues (Fig. 12.7). As we have seen above, the mechanisms that generate pattern and form operate in the early period of embryogenesis. We shall now take into account the factors affecting the movement of fibroblasts. These factors include convection (passive transport), chemotaxis (chemical gradient), contact guidance (preferred direction for the

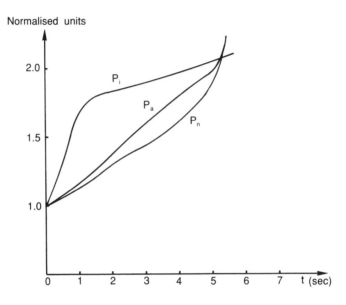

Fig. 12.6.   Variation of the internal pool P_i, the amino acid pool P_a, and the nucleotide pool P_n in function of time (normalised units) according to Davison (1975).

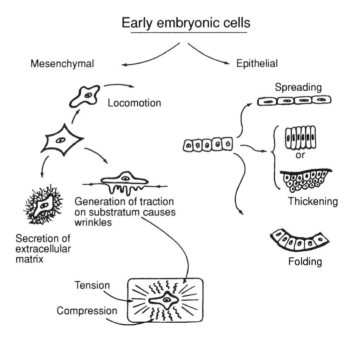

Fig. 12.7.   Early embryonic cells (after J. D. Murray (1989)). Mesenchymal cells are motile and generate large traction forces. Epithelial cells do not move about, but can spread or thicken under the influence of force. Force generation affects cell division.

cells), contact inhibition, haptotaxis (the cells move up an adhesive gradient), diffu-sion (the movement is random but generally follows a cell density gradient), galvanotaxis (movement generated by electric potentials).

The mechano-chemical model of mesenchymal morphogenesis is based on the mathematical description of the spatiotemporal variation of some of these mechan-isms. The equations govern: (i) the conservation law for the cell population density n ($\mathbf{r}$, t); (ii) the mechanical balance of the forces between the cells and the extracellular matrix (ECM); (iii) the conservation law for the ECM density $\rho(\mathbf{r}, t)$ at point $\mathbf{r}$ and at time t. The mechanisms involved may be formulated as follows:

Cell population density: Let J be the flux of cells (the number of cells crossing a unit area in unit time), and M the source (the cell proliferation rate equals $rn(N - n)$, according to the logistic model of cell growth, where r is the proliferative rate at initial time and N is the maximum cell density when other effects are absent). Then, from Eq. (2.36), the conservation equation is:

$$\frac{\partial n}{\partial t} = -\nabla \cdot \mathbf{J} + M. \tag{12.20}$$

Convection: Let $\mathbf{u}(\mathbf{r}, t)$ be the displacement vector of the ECM. The convective flux component is then given by:

$$\mathbf{J}_c = n \frac{\partial \mathbf{u}}{\partial t} \tag{12.21}$$

which is the product of the number of cells and the velocity of the deformation of the ECM.

Diffusion: The flux of cells $\mathbf{J}_D$ due to diffusion is considered to have two origins: a classical Fickian diffusion with a coefficient D_1, and a *non-local* effect on diffusive dispersal due to the more distant densities that have a neighbouring average effect on the cells considered at ($\mathbf{r}$, t). This long-range diffusion has a coefficient D_2 (Murray *et al.*, 1983), after Othmer (1969). Thus:

$$\mathbf{J}_D = -D_1\nabla n + D_2\nabla(\nabla^2 n). \tag{12.22}$$

If D_2 is positive, the long-range diffusion has a stabilising effect.

Haptotaxis: Gradients in the matrix density are created by the traction exerted by the cells on the ECM. This haptotactic flux is given by:

$$\mathbf{J}_h = n \, (a_1\nabla\rho - a_2\nabla^3\rho) \tag{12.23}$$

where a_1 and a_2 are two positive coefficients. The second term sum represents a non-local sensing property of the cells. The influence of cell-haptotaxis on the generation of spatial and spatiotemporal patterns has been studied by Maini (1989).

Finally, the cell conservation equation is:

$$\frac{\partial n}{\partial t} = -\nabla \cdot \left[n\frac{\partial \mathbf{u}}{\partial t} \right] + \nabla \cdot [D_1\nabla n - D_2\nabla(\nabla^2 n)] - \nabla \cdot [a_1\nabla\rho - a_2\nabla^3\rho] + r\,n(N - n). \tag{12.24}$$

Mechanical cell–ECM interactions

It should be possible to take into account the elastic strain in the ECM which results
in aligned fibres. The directions of the strain facilitate the movement of cells. This
effect can be included in the elastic strain tensor. However, the ECM is a complex
medium which is difficult to model. Some mechanical deformations can be described
by a stress tensor $\sigma(\mathbf{r}, t)$ acting on a linear, isotropic viscoelastic continuum. The
mechanical interactions between the cells and the ECM may be described by the
stress tensor such that:

$$\sigma = \sigma_{\text{ECM}} + \sigma_{\text{cell}} \tag{12.25}$$

which satisfies the equation:

$$\nabla \cdot \sigma + \rho \mathbf{F} = 0 \tag{12.26}$$

describing the mechanical equilibrium between the traction forces generated by the
cells and the elastic restoring forces developed in the ECM. The expression of the
two contributions to the stress tensor is a difficult task, but it allows the description
of an anisotropic medium.

ECM conservation equation

The matrix conservation equation is, as usual:

$$\frac{\partial \rho}{\partial t} + \nabla \cdot (\rho \mathbf{u}_t) = S(n, \rho, \mathbf{u}) \tag{12.27}$$

where the source term S is the rate of secretion of matrix by the cells.

The final system of equations for the density fields $n(\mathbf{r}, t)$ $\rho(\mathbf{r}, t)$ and the displacement field $\mathbf{u}(\mathbf{r}, t)$ contains 14 parameters which have a physical meaning and are, in
principle, measurable. The non-dimensionalisation leads to a reduced system with 12
parameters:

$$\frac{\partial n}{\partial t} = \nabla \cdot [D_1 \nabla n - D_2 \nabla(\nabla^2 n)] - \nabla \cdot [a_1 \nabla \rho - a_2 \nabla^3 \rho] + -\nabla \cdot \left[n \frac{\partial \mathbf{u}}{\partial t} \right] + r\, n(N - n)$$

$$\nabla \cdot \left[\left(\mu_1 \frac{\partial \varepsilon}{\partial t} + \mu_2 \frac{\partial \theta}{\partial t} \mathbf{I} \right) + (\varepsilon + v'\theta \mathbf{I}) + \frac{\tau n}{1 + \lambda n^2}(\rho + \gamma \nabla^2 \rho)\mathbf{I} \right] = s\rho \mathbf{u}$$

$$\frac{\partial \rho}{\partial t} + \nabla \cdot (\rho \mathbf{u}_t) = S(n, \rho, \mathbf{u}). \tag{12.28}$$

The 12 parameters are a_1, a_2, D_1, D_2, r, τ, λ, associated with the cell properties,
and μ_1, μ_2, v', γ, s, associated with the ECM properties. Figure 12.8 shows the
conceptual framework for the mechanical models. The mathematical study of the
above system is quite difficult but it is possible to use the techniques described in
Chapter 10 to find the dispersion relation. The traction parameter τ plays an important role in the qualitative variation of these dispersion relations. The Oster–Murray
theory leads to a remarkable variety of dispersion relations from relatively simple

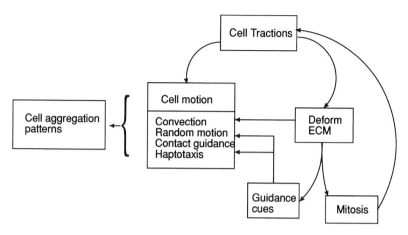

Fig. 12.8. Mechanical models: cell tractions play a central role in pattern formation (after J. D. Murray (1989)).

mechanisms, thus predicting complex growth behaviour. An interesting application concerns the morphogenesis of limb cartilage and the morphogenetic rules that can be deduced. Figure 12.9 shows the three basic types of cell condensation which generate cartilage patterns: focal condensation, branching bifurcation, and segmentation condensation. These bifurcating pattern elements can be combined to construct all limb cartilages (Fig. 12.10).

Morphogenetic rules have been discussed in relation to evolution (Oster *et al.*, 1988). Murray (1989) observes that whereas, on a macroscopic scale, morphogenesis appears to be deterministic, cellular activities on a microscopic scale during the formation of the limb involve considerable randomness. Although not strictly forbidden mathematically, some morphogenetic events, such as trifurcations from a single chondrogenetic condenstation, are highly unlikely since they correspond to a 'developmental bias' involving a delicate choice of conditions and parameter tuning. Beloussov and Lakirev (1991) have discussed generative rules for the morphogenesis of epithelial tubes using a finite element model, based upon the lateral pressure between adjacent epithelial cells. Biologically realistical shapes are obtained, almost all of them belonging to the same basic archetype. This establishes the morphogenetic role of changes in viscoelastic coefficients. Geometry has proved to be important in models of this type (Briere and Goodwin, 1988). The possibility that a reacting–diffusing morphogen might be the direct agent in the control of the morphogenetic movements of cell sheets has been studied by Cummings (1990) who has determined the minimum number of cellular parameters, related to the concentration of morphogen, necessary for the definition of the morphogen-coupled surface geometry.

Although the actual morphogens have not yet been identified, Murray's use of reaction–diffusion equations to explain the phenomena of morphogenesis appears to be a promising approach. A principle of continuity has emerged from morphogenesis for many developing systems since a morphogenetic field, when perturbed, tends to restore the normal pattern of structures in its organ region (Clarke *et al.*, 1988).

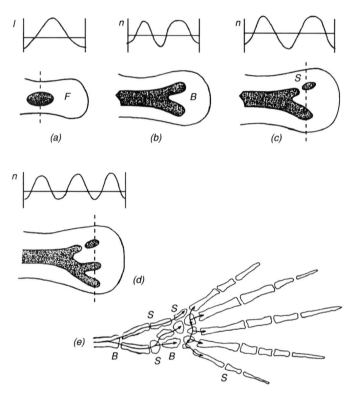

Fig. 12.9. The three basic types of cell condensation which generate cartilage patterns in the developing vertebrate limb (after J. D. Murray (1989)): (a) focal condensation; (b) branching bifurcation; and (c) segmentation condensation. (d) shows the formation of more patterns occurs by further branching or independent foci; and (e) represents a branching sequence showing how the cartilage patterns in the limb of a salamander can be built up from a sequence of F, B, and S bifurcations.

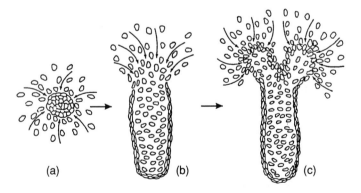

Fig. 12.10. Cell condensation process (after J. D. Murray (1989)). (a) Initial aggregation of cells into a central focus. (b) Development of the cartilage at the distal end of the condensation. (c) With appropriate conditions, the aggregation undergoes a Y-bifurcation.

These authors have investigated physical field theories for a morphogenetic field, seeking constraints which would make a field theory produce the principle of continuity. This approach has led to an extremum criterion for epimorphic regeneration. *However, a major problem remaining to be solved is that of the mechanism by which the physiological function is maintained during the structural transformations of the organism during development.* Our main objective is to define an extremum criterion for the organisation of physiological functions during development. Let us now consider this important issue.

IV.   Topological description of developmental dynamics: potential of functional organisation

1.   *Introduction: variational principles in biology*

A theory of functional organisation, based on five definitions, in terms of the functional interactions of a Formal Biological System (FBS), was presented in Chapter 4. A functional interaction, identified with an elementary physiological function, was defined as the action of a product emitted by one structural unit to another. The emitted product $P_{\alpha,1}$, synthesised by the source u_1, is transferred towards a sink u_2. This process is described by the relation: $P_{\alpha,2} = \psi_{12}^{\alpha}(P_{\alpha,1})$. For example, a molecule emitted by a specific cell acts on another cell within which it induces a series of transformations. The FBS concept was introduced to provide a mathematically defined biological system with specific properties as similar as possible to those of a real biological system. An FBS is characterised by two such properties: (i) it is an '*equipotent system*', i.e. all the potentialities of the genome are identical at each level of organisation; and (ii) it is a '*mutational system*', in which the structural units undergo independent modifications. The construction of the set of functional interactions, constituting the topology of the biological system, is based on a very simple hypothesis, called the self-association hypothesis. According to this non-trivial hypothesis, an interaction between two structural units becomes necessary when the elementary function of one of the units is destroyed, i.e. when a source becomes a sink. Indeed, each elementary function can be potentially executed by a structural unit for its own survival. Thus, if such a function is lost, for example due to the occurrence of micromutations, the survival condition for the structural unit (the cell for instance) implies that the missing product be made available by another unit synthesising the product. The topological system, or the observed functional biological system (O-FBS), composed of such functional interactions represents the functional organisation of the FBS. The (O-FBS) is constructed according to what we have called the *principle of vital coherence*.

The hypothesis of self-association was shown to be valid, at least in the case of an FBS with two fundamental properties of living organisms: metabolism and self-reproduction. We demonstrated that *self-association increases the domain of stability of the dynamic functional biological system (D-FBS) describing the physiological function*, represented by the hierarchical system of functional interactions. The dynamic system (D-FBS) is associated with the topological system (O-FBS) defined by the mathematical graph representing the set of functional interactions. The property of the self-association that increases the domain of stability may then be formulated by stating that the increase in the complexity of the (O-FBS) by self-association is a natural tendency of the biological system since it accompanies an increase in the domain of stability of the corresponding (D-FBS). Thus, there exists a natural tendency for a biological system to move towards greater complexity as soon as the conditions for self-association are satisfied, for example after the occurrence of micromutations.

An integrated description of physiological phenomena from the cell level to the organ level must include a unique, formalised definition of the biological subsystem within a mathematical framework. Such a description will have to be able to deal simultaneously with several levels of organisation. Indeed, the nuclear and cytoplasmic subsystems of the cell and, for example, the renal or respiratory subsystems of living organisms can all be considered as hierarchical systems based on the same organisational principles (Chauvet, 1987, 1993b). At the higher levels of functional organisation, we have the intuitive impression of an increase in complexity, which may be thought of as an increase in the number of state variables of the biological system. However, in reality, the hierarchy of the system leads to a decrease in the number of state variables.

In the conceptual framework proposed here, we shall see that because of the relationship between the topology and the dynamics of a biological system, there is a simultaneous increase in the complexity of the (O-FBS) and the (D-FBS). The hypothesis of self-association will be used to express functional complexity in terms of combinations of functional interactions represented by oriented digraphs.

The measurement of the complexity of a system poses a fascinating challenge in all branches of science and has stimulated much research. For example, Ferdinand (1974) has used the concept of 'default entropy' to evaluate the complexity of computer circuitry, and Walter (1980; 1983) has defined a family of indices of complexity linked to the stability of compartmental systems. In the case of the biological functional organisation, the advantage of our method stems from the double description: topological and dynamical. The former describes the existence of functional interactions, whereas the latter describes the spatiotemporal variations of the processes associated with these interactions. Consequently, *the stability of the biological system will correspond to that of the two systems when the number of structural units varies.*

In a variational approach, a fundamental theoretical question arises concerning the existence of an optimum principle, similar to optimum principles known in the physical sciences, capable of explaining the stability of a formal biological system. Such a principle could be used as a criterion for the evolution of biological systems and would provide a model for the comparison of biological and physical systems. We shall therefore address this important question. The topological stability of the graph of the system (O-FBS) will be expressed as the redistribution of the edges between the vertices of the graph, i.e. as the redistribution of the sources and sinks when one of these is destroyed, taking into account the constraint of the invariance of the physiological function. This constraint, expressed in terms of the principle of vital coherence, implies that the functional interactions are reorganised by self-association in such a way that the collective function subsists after the reorganisation. We shall demonstrate that *just one hypothesis, which is experimentally falsifiable, is sufficient for the basis of this theory.*

2. The potential of functional organisation

a. The nature of the concept: the combinatorial approach and non-symmetry

If an FBS is considered to be an equipotent system, i.e. made up of structural units that can have the same elementary physiological functions, then the self-association hypothesis accounts for the functional organisation of these structural units, and subsequently for the hierarchical system (Chauvet, 1993a). As mentioned above, all the cells of a given organism have the same potentialities of expression at the lowest gene level of organisation. We shall see that this property can be used for the construction of the FBS corresponding to the hierarchical system. The (O-FBS) consists of functional interactions, i.e. structural units and elementary functions, the topology of which is the functional organisation. More precisely, the functional organisation at level l is defined by the distribution $(n_\alpha^{(l)})_{\alpha=1,\mu^{(l)}}$ of functional links between structural units at this level. This distribution will be called *a state of organisation*. $n_\alpha^{(l)}$ is also the number of zeros in the row α of the matrix M, i.e. the number of sinks for the function P_α of the system (see Definition IV in Section III.3, Chapter 4).

In the observed state of organisation, the FBS, because of the property of idempotence, presents other non-observed, potential states of organisation. *These states exist as potentialities in the system, but are not expressed.* For example, the passage of organisms through particular stages during morphogenesis, and the capacity of regeneration in certain species, show that such potential states of organisation can be observed during the life of the system.

Let (S) be a set of v structural units all having the same potentialities. According to the self-association hypothesis, each unit is either a source or a sink. If a source could not become a sink it would die following an alteration in the function. Suppose that n of the v units do not yield the product necessary for their survival. These n units will therefore have to be coupled to the other $(v - n)$ units. Several possibilities can be envisaged to account for the associations that create the topology of an (O-FBS). We have proposed a function $\Pi(n)$, called the *potential of functional organisation*, representing the reservoir of possibilities during the lifetime of the system. This mathematical function is determined if the following properties are satisfied:

(i) Π gives a measure of the number of potential functional interactions;
(ii) Π leads, under some conditions, to a hierarchy of the system;
(iii) the hierarchy found is such that the value of Π^l, calculated *for a level* l, decreases from this level l to the next higher $(l + 1)$;
(iv)   the value of Π, calculated *for the total hierarchical system*, increases, whereas a reorganisation of the system, as defined below, implies a decrease in this value.

These four properties represent general facts about biological systems. However, the problem of the evolution of the functional organisation is difficult to solve. Indeed, if a function such as Π admittedly describes a certain biological reality, further questions immediately arise. Among all the possible organisations available to the system during its evolution, why was one particular organisation chosen?

b.   Definition and formulation

Definition: Potential of functional organisation

The potential Π^l at level l is the logarithm of the number o^l of all functional organisations that are possible for the observed state of organisation, at its ith level. All the structural units are assumed to have identical potentialities.

Let o^l be the number of organisations available for the observed state, i.e. a sequence (n_α^l), $\alpha = 1, \mu^l$, where n_α^l is the number of sinks for the P_α-function at level l. Then:

$$o^l = \prod_{\alpha=1}^{\mu^l}(v^l - n_\alpha^l)^{n_\alpha^l} \tag{12.29}$$

since n_α^l units can be coupled to the $(v^l - n_\alpha^l)$ units which have the P_α-function, and all organisations can be associated to one another such that $1 \leq \alpha \leq \mu^l$. The number of structural units at level l is v^l, and is called the *degree of organisation* of the system at this level because it corresponds to the number of classes of structural equivalence. Then:

$$\Pi^l = \ln o^l = \sum_{\alpha=1}^{\mu l} n_\alpha^l \ln(v^l - n_\alpha^l)$$

$$\Pi = \sum_{l=1}^{L} \ln o^l = \sum_{l=1}^{L} \sum_{\alpha=1}^{\mu^l} n_\alpha^l \ln(v^l - n_\alpha^l) \tag{12.30}$$

where L is the number of levels and μ^l is the number of products at level l. A simplified form would be:

$$\Pi = \sum_{levels} \sum_{products} [sinks] \ln [sources] \tag{12.31}$$

Note that $v^l - n_\alpha^l > 0$ and $\Pi = 0$, if $n_\alpha^l = 0$ or $v^l - n_\alpha^l = 1$, i.e. if there exists no sink or only one source (see Fig. 12.11). These two eventualities are the simplest; all the others correspond to combinations of interactions. Thus, Π could be interpreted to represent functional complexity. However, as we shall see, the concept of potential can be used as the basis for the development of a variational theory.

3 sinks   2 sinks   1 sink

1 source   2 sources   3 sources

Potential=0   Potential=ln 4   Potential=ln 3

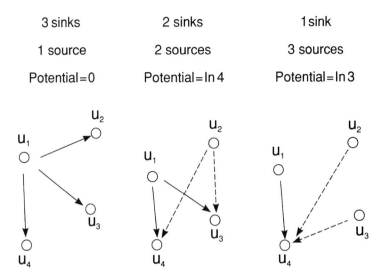

Fig. 12.11. Combinatoric representation of the functional organisation as relations between sources and sinks. In this example, with four structural units and one product, there exist three values of the potential of organisation when the number of sinks equals 3, 2, 1. The condition is that a source cannot be a sink (after G. A. Chauvet (1993b)).

3.   Criterion of maximality for the potential of organisation: a class of biological systems

a.   State of maximum organisation

Can the Π-function describe the dynamics of an (O-FBS)? The existence of a particular organisation (n_M), for which $\Pi(n_M)$ is maximum (Fig. 12.12), suggests a positive answer. The following properties may be readily deduced from the definition above:

Property I:

The potential of organisation Π increases with both the degree of organisation and the number of levels for a given distribution (n_α^l), $\alpha = 1$, μ^l.

Property II:

For a given level l and a given degree of organisation v^l, there exists a maximum for the potential of organisation obtained for the value (n_M) of the organisation. This particular value $\Pi_{max}(n_M)$ is called the maximal potential of organisation and describes a compromise between the complete specialisation and the complete independence of the units at this level.

Potential Π(x)

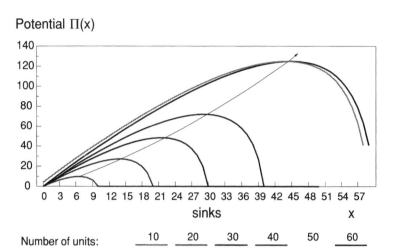

Fig. 12.12. Potential of functional organisation for six values of the degree, from 10 to 60. The maximum of the potential determines a particular organisation that is assumed to define the biological system (after G. A. Chauvet (1993b)).

Proof: Let $n_\alpha^l = E(x_\alpha^l)$ be the integer part of the real x_α^l. Then, Π^l is now the real function: $x \rightarrow \Pi^l(x)$ defined in $\mathbb{R}^\mu$. As mentioned in the introduction, this Darwinian system is characterised by random suppressions or appearances of an individual product P. Mathematically, this property corresponds to the independent variations dx_α of x_α that occur in a given unit. Then, the maximum of Π (which is not a bound maximum) is obtained for $(x_\alpha)_{\alpha=1,\mu^l}$, which satisfies the equation:

$$d\Pi = \sum_{\alpha=1}^{\mu} d\Pi_\alpha = \sum_{\alpha=1}^{\mu} \left(\ln(v - x_\alpha) - \frac{x_\alpha}{v - x_\alpha} \right) dx_\alpha = 0 \qquad (12.32)$$

where $\Pi_\alpha = x_\alpha \ln(v - x_\alpha)$ is the potential corresponding to product P_α, and where the superscript l has been suppressed for clarity.

Therefore, $\forall \alpha \in [\![1, \mu]\!]$ a necessary condition for the extremum is:

$$\ln(v - x_\alpha) - \frac{x_\alpha}{v - x_\alpha} = 0. \qquad (12.33)$$

It can be shown that (i) $\Pi(x)$ really has a maximum at one point in $[0, v - 1]^\mu$; and (ii) this maximum is unique (Chauvet, 1993b).

All structural units involved in the generation of a physiological function are supposed to have identical elementary physiological functions. This is the case in a non-differentiated tissue, with all cells having the same individual potentialities, and which is being transformed into a differentiated tissue, i.e. one in which some cells produce individual products for all the others. We could say that the graph G and the matrix M come into being when the 'functional isotropy' has disappeared. The two limit cases correspond to 'functional isotropy': (1) when $n_\alpha = 0$, and all units are independent; and (2) when $n_\alpha = v - 1$, and only one unit is specialized for the elementary function P_α. Then, a possible interpretation of Π could be the maximum of functional anisotropy associated with potential organisations.

b. The extremum hypothesis: a class of biological systems

Two important properties may be deduced from the definition of $\Pi(x)$.

α. The organisational state is an attractor.

Property III

The organisational state, i.e. the state of organisation for the maximum potential $y_M = \Pi(x_M)$, is an attractor for the dynamics of the organisation which tends to x_M, either by decreasing or by increasing values of x when the time t tends towards t_M.

Proof: $D(x) = \Pi_{\text{max}} - \Pi(x)$, where Π_{max} is the state of maximal organisation, is a Lyapunov function when $x(t)$, considered as a dynamical system, tends towards x_M either by decreasing or by increasing values, according to the initial value of x, i.e.:

$$\forall \alpha \in [1, \mu] \qquad \forall x_\alpha \qquad D(x_\alpha) > 0. \tag{12.34}$$

The time derivative of $D(x(t))$ is negative or null:

$$\frac{dD}{dt} = \sum_\alpha \frac{\partial D}{\partial x_\alpha} \frac{dx_\alpha}{dt}$$

if the two derivatives:

$$\frac{\partial D}{\partial x_\alpha}, \frac{dx_\alpha}{dt} \tag{12.35}$$

have opposite signs. Then the dynamical system which describes the time variation of the graph:

$$\frac{dx_\alpha}{dt} = f_\alpha(x_1, x_1, \dots, x_\mu) \tag{12.36}$$

has an equilibrium point at $X_M \equiv (x_M, x_M, \dots, x_M) \equiv (x_1, x_2, \dots, x_\mu)_M$ which is reached at a time t_M according to increasing values when $x < X_M$, and decreasing values when $x > X_M$. In the former case ($x < X_M$), the partial derivative $\partial D/\partial x_\alpha$ is negative and the time derivative of x_α must be positive, whereas in the latter case, $\partial D/\partial x_\alpha$ is positive ($x > X_M$), and the time derivative of x_α must be negative (Fig. 12.13).

β. Consequence: the extremum hypothesis for the time-variation of the number of sinks.
There are two possibilities for the functional organisation of the biological system for which the potential of organisation is given by the function $\Pi(x)$: before

t_M, the number of sinks either decreases or increases towards a stable limit. This *monotonic* property of the variation of the number of sinks is important in the characterisation of the potential of organisation. An interpretation of this result in the second case could be the special role given to the sources of a system for a maximum redundancy among the structural units: the Darwinian biological system would vary with time so that the number of production sites (sources) remains maximum.

We may now state the extremum hypothesis as follows: A biological system evolves from initial conditions such that the number of sinks either decreases or increases, i.e. it corresponds to a monotonic time function, and therefore reaches a state where the potential of organisation is maximum. This extremum hypothesis defines a class of formal biological systems.

The validity of the proposed theory, which is finally based on the extremum hypothesis, can be experimentally tested. The definition of functional interactions makes it possible to identify sources and sinks, and to verify the decreasing monotonic property of the number of sinks, *even if the explicit dynamics of the system is unknown*. Another possibility would be to check the validity of the consequences of the theory developed below.

4. Criterion of evolution for the functional organisation: orgatropy

a. The concept of 'orgatropy'

In the preceding section, the influence of the variation of the number of sinks on the functional organisation was studied by means of the potential of organisation. Here, we shall explore the effect of a variation of the number of structural units in the system, i.e. the effect of the degree of organisation. Some interesting results regarding the development of biological systems may be deduced from such variations. For example, when the degree of organisation increases, the topology of the system is transformed such that the constraint of maximum potential is satisfied. We shall see that the evolution of the system is governed by a function called *'orgatropy'*, deduced from the function Π.

Fig. 12.13. This set of three figures illustrates the fundamental property III of the existence of an attractor for an organisation having a maximum potential. Here we have a specific case with two state variables and a degree of 300. (a) The dynamics of the organisation $(x_1(t), x_2(t))$, i.e. the number of sinks for two products P_1 and P_2, tends asymptotically towards x_M = 160 either by decreasing values from initial conditions (0,0), or by increasing values from initial conditions (290, 290). (b) Corresponding variation in time of the potential $\Pi(x_1(t), x_2(t))$ = $x_1 \ln(v - x_1) + x_2 \ln(v - x_2)$ for the dynamics shown in (a). (c) Potential and related Lyapunov function of the dynamics shown in (a) and (b), with degree $v = 30$ (after G. A. Chauvet (1993b)).

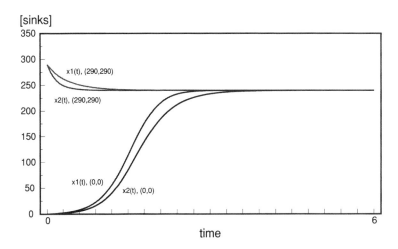

[sinks]

x1(t), (290,290)

x2(t), (290,290)

x1(t), (0,0)

x2(t), (0,0)

time

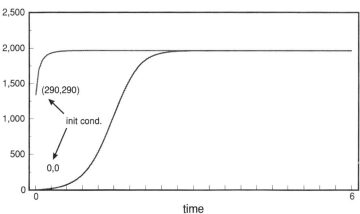

Potential

(290,290)

init cond.

0,0

time

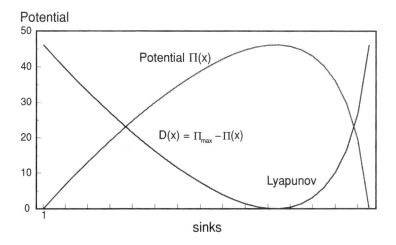

Potential

Potential $\Pi(x)$

$D(x) = \Pi_{max} - \Pi(x)$

Lyapunov

sinks

Property IV: The concept of orgatropy

If the degree of organisation v^l changes with a variation in the number of structural units at level 1 without reorganisation, then the system with maximum potential satisfies the criterion of evolution for the function F, called orgatropy:

$$dF^{\,l}(v^l) \geq 0 \tag{12.37}$$

and defined from the following functions h and Π_{max} as:

$$F : v \rightarrow x_M \rightarrow y_M = F(v) = \Pi_{max} \circ h(v).$$

This function, which is important in considering the time-variation of the (O-FBS), has been called orgatropy because of its similarity in form to physical entropy, as discussed below: the time variation of $\Pi(n_M^l)$ for level l is given by the function F when the degree of organisation at that level varies without a re-organisation of the system.

Proof: The maximum of Π is given by equation (12.32):

$$d\Pi(x_M) = 0 \Leftrightarrow x_M = (v - x_M)\ln(v - x_M). \tag{12.38}$$

Let h be the implicit function of v: $v \rightarrow x_M = h(v)$. Then the equation for the maximum is obtained by eliminating v between $\Pi(x) = x \ln(v - x)$ at x_M and $d\Pi(x_M) = 0$:

$$y_M = \Pi_{max}(x_M) = x_M^2/(v - x_M).$$

Then:

$$y_M = 2x_M \ln x_M - x_M \ln y_M. \tag{12.39}$$

For the function $F(v) = \Pi_{max} \circ h(v)$, as defined above, it is easy to show that F is the product of two increasing functions, h and Π_{max}. Indeed, by differentiating we have:

$$\frac{dy_M}{dx_M} = \frac{2 + \ln(x_M^2/y_M)}{1 + (x_M/y_M)} \tag{12.40}$$

and:

$$\frac{dx_M}{dv} = \frac{1 + \ln(v - x_M)}{2 + \ln(v - x_M)} > 0. \tag{12.41}$$

Since:

$$\frac{x_M^2}{y_M} = v - x_M > 1, \quad \frac{x_M}{y_M} > 0$$

the first ratio (12.40) is positive. Thus, Π_{max} and h increase, and therefore the product F also increases. Finally, $F(v)$ is obtained from the solution of:

$$x_M^2 + x_M y_M - v y_M = 0$$

which is put in $y_M = \Pi_{max}(x_M)$.

b.   Does orgatropy provide a criterion for the time-variation of the (O-FBS)?

When the system grows by an increase in the number of structural units $v(t)$, the concentration of sources and sinks varies such that the potential of organisation remains maximum. However, the quality of the structural units is conserved, i.e. a source remains a source, and a sink remains a sink. In this case, the (O-FBS) develops in time without reorganisation. We have shown that orgatropy cannot decrease, and thus indicates the direction of the time-variation. *The biological system develops such that the orgatropy, which represents the most developed combination of potentialities, increases.*

Nevertheless, reorganisations are observed during the developmental process. Such reorganisations may be formally described as a sequence of specialisations (Chapter 10, Section V). Functional interactions are created during development, i.e. the sources and sinks produce a hierarchical system. A major problem is to determine the principle governing this evolution. The orgatropy function does not include the process of specialisation and therefore cannot describe the development of the system as a hierarchical system. It describes only a part of the time-variation, i.e. the fact that functional interactions are added to the system without reorganisation, therefore without the emergence of a hierarchy. But *orgatropy includes non-symmetry which is the main property of the functional interaction.* This result, which is important in the framework of the theory because of its self-coherence, brings out the fundamental distinction between orgatropy and physical entropy $S = \rho \ln \rho$ which describes symmetrical structural interactions (see below).

5.   Criterion of specialisation and reorganisation of the (O-FBS) during development

a.   Criterion of specialisation

α.   The concept of specialisation.   The mathematical definition of specialisation and the emergence of levels of organisation in a hierarchical system raises a major difficulty. If the concept of functional interaction is accepted, then we may assign the following meaning to the notion of specialisation. Let us assume that the level l contains n_1 sinks for the product P_1, and n_2 sinks for the product P_2 such that

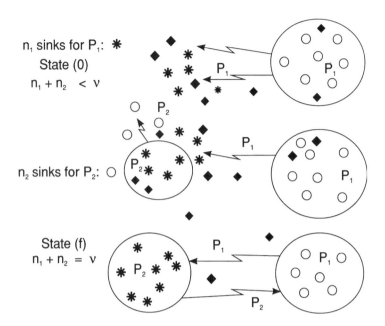

Fig. 12.14. An interpretation of the mechanism of specialisation with two products. From the top to the bottom, successive self-associations lead to two sets of structural units that produce only P_1 or P_2. At the beginning of the process, in the state (0), there are n_1 sinks for P_1, and n_2 other sinks such that: $n_1 + n_2 < v$. In the final state (f), there are only two sets, one producing P_1 (on the right), the other producing P_2 (on the left), and therefore: $n_1 + n_2 = v$, which is the condition of specialisation (after G. A. Chauvet (1993b)).

$n_1 + n_2 < v$ (Fig. 12.14). Thus, at level l, some structural units synthesise a product P_1, some synthesise P_2, some synthesise both, and some synthesise products other than P_1 and P_2, i.e.:

$$v = n_{P_1 P_2} + n_{\bar{P}_1 P_2} + n_{P_1 \bar{P}_2} + n_{\bar{P}_1 \bar{P}_2}$$

In this equation, $n_{\bar{P}_1 P_2}$ for example is the number of units that synthesise P_2 and not P_1. Therefore:

$$n_1 = n_{\bar{P}_1 P_2} + n_{\bar{P}_1 \bar{P}_2}, \quad n_2 = n_{P_1 \bar{P}_2} + n_{\bar{P}_1 \bar{P}_2}.$$

If $n_1 + n_2 < v'$ then: $n_{P_1 P_2} \neq 0$, i.e. some units synthesise P_1 and P_2. Thus: $n_{P_1 P_2} > n_{\bar{P}_1 \bar{P}_2}$.
Therefore, we define specialisation as follows:

Definition:

Given, at initial time $t^{(0)}$, $n_1^{(0)}$ *sinks for the product* P_1, *and* $n_2^{(0)}$ *sinks for the product* P_2, *such that* $v^{(0)} > n_1^{(0)} + n_2^{(0)}$, *the transformation at this l-level, called the specialisation of the l-level at final time* $t^{(f)}$, *corresponds to a partition into two subsets of* $n_1^{(f)}$ *and* $n_2^{(f)}$ *units such that:*

$$v^{(f)} = n_1^{(f)} + n_2^{(f)} \text{ with } n_{P_1 P_2}^{(f)} = n_{\bar{P}_1 \bar{P}_2}^{(f)} = 0. \tag{12.42}$$

β.   The relation between specialisation and hierarchisation.   Thus, according to the definition of a structural unit (Definition I, Chapter 4, Section III), the subset constitutes a class of equivalence, and therefore a new structural unit: the transformation from initial time $t^{(0)}$ to final time $t^{(f)}$ corresponds to the emergence of a level of organisation defined by new structural units that are specialised in the dynamics of $\{P_1, P_2\}$. This transformation of the 'quality' of the units, which implies a different number of these units with a given quality, will be indicated by the specific operator δ. When, at time $t^{(0)}$, $n_1^{(0)}$ sinks for the product P_1, and $n_2^{(0)}$ sinks for the product P_2 constitute a partition of $v^{(0)}$ at $t^{(0)}$, we have the following property:

Property V: Hierarchisation operator

Let 1 *be a level of organisation with* $v^{(0)}$ *structural units whose* $n_1^{(0)}$ *and* $n_2^{(0)}$ *are sinks for the products* P_1 *and* P_2 *respectively. A necessary and sufficient condition for the reorganisation of the system through the emergence of a higher level of organisation is the maximisation of* Π^1:

$$d\Pi^1(n_1^{(0)}, n_1^{(0)}) = 0 \tag{12.43}$$

where $n_1{}^{(0)}$ *and* $n_2{}^{(0)}$, *specialised respectively in the production of* P_2 *and* P_1, *are two sets of structural units such that* $v^{(0)} \neq n_1^{(0)} + n_2^{(0)}$. *Then:*

$$\delta v^l = v^{(f)} - v^{(0)} < 0, \quad \delta F^l \leq 0 \tag{12.44}$$

and the system, which reaches its maximum is stabilised for P_1 *and* P_2.

We may note that if $dv^l > 0$ without reorganisation, then $v^{(f)} - v^{(0)} < 0$ after reorganisation. This transformation in the hierarchical system is indicated by the operator δ: $\delta v^l < 0$. With the same notation, $\delta F^l \leq 0$, whereas $dF^l > 0$

Proof: The condition of maximality of Π is:

$$\frac{\partial \Pi}{\partial x_1} = \frac{\partial \Pi}{\partial x_2} = 0 \tag{12.45}$$

In the initial state (0), there are $n_1^{(0)}$ and $n_2^{(0)}$ sinks in the l-level of organisation, such that: $v^{(0)} \neq n_1^{(0)} + n_2^{(0)}$. In the final state (f), the partition of the system is expressed by:

$$v^{(f)} = n_1{}^{(f)} + n_2{}^{(f)}, \; v^{(f)} \neq v^{(0)} \tag{12.46}$$

Conditions (12.34) give two relations in x_1 and x_2:

$$\ln(v - x_1) - \frac{x_1}{v - x_1} = 0 \Rightarrow \ln x_2 = \frac{x_1}{x_2}$$

$$\ln(v - x_2) - \frac{x_2}{v - x_2} = 0 \Rightarrow \ln x_1 = \frac{x_2}{x_1} \tag{12.47}$$

which determine $x_1^{(f)}$ and $x_2^{(f)}$, and then $v^{(f)}$. The vector $(x_1^{(f)}, x_2^{(f)}) = (e, e)$ is the only solution of the system (12.47):

$$x_1 = x_2 \ln x_2$$
$$x_2 = x_1 \ln x_1. \tag{12.48}$$

A mathematical difficulty arises for the solution in integer numbers. We have shown (Chauvet, 1993b) that the partition of the system with a maximum potential can also be obtained for the vectors (3,3) and (4,4). An interpretation of this strange property will be given below. Assuming that these three solutions are possible, let us consider the solution (2,2) as representative. Then:

$$(n_1^{(f)}, n_2^{(f)}) = (2, 2) \tag{12.49}$$

is the integer solution of this problem. Inversely, if e is a solution of (12.47), then Π is maximum. Thus, we see that if there is a partition of $v^{(f)}$ units into $x_1^{(f)}$ sinks for P_1 and $x_2^{(f)}$ sinks for P_2, then the potential Π will be maximum if $x_1^{(f)} = x_2^{(f)} = e$.

Finally, $dF = 0$, because $d\Pi(x)/dx = 0$ for $x = e$, and $\Pi_{max}(e) = e$. The transformation from $v^{(0)}$ to $v^{(f)} < v^{(0)}$, which corresponds to an increase of units $v(t_f) > v(t_0)$ without reorganisation is such that $dF(v) \geq 0$: the criterion of evolution $F(v(t_f)) \geq F(v(t_0))$ is satisfied, and emergence of a level of organisation represented by the operator δ is described in terms of the orgatropy function by the relation: $\delta F(v) = F(v^{(f)}) - F(v^{(0)}) \leq 0$ which is therefore satisfied with $e = F(v^{(f)})$. For this reason, δ will be called the '*hierarchisation operator*'.

b. *Consequence: mathematical expressions of specialisation and emergence of a level of organisation*

When the l-level of organisation contains exactly $v = N$ structural units of which n_1 receive P_1 (or equivalently, $N_1 = N - n_1$ emit P_2), and n_2 receive P_2 ($N_2 = N - n_2$ emit P_1), then the constraint:

$$v^{(f)} = n_1^{(f)} + n_2^{(f)} \tag{12.50}$$

implies the creation of a higher $(l + 1)$-level, so that N_1 structural units synthesise P_2, and N_2 units synthesise P_1. According to Definition I (see Chapter 4, Section III), a structural equivalence class results from the specialisation of units, and a higher level of organisation is obtained (Fig. 12.15). From Property III, this system passes from one stable extremum of the organisation to another when the degree of organisation varies. Thus, for a given state of organisation, we have a mathematical expression of specialisation:

$$N_1(t; P_1) + N_2(t; P_2) = N_1^2 = \text{constant}$$

or:

$$dN_1^{2,l}(t) = 0 \tag{12.51}$$

which describes the time invariance of the partition at level l for products P_1 and P_2.

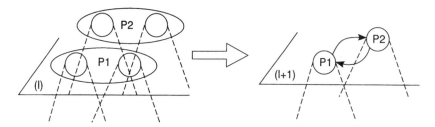

Fig. 12.15. Emergence of a level of organisation. Assuming that the conditions of specialisation are satisfied, as in Fig. 12.14, then we see that two new structural units have been
created at the upper level, because they are respectively specialised in the production of P_1 and
P_2. On the left, a similar figure to Fig. 12.14; on the right, the new level of organisation (after
G. A. Chauvet (1993b)).

Both conditions: Π maximum and partition of v, lead to a reduction of the number
of structural units, and correlatively to the creation of a level for the corresponding
hierarchical system:

$$n_1^{(f)} = 2 \ll n_1^{(0)} \quad n_2^{(f)} = 2 \ll n_2^{(0)}. \tag{12.52}$$

These two new sets of units collectively execute the elementary physiological functions P_2 and P_1, respectively. The variation of the potential of organisation from the
initial state (0) to the final state (f) is expressed by:

$$^{(0)}\Pi^l = n_1^{(0)} \ln n_2^{(0)} + n_2^{(0)} \ln n_1^{(0)} \tag{12.53}$$

$$^{(f)}\Pi^l = n_1^{(f)} \ln x_2^{(f)} + n_2^{(f)} \ln x_1^{(f)} = 4.$$

Then:

$$^{(0)}\Pi^l > {}^{(f)}\Pi^l = \Pi^{l+1}. \tag{12.54}$$

This relation corresponds to the required property (iii) of Π. Expression (12.54)
mathematically describes the emergence of a level of organisation.

c.   Functional order

In Subsection 4 above, a function $F(v)$ was used to describe the time-variation of an
FBS without reorganisation. Here, we shall determine a similar function to describe
the evolution of the FBS when reorganisation (with specialisation) is assumed. We
shall show that a function $\Omega(t)$, called the *functional order*, describes the time-
variation of the hierarchical system.

Definition:

*The functional order of a hierarchical biological system is defined by the state
function:*

$$\Omega(t) = -\sum_{l=1}^{L} F^l(v^l).$$ (12.55)

Theorem: On the time variation of an (O-FBS)

Let a biological system be defined by the extremum hypothesis:

$$\Pi(n_M^l)_{l=1,L} \ maximum \Leftrightarrow \forall \alpha = 1, \mu^l \quad (dn_\alpha^l(t) < 0 \ if \ n_\alpha^l > n_M^l)$$

$$or \ (dn_\alpha^l(t) = 0 \ if \ n_\alpha^l \neq n_M) \quad (12.56)$$

$$or \ (dn_\alpha^l(t) > 0 \ if \ n_\alpha^l < n_M)$$

for all $l = 1$, L. *If the degree of organisation,* v^l, *is modified such that the criterion of specialisation is satisfied for products* P_α *and* P_β:

$$dN_\alpha^\beta(t) = 0$$ (12.57)

then the self-organisation of the system is such that:

(1)   At most two structural equivalence classes are created, each being specialised in the production of a given product, according to a bi-unitary process, i.e. by the association of two units:

$$\delta v^l < 0.$$ (12.58)

(2)   Its functional order increases and corresponds to the emergence of one level of organisation:

$$\delta \Omega \geq 0$$ (12.59)

and simultaneously its orgatropy decreases:

$$\delta F \leq 0.$$

(3)   The potential of organisation Π *decreases:*

$$^{(0)}\Pi^l \rightarrow \Pi^{l+1} = {}^{(f)}\Pi^l.$$ (12.60)

(4)   At the limit, in the state of maximum specialisation, Π *is minimum and its value is:*

$$\Pi^{l+p} = 2\mu^l$$ (12.61)

with $p = \mu/2$ *if* μ^l *is even, and:*

$$\Pi^{l+p} = 2(\mu^l - 1) + \Pi_\omega^l$$ (12.62)

with $p = (\mu^l - 1)/2$ *if* μ^l *is odd.*

Here, μ^l is the number of products in the *l*-level before the complete specialisation of the units, and p is the step of the last product synthesised.

Proof:

(1) The criterion of specialisation (Property V) can be generalised as follows. In a system with v^l structural units that exchange their products, is it possible to obtain a partition of v^l according to the numbers $(n_\alpha^l)_{\alpha=1,\mu^l}$? In other words, can the number of units $N_\alpha = N_\alpha^\beta - n_\alpha$ evolve towards the number of units $N_\alpha^{(f)}$ which are specialised in the production of P_α, i.e. which emit only the product P_α consumed by all the other units? The extremum condition (12.56) leads to a system that is similar to (12.47):

$$n_\alpha = S_\alpha \ln S_\alpha \quad \forall \ \alpha = 1, \mu^l$$

and:

$$S_\alpha = \sum_{\beta=1}^{\mu} n_\beta \quad \beta \neq \alpha$$

whose solution in $\mathbb{R}^\mu$ is:

$$x_\alpha = \frac{1}{\mu - 1} \exp \frac{1}{\mu - 1}$$

for all α. In $\mathbb{N}$, $n_\alpha = 0$ when $\mu \geq 3$. Therefore, the partition applies only to a maximum of two sets of units.

(2) The functional order is defined by (12.55), and its time derivative is:

$$\frac{d\Omega}{dt} = -\sum_{l=1}^{L} \frac{\partial F^l}{\partial v^l} \frac{\delta v^l}{\partial t} \geq 0$$

due to the signs of $dF/dv \geq 0$ (Property IV), and $\delta v < 0$ (Property V for two given products at this level).

(3) The potential of organisation in the initial state (0) is:

$$^{(0)}\Pi^l = \sum_{\alpha=1}^{\mu^l} (v^l - N_\alpha^{\ l}) \ln N_\alpha^l. \tag{12.63}$$

The condition of specialisation (12.57) implies:

$$N_\alpha(t; P_\alpha) + N_\beta(t; P_\beta) = N_\alpha^\beta = K \tag{12.64}$$

where K is a constant. It describes the invariance of the partition at this level and indicates whether a variation occurs in the topology of the system. Therefore, with $\alpha = 1$ and $\beta = 2$, following Property V, we have:

$$^{(0)}\Pi^l = N_1^{(0)} \ln N_2^{(0)} + N_2^{(0)} \ln N_1^{(0)} + \sum_{\alpha=3}^{\mu^l} \Pi_\alpha^l \tag{12.65}$$

and:

$$N_1 = N_2 \ln N_2$$
$$N_2 = N_1 \ln N_1$$
$$N_1^{(f)} = N_2^{(f)} = 2.$$

Thus, the reorganisation of the l-level leads to:

$$^{(f)}\Pi^l = 2 + 2 + \sum_{\alpha=3}^{\mu^l} \Pi_\alpha^l < {}^{(0)}\Pi^l \tag{12.66}$$

assuming that Π^{l+1} is defined by:

$$\Pi^{l+1} = {}^{(f)}\Pi^l. \tag{12.67}$$

(4) This condition of specialisation can be repeated for two elementary functions P_α and P_β, α, $\beta = 3$ to μ^l if μ^l is even, and α, $\beta = 3$ to $\mu^l - 1$ if μ^l is odd. Then, the decomposition of this expression is obtained $p = \mu^l/2$ times, if μ^l is even, and $p = (\mu^l - 1)/2$ times if μ^l is odd.

d. Time-variation of an (O-FBS) during development

The FBS studied here has two characteristic biological properties: (i) identical potentialities for all structural units (equipotent system) at each level, and (ii) variations within a given unit for a given product (Darwinian mutational system). It is important to determine whether these biological properties lead to the interpretation of other biological properties, and whether they provide the means of transformation into a more complex biological system, closer to a real biological system. The criterion for the development of functional organisation leads to some formalised properties of biological systems that will have to be experimentally tested.

For example, let us consider a formal tissue having the following organisation: $N_1^{(0)}$ units $u_1(P_1, P_2, P_3, P_4, \ldots)$ which supply $P_1, P_2, P_3, P_4, \ldots$. With the same notation, $N_2^{(0)}$ units $u_2(P_1, P_2, P_3, P_4, \ldots)$, $N_3^{(0)}$ units $u_3(P_1, P_2, P_3, P_4, \ldots)$, $\ldots$ supply $P_1, P_2, P_3, P_4, \ldots$. According to the theorem above, the l-level will move towards a new organisation $(N_1^{(f)}, N_2^{(f)}, N_3^{(0)}, N_4^{(0)}, \ldots)$ that may be described by the sequence:

$$2*u_1(P_1, P_3, P_4, \ldots), \; 2*u_2(P_2, P_3, P_4, \ldots),$$
$$N_3^{(0)}*u_3(P_3, P_4, \ldots), \; N_4^{(0)}*u_4(P_3, P_4, \ldots), \ldots \tag{12.68}$$

We may therefore say that, in the final state (f), the two units $u_1(P_1, P_3, P_4, \ldots)$ are two structural equivalence classes containing all the units supplying P_1, which is considered to be in state (0) before the reorganisation of the entire system. They constitute the highest level of organisation of the hierarchical system, which is specialised in the production of P_1. The same result is obtained for $u_2(P_2, P_3, P_4, \ldots)$. Thus, two parallel hierarchical systems, coupled by the two functional interactions P_1 and P_2, appear in the final state (f) (Fig. 12.15).

In the previous section, we have shown that a system with a maximum potential can increase its specialisation by a partition into two new structural units. The analysis with real numbers, which gives a unique solution (e, e) leads to three possibilities with integer numbers: $(2, 2)$, $(3, 3)$ and $(4, 4)$. It is interesting to look for some physiological functions having such a number of structural units. At the highest level of functional organisation, most of the physiological functions are carried out by two structural units, e.g. eyes, lungs, and kidneys, but only one function, calcium control, involves four structural units: the parathyroids.

V.   A comparison between biological and physical systems

1.   *Structural entropy and functional orgatropy*

For a molecular gas described by a velocity distribution function $f(v)$, Boltzmann (1877) defined the quantity H:

$$H = \iiint f(v)\ln f(v)dv$$

such that $dH \leq 0$. Later (1877), Boltzmann identified the quantity H with the entropy S:

$$S = -k\,H$$

such that $dS \geq 0$. In this expression, k is the Boltzmann constant. Here, the definition is of a statistical nature since it is based on the statistical mechanical description of a molecular gas. More generally, a statistical mechanical definition of S, valid for an arbitrary system obeying the laws of classical mechanics in canonical form versus phase variables (p, q), is given by:

$$S = -k\iint f(p, q)\ln f(p, q)dpdq. \tag{12.69}$$

Then Gibbs (1902), with his formulation in terms of statistical ensembles, conceived of entropy as an ensemble property. This definition of entropy provides a link between information and entropy.

Our definition of orgatropy F is of a different nature:

(i) $F(v)$ describes the maximum potential of organisation Π_{max} for a *non-statistical ensemble*.

(ii)   The definition of Π elicits a property of *non-symmetry* expressing the fact that a source is not equivalent to a sink. This property appears to be fundamental: in physics, the interaction is a force which couples two elements (action and reaction imply interaction); whereas in biology, the functional interaction describes a non-symmetrical effect of one element on another. Thus, distinctive roles are attributed to the source and the sink.

Table 12.3.

		Function	Derivative
Structural organisation (in isolated physical systems and biological systems)	**Thermodynamical entropy**	S	$dS \geq 0$
	Neguentropy	$N = -S$	$dN \leq 0$
	Lyapunov function	$P = d_i S \quad P \geq 0$	$dP \leq 0$
Functional organisation (in biological systems only)	**Orgatropy**	F	$dF(v) > 0 \quad \delta F(v) \leq 0$
	Functional order	$\Omega = -F$	$\delta\Omega \geq 0$
	Lyapunov function	$\Omega < 0$	$\delta\Omega \geq 0$

(iii) The definition of Π calls for an *equipotence principle* at each level of organisation. In contrast, a physical system is intrinsically supposed to be in an equilibrium state, or at least a steady state far from equilibrium, and therefore satisfies a *principle of energy equipartition*.

(iv) Orgatropy is a *global concept* regarding sources and sinks.

A physical system is defined by the large number of states in the phase space through which it can pass. Thus, a definition of the equilibrium state is the condition of greatest probability with present constraints. A biological system is defined, in part, by its functional topology. The related organisational system, represented by the (O-FBS), is not statistical in character, although the dynamic system (D-FBS) has this property. One important problem is connected with the possible effect of a perturbation of the (D-FBS) on the (O-FBS), i.e. the influence of a structural perturbation on the functional organisation.

2. *The consequence of the optimum principle*

What is the major result of the theorem on the evolution of the (O-FBS)? The criterion (12.37) and the extremum hypothesis (12.56) imply an increase in the complexity at the l-level. At the same time, the system decreases this complexity by means of a reorganisation which, in turn, increases the number of levels. With the definitions proposed, it is possible to conclude that an organism, i.e. a system that possesses both the basic qualities of being *equipotent* and *mutational*, develops according to the principles of *a decrease in orgatropy and an increase in functional order*.

This consequence of the optimum principle provides a definition for living systems by clearly distinguishing between biological and physical systems (Table 12.3). The evolution of a physical system is characterised by an increase in the thermo-dynamical entropy, i.e. by an increase in the molecular disorder (second principle of thermo-dynamics). Regarding its structure, i.e. its physical, molecular structure, a biological system obviously satisfies the second principle, which states that the production of entropy $P = \mathrm{d}_i S$ is a Lyapunov function. In contrast, the functional organisation, which is the critical feature of a biological system, evolves with a decrease in orgatropy and an increase in functional order.

In summary, *a living system is characterised by two complementary sets of mathematical laws: the first set governs the physical structure with an optimum principle concerning thermodynamical entropy (Prigogine's criterion of evolution (Glansdorff and Prigogine, 1970)) including living organisms (Prigogine et al., 1972); whereas the second set governs the organisation of physiological functions with an optimum principle concerning the potential of organisation.* A law concerning the functional order results from this second optimum principle. The state functions, and their criterion of evolution, for both structural organisation and functional organisation are summarised in Table 12.3.

3. On the meaning of the optimum principle

Our theory of functional organisation is based on the extremum hypothesis: a biological system evolves such that its potential of organisation remains maximum. A sufficient condition of this extremum hypothesis is the monotonic trend of the dynamics of the sinks towards an asymptotical limit. Even without the complete knowledge of the dynamics, such a property can be experimentally observed and constitutes evidence if it is observed for a specific biological system, as in the case of the nervous system. Therefore, the problem in biological terms is to find out if such behaviour can be observed during development, and, in particular if, during embryogenesis, an organism has an optimum number of sources which emit chemical products, signals, or execute any kind of elementary function, necessary for the maintenance of life. Then, it could be said that the number of sinks as a function of time evolves toward a minimum.

Probably a living system has to have an optimum number of sources, i.e. defined as implying Π maximum, at least at the beginning of life, before some adjustments are made by intrinsic and/or extrinsic controls. For example, we know that spontaneous neuronal death occurs during neuroembryogenesis and that the number of nervous endings decreases. As we have shown (Chauvet, 1993b), the latter process presumably corresponds to 'learning', or adaptation to specific circumstances. At this point, experimentation on living systems would be required to determine whether the number of sinks evolves, monotonically, towards a minimum or a maximum during its development.

Our formulation of a biological system was made in terms of functional interactions in the representation (ψ, ρ). This is very different from the representation (N, a) (see Chapter 4), where N (generally a large number) is the occupation number of a functional equivalence class, and *a* the rate constant between classes. Statistical mechanics have been transposed to population models (Kerner, 1957; 1972; Cowan, 1968). Demetrius (1983; 1984) has extended and found a variation principle for evolutionary models, and Auger (1986) has determined the conditions for the emergence of a hierarchy in such systems. All of these methods apply to structural organisations.

In our approach to the study of the physiology of biological systems, we have distinguished between the structural and the functional organisations. The distinction between the two types of organisation will appear to be even sharper when we consider the major physiological systems (Volume II). The mathematical formulation of the ancient problem of the relationship between structure and function, which is in fact that of the relationship between the two types of organisation, has allowed us to deduce the novel results presented in Volume III. Biological development, i.e. cell growth, division and differentiation, involves not only the spatiotemporal variation of the structural organisation, which of course corresponds to morphogenesis, but also the spatiotemporal variation of the functional organisation, which we may call '*physiogenesis*'. Both these aspects have been considered in this chapter. According to Oster and Murray, morphogenesis appears to be characterised by a set of rules obtained from the dispersion relation of the reaction–diffusion equations representing local phenomena. Physiogenesis appears to depend on a maximum potential of organisation which maintains the permanency of the physiological functions during development. The phenomena involved are non-symmetric and non-local. These concepts, which are of a strictly biological origin, give biological systems a specific status quite distinct from that of physical systems. In particular, the hierarchisation of a biological system which is accompanied by an increase in functional order yields a principle that may be termed *the principle of functional order through hierarchy*. This leads to an experimentally falsifiable extremum hypothesis which may be expressed as the monotonous variation of the number of sinks for a given physiological function. This aspect of the theory concerning the topology of the organisation of the functional biological system (O-FBS), which is due to the non-symmetry of the interactions, will be completed by the study of the dynamics of the functional biological system under the form of non-local fields (D-FBS) in Volume III.

Summary of Part III

Organisation at the cellular level involves complex processes of control and regulation. Currently the best known is the mechanism of regulation discovered by Jacob and Monod in *Escherichia coli*. It has the advantage of simplicity, which appears not to be the case in eucaryotes. We examine mathematical models of this mechanism, from the simplest model with *two coupled operons* to the most complex with *time lags for certain chemical reactions with respect to others*, as found in reality. Unfortunately, this latter method leads to a mathematical complexity which is rather difficult to handle.

The concepts of *cell growth, development, differentiation* and *morphogenesis* are introduced and applied within the context of existing theories. We see how the morphogenesis in the Acrasiales (*Myxomycete amoebae*) may be explained by the double mechanism of diffusion and chemotaxis. In higher organisms, these notions are more difficult to interpret. Cell differentiation depends on a mechanism of *regulation* (a double gradient, animal and vegetal) and on the existence of a *morphogenetic field*. The cell recognises its state of development through *positional information*. It is believed that an invisible *primary pattern* leads to the visible *morphological pattern*. Several mathematical models have been constructed to test the validity of the hypotheses proposed on the basis of known experimental results. This delicate problem is tackled both theoretically, by the definition of the concepts involved, and experimentally. The number of morphogens and the conditions required for the process of morphogenesis are discussed within the framework of Turing's theory as well as that of Thom. Thus, the phenomenon of organogenesis is interpreted as the evolution of several metabolic fields corresponding the local potentials of biochemical kinetics.

The mechanisms of cell division are considered in terms of a model of the cell cycle called the *causal loop model*. A limit cycle of biochemical oscillations may be obtained by the use of justified constraints on the mitotic system. The evolution of a cell population is analysed for two types of variable, *age* and *maturity*. The cell cycle is described using the formalism of the *Leslie matrices* which leads to a simple interpretation of the fraction labelled mitoses curves.

Finally, the relationship between cell growth, division and differentiation is investigated by analysing the evolution of the cell mass and volume during successive mitoses. Here, important concepts such as the *elongation factor* and the *mass factor* are introduced. Although, in view of the present limited knowledge of cell phenomena as well as of the complexity of cell properties, a global description of cell behaviour may today seem to be of but little interest, the methodology used is well worth the study for future application when more reliable experimental results will surely be available. This is why we describe a method of *numerical simulation*, with a criterion of minimisation ensuring mitotic division.

Cell growth, division and differentiation play a major role in the development of an organism. In the framework of our theory of the functional organisation of biological systems, we consider two types of organisation, one structural and the other functional. These organisations, which are quite distinct, are nevertheless coupled. The variation of the structural organisation during development, or morphogenesis, has been the object of several communications, in particular by Oster and Murray. It has been shown that morphogenetic phenomena can be explained on the basis of a fairly restricted set of rules. The variation of the functional organisation during development, or '*physiogenesis*', depends on a potential of functional organisation leading to an optimum principle. A mathematical function, called the *orgatropy*, allows us to describe the variation of the biological system from a functional point of view, showing how the various levels of organisation emerge by specialisation. We also see how the biological system decreases its complexity through reorganisation. The criterion of variation which corresponds to the increase of *functional order* by stabilisation, also provides a criterion for the comparison of biological and physical systems. This finally leads to the principle of functional order through hierarchy.

Conclusion to Volume I:
Unity at the Gene Level

The idea of biological unity, possibly situated at the gene level, arises from the molecular and cellular description of physiological phenomena. This is not surprising since the chemical processes involved are directly controlled by the dynamics of the genome. A good indication is the recent creation by geneticists of a new terminology analogous to that developed by physicists. For example, the protons, nucleons, bosons and fermions of physics seem to be echoed by the cistrons, recons, operons and codons of genetics. However, the difference is crucial, *for whereas physicists describe structural units biologists have to consider functional units*. So, to conclude this volume, let us take a look at the current picture of molecular genetics with its new functional units.

The basic biological structure is the sequence of nucleotides — guanine, thymine, adenine and cytosine — attached to the sugar–phosphate chain to form the DNA double helix. This molecular structure explains two fundamental genetic characteristics, *heredity* and *variation*, heredity being due to the replication of the molecule, generation after generation, and variation being caused by errors in the replication of certain nucleotides. The essential aspects of this phenomenon, as discussed in Chapter 5, are: bidirectional DNA replication with the initiation occurring at various, apparently definite, points along the length of the molecule; DNA synthesis consisting in the formation of new, short fragments of strands, $5'$–$3'$, by the action of an enzyme, polymerase III; and finally, the assembly of these fragments, called Okasaki fragments, by a ligase. It has been estimated that human DNA replication proceeds at a rate of about 1 μm/min so that, if the replication occurred linearly from one end to the other of the DNA molecule, the whole operation would take more than 10 000 hours whereas it is known to be carried out in as little as 6 to 8 hours in some cells. In fact, the extraordinary speed of replication is due to the presence of several thousand functional units called *replicons*, each being triggered at a specific time and remaining active for less than an hour (Jacob and Brenner, 1963).

Other biological phenomena, currently under active investigation, provide the bases for the definition of new functional units. What were yesterday merely tentative hypotheses are now being transformed into rigorous definitions in molecular terms. Obviously, this is the first step towards the precise formalisation of the processes involved. Since the discovery of the DNA double helix in 1953, the new techniques of genetic engineering — the snipping, displacement, recombination and splicing of DNA fragments — have produced fantastic advances. In Chapter 5, protein biosynthesis was roughly formalised as a stochastic process but in fact it was the *operon*, the functional unit introduced by Jacob and Monod (Chapter 9) to explain the regulation of synthesis in procaryotes, which led to the construction of really interesting formalised models. In general terms, the gene, considered as a functional unit, is called a *cistron* which, according to Benzer (1957), is a segment of genetic material, DNA or RNA, within which the mutant pairs in the *trans* position are deficient in a specific enzyme and will therefore only synthesise a structurally abnormal enzyme. In other words, the sequence of the amino acids in a polypeptide chain is determined by the nucleotide pairs according to a one-to-one *cistron–polypeptide* relationship. Only the genes in the *cis* position are functional, whence the term *cistron*. A mutation may lead to a loss of function that can be directly observed through some change in protein structure. Mutagenesis may be produced by the *transition* or *transversion* of nucleotides (through the action of base analogues or the modification of certain bases) or else by the *insertion* or *deletion* of nucleotides. Thus, the *muton*, the basic unit of gene mutation, corresponds to the nucleotide pair within the cistron which is *the smallest alterable element* of the one-dimensional structure of genetic material (Lints, 1981). Finally, the working of a chromosome involves a complex mechanism of gene displacement and association termed genetic recombination. Although this mechanism has not been fully elucidated, it may be considered to consist of two successive events, cutting and splicing. DNA replication is believed to precede recombination which occurs at certain points when homologous chromosomes are paired off. The splicing does not take place by simple end-to-end binding of the DNA chain but by the addition of complementary fragments from homologous regions. This corresponds to the *polarised hybrid* DNA model, whence the term *polaron* to describe the DNA segment in which genetic recombination is non-reciprocal — thus a conversion — and polarised. In other words, the recombination frequency of the converted allele depends on its position with respect to the other alleles. Therefore there must exist a smaller unit of genetic material, which may be exchanged but not divided, by intra-genic recombination between homologous chromosomes. This hypothetical unit is called the *recon*.

How do these functional units intervene in cell growth, division and differentiation? Very little is known about this although it has been possible to define the notion of *morphogens* (Crick) and *positional information* (Wolpert). Indeed, from a mathematical point of view, it is here that the determinism of development appears most unequivocally. Perhaps the answers will be found at the gene level. During development and differentiation, gene action appears to be a result of three possible processes: (i) the *irreversible suppression* of genes when they are no longer required; (ii) the *irreversible blocking* of genes which will play no further part in the development after a specific instant; and (iii) the *reversible blocking* of genes through a control process. In Chapter 7 we have shown mathematically that such a structure, obtained by the association of substructures (or subunits) with different properties, could be physiologically stable. In fact, these three types of process have been experimentally observed. Recent research, based on the construction of *fate maps* for blastoderms, has demonstrated the precise destiny of embryonic cells. Structures such as the humerus, the abdomen, and so on, are localised on a fate map at blastodermal sites situated at experimentally determined distances, expressed in *sturts*. Thus, a *compartmentalisation* of structures may be visualised, *each compartment representing a unit of determination.*

Another way of formulating this important issue may be deduced from the above description, and summarises the contents of this volume: *what is the relation between the structural and the functional organisations, and how do these organisations simultaneously evolve satisfying certain properties such as idempotence?* In fact, the functional units enumerated above are structural units with a certain collective behaviour called the physiological function. The structural units are at the highest level of the hierarchy for the function considered, i.e. composed of structural units with functional interactions between them. It is possible that the compartmentation of structures corresponds to an increase in the stability of the dynamics by association of structures, as we have shown for the phenomenon of channelling. Thus, the relation between the structural organisation and the functional organisation would be included in the relation between the structural units and the functional interactions. Due to the property of idempotence of structural units, the functional organisation has potentialities that are represented by the potential of organisation. Using a variational approach, we have determined a criterion of evolution for the functional organisation which is deduced from the potential, and thus *a functional order from hierarchy* that can be compared with (and opposed to) the well-known principles of order recalled in Chapter 4, i.e. the principle of order from noise related to the theory of information, the principle of order from order deduced from thermodynamics (Schrodinger, 1945), and the principe of order from fluctuations (Prigogine *et al.*, 1972).

In contrast, our theory concerns functional organisation. It shows that the observed organisation, among all the possible organisations that could lead to the correct couplings, is the one that implies an increase in the functional order. This is an optimum constraint for the physiological mechanisms of an individual system subject to micromutations. Another constraint is given by external, i.e. environmental pressure, which causes the micromutations in the population of such interactive systems. Some results concerning the influence of defined parameters on the dynamics of the (D-FBS) will be presented in Volume III. Now, the problem is to determine whether this second constraint at the level of the population leads to the selection of the organisation that has in fact been chosen during the course of the evolution of the species. This problem may be formalised as follows: Does there exist a similar optimum principle, unifying physiological mechanisms and behaviour for the population of biological systems, that could describe the evolution of functional organisation at the highest level of organisation? If the answer is positive, then the highest level of organisation in the species population would be ecological, because the 'biological system' to be considered is the environment (including all organisms) and the given organism. In this case, the set of biological systems in its environment, with the functional interactions between them describing their ethological behaviour, would constitute just one biological system.

This is all in favour of the existence of a great number of functional units which, even when situated in the cytoplasm, are the expression of gene activity. As a cell receives information from multiple external sources it cannot, of course, be dissociated from its environment. Groups of cells will thus be determined by their relative position at a given time. Here, as we have seen, a considerable difficulty arises since we have to pass, mathematically, from the local to the global situation. How does a developing organism simultaneously take into account, on the one hand, the local intracellular genetic and cytoplasmic constraints, and, on the other, the global constraints due to neighbouring groups of cells and the environment? This is indeed the great mystery of developmental biology.

Mathematical Appendices

The sections that follow recall, as simply as possible, some of the mathematical notions used in the text. The approach is deliberately intuitive rather than rigorous and should be of some help to biologists whose mathematical training may be far behind them.

Appendix A: Vector analysis

1. *Gradient of a function* U(x, y, z)

The vectors of ordinary space ($\mathbb{R}^3$) are often necessary for the representation of physical dimensions such as speed, or the force applied to a solid. We have a *vector field* when the vector is defined in all points of a part (*D*) of the space. Let $M \in (D)$ and *U* be a function of *M* with coordinates *x*, *y* and *z*. This is a three-variable function with a real value: $\mathbb{R}^3 \rightarrow \mathbb{R}$. When the displacements of *x*, *y* and *z* are infinitely small (respectively d*x*, d*y* and d*z* from x_0, y_0 and z_0), then *U* varies by a certain quantity d*U* which depends on the *partial derivatives*: $\partial U/\partial x$, $\partial U/\partial y$ and $\partial U/\partial z$. We know that for a function with a single variable $f: \mathbb{R} \rightarrow \mathbb{R}$, so that:

$$f(x_0 + \mathrm{d}x) = f(x_0) + f'(x)\big|_{x=x_0}\mathrm{d}x$$

which may be written in differential form as:

$$\mathrm{d}f = f'(x)\big|_{x=x_0}\mathrm{d}x.$$

It is easy to show that for three variables (and the same reasoning will hold good for any number of variables) we have:

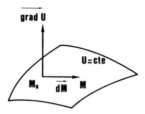

Fig. A.1.   The vector **grad** U is normal to the surface $U = 1$.

$$dU = \frac{\partial U}{\partial x}\, dx + \frac{\partial U}{\partial y}\, dy + \frac{\partial U}{\partial z}\, dz \tag{A.1}$$

where $\partial U/\partial x$ is the derivative of U with respect to x alone, y and z being considered as constants, equal to y_0 and z_0. $\partial U/\partial y$ and $\partial U/\partial z$ are defined in the same way. In fact, dU is the *scalar product* of two vectors: $(\partial U/\partial x, \partial U/\partial y, \partial U/\partial z)$, called the *gradient* of U, written **grad** U, and (dx, dy, dz), the increment of M_0 at M, written d**M**. Then:

$$dU = \mathbf{grad}\ U \cdot d\ \mathbf{M}. \tag{A.2}$$

Properties of the gradient vector

Any function U with a first derivative will have a gradient. The vector **grad** U is oriented in the sense of increasing U values. The significance is that *it indicates the direction of the greatest slope*, i.e. the direction along which the variation will be the fastest. The direction of the displacement corresponds to increasing values of U (Fig. A.1). Conversely, for any vector $\mathbf{A}(X, Y, Z)$ to be the gradient of a function $U(\mathbf{A} = \mathbf{grad}\ U)$, the necessary and sufficient conditions are:

$$\left.\begin{array}{c} \dfrac{\partial Z}{\partial y} - \dfrac{\partial Y}{\partial z} = 0 \\[2mm] \dfrac{\partial X}{\partial z} - \dfrac{\partial Z}{\partial x} = 0 \\[2mm] \dfrac{\partial Y}{\partial x} - \dfrac{\partial X}{\partial y} = 0 \end{array}\right\}. \tag{A.3}$$

If $\mathbf{A}(X, Y, Z)$ is such that $\mathbf{A} = \mathbf{grad}\ U$, $\mathbf{A}$ is said to *derive from a potential* $V = -U$.

2.   Divergence of a vector

The vector **A** may represent a dimension which itself represents some kind of flow, for example that of energy, across a surface (S) delimiting a volume (V). This

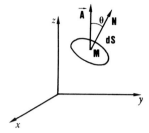

Fig. A.2. Flux of vector **A** across the surface element dS. **MN** is the normal at this element.

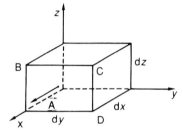

Fig. A.3. Flux of vector **A** across the face $ABCD$ of the parallelepiped with sides dx, dy and dz.

volume is 'immersed' in a field of vectors **A**. Let $M \in (S)$; the surface element dS in M is a vector of which, by definition, the modulus |dS| is equal to the area of the element, and the direction is in the positive sense of the normal **MN** at the surface in M. The *flux* of the vector **A** across the surface element dS is, by definition, the scalar product:

$$d\Phi = \mathbf{A} \cdot d\mathbf{S}. \tag{A.4}$$

If $\theta = (A, MN)$, then d$\Phi = A.$ d$S \cos \theta$ (Fig. A.2). The flux is maximum when **A** is perpendicular to the surface (cos $\theta = 1$ and $\theta = 0$), which is intuitively evident. The flow across the surface dS is maximum when it is perpendicular to the surface. dS may be assimilated to the surface of a parallelepiped of sides dx, dy and dz.

If A has X, Y and Z components, then:

$$d\Phi = X \, dy \, dz + Y \, dx \, dz + Z \, dx \, dy$$

is the flux of **A** across the faces of the parallelepiped since the flux is additive (according to the additive property of a scalar product). For example, dΦ_X, the value of the flux of the component X of **A** on Ox, is written as the scalar product of $\mathbf{A}_x(X, 0, 0)$ and d$\mathbf{S}$(dy dz, 0, 0) normal to the surface $ABCD$ (Fig. A.3). When (S) is a closed surface, the positive direction is always from the inside to the outside. Finally, we may use the additive property of fluxes to integrate $d\Phi$ to obtain the resultant flux across any surface (S):

$$\Phi = \int\int_{(S)} \mathbf{A} \cdot d\mathbf{S} = \int\int_{(S)} (X \, dy \, dz + Y \, dx \, dz + Z \, dx \, dy). \qquad (A.5)$$

3.   Green's theorem

A closed surface (S) delimits a volume V. Can the flux Φ be expressed for the field of vectors of the entire volume V and not just for the direction of $\mathbf{A}$ at the frontier of the volume? Green's theorem gives a positive answer to this question.

It can be shown that:

$$d\Phi = \left[\frac{\partial X}{\partial x} + \frac{\partial Y}{\partial y} + \frac{\partial Z}{\partial z} \right] dV \qquad (A.6)$$

where $dV = dx \, dy \, dz$ is the volume of the parallelepiped as above. We may write:

$$\text{div } \mathbf{A} = \frac{\partial X}{\partial x} + \frac{\partial Y}{\partial y} + \frac{\partial Z}{\partial z}.$$

Contrary to the gradient, the divergence of a vector is scalar. Thus:

$$d\Phi = \text{div } \mathbf{A} \, dV \qquad (A.7)$$

so that (if the components of $\mathbf{A}$ have derivatives at all points):

$$\int\int_{(S)} \mathbf{A} \cdot d\mathbf{S} = \int\int\int_{(V)} \text{div } \mathbf{A} \, dV. \qquad (A.8)$$

This expression of Green's formula allows the calculation of a flux across a surface (S) as an integral of either the surface or the volume delimited by the surface.

Properties of the divergence

The positive or negative sign of the divergence at a point M is characteristic of the divergence or the convergence of the field in the neighbourhood of M, whence the term. When the field is a vortex, the divergence is null at M (Fig. A.4).

The flux $\mathbf{A}$ across any closed surface (S) is null if, and only if, the divergence is null at all points of the field, under the conditions in which Green's theorem is applicable. In this case, the flux of $\mathbf{A}$ is said to be *conservative*.

4.   The Laplace function: the second-order scalar operator

Since the gradient of a function U is a vector, the divergence may be calculated:

$$\text{div } (\mathbf{grad} \, U) \equiv \text{div} \cdot \mathbf{grad} \cdot U \qquad (A.9)$$

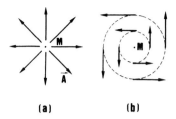

(a)   (b)

Fig. A.4.   (a) Field of divergent vectors **A** (positive divergence); (b) vortex field (divergence null).

where the dot notation indicates the operation as the action of an *operator*. We may then put:

$$\Delta = \text{div} \cdot \mathbf{grad}.$$

As the *vectorial* operator **grad** is generally noted ∇, we obtain:

$$\Delta = \text{div} \cdot \nabla = \nabla \cdot \nabla. \tag{A.10}$$

The operator Δ is called the *Laplacian* and, under the *operator form*, it may be considered as the scalar product of the operator ∇ multiplied by itself.
We have:

$$\text{div} \, (\mathbf{grad} \, U) = \frac{\partial}{\partial x}\left(\frac{\partial U}{\partial x}\right) + \frac{\partial}{\partial y}\left(\frac{\partial U}{\partial y}\right) + \frac{\partial}{\partial z}\left(\frac{\partial U}{\partial z}\right)$$

$$\tag{A.11}$$

$$\Delta U = \frac{\partial^2 U}{\partial x^2} + \frac{\partial^2 U}{\partial y^2} + \frac{\partial^2 U}{\partial z^2}.$$

5.   *Summary*

If:

$$U = U(x, y, z) \text{ and } \mathbf{A} = \mathbf{A}(X, Y, Z)$$

we have:

$$\nabla U = \left(\frac{\partial U}{\partial x}, \frac{\partial U}{\partial y}, \frac{\partial U}{\partial z}\right), \quad \text{div} \, \mathbf{A} = \frac{\partial X}{\partial x} + \frac{\partial Y}{\partial y} + \frac{\partial Z}{\partial z}$$

$$\Delta U = \frac{\partial^2 U}{\partial x^2} + \frac{\partial^2 U}{\partial y^2} + \frac{\partial^2 U}{\partial z^2}.$$

Appendix B: Dynamic systems

1. *Notion of a dynamic system*

Let us consider a material system of which the movement in the ordinary space depends on n variables, such as the temperature, the concentration of certain molecules, the velocity of certain elements, and so on. These are the n *state variables* x_1, x_2, ..., x_n. The element $\mathbf{X} = (x_1, x_2, ..., x_n) \in \mathbb{R}^n$ describes the *space of the states* or the *phase space* (D) according to a particular trajectory and each coordinate depends on time. If $x_i \equiv x_i(t)$ is known $\forall i = 1, 2, ..., n$, then the 'movement' of the system is completely determined. Obviously, the question we are interested in is just the converse. A *dynamic system* is characterised by the hypothesis that a state variable $x_i(t)$ undergoes, in the neighbourhood of a given state $(x_1(t), x_2(t), ..., x_n(t))$, a variation that depends only on this state, being equal to its velocity at this point.

In other words:

$$\frac{\mathrm{d}x_i}{\mathrm{d}t} = f_i(x_1, x_2, ..., x_n) \quad i = 1, 2, ..., n \tag{B.1}$$

or:

$$\frac{\mathrm{d}\mathbf{X}}{\mathrm{d}t} = \mathbf{F}(\mathbf{X}).$$

But certain systems may depend *explicitly* on time. These are said to be *non-autonomous* and may be written in the form:

$$\frac{\mathrm{d}\mathbf{X}}{\mathrm{d}t} = \mathbf{F}(\mathbf{X}, t).$$

This is the case of material systems for which the forces acting on the system vary with time. Very often, dynamic systems depend on external *parameters* or *constraints* which strongly influence their solutions. For example:

$$\frac{\mathrm{d}\mathbf{X}}{\mathrm{d}t} = F(\mathbf{X}, t, \boldsymbol{\lambda}) \quad \text{where} \quad \boldsymbol{\lambda} = (\lambda_1, \lambda_2, ..., \lambda_p) \in \mathbb{R}^p. \tag{B.2}$$

This non-autonomous system further depends on p parameters λ_i, $i = 1$ to p. A process is said to be *adiabatic* when the variation of the external conditions of a body is sufficiently slow. Historically, the term is of thermodynamic origin: during an adiabatic process the entropy of a material body remains unchanged, i.e. the process is reversible. Today the term is used to describe any process during which $\mathrm{d}\lambda/\mathrm{d}t$ is small.

The first dynamic systems to be investigated were mechanical systems and it was Poincaré who, working on the three-body problem in celestial mechanics, laid down the bases for the study of non-linear systems.

Thus Newton's equation:

$$\mathbf{F} = m\gamma \Rightarrow F = m\frac{d^2x}{dt^2}$$

where x is the coordinate of a material point along the axis Ox, is applied to a harmonic oscillator defined by the force (as in the case of a spring):

$$F = -kx$$

where k is a positive constant, so that the corresponding dynamic system is:

$$\left. \begin{aligned} m\frac{dx}{dt} &= p \\ \frac{dp}{dt} &= -kx \end{aligned} \right\} \tag{B.3}$$

where p, the quantity of movement, is the second state variable in the phase space:

$$x_1 \equiv x \quad \text{and} \quad x_2 \equiv p.$$

This is a dynamic system of the form:

$$\left. \begin{aligned} \frac{dx_1}{dt} &= f_1(x_1, x_2) \\ \frac{dx_2}{dt} &= f_1(x_1, x_2) \end{aligned} \right\}.$$

2. The Hamiltonian form. Conservative systems

We know that in mechanics there exists a fundamental dimension, the *energy*, the sum of the kinetic and potential energies, which is conserved all along the trajectory. The Hamiltonian formulation provides an elegant determination of the conditions for the existence of such an observable dimension. The system is then said to be *conservative*. Let $H = H(x, p)$ be this observable dimension. According to the theorem for the derivation of composite functions, we have:

$$\frac{dH}{dt} = \frac{\partial H}{\partial x}\frac{dx}{dt} + \frac{\partial H}{\partial p}\frac{dp}{dt}.$$

If $H(x, p)$ is constant over the trajectory of the material system, then:

$$\frac{dH}{dt} = 0.$$

Thus a *sufficient* condition for this is:

$$\left. \begin{aligned} \frac{dx}{dt} &= \frac{\partial H}{\partial p} \\ \frac{dp}{dt} &= -\frac{\partial H}{\partial x} \end{aligned} \right\}. \tag{B.4}$$

These are the equations of movement for a system in the Hamiltonian form. If the system can be integrated, then the function H is called the *Hamiltonian*. The function has the physical property of being conserved all along the trajectories. However, in some cases, the identity between the Hamiltonian and the total energy of the system is not complete. This occurs only if the system is *conservative*, and it can then be shown that there exists a *potential* function V such that:

$$\frac{dp}{dt} = -\frac{\partial V}{\partial x}.$$ (B.5)

This *condition is necessary and sufficient*. The equations of movement can then be deduced in Hamiltonian form with:

$$H = T + V = \text{constant}$$

where T is the kinetic energy:

$$T = \frac{1}{2} m\dot{x}^2 = \frac{p^2}{2m} \quad \text{with} \quad p = mv.$$ (B.6)

More generally, a *conservative* dynamic system is a dynamic system for which a *constant* function f has been defined for each trajectory. This function is called the *first integral* of the dynamic system.

3.   Stability: Lyapunov functions

The problem of the *stability* of a dynamic system is that of the evolution of a system subjected to small perturbations. This involves a comparison of the properties of the trajectories of systems with and without perturbations. Such studies are generally based on the theorems due to Lyapunov.

Starting with the dynamic system (B.1):

$$\frac{dx_i}{dt} = f_i(x_1, x_2, \ldots, x_n) \quad i = 1, 2, \ldots, n$$

we may study the *trajectories* $x_i(t) = \text{constant}$, represented in the phase space by a single point called the *critical point*, the *singular point*, the *equilibrium point* or the *stationary state*. These terms should be used with discernment according to the field of application (see, for example, the theory of thermodynamics and the theory of catastrophes in Chapter 3).

The stability of the critical points is given by the *Lyapunov functions* $U(x_1, x_2)$ according to the following theorems:

Theorem B.I: *Let a dynamic system*:

$$\left.\begin{aligned}
\frac{dx_1}{dt} &= f_1(x_1, x_2) \\
\frac{dx_2}{dt} &= f_2(x_1, x_2)
\end{aligned}\right\}$$ (B.7)

be defined in an open interval Ω containing the origin and possessing a single trajectory such that $f_1(0, 0) = f_2(0, 0) = 0$. If there exists a function $U(x_1, x_2)$ defined in Ω and such that:

(i) $U(0, 0) = 0$ and $U(x_1, x_2) > 0$ elsewhere; and

(ii) $\dfrac{dU}{dt}(0, 0) = 0$ and $\dfrac{dU}{dt}(x_1, x_2) \leq 0$ elsewhere,

then the origin is a critical stable point.

Theorem B.II: *If condition (ii) is replaced by:*

$$\frac{dU}{dt}(0, 0) = 0 \ and \ \frac{dU}{dt}(x_1, x_2) < 0 \ elsewhere,$$

then the origin is an asymptotically unstable critical point.

Theorem B.III: *If condition (ii) is replaced by:*

$$\frac{dU}{dt}(0, 0) = 0 \ and \ \exists(x_1, x_2) \ such \ that \ \frac{dU}{dt} > 0$$

then the origin is unstable.

These theorems may be used for critical points beyond the origin by simply changing the variable $y_i(t) = x_i(t) - x_i^0(t)$ where x_i^0 is a critical point of the system (B.7). Then $(0, 0)$ is obviously a critical point of the (y_i).

Let us illustrate this in the field of mechanics, again considering the example of the harmonic oscillator. The Hamiltonian may be written (Eqs (B.3) and (B.6)):

$$H(x, p) = -\int (-kx)\,dx + \frac{p^2}{2m} = \frac{kx^2}{2} + \frac{p^2}{2m}.$$

This is the total energy of the system if it is conservative. It can be shown that H is a Lyapunov function if, and only if, the potential energy V passes through a minimum at the origin. For the harmonic oscillator we clearly have:

$$H(0, 0) = 0 \quad and \quad H(x, p) > 0 \ \forall(x, p) \neq (0, 0)$$

and following (B.3):

$$\frac{dH}{dt} = \frac{p}{m}(kx) - (kx)\frac{p}{m} = 0.$$

Thus the conditions of theorem B.I are satisfied. This theorem finally shows that the critical point will be an *attractor* (a stable point) if U and dU/dt are of opposite signs, which is geometrically interpreted as a return to equilibrium after perturbation and, in mechanics, implies a minimum potential (see Fig. B.1).

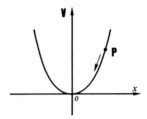

Fig. B.1.   The potential function V is a Lyapunov function for a conservative system: $V \geq 0$, and $dV/dt \leq 0$.

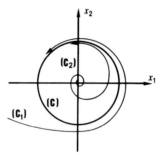

Fig. B.2.   *Stable* limit cycle. All trajectories, external (C_1) or internal (C_2), wind on to the limit cycle (C).

4.   Limit cycles, critical points, Jacobian, Hessian

A dynamic system of the type:

$$\frac{dx_1}{dt} = f_1(x_1, x_2)$$

$$\frac{dx_2}{dt} = f_2(x_1, x_2)$$

(B.7)

may admit solutions represented by closed curves in the phase plane (phase space $\mathbb{R}^2$), called *limit cycles*. A limit cycle is *an isolated closed trajectory*. It is stable if all the trajectories 'not too distant' approach it when $t \to \infty$. It is unstable if such trajectories move further away. When a trajectory *exterior* to the limit cycle approaches it (resp. moves away) whereas a trajectory *interior* to the limit cycle moves away from it (resp. approaches it), the cycle is said to be semi-stable. In practice, however, such a cycle will be considered to be unstable (Fig. B.2).

Limit cycles are of fundamental importance since: (i) they occur only in *non-*

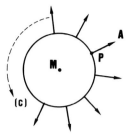

Fig. B.3. Index of a closed curve (C). The point P of the curve is displaced and the associate vector **A** rotates continuously in space. When P comes back to the starting point, **A** may have completed a number of rotations $I(C)$. The sign of I depends on the orientation of the plane (positive if the vector rotates in the direction of the orientation of the plane).

linear non-conservative systems; and (ii) if they are stable, they represent a *stable stationary* oscillation of a physical system in exactly the same way as a stable critical point represents a stable *equilibrium*. In fact, as the ultimate stationary movement established on the closed curve, the limit cycle may be considered to be independent of initial conditions.

A difficult problem is posed by the search for limit cycles. Poincaré introduced the notion of an *index* to establish a *necessary* criterion for the existence of a limit cycle. Details of this approach will be found in any good treatise on dynamic systems, but here let us just give the result: *the index* I(C) *of a closed trajectory* (C) *is given by*:

$$I(C) = \frac{1}{2\pi} \int_{(C)} \frac{f_2 \mathrm{d}f_1 - f_1 \mathrm{d}f_2}{f_1^2 + f_2^2}. \tag{B.8}$$

It can be shown that, if (C) has a tangent vector able to rotate continuously, then $I(C) = 1$ (Fig. B.3). It can then be shown that the index of an elementary critical point is equal to ± 1. This result is obtained from the calculation of the *Jacobian*:

$$\Delta = \begin{vmatrix} \dfrac{\partial f_1}{\partial x_1} & \dfrac{\partial f_1}{\partial x_2} \\ \dfrac{\partial f_2}{\partial x_1} & \dfrac{\partial f_2}{\partial x_2} \end{vmatrix} \tag{B.9}$$

which, by definition, is the determinant of the Jacobian matrix of the first-order partial derivatives of f_i with respect to x_j, $\forall i, j$.

The *Hessian* is the determinant of the matrix of the second-order partial derivatives. It is more rarely used as, in general, the first order will suffice for the characterisation of the singularities. Thus:

$$|\mathcal{H}(f)| = \begin{vmatrix} \dfrac{\partial^2 f}{\partial x_1^2} & \dfrac{\partial^2 f}{\partial x_1\,\partial x_2} \\[2ex] \dfrac{\partial^2 f}{\partial x_2\,\partial x_1} & \dfrac{\partial^2 f}{\partial x_2^2} \end{vmatrix}.$$

If a is a critical point of f and if $|\mathcal{H}(f)(a)| \neq 0$, then a is a non-degenerate critical point of f.

There are four kinds of singularities, called *elementary singularities*, with real values, which can be identified as a *node*, a *saddle-point*, a *focal point* and a *centre*. To find these singularities, the dynamic system (B.7) is transformed into:

$$\frac{dx_1}{dt} = a_{11}x_1 + a_{12}x_2 + P_1(x_1,x_2)$$

$$\frac{dx_2}{dt} = a_{21}x_1 + a_{22}x_2 + P_2(x_1,x_2)$$

(B.10)

where P_1 and P_2 are polynomials of at least the second degree in x_1 and x_2 with:

$$a_{11} = \frac{\partial f_1}{\partial x_1}, \; a_{12} = \frac{\partial f_1}{\partial x_2}, \; a_{21} = \frac{\partial f_2}{\partial x_1}, \; a_{22} = \frac{\partial f_2}{\partial x_2}$$

elements of the Jacobian matrix of which the determinant is the Jacobian Δ (Eq. (B.9)). The linearised system (where $P_1 = P_2 = 0$) has solutions of the form:

$$x_1 = x_1(t_0)\, e^{\omega t} \quad x_2 = x_2(t_0)\, e^{\omega t}$$

(B.11)

called *normal modes*. The algebraic system obtained by substitution of $x_1(t)$ and $x_2(t)$ has non-trivial solutions if:

$$|a_{ij} - \omega \delta_{ij}| = 0 \Leftrightarrow \omega^2 - S\omega + P = 0.$$

(B.12)

This is the characteristic equation of which the solutions may be discussed in function of the discriminant $(S^2 - 4P)$, where:

$$S = a_{11} + a_{22} = \frac{\partial f_1}{\partial x_1} + \frac{\partial f_2}{\partial x_2}$$

$$P = a_{11}a_{22} - a_{12}a_{21}.$$

(B.13)

If ω_1 and ω_2 are the two (distinct) solutions of (B.12), then the solutions of (B.11) may be written:

$$x_1(t) = C_1 e^{\omega_1 t} + C_2 e^{\omega_2 t}$$

$$x_2(t) = C_1 K_1 e^{\omega_1 t} + C_2 K_2 e^{\omega_2 t}$$

where the constants C_1, C_2, K_1 and K_2 are determined by the initial conditions. We then have the following classification of singular points (illustrated in Fig. B.4):

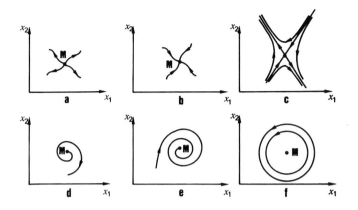

Fig. B.4. (a) Unstable *node*; (b) stable *node*; (c) unstable *saddle*; (d) unstable *focus*; (e) stable *focus*; (f) *centre*.

Discriminant	Sign of ω_1 and ω_2	Singularity	Figure
$S^2 - 4P \geq 0$			
$(\omega_1, \omega_2 \in \mathbb{R})$			
$\quad P > 0$	same sign	node	
$\quad\quad S > 0$	positive	unstable node	B.4(a)
$\quad\quad S < 0$	negative	stable node	B.4(b)
$\quad P < 0$	opposite signs	unstable saddle	B.4(c)
$S^2 - 4P < 0$			
$(\omega_1, \omega_2 \in \mathbb{C})$			
Conjugates			
$\quad S > 0$	real positive part	unstable focus	B.4(d)
$\quad S < 0$	real negative part	stable focus	B.4(e)
$\left.\begin{array}{l} P > 0 \\ S = 0 \end{array}\right\}$	purely imaginary	centre	B.4(f)

5. *Partial differential equations*

Partial differential equations are of the form:

$$f\left(x,\, y,\, u,\, \frac{\partial u}{\partial x},\, \frac{\partial u}{\partial y},\, \frac{\partial^2 u}{\partial x^2},\, \frac{\partial^2 u}{\partial y^2},\, \frac{\partial^2 u}{\partial x \partial y},\, \frac{\partial^3 u}{\partial x^3},\, \ldots \right) = 0.$$

They contain partial derivatives of a function u with respect to independent variables, and can only be solved in particular cases. One example is the diffusion equation of the type:

$$\frac{\partial u}{\partial t} = D\nabla^2 u$$

where D is the coefficient of diffusion. The methods of resolution of such equations are highly specific, and each case will require special treatment.

   Ordinary differential equations, on the contrary, depend only on the variables x_1, x_2, ..., x_n, as in the case of the dynamic systems considered above.

6.   Some notes on the terminology of ordinary differential equations. Compact differential manifolds

A rigorous presentation of dynamic systems of the type (B.1) calls for the introduction of abstract notions. Thus, the correspondence:

$$t \rightarrow x_i(t) \rightarrow f_i(x_1, x_2, ..., x_n) = \frac{d x_i}{d t}$$

implies several spaces:

(a)   The *phase space (D)* is the set of all the states of the process.
(b)   The *dimension* of the process is that of its phase space.
(c)   A process is *differentiable* if its phase space has a structure of the *differential manifold*.
(d)   A *differential variety* of $\mathbb{R}^n$ is a part $V \subset \mathbb{R}^n$ on which a *structure of a differential variety* has been defined. Here we seek to *cover* V with *similar* parts. This is accomplished by a union of *diffeomorphisms* of open sets to the open sets U_i of $\mathbb{R}^p$ ($p \leq n$). Let $\phi_i : W_i \rightarrow U_i$, as well as its reciprocal ϕ_i^{-1}, be such a diffeomorphism, i.e. a differential homeomorphism. U_i is called a *map* and $\phi_i(x)$ is the *image* of $x \in W_i$ on the map U_i. These concepts lead to the representation of a complex space in a simpler space. For example, the three-dimensional space in which we live may be represented on a geographical map, the set of land surfaces, roads, and so on, appearing on a map constructed by the union of smaller, juxtaposed maps. This is just what the diffeomorphism ϕ_i provides: an application that is bijective (the relief appears) but smooth in terms of continuous variation of distances (through the existence of derivatives) on the earth and on the paper. However, even this is not sufficient. Indeed, what if the land surfaces have a non-null intersection, as might be the case, for example, if the geographical work were done by different surveyors? Common rules would have to be applied, in other words we would require similar or compatible maps. Mathematically, these could be obtained as follows:

   With $W_i \cap W_j \neq \varnothing$, let us consider:

$$U_{ij} = \varphi_i(W_i \cap W_j)$$
$$U_{ji} = \varphi_j(W_j \cap W_i).$$

Suppose surveyor i works on a part $W_i \cap W_j$ which also belongs to W_j, studied by surveyor j. It should be possible to go from one map to the other. Let ϕ_{ij} be this application, i.e. the rule followed by the two surveyors:

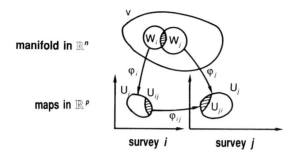

Fig. B.5. Structure of a differential manifold: ϕ_{ij} is defined by $\phi_{ij}(x) = \phi_j(\phi_i^{-1}(x))$ and must be differentiable.

$$\phi_{ij}: U_{ij} \to U_{ji} \quad \text{such that} \quad \phi_{ij}(x) = \phi_j \circ \phi_i^{-1}(x).$$

The applications ϕ_{ij} are the diffeomorphisms between the open sets of $\mathbb{R}^p$. The class of differentiability of these applications determines the class of manifolds. The variety will be said to be of the class $C^r(1 \le r \le \infty)$ if the applications ϕ_{ij} are differentiable up to order r.

A set of maps U_i, sufficiently *compatible* for all points of V to be located, is called an *atlas*. Obviously, if the set of maps is to be of any use, each place that can be visited must be represented on the atlas. Finally, if the maps of different atlases, edited by different publishers, are to be compatible, there must be an *equivalence* between the atlases such that the union of atlases on V is again an atlas on V.

A *structure of a differential manifold on* V *is a class of equivalent atlases.* This is summed up in Fig. B.5. A differential manifold possesses all the 'right' properties that allow the study of dynamic systems and their singularities.

A *compact differential manifold* is a differential manifold $V \subset \mathbb{R}^n$ such that the whole surface of V can be covered by a *finite* number of open sets, according to the topological definition of *compacity*.

(e) *Dynamic systems; flow.* We have seen that the notion of a dynamic system involves a *field of vectors*.

Definition I: *Let* V *be a differential manifold of dimension* n, *and* T_bV *the tangent 'plane' at any point* $b \in V$, *homeomorph at* $\mathbb{R}^n$. *The set:*

$$TV = \left(\bigcup_{b \in V} \right) T_bV$$

is called the tangent bundle TV *at* V.

Definition II: *A field of vectors* **X** *on a variety* V *is a* C^r *differential application s:* $V \to TV$, *which associates a vector* $\mathbf{v}_b = s(b)$ *of the tangent plane* T_bV *to all points* $b \in V$.

Fig. B.6.   Field of vectors on the circle S^1.

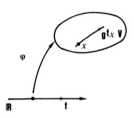

Fig. B.7.   Movement of a point in phase space V.

With these definitions we obtain a continuous variation of the tangent vector $\mathbf{v}_b$ (situated in the tangent plane) with b (Fig. B.6).

<u>Definition III</u>: *A dynamic system* ($\mathcal{D}$) *is produced by a variety* V *and a field of vectors* **X** *defined on this variety:* $\mathcal{D} = (V, \mathbf{X})$.

<u>Definition IV</u>: *A singularity of a field of vectors is a point* $a \in$ V *where the vector* $\mathbf{v}_a \in \mathbf{X}$ *is null.*

<u>Definition V</u>: *Let* $x \in$ V *be any initial state of the process, and* $g^t x$ *the state of the process at time* t. *A flow* $(V, \{g^t\})$ *is the couple formed by the variety* V *and the one-parameter group* $\{g^t\}$, *i.e. the family of applications* $\{g^t \colon V \to V\}$ *such that* $\forall s$, $\forall t \in \mathbb{R}$, *we have* $g^{t+s} = g^t g^s$, *where* g^0 *is the identical application.*
    The *movement* of $x \in$ V under the action of the flow $(V, \{g^t\})$ is the application $\phi \colon \mathbb{R} \to V$, $\phi(t) = g^t x$ (Fig. B.7).
    The *orbit* of the flow $(V, g^t x)$ is the image of $\phi \colon \mathbb{R} \to V$.
    This is a subspace of V (Fig. B.8). The graph of the movement $\phi(t) = g^t x$ is an *integral curve* of the flow (Fig. B.9).
    In conclusion, the fundamental problem of the theory of dynamic systems consists in the study of one-parameter groups $\{g^t\}$ of diffeomorphisms of a variety V, the fields of vectors defined over V, and the relationships existing between them. *A group* $\{g^t\}$ *defines a field of vectors* **X**, *which is the field of velocities* **v** *of the process:*

$$\left. \frac{d}{dt} \right|_{t=0} g^t x = v(x).$$

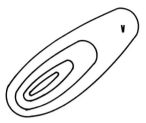

Fig. B.8. Orbits.

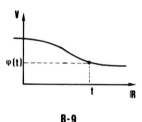

Fig. B.9. Integral curve.   **B-9**

which is generally written:

$$\dot{x} = \mathbf{v}(x).$$

This is the differential equation defined for a field of vectors **v**.

Appendix C: Notations in matrix algebra

Let E be a vector space of dimension n over a field $\mathbb{K}$; f an endomorphism, and $\mathbf{A}$ the matrix representing the linear application f, with elements (a_{ij}), i and $j = 1$ to n. $\mathbf{A}$ is a square matrix.

The *transposed* matrix $\mathbf{A}^t$ is made up of elements (a_{ji}) where the rows of $\mathbf{A}$ are the columns of $\mathbf{A}^t$.

The square matrix $\mathbf{A}$ is *symmetric* if $\mathbf{A}^t = \mathbf{A}$, i.e. if $a_{ij} = a_{ji}$ $\forall i, j = 1$ to n.

If $\mathbb{K} = \mathbb{C}$, the elements a_{ij} are complex numbers. The *adjoint* matrix $\mathbf{A}^*$ is defined by:

$$(a^*)_{ji} = (a_{ij})^*.$$

It is obtained by transposing $\mathbf{A}$ and by taking the conjugate of each complex term.

The matrix $\mathbf{A}$ is *hermitian* if $\mathbf{A}^* = \mathbf{A}$, i.e. if it is equal to its adjoint.

The matrix $\mathbf{A}$ is *unitary* if $\mathbf{A}^* \mathbf{A} = \mathbf{I}$, where $\mathbf{I}$ is the unit matrix.

These definitions are of interest as they allow the construction of spaces with structures giving them the property of hermiticity, of unitarity, and so on. Further details will be found in standard works. The relationship between the matrix and the operator will be discussed in Volume III, Appendix A.

Appendix D: Probability and information theory

1. *Probability*

Let Ω be the universe of possible events and E an event belonging to the set of parts $\mathcal{P}(\Omega)$. The probability is an application:

$$\text{Pr: } \mathcal{P}(\Omega) \rightarrow [0,\ 1]$$

which associates a real number in the interval $[0,\ 1]$ to $E \subset \Omega$ and which satisfies the following axioms:

(i) $\text{Pr}[\Omega] = 1$;
(ii)   If E_1 and E_2 are two events such that $E_1 \cap E_2 = \varnothing$ (incompatible events), then:

$$\text{Pr}[E_1 \cup E_2] = \text{Pr}[E_1] + \text{Pr}[E_2].$$

This is the *axiom of the sum of the probabilities.*

An arbitrary value, generally real, is associated by X to each possible event $\omega \in \Omega$; this variation of x is represented by a random variable X (which takes the value x). Then:

$$\text{Pr}[E] = \text{Pr}[X = x]$$

where $E = X^{-1}(x)$. If Ω is continuous, then X is a continuous random variable.

The *mathematical expectation* $E(X)$ is defined by:

$$E(X) = \int_{-\infty}^{\infty} p(x)\,x\,\mathrm{d}x$$

where:

$$p(x)\,\mathrm{d}x = \text{Pr}[x \leq X \leq (x + \mathrm{d}x)]$$

is the density of probability for $X \in\ ]-\infty,\ +\infty[$. In the case of a *discrete* random variable, where:

$$\Omega = \{E_1,\ E_2,\ ...,\ E_i...\} \quad \text{and} \quad \text{Pr}[E_i] = \text{Pr}[X = x_i] = p_i.$$

$\forall i$, we have:

$$E(X) = \sum_i p_i\, x_i.$$

If Ω is a *finite* universe, and if the elementary probabilities p_i are equal, then, by the application of the law of large numbers, $E(X)$ represents the usual arithmetical mean:

$$\bar{x} = \frac{\displaystyle\sum_{i=1}^{N} n_i\, x_i}{N} = \sum_{i=1}^{N} p_i\, x_i$$

where $p_i = n_i/N$ is the relative frequency of the event and x_i is the value of the statistical series of which the occurrence is equal to n_i.

The mathematical expectation is also the *moment* of order 1. In a general way, the moment of order k of a continuous random variable is defined by:

$$E(X^k) = \int_{-\infty}^{+\infty} x^k p(x)\, dx.$$

For a discrete random variable we have:

$$M_k = E(X^k) = \sum_{i=1}^{\infty} x_i^k p_i.$$

The moment of order k, about the centre, is then:

$$m_k = E(X - EX)^k$$

where $m_1 = 0$, $m_2 = \text{Var}(X)$, the variance of X, and so on. An important relationship used in Chapter 4 is:

$$m_2 = M_2 - M_1^2.$$

2.   The Shannon function of information

The Shannon function H as defined by:

$$H(x) = - \sum p_i \log_2 p_i \tag{D.1}$$

where p_i represents the probability of the symbol x_i in the message x, applies to a set of messages that all use the same number of different symbols with the same distribution of probability for these symbols. The function H will be the same whatever the content of the messages. For a message y we will have:

$$H(y) = - \sum p_j \log_2 p_j$$

and for two dependent messages x and y we have:

$$H(x, y) = - \sum_{i,j} p_{ij} \log_2 p_{ij}. \tag{D.2}$$

This measures the incertitude associated with the couple of events. If the events are *independent*, we have:

$$p_{ij} = p_i \cdot p_j$$

so that:

$$H(x,y) = - \sum_{i,j} p_{ij} \log_2 p_i - \sum_{i,j} p_{ij} \log_2 p_j$$

$$= - \sum_j p_j \sum_i p_i \log_2 p_i - \sum_i p_i \sum_j p_j \log_2 p_j$$

$$H(x,y) = H(x) + H(y)$$

as:

$$\sum_j p_j = \sum_i p_i = 1.$$

When the messages are not independent, conditional probabilities will have to be introduced:

$$P_{i/j} = \Pr[x = i/y = j]$$

and

$$P_{j/i} = \Pr[y = j/x = i].$$

The probability:

$$\Pr[(x = i) \text{ and } (y = j)] = \Pr[x = i] \cdot \Pr[y = j/x = i]$$

or:

$$p_{ij} = p_i p_{j/i}.$$

The conditional incertitude may be deduced from the conditional probabilities:

$$H(y/i) = - \sum_j p_{j/i} \log_2 p_{j/i} \tag{D.3}$$

when the symbol i is known. Thus, the *mean incertitude* $H(y/x)$ is given by:

$$H(y/x) = \sum_i p_i H(y/i)$$

$$\tag{D.4}$$

$$H(y/x) = - \sum_{i,j} p_i p_{j/i} \log_2 p_{j/i}.$$

From this, we may deduce $H(x, y)$.

In fact:

$$p_{j/i} = p_{ij}/p_i$$

leads to:

$$H(y/x) = - \sum_{i,j} p_i \frac{p_{ij}}{p_i} \log_2 \frac{p_{ij}}{p_i}$$

$$= - \sum_{i,j} p_{ij} \log_2 p_{ij} + \sum_{i,j} p_{ij} \log_2 p_i$$

$$= H(x,y) + \sum_i \left(\sum_j p_{ij} \right) \log_2 p_i$$

$$= H(x,y) + \sum_i p_i \log_2 p_i$$

$$H(y/x) = H(x,y) - H(x).$$

Thus:

$$H(x, y) = H(y/x) + H(x).$$

This result means that the incertitude concerning the two messages x and y is equal to the incertitude of one of them *plus* the conditional incertitude of the other. Under the additive form, this result is analogous to that concerning conditional probabilities, the logarithms having transformed the product of terms into a sum.

These results have been applied to the transmission of information in a transmission channel: $H(x)$ is the incertitude on the *source*; $H(y)$ is the incertitude on the receiver; $H(y/x)$ is the quantity of information at the output when the input is known (*ambiguity*); $H(x/y)$ is the quantity of information at the input when the output is known (*equivocation*). If the transmission along a channel were perfect, we would have:

$$H(x) = H(y) = H(x, y)$$

and

$$H(x/y) = H(y/x) = 0.$$

Symbols and Constants (Mainly in CGS Units)

A = Area (cm^2)

a = Affinity (ergs/mole)

$A, B, C, \ldots, X, Y, Z$ = Chemical species

AA_i = i-rank amino acid in a protein chain

AQ = Rate of formation of carriers (sec^{-1})

$\mathcal{A}$ = Recurrent matrix

C = Factor of conversion

D = Constant of diffusion (cm^2/sec)

D = Rate of destruction of carriers (sec^{-1})

d_t = Topological dimension

E = Enzyme

E = Internal energy (ergs)

E = Number of enlacements between two curves

E = Excess production (sec^{-1})

E = Average production (sec^{-1})

ES = Enzyme–substrate complex

E_j = Class of equivalence

F = Factor of mass (dimensionless)

F = Helmhlotz's free energy (ergs)

F = Orgatropy

F^l $(l = 1, \ldots, n)$ = Elementary physiological function

G = Gibb's free energy (ergs)

$g_i(P_i, P_i)$ = Transport function between two different locations

g_{ij} = Transport function from a structural unit u_i onto a structural unit u_j

H = Shannon's average quantity of information (bits)

$\mathbf{H}$ = Matrix of membrane permeabilities

H_M = Quantity of information in a message supposing the equiprobability of symbols (bits)

H_R = Quantity of information in a message taking into account the conditional probability of the symbols (bits)

$H_{\max}$ = Shannon's maximum quantity of information (bits)

I = Quantity of information stored per symbol (bits)

I_1 = Deviation from the equiprobable state (bits)

I_2 = Deviation from independence (bits)
J = Flux, rate of flow per unit surface (moles/(cm²/sec))
J_m = Metabolic flux
J_t = Transport flux (number of moles per unit time)
K = Equilibrium constant of reaction (dimensionless)
L   = Leslie matrix
L = Cell length (cm)
Λ = Space of parameters
ln   = Natural logarithm
$\log_2$ = Logarithm to the base 2
W = Writhing: $E - T$
X_i = Number of carriers of type i
$\overline{Y}_s$ = Saturation function of a macromolecule for a ligand S
Z_i = Integral of configuration

a = Age of a cell (sec)
α = Aspect ratio
c = Volume concentration (moles/cm³)
δ = Hierarchisation operator
∂_{ij} = Partial derivative of v_{x_i} with respect to x_j (Higgins)
η = Total quantity of ligand
i = Information carrier
k_+ = Direct reaction constant (sec⁻¹)
k_- = Inverse reaction constant (sec⁻¹)
k_B = Boltzmann's constant: $R/N_0 = 1.38 \times 10^{-16}$ ergs/deg K
λ = Parameter
λ_i = Probability that the length of a growing chain be i
m = Mass (g)
m^* = $\frac{1}{2}$(Magic mass (g))
m_i = Probability that an element of E_i dies
μ = Chemical potential (ergs/mole)
μ = Cell maturity (dimensionless)
N_j = Number of elements in a class of equivalence
N_0 = Avogadro's number: 6.0234×10^{23}/mole
n_α^l = Number of sinks for the P_α-function at level l
$n_{P_1 P_2}$ = Number of units that synthesise P_2 and not P_1
o^l = Number of organisation available for an observed state
P = Product
P = Total production of entropy (entropy/s = ergs (deg K.sec))
P = Binding polynomial
Pr   = Probability
Q = Flow (moles/sec)
Q = Factor of quality
R = Von Foerster's self-organising function
R = Redundance

R = Perfect gas constant: 8.31434×10^7 ergs/deg K

Re = Repressor

$\overline{R}_j$ = Quantity of proteins in state j

S = Substrate

S = Entropy (ergs/deg K)

T = Temperature (deg K)

T = Twist (a measure of the torsion)

T = Inverse of the time constant of cell kinetics (sec)

V = Volume (cm^3)

V, E = Energy (1 joule = 10^7 ergs)

υ = Stoichiometric coefficient (< 0 on the left, > 0 on the right)

$\Omega(t)$ = Functional order

Ω = Solid angle (steradians)

ω = Mobility (cm^2/(sec.erg))

ϕ = Probability that the j successive pairs be dissociated

$\Pi(n)$ = Potential of functional organisation

π = Elementary probability

ψ = Probability of moving from one site to the next

ψ_{ij}^{α} = Functional interaction from the i-unit to the j-unit

r = Vector radius (cm)

r_i = Probability of an element of E_i disappearing to give birth to β daughter cells

ρ = Density of source activators (moles/cm^3)

$\overline{\rho}$ = Density of source inhibitors (moles/cm^3)

ρ_E = Total quantity of protein

s = Density of entropy (ergs/(deg K cm^3))

σ = Production of entropy (internal entropy per unit time and unit volume)

t = Time (sec)

τ = Generation time (sec)

$\overline{\tau}$ = Time at which the mass is doubled

$\tilde{\tau}$ = Time at which the factor of mass F is doubled

u = Self-replicating unit

v = Rate of chemical reaction (concentration/sec)

v_i = Probability that an element of E_i pass into E_{i+1} by ageing

v^l = Degree of organisation

$\overline{\omega}$ = Stoichiometry of the interaction

General Reading

Mathematics

Arnold V. I. *Mathematical Methods of Classical Mechanics*, Springer-Verlag, Berlin (1989).
Hirsch W., Smale S. *Differential Equations, Dynamical Systems and Linear Algebra*, Academic Press, New York (1974).
Murray J. D. *Mathematical Biology*, Springer-Verlag, Berlin (1989).
Skelton R. E. *Dynamic System Control: Linear System Analysis and Synthesis*, John Wiley & Sons, New York (1988).
Zauderer E. *Partial Differential Equations of Applied Mathematics*, Wiley Interscience, New York (1989).

Physics

Feynman R. *et al. Feynman Lectures on Physics*, Addison-Wesley, Reading, MA (1969).
Goldstein H. *Classical Mechanics*, Addison-Wesley, Reading, MA (1980).
Keizer J. *Statistical Thermodynamics of Nonequilibrium Processes*, Springer-Verlag, Berlin (1987).
Landau L. D. *et al. Course of Theoretical Physics*, Pergamon Press, New York (1982).
Morse P. M., Feshbach H. *Methods of Theoretical Physics*, McGraw-Hill, New York (1953).

Physiology

Alberts B. *et al. Molecular Biology of the Cell*, Garland Publishing, New York (1983).
Giese A. C. *Cell Physiology*, Saunders, Philadelphia, PA (1979).
Guyton A. C. *Textbook of Medical Physiology*, Saunders, Philadephia, PA (1986).
Lehninger A. L. *Principles of Biochemistry*, Worth, New York (1982).
Mountcastle V. B. (Ed.) *Medical Physiology*, Mosby, St Louis, MO (1980).
Segel L. A. *Modeling Dynamic Phenomena in Molecular and Cellular Biology*, Cambridge University Press, Cambridge (1984).
Vander A. G., Sherman J. H. *Human Physiology*, McGraw-Hill, New York (1985).
Watson J. D. *et al. Molecular Biology of the Gene*, Benjamin-Cummings, Menlo Park, CA (1987).

Anatomy

Hollinshead W. H., Ross C. *Textbook of Anatomy*, Lippincott, Washington (1985).
Woodburne R. T., Ross C. *Essentials of Human Anatomy*, Oxford University Press, New York (1988).

Bibliography

Acerenza L., Sauro H. M., Kacser H. Control analysis of time-dependent metabolic systems, *J. Theoret. Biol.*, **137**, 423–444 (1989).

Allison L., Yee C. N. Minimum message length encoding and the comparison of macro-molecules, *Bull. Math. Biol.*, **52**, 431–453 (1990).

Arbib M. A. Organizational principles for theoretical neurophysiology, in *Towards a Theoretical Biology*, Vol. 4 (Essays), Ed. Waddington C. H., Edinburgh University Press, Edinburgh (1972).

Arino O., Kimmel M. Asymptotic behavior of a nonlinear functional-integral equation of cell kinetics with unequal division, *J. Math. Biol.*, **27**, 341–354 (1989).

Arino O., Mortabit A. A periodicity result for a nonlinear functional integral equation, *J. Math. Biol.*, **30**, 437–456 (1992).

Arnold V. *Equations différentielles ordinaires*, Mir, Moscow (1974).

Atlan H. *L'Organisation biologique et la théorie de l'information*, Hermann, Paris (1972).

Atlan H. Les modèles dynamiques en réseaux et les sources d'information en biologie, in *Structure et dynamique des systèmes*, Maloine, Paris (1976).

Auger P. Dynamics in hierarchically organized systems: A general model applied to ecology, biology and economics, *Syst. Res.*, **3**, 41–50 (1986).

Babloyantz A., Hiernaux J. Models for cell differentiation and generation of polarity in diffusion-governed morphogenetic fields, *Bull. Math. Biol.*, **37**, 637 (1975).

Babloyantz A., Kaczmarek L. K. Self-organization in biological systems with multiple cellular contacts, *Bull. Math. Biol.*, **41**, 193 (1979).

Babloyantz A., Nicolis G. Chemical instabilities and multiple steady-state transitions in Monod–Jacob type models, *J. Theoret. Biol.*, **34**, 185 (1972).

Bak A. L., Bak P., Zeuthen J. Higher levels of organisation in chromosomes, *J. Theoret. Biol.*, **76**, 205–217 (1979).

Bardsley W. G., Waight R. D. Factorability of the Hessian of the binding polynomial. The central issue concerning statistical ratios between binding constants, Hill-plot slope and positive and negative co-operativity, *J. Theoret. Biol.*, **72**, 321–372 (1978).

Barrett J. C. A mathematical model of the mitotic cycle and its application to the interpretation of percentage labelled mitoses data, *J. Natl Cancer Inst.*, **37**, 4, 443 (1966).

Bartholomay A. F. Chemical kinetics and enzyme kinetics, in *Foundations of Mathematical Biology*, Vol. I, Ed. Rosen R., Academic Press, New York (1972).

Batke J. Channelling by loose enzyme complexes *in situ* is likely, though physiological significance is open for speculation, *J. Theoret. Biol.*, **152**, 41–46 (1991).

Bauer W. R. Structure and reactions of closed duplex DNA, *Ann. Rev. Biophys. Bioeng.*, **7**, 287–313 (1978).

Bauer W. R., Crick F., White J. Supercoiled DNA, *Sci. Am.*, **243** (1), 100–113 (1980).

Beloussov L. V., Lakirev A. V. Generative rules for the morphogenesis of epithelial tubes, *J. Theoret. Biol.*, **152**, 455–468 (1991).

Benzer S. The elementary units of heredity, in *The Chemical Basis of Heredity*, Eds McElroy W. D., Glass B., Johns Hopkins Press, Baltimore, MD, pp. 70–93 (1957).

Bernard-Weil E., Duvelleroy M., Droulez J. Analogical study of a model for the regulation of ago-antagonist couples. Applications to adrenal–postpituitary interrelationships, *Math. Biosci.*, **27**: 333–348 (1975).

Berthalanfy L. *Théorie générale des systèmes*, Dunod, Paris (1973).

Bodnar J. W. A domain model for eukaryotic DNA organization: a molecular basis for cell differentiation and chromosome evolution, *J. Theoret. Biol.*, **132**, 479–507 (1988).

Bodnar J. W., Jones G. S., Ellis C. H. Jr. The domain model for eukaryotic DNA organization. 2: a molecular basis for constraints on development and evolution, *J. Theoret. Biol.*, **137**, 281–320 (1989).

Bois F. Y., Compton-Quintana P. J. E. Sensitivity analysis of a new model of carcinogenesis, *J. Theoret. Biol.*, **159**, 361–375 (1992).

Boncinelli E. A model of post-transcriptional control in eukaryotic cell, *J. Theoret. Biol.*, **72**, 75–79 (1978).

Botts J., Morales M. Analytical description of the effects of modifiers and of enzyme multivalency upon the steady state catalyzed reaction rate, *Trans. Faraday Soc.*, **49**, 696–707 (1953).

Bouligand Y. *Morphogenèse: de la biologie aux mathématiques*, Maloine, Paris (1980).

Bradbury E. M., Inglis R. J., Matthews H. R., Sarner N. Phosphorylation of very lysine-rich histone in Physarum polycephalum. Correlation with chromosome condensation, *Eur. J. Biochem.*, **33**, 131–139 (1973).

Bremermann H. J. Unlinked strands as a topological constraint on chromosomal DNA, Plasmid integration and DNA repair, *J. Math. Biol.*, **8**, 393–401 (1979).

Briere C., Goodwin B. Geometry and dynamics of tip morphogenesis in Acetabularia, *J. Theoret. Biol.*, **131**, 461–475 (1988).

Brillouin L. *La science et la théorie de l'information*, Masson, Paris (1959).

Britten R. J., Davidson E. H. Gene regulation for higher cells: a theory, *Science*, **165**, 349–357 (1969).

Britton N. F. A singular dispersion relation arising in a caricature of a model for morphogenesis, *J. Math. Biol.*, **26**, 387–403 (1988).

Brock Fuller F. The writhing number of a space curve, *Proc. Natl Acad. Sci. USA*, **68**, 4, 815–819 (1971).

Bruter C. P. *Topologie et Perception*, Tome I, *Bases mathématiques of philosophiques*, Maloine, Paris (1974).

Bruter C. P. *Les architectures du feu, Considérations sur les modèles*, Flammarion, Paris (1982).

Budriene E. O., Polezhaev A. A., Ptitsyn M. O. Mathematical modelling of intercellular regulation causing the formation of spatial structures in bacterial colonies, *J. Theoret. Biol.*, **135**, 323–341 (1988).

Burger J., Machbub C., Chauvet G. Numerical stability of interactive biological units, *Syst. Anal. Model. Simul.*, **13**, 197–207 (1993).

Cesarini G., Banner, D. W. Regulation of plasmid copy number by complementary RNAs, *Trends Biochem. Sci.*, **10**, 303–306 (1985).

Chambon P. Split genes, *Sci. Am.*, **244** (5), 60–71 (1981).

Chao K.-M., Hardison R. C., Miller W. Constrained sequence alignment, *Bull. Math. Biol.*, **55**, 503–524 (1993).

Chauvet G. Les concepts de complexité, autonomie et auto-organisation en biologie. Application à la spécialisation tissulaire, *Actes du deuxième séminaire de l'École de Biologie théorique*, CNRS, Université de Rouen (1982).

Chauvet G. Hierarchical functional organization of formal biological systems: a dynamical approach I. An increase of complexity by self-association increases the domain of stability of a biological system, *Phil. Trans. R. Soc. Lond. B.*, **339**, 425–444 (1993a).

Chauvet G. Hierarchical functional organization of formal biological systems: a dynamical approach. II. The concept of non-symmetry leads to a criterion of evolution deduced from an optimum principle of the (O-FBS) sub-system, *Phil. Trans. R. Soc. Lond. B.*, **339**, 445–461 (1993b).

Chauvet G. Hierarchical functional organization of formal biological systems: a dynamical approach. III. The concept of non-locality leads to a field theory describing the dynamics at each level of organization of the (D-FBS) sub-system, *Phil. Trans. R. Soc. Lond. B.*, **339**, 463–481 (1993c).

Chauvet G. Non-locality in biological systems results from hierarchy: Application to the nervous system, *J. Math. Biol.*, **31**, 475–486 (1993d).

Chauvet G., Costalat R. On the functional organization in a biological structure: the example of enzyme organization C. R. Acad. Sci. Paris, Série C. **318**, 529–535 (1995).

Chauvet G., Girou D. On the conservation of physiological function in a population of inter-active and self-replicative units, in *Rhythms in Biology and their Fields of Application, Lecture Notes in Biomathematics*, No. 49, Eds Cosnard M., Demongeot J., Le Breton A., Springer-Verlag, Berlin, pp. 101–113 (1983).

Chauvet G., Taravel B., Lorenzelli V. A preliminary method of frequency assignment to molecular vibrations, *J. Mol. Struct.*, **20**, 189–196 (1974).

Chay T. R. Kinetic modeling for the channel gating process from single channel patch clamp data, *J. Theoret. Biol.*, **132**, 449–468 (1988).

Chernavskii D. S., Ruijgrok T. W. Dissipative structures in morphogenetic models of the Turing-type, *J. Theoret. Biol.*, **73**, 585–607 (1978).

Chevalet C., Gillois M., Micali A. Sur un modèle en cinétique enzymatique in *Modèles mathématiques en biologie, Lecture Notes in Biomathematics* No. **41**, Springer-Verlag, Berlin (1981).

Chevalet C., Corpet F., Gillois M., Micali A. Modélisation dynamique des systèmes génétiques de régulation, *École d'Automne de Biologie Théorique*, CNRS, Solignac (1983).

Clarke B. S., Mittenthal J. E., Arcuri P. A. An extremal criterion for epimorphic regeneration, *Bull. Math. Biol.*, **50**, 595–634 (1988).

Cohen M. H., Robertson A. Differentiation for aggregation in the cellular slime molds, in *Cell Differentiation*, Eds Harris R., Allin P., Viza D., Munksgaard, Copenhagen (1972a).

Cohen M. H., Robertson A. Cell migration and the control of development, in *Statistical Mech⁻nics*, Eds Rice S. A., Freed K. F., Light J., University of Chicago Press, Chicago, Ill (1972b).

Collado-Vides J. A transformational-grammar approach to the study of the regulation of gene expression, *J. Theoret. Biol.*, **136**, 403–425 (1989).

Comorosan S., Platica O. On the biotopology of protein syntheses, *Bull. Math. Biophys.*, **29**, 665–676 (1968).

Cooper S. The continuum model and c-myc synthesis during the division cycle, *J. Theoret. Biol.*, **135**, 393–400 (1988).

Cornish-Bowden A. 1989 Metabolic control theory and biochemical systems theory: different objectives, different assumptions, different results, *J. Theoret. Biol.*, **136**, 365–377 (1989).

Cosmi C., Cuomo V., Ragosta M., Macchiato M. F. Characterization of nucleotidic sequences using maximum entropy techniques, *J. Theoret. Biol.*, **147**, 423–432 (1990).

Cowan J. D. *Statistical Mechanics of Neural Nets*, Springer-Verlag, Berlin (1968).

Cramp D. G., Carson E. R. Dynamics of blood glucose and its regulating hormones, in *Biological Systems, Modelling and Control*, Ed. Linkens D. A., Peter Peregrinus, Stevenage, (1979).

Crick F. Linking numbers and nucleosomes, *Proc. Natl Acad. Sci. USA*, **73**, 2639–2643 (1976).

Crick F. H. C. Codon–anticodon pairing: the wobble hypothesis, *J. Mol. Biol.*, **19**, 548–555 (1966).

Cummings F. W. A model of morphogenetic pattern formation, *J. Theoret. Biol.*, **144**, 547–566 (1990).

Curnow R. N. 1988 The use of Markov chain models in studying the evolution of the proteins, *J. Theoret Biol.*, **134**, 51–57 (1988).

Czinak G., Linhart J. Computer modeling of pole formation in cell division, *Math. Biosci.*, **110**, 175–180 (1992).

Dancoff S. M., Quastler H. The information content and error rate of living things, in *Information Theory in Biology*, Ed. Quastler H., University of Illinois Press, Urbana (1953).

Darvey G., Staff P. J. The application of the theory of Markov processes to the reversible one substrate — one intermediate — one product enzyme mechanism, *J. Theoret. Biol.*, **14**, 157–172 (1967).

Davison E. J. Simulation of cell behavior: normal and abnormal growth, *Bull. Math. Biol.*, **37**, 427–458 (1975).

Day W. H. E., McMorris F. R. Cosensus sequences based on plurality rule, *Bull. Math. Biol.*, **54**, 1057–1068 (1992).

Dayhoff M. O. *Atlas of Protein Sequence and Structure*, Natl. Biomed. Res. Found, Washington, DC (1976).

Debru C. *L'Esprit des protéines: histoire et philosophie biochimiques*, Hermann, Paris (1983).

Delattre P. Topological and order properties in transformation systems, *J. Theoret. Biol.*, **32**, 269–282 (1971).

Delattre P. *L'Evolution des systèmes moléculaires*, Maloine, Paris (1971a).

Delattre P. *Système, structure, fonction, évolution*, Maloine, Paris (1971b).

Delattre P. Direct and inverse regulation control in transformation systems, *Math. Biosci.*, **34**, 303–323 (1977).

Delattre P. Une définition formalisée de l'autocatalyse, *C. R. Acad. Sci. Paris, Série C* **290**, 101–104 (1980).

Delattre P. *La théorie des systèmes de transformations et ses applications*, Cours de Biologie théorique, Faculté de Médecine, Angers (1981).

De Lisi C. *Antigen–Antibody Interactions*, Lecture Notes in Biomathematics, Vol. 8, Springer-Verlag, Berlin (1975).

Demetrius L. La valeur adaptive et la sélection naturelle, *C. R. Acad. Sci. Paris, Série D* **290**, 1491–1494 (1980).

Demetrius L. Statistical mechanics and population biology, *J. Stat. Phys.*, **30**, 709–753 (1993).

Demetrius L. Self-organization in macromolecular systems: The notion of adaptive value, *Proc. Natl. Acad. Sci. USA*, **81**, 6068–6072 (1984).

Demetrius L., Ziehe M. The measurement of Darwinian fitness in human populations, *Proc. R. Soc. Lond. B*, **222**, 33–50 (1984).

Di Giulio M. On the origin of the transfer RNA molecule, *J. Theoret. Biol.*, **159**, 199–214 (1992).

Dobzhansky T., Boesiger E. *Essais sur l'évolution*, Masson, Paris 1968.

Donachie W. D., Masters M. Temporal control of gene expression in bacteria, in *The Cell Cycle: Gene–Enzyme Interactions*, Eds Padilla G. M., Whitson G. L., Cameron I. L., Academic Press, New York (1969).

Doubabi S., Morillon J. P., Costalat R. Effect of coupling on the oscillations of a biochemical pathway, in *Proceedings of the 1st MATHMOD Symposium*, Eds. Troch I., Breitenecker F., IMACS, February 2–4, 1994, Vienna, Austria (1994).

Dowse H. B., Ringo J. M. The search for hidden periodicities in biological time series revisited, *J. Theoret. Biol.*, **139**, 487–515 (1989).

Dumas J. P., Ninio J. Efficient algorithms for folding and comparing nucleic acid sequences, *Nucleic Acids Res.*, **10** (1), 197–206 (1982).

Dvorak I., Siska J. Analysis of metabolic systems with complex slow and fast dynamics, *Bull. Math. Biol.*, **51**, 255–274 (1989).

Eddington A. *The Nature of the Physical World*, Dent & Sons, London (1958).

Edelstein B. B. The dynamics of cellular differentiation and associated pattern formation, *J. Theoret. Biol.*, **37**, 221–243 (1972).

Eigen M. Self-organization of matter and the evolution of biological macromolecules, *Naturwissenschaften*, **58**, 465–523 (1971).

Eigen M., Schuster P. *The Hypercycle, a Principle of Natural Self-organization*, Springer-Verlag, Berlin (1979).

Eigen M., Gardiner W., Schuster P., Winkler-Oswatitsch R. The origin of genetic information, *Sci. Am.*, **244** (4), 88–118 (1981).

Ekeland I. La théorie des catastrophes, *La Recherche*, Paris, **81** (8), 745–754 (1977).

Ferdinand A. E. A theory of system complexity, *Int. J. Gen. Syst.,* **1**, 19–33 (1974).

Fernandez A. Stochastic dynamical constraints in de novo RNA replication, *J. Theoret. Biol.,* **134**, 419–430 (1988).

Frenzen C. L., Maini P. K. Enzyme kinetics for a two-step enzymic reaction with comparable initial enzyme–substrate ratios, *J. Math. Biol.,* **26**, 689–703 (1988).

Friedrich P. Protein structure: the primary substrate for memory, *Neuroscience,* **35**, 1–7 (1990).

Friedrich P. Metabolic compartmentation via 'quenching enzymes', *J. Theoret. Biol.,* **152**, 115–116 (1991).

Fukuchi S., Otsuka J. Evolution of metabolic pathways by chance assembly of enzyme proteins generated from sense and antisense strands of pre-existing genes, *J. Theoret. Biol.,* **158**, 271–291 (1992).

Fuxman Y. L. Reproduction rate, feeding process, and Leibich limitations in cell populations: Part 1. Feeding stochasticity and reproduction rate, *Bull. Math. Biol.,* **55**, 175–195 (1993).

Galar R. Evolutionary search with soft selection, *Biol. Cyber.,* **60**, 357–364 (1989).

Gatlin L. C. The information content of DNA, *J. Theoret. Biol.,* **18** (2), 181–194 (1968).

Gatlin L. C. The entropy maximum of proteins, *Math. Biosci.,* **13**, 213–227 (1972).

Georgiev G. P. On the structural organization of operon and the regulation of RNA synthesis in animal cells, *J. Theoret. Biol.,* **25**, 473–490 (1969).

Ghosh A. K., Chance B., Pye E. K. Metabolic coupling and synchronization of NADH oscillations in yeast cell populations, *Arch. Biochem. Biophys.,* **145**, 319–331 (1971).

Gibbs J. W. *Elementary Principles of Statistical Mechanics,* Yale University Press, New Haven (1902).

Giebisch G., Tosteson D. C., Ussing H. H., Tosteson M. T. *Membrane Transport in Biology,* Springer-Verlag, Berlin (1978).

Gierer A., Meinhardt H. A theory of biological pattern formation, *Kybernetik,* **12**, 30–39 (1972).

Giersch C. 1988 Control analysis of biochemical pathways: a novel procedure for calculating control coefficients, and an additional theorem for branched pathways, *J. Theoret. Biol.,* **134**, 451–462 (1988).

Glansdorff P., Prigogine I. *Stabilité, structure et fluctuations,* Masson, Paris (1971).

Goad W. B., Kaneshisa M. I. Pattern recognition in nucleic acid sequences I. A general method for finding local homologies and symmetries, *Nucleic Acids Res.,* **10** (1), 247–263 (1982).

Goldbeter A. Kinetic negative co-operativity in the allosteric model of Monod, Wyman and Changeux, *J. Mol. Biol.,* **90**, 185–190 (1974).

Goodwin B. C. *Analytical Cell Physiology,* Academic Press, New York (1978).

Goodwin B. C., Cohen M. S. A phase shift model for the spatial and temporal organization of developing systems, *J. Theoret. Biol.,* **25**, 49 (1969).

Goodwin B. C., Kauffman S. A. Spatial harmonics and pattern specification in early Drosophila development. I. Bifurcation sequences and gene expression, *J. Theoret. Biol.,* **144**, 303–319 (1990).

Gotoh O. Consistency of optimal sequence alignments, *Bull. Math. Biol.,* **52**, 509–525 (1990).

Grasse P. P. *L'évolution du vivant,* Albin Michel, Paris (1973).

Greenspan H. P. On fluid-mechanical simulations of cell division and movement, *J. Theoret. Biol.,* **70**, 125–134 (1978).

Gutfreund H., Chock P. B. Substrate channeling among glycolytic enzymes: Fact or fiction, *J. Theoret. Biol.,* **152**, 117–121 (1991).

Gyllenberg M., Webb G. F. A nonlinear structured population model of tumor growth with quiescence, *J. Math. Biol.,* **28**, 671–694 (1990).

Haken H. *Synergetics. An Introduction,* Springer-Verlag, Berlin (1978).

Hanusse P. De l'existence d'un cycle limite dans l'évolution des systémes chimiques ouverts, *C. R. Acad. Sci., Paris, Série C,* **274**, 1245–1247 (1972).

Hariri A., Weber B., Olmsted J. III. On the validity of Shannon-information calculations for molecular biological sequences, *J. Theoret. Biol.,* **147**, 235–254 (1990).

Harris H. *Cell Fusion*, Oxford University Press, Oxford (1970).

Harrison L. G., Kolar M. Coupling between reaction–diffusion prepattern and expressed morphogenesis, applied to desmids and dasyclads, *J. Theoret. Biol.*, **130**, 493–515 (1988).

Hartwell L. H., Culotti J., Pringle J. R., Reid B. J. Genetic control of the cell division cycle in yeast, *Science*, **183**, 46–51 (1974).

Havsteen B. H., Garcia-Moreno M., Valero E., Manjabacas M. C., Varon R. The kinetics of enzyme systems involving activation of zymogens, *Bull. Math. Biol.*, **55**, 561–583 (1993).

Heinmets F. *Analysis of Normal and Abnormal Cell Growth*, Plenum Press, New York (1966).

Hejblum G., Costagliola D., Valleron A.-J., Mary J.-Y. Cell cycle models and mother–daughter correlation, *J. Theoret. Biol.*, **131**, 255–262 (1988).

Henri V. Théorie de l'action des diastases, *C. R. Soc. Biol. Paris*, **58**, 610–613 (1905).

Herrick J., Bensimon D. Gene regulation under growth conditions. A model for the regulation of initiation of replication in *Escherichia coli*, *J. Theoret. Biol.*, **151**, 359–365 (1991).

Herschkovitz-Kaufman M. Bifurcation analysis of non linear reaction–diffusion equations, *Bull. Math. Biol.*, **37**, 589 (1975).

Hickey D. A., Benkel B. F., Abukashawa S. M. A general model for the evolution of nuclear pre-mRNA introns, *J. Theoret. Biol.*, **137**, 41–53 (1989).

Higgins J. Dynamics and control in cellular reactions, in *Control of Energy Metabolism*, Eds Chance B., Estabrook, Williamson J. R., Academic Press, New York (1965).

Hinshelwood C. N. Quasi-unimolecular reactions. The decomposition of di-ethyl ether in the gaseous state, *Proc. R. Soc. Lond. A* **114**, 84–97 (1927).

Hopfield J. J. Neural networks and physical systems with emergent collective computational abilities, *Proc. Natl. Acad. Sci. USA*, **79**, 2554–2558 (1982).

Houillon C. *Embryologie*, Hermann, Paris (1969).

Hyver C. Impossibility of existence of undamped oscillations in linear chemical systems, *J. Theoret. Biol.*, **36**, 133–138 (1972).

Hyver C. Valeurs propres des systèmes de transformation représentatives par des graphes en arbre, *J. Theoret. Biol.*, **42**, 397–409 (1973).

Igo-Kemenes T., Zachau H. G. Domains in chromatin structure, *Cold Spring Harbor Symp. Quant. Biol.*, **XIII**, 109–118 (1978).

Iwasa Y. Free fitness that always increases in evolution, *J. Theoret. Biol.*, **135**, 265–281 (1988).

Jacob F. *La logique du vivant*, Gallimard, Paris (1970).

Jacob F., Brenner S. Sur la régulation de la synthèse du DNA chez les bactéries: l'hypothèse du replicon., *C. R. Acad. Sci., Paris, Série D*, **256**, 298–300 (1963).

Jacob F., Monod J. Genetic regulatory mechanisms in the synthesis of proteins, *J. Mol. Biol.*, **3**, 318–356 (1961).

Jacquez J. A. *Compartmental Analysis in Biology and Medicine*, The University of Michigan Press, Ann Arbor (1985).

Jejedor A. G., Sanz-Nuno J. C., Olarrea J., de la Rubia F. J., Montero F. Influence of the hypercycle on the error threshold: a stochastic approach, *J. Theoret. Biol.*, **134**, 431–443 (1988).

Job D., Soulié J. M., Job C., Shire D. Potential memory and hysteretic effects in transcription, *J. Theoret. Biol.*, **134**, 273–289 (1988).

Jones B. L., Enns R. H., Rangnekar S. S. On the theory of selection of coupled macromolecular systems, *Bull. Math. Biol.*, **38**, 15–28 (1976).

Kaden F., Koch I., Selbig J. Knowledge-based prediction of protein structures, *J. Theoret. Biol.*, **147**, 85–100 (1990).

Kanavarioti A. Self-replication of chemical systems based on recognition within a double or a triple helix: a realistic hypothesis, *J. Theoret. Biol.*, **158**, 207–219 (1992).

Kanehisa M. I., Goad W. B. Pattern recognition in nucleic acid sequences. II. An efficient method for finding locally stable secondary structures, *Nucleic Acids Res.*, **10**, 265–278 (1982).

Kaufman S. Dynamic models of the mitotic cycle: evidence for a limit cycle oscillator, in *Mathematical Models in Biological Discovery, Lecture Notes in Biomathematics*, Vol. 13, Eds Solomon D. L., Walter C. F., Springer-Verlag, Berlin (1977).

Kaufman S. A. Boolean systems, adaptive automata, evolution, in *Disordered Systems and Biological Organization*, Eds Bienenstock E., Fogelman-Soulié F., Weisbuch G., F20 NATO Series (1985).

Kaufman S. A., Goodwin B. C. Spatial harmonics and pattern specification in early Drosophila development. II, The four colour wheels model, *J. Theoret. Biol.*, **144**, 321–345 (1990).

Keasling J. D., Palsson B. O. Letter to the editor: On the kinetics of plasmid replication, *J. Theoret. Biol.*, **136**, 487–492 (1989).

Keller E., Segel L. Initiation of slime mold aggregation viewed as an instability, *J. Theoret. Biol.*, **26** (3), 399–415 (1970).

Kempner E. S. *Cell Compartmentation and Metabolic Channelling*, Eds Nover L., Lynen F., Mothes K., Elsevier/North Holland, Amsterdam (1980).

Kerner E. H. A statistical mechanics of interacting biological species, *Bull. Math. Biophys.*, **19**, 121–146 (1957).

Kerner E. H. *Gibbs Ensemble: Biological Ensemble*, Gordon and Breach, New York (1972).

Kimmel M., Arino O. Cell cycle kinetics with supramitotic control, two cell types, and unequal division: A model of transformed embryonic cells, *Math. Biosci.*, **105**, 47–79 (1991).

Ko M. S. H. A stochastic model for gene induction, *J. Theoret. Biol.*, **153**, 181–194 (1991).

Kornberg R. Chromatin structure: a repeating unit of histones and DNA, *Science*, **184**, 868–871 (1974).

Kornberg R., Klug A. The nucleosome, *Sci. Am.*, **244** (2), 52–64 (1981).

Koshland D. E. Application of a theory of enzyme specificity to protein synthesis, *Proc. Natl Acad. Sci. USA*, **44**, 98–104 (1958).

Koshland D. E., Nemethy G., Filmer D. Comparison of experimental binding data and theoretical models in proteins, *Biochemistry*, **5** (1), 365–385 (1966).

Koziol J. A. On the prevalence of transcriptional regions in human genomic DNA, *J. Theoret. Biol.*, **149**, 377–380 (1991).

Krasnow R. A. Mass, length and growth rate in single cells, *J. Theoret. Biol.*, **72**, 659–699 (1978).

Lacalli T. C. Modeling the Drosphila pair-rule pattern by reaction–diffusion: gap input and pattern control in a 4-morphogen system, *J. Theoret. Biol.*, **144**, 171–194 (1990).

Lahav N. Prebiotic co-evolution of self-replication and translation or RNA world?, *J. Theoret. Biol.*, **151**, 531–539 (1991).

Laiken N., Nemethy G.-J. A statistical–thermodynamic model of aqueous solutions of alcohols, *J. Phys. Chem.*, **74**, 3501–3509 (1970).

Landauer R. Inadequacy of entropy and entropy derivatives in characterising the steady-state, *Phys. Rev.*, **12**, 636–638 (1975).

La Salle J., Lefchetz S. *Stability by Lyapunov's Direct Method with Applications*, Academic Press, New York (1961).

Lasota A., MacKey M. C., Tyrcha J. The statistical dynamics of recurrent biological events, *J. Math. Biol.*, **30**, 775–800 (1992).

Latt S. A., Davidson R. L., Lin M. S., Gerald P. S. Lateral asymmetry in the fluorescence of human Y chromosomes stained with 33258 Hoechst, *Exp. Cell. Res.*, **87**, 425–429 (1974).

Lehman N., Jukes T. H. Genetic code development by stop codon takeover, *J. Theoret. Biol.*, **135**, 203–214 (1988).

Lewontin R. C. Population genetics, *Ann. Rev. Genet.*, **1**, 37–70 (1967).

Li Y.-X., Goldbeter A. Oscillatory isozymes as the simplest model for coupled biochemical oscillators, *J. Theoret. Biol.*, **138**, 149–174 (1989).

Lindemann F. A. Discussion on: "The radiation theory of chemical action", *Trans. Faraday Soc.*, **17**, 598–606 (1922).

Lints F. *Génétique*, Office International de Librairie, Bruxelles, et Technique et Documentation, Paris (1981).

Liu C. Q., Wang Y., Huang J. F., Zhang H. A quantum biological approach to the relations of DNA methylation with gene transcription and mutation, *J. Theoret. Biol.*, **148**, 145–155 (1991).

Lloyd D., Lloyd A. L., Olsen L. F. The cell division cycle: a physiologically plausible dynamic model can exhibit choatic solutions, *Biosystems*, **27**, 17–24 (1992).

Loewy A. G., Siekevitz P. *Cell Structure and Function*, Reinhart and Winston, New York (1974).

López-Quintela M. A., Casado J. Revision of the methodology in enzyme kinetics: a fractal approach, *J. Theoret. Biol.*, **139**, 129–139 (1989).

Lumry R., Biltonen R. Thermodynamic and kinetic aspects of protein conformations in relation to physiological function, in *Structure and Stability of Biological Macromolecules*, Eds Timasheff S. N., Fasman G. D., Marcel Dekker, New York (1969).

Luo L., Li H. The statistical correlation of nucleotides in protein-coding DNA sequences, *Bull. Math. Biol.*, **53**, 345–353 (1991).

Luo L., Trainor L. E. H. A stochastic evolutionary model of molecular sequences, *J. Theoret. Biol.*, **157**, 83–94 (1992).

Luo L. F., Tsai L., Zhou M. Y. Informational parameters of nucleic acid and molecular evolution, *J. Theoret. Biol.*, **130**, 351–361 (1988).

MacDonald C. T., Gibbs J. H., Pipkin A. C. Kinetics of biopolymerization on nucleic acid templates, *Biopolymers*, **6**, 1–25 (1968).

Machbub C., Burger J., Chauvet, G. Numerical study of interactive of biological units, in *Computational Systems Analysis*, Ed. Sydow A. Elsevier Science Publishers, New York (1992).

McMahon D. A cell-contact model for cellular position determination in development, *Proc. Nat. Acad. Sci. USA*, **70** (8), 2396–2400 (1973).

MacWilliams H. K., Papageorgiou S. A model of gradient interpretation based on morphogen binding, *J. Theoret. Biol.*, **72**, 385–411 (1978).

McElwain D. L. S., Pettet G. J. Cell migration in multicell spheroids: swimming against the tide, *Bull. Math. Biol.*, **55**, 655–674 (1993).

McLean N. *The Differentiation of Cells*, University Park Press, Baltimore, Md (1975).

Mahaffy J. M., Zyskind J. W. A model for the initiation of replication in *Escherichia coli*, *J. Theoret. Biol.*, **140**, 453–477 (1989).

Maini P. K. Spatial and spatio-temporal patterns in a cell-haptotaxis model, *J. Math. Biol.*, **27**, 507–522 (1989).

Mandelbrot B. B. *The Fractal Geometry of Nature*, W. H. Freeman, New York (1983).

Mani G. S. Correlations between the coding and non-coding regions in DNA, *J. Theoret. Biol.*, **158**, 429–445 (1992a).

Mani G. S. Long-range doublet correlations in DNA and the coding regions, *J. Theoret. Biol.*, **158**, 447–464 (1992b).

Mano Y. Cytoplasmic regulation and cyclic variation in protein synthesis in the early cleavage stage of the sea urchin embryo, *Develop. Biol.*, **22**, 433–460 (1970).

Marcelpoil R., Usson Y. Methods for the study of cellular sociology: Voronoi diagrams and parametrization of the spatial relationships, *J. Theoret. Biol.*, **154**, 359–369 (1992).

Marmillot P., Hervagault J.-F., Welch G. R. Patterns of spatiotemporal organization in an 'ambiquitous' enzyme model, *Proc. Natl Acad. Sci. USA*, **89**, 12103–12107 (1992).

Martinez H. M. Morphogenesis and chemical dissipative structures, *J. Theoret. Biol.*, **36**, 479–501 (1972).

Mayr E. *Populations, espèces et évolution*, Hermann, Paris (1974).

Meinhardt H. Space-dependent cell determination under the control of a morphogen gradient, *J. Theoret. Biol.*, **74**, 307–321 (1978).

Michaelis L., Menten M. L. Die Kinetic der Invertinwerkung, *Biochem Zeitschrift*, **49**, 333–369 (1913).

Mikulecky D. C. *Applications of Network Thermodynamics to Problems in Biomedical Engineering*, New York University Press, New York (1993).

Mikulecky D. C., Wiegand W. A., Shiner J. S. A simple network thermodynamics method for modeling series–parallel coupled flows, *J. Theoret. Biol.*, **69**, 471–510 (1977).

Miller S. C. Production of some organic compounds under possible primitive earth conditions, *J. Am. Chem. Soc.*, **77**, 2351–2361 (1969).

Monod J. *Le hasard et la nécessité*, Seuil, Paris (1970).

Monod J., Jacob F. General conclusions: teleonomic mechanisms in cellular metabolism, growth and differentiation, *Cold Spring Harbor, Symp. Quant. Biol.*, **26**, 389–401 (1961).

Monod J., Wyman J., Changeux J. P. On the nature of allosteric effect, a plausible model, *J. Mol. Biol.*, **12**, 88–118 (1965).

Morgan A. R., Severini A. Interconversion of replication and recombination structures: implications for terminal repeats and concatemers, *J. Theoret. Biol.*, **144**, 195–202 (1990).

Morillon J.-P., Costalat R., Burger N., Burger J. Modelling two associated biochemical pathways, in *Proceedings of the 1st MATHMOD Symposium*, IMACS, Eds. Trach I., Breitenecker F., February 2–4, 1994, Vienna, Austria (1994).

Murray J. D. A pattern formation mechanism and its application to mammalian coat markings. In: *Vito Volterra Symposium on Mathematical Models in Biology, Lecture Notes in Biomathematics, Vol. 39*, Ed. Barogozzi C., Springer-Verlag, Berlin (1980).

Murray J. D. On pattern formation mechanisms for lepidopteran wing patterns and mammalian coat markings, *Phil. Trans. R. Soc. Lond.*, **B 295**, 473–496 (1981).

Murray J. D. *Mathematical Biology*, Springer-Verlag, Heidelberg (1989).

Murray J. D., Oster G. F. Generation of biological pattern and form, *IMA J. Math. Appl. Med. Biol.*, **1**, 51–75 (1984a).

Murray J. D., Oster G. F. Cell traction models for generating pattern and form in morphogenesis, *J. Math. Biol.*, **19**, 265–279 (1984b).

Murray J. D., Oster G. F., Harris A. K. A mechanical model for mesenchymal morphogenesis, *J. Math. Biol.*, **17**, 125–129 (1983).

Myers E. W., Miller W. Approximate matching of regular expressions, *Bull. Math. Biol.*, **51**, 5–37 (1989).

Nagorcka B. N. Wavelike isomorphic prepatterns in development, *J. Theoret. Biol.*, **137**, 127–162 (1989).

Needleman S. B., Wunsch C. D. A general method applicable to the search for similarities in the amino acid sequence of two proteins, *J. Mol. Biol.*, **48**, 443–453 (1970).

Nemethy G. Molecular interactions and allosteric effects, in *Sub-units in Biological Systems*, Part C, Eds Timasheff S. N., Fasman G. D., Marcel Dekker, New York (1975).

Newman S. A., Frisch H. L., Percus J. K. On the stationary state analysis of reaction–diffusion mechanisms for biological pattern formation, *J. Theoret. Biol.*, **134**, 183–197 (1988).

Nicolis G., Prigogine I. *Self-organisation in Non-equilibrium Systems*, John Wiley and Sons, New York (1977).

Ninio J. *Approches moléculaires de l'évolution*, Masson, Paris (1979).

Odell G. M., Oster G. F., Burnside B., Alberch P. The mechanical basis for morphogenesis, *Dev. Biol.*, **85**, 446–462 (1981).

Orengo C. A., Taylor W. R. A rapid method of protein structure aligment, *J. Theoret. Biol.*, **147**, 517–551 (1990).

Ortoleva P., Ross J. A chemical instability mechanism for asymmetric cell differentiation, *Biophys. Chem.*, **1**, 87–96 (1973).

Oster G. F. Lateral inhibition models of developmental processes, *Math. Biosciences*, **90**, 265–286 (1988).

Oster G. F., Murray J. D., Harris A. K. Mechanical aspects of mesenchymal morphogenesis, *J. Embryol. Exp. Morphol.*, **78**, 83–125 (1983).

Oster G. F., Perelson A., Katchalsky A. Network thermodynamics: dynamic modelling of biological systems, *Quart. Rev. Biophys.*, **6** (1), 1–134 (1973).

Oster G. F., Shubin N., Murray J. D., Alberch P. Evolution and morphogenetic rules. The shape of the vertebrate limb in ontogeny and phylogeny, *Evolution* **45**, 862–884 (1988).

Othmer H. *Interactions of Reaction and Diffusion in Open Systems*, University of Minnesota Press (1969).

Ovádi J. Physiological significance of metabolic channelling, *J. Theoret. Biol.*, **152**, 1–22 (1991).

Palsson B. O., Lightfoot E. N. Mathematical modeling of dynamics and control in metabolic networks. IV. Local stability analysis of single biochemical control loops, *J. Theoret. Biol.*, **113**, 261–277 (1985).

Pattee H. H. The problem of biological hierarchy, in *Towards a Theoretical Biology*, Vol. 3 (Drafts), Ed. Waddington C. H., Edinburgh University Press, Edinburgh (1970).

Perutz M. F. Submicroscopic structure of the red cell, *Nature*, **161**, 204–205 (1948).

Pettersson G. No convincing evidence is available for metabolite channelling between enzymes forming dynamic complexes, *J. Theoret. Biol.*, **152**, 65–69 (1991).

Pipkin A. C., Gibbs J. H. Kinetics of synthesis and/or conformational changes of biological macromolecules, *Biopolymers*, **4**, 3–15 (1966).

Pohl W. F. The self-linking numbers of a closed space curve, *J. Math. Mech.*, **17**, 975–985 (1968).

Pohl W. F., Roberts G. W. Topological considerations in the theory of replication of DNA, *J. Math. Biol.* **6**, 383–402 (1978).

Polozov R. V., Yakushevich L. V. Nonlinear waves in DNA and regulation of transcription, *J. Theoret. Biol.*, **130**, 423–430 (1988).

Prigogine I. L'ordre par fluctuations et le systéme social, in *L'idée de régulation dans les sciences*, Eds Lichnerowicz A., Perroux F., Gadoffre G., Maloine, Paris (1977).

Prigogine I. *Physique, temps et devenir*, Masson, Paris (1980).

Prigogine I., Lefever R. Symmetry breaking instabilities in dissipative systems II, *J. Chem. Phys.*, **48**, 1695–1700 (1968).

Prigogine I., Mayné F., George C., De Hann M. Microscopic theory of irreversible processes, *Proc. Natl Acad. Sci. USA*, **74**, 4152–4156 (1977).

Prigogine I., Nicolis G. Biological order, structure and instabilities, *Q. Rev. Biophy*, **4**, 107–148 (1971).

Prigogine I., Nicolis G., Babloyantz A. Thermodynamics of evolution. The functional order maintained within living systems seems to defy the Second Law; nonequilibrium thermo-dynamics describes how such systems come to terms with entropy, *Phys. Today*, November, 23–28 (1972).

Prigogine I., Nicolis G., Babloyantz A. Thermodynamics of evolution. The ideas of nonequilibrium order and of the search for stability extend Darwin's concept back to the prebiotic stage by redefining the 'fittest', *Physics Today* December, 38–44 (1972).

Quastler H. The Analysis of cell population kinetics, in *Cell Proliferation*, Eds Lamerton L. F., Fry R. J. M. Blackwell, Oxford (1963).

Rao B. R., Talwalker S. Bounds on life expectancy for the Rayleigh and Weibull distributions, *Math. Biosci.*, **96**, 95–115 (1989).

Rapp P. Analysis of biochemical phase shift oscillators by a harmonic balancing technique, *J. Math. Biol.*, **3**, 203–224 (1976).

Rapp P. E., Berridge M. J. Oscillations in calcium AMP control loops form the basis of pacemaker activity and other high frequency biological rhythms, *J. Theoret. Biol.*, **66**, 497–525 (1977).

Rashevsky N. *Mathematical Principles in Biology and their Applications*, Charles C. Thomas, Springfield, Illinois (1961).

Reder C. Metabolic control theory: a structural approach, *J. Theoret Biol.*, **135**, 175–201 (1988).

Reiner J. M. Molecular biology and the kinetics of cell growth. II. A note on mass–volume relations, *Bull. Math. Biol.*, **35**, 109–113 (1973).

Revel M., Groner Y., Pollack Y., Schepos R., Berissi H. Protein synthesis machinery and the regulation of messenger RNA translation, in *Polymerisation in Biological Systems*, Ciba Foundation Symposium 7, Eds Wolstenholme G. E. W., O'Connor M., Associated Scientific Publishers, Amsterdam (1972).

Ricard J., Kellershohn N., Mulliert G. Dynamic aspects of long distance functional interactions between membrane-bound enzymes, *J. Theoret. Biol.*, **156**, 1–40 (1992).

Rinsma I., Hendy M., Penny D. Distribution of the number of matches between nucleotide sequences, *Bull. Math. Biol.*, **52**, 349–358 (1990).

Rodley G. A., Scobie R. S., Bates R. H. T., Lewitt R. M. A possible conformation for double stranded polynucleotides, *Proc. Natl Acad. Sci. USA*, **73**, 2959–2963 (1976).

Rosen R. A relational theory of biological systems, *Bull. Math. Biophys*, **20**, 245–260 (1958).

Rosen R. The representation of biological systems from the standpoint of the theory of categories, *Bull. Math. Biophys.*, **20**, 317–341 (1958a).

Rubinow S. I. A maturity time representation for cell populations, *Biophys. J.*, **8**, 1055–1073 (1968).

Rumer Y. B. *Proc. Acad. Sci. USSR*, **167**, 1393 (1966).

Sachs T. Epigenetic selection: an alternative mechanism of pattern formation, *J. Theoret. Biol.*, **134**, 547–559 (1988).

Salerno C. Misleading interpretations of coupled enzyme kinetics, *J. Theoret. Biol.*, **152**, 73–75 (1991).

Sankoff D. Matching sequences under deletion/insertion constraints, *Proc. Natl Acad. Sci. USA*, **69**, 4–6 (1972).

Sarai A. Molecular recognition and information gain, *J. Theoret. Biol.*, **140**, 137–143 (1989).

Sattinger D. *Topics in Stability and Bifurcation Theory*, Springer-Verlag, Berlin (1973).

Saunders P. T. *An Introduction to Catastrophe Theory*, Cambridge University Press, Cambridge (1980).

Savageau M. A. Growth of complex systems can be related to the properties of their underlying determinants, *Proc. Natl Acad. Sci. USA*, **76**, 5413–5417 (1979).

Savageau M. A. Biochemical systems theory: Operational differences among variant representations and their significance, *J. Theoret. Biol.*, **151**, 509–530 (1991a).

Savageau M. A. Metabolite channeling: Implications for regulation of metabolism and for quantitative description of reactions *in vivo*, *J. Theoret. Biol.*, **152**, 85–92 (1991b).

Savageau M. A., Sands P. J. Completely uncoupled or perfectly coupled circuits for inducible gene regulation, in *Canonical Non-linear Modeling*, Ed. Voit E. O., Van Nostrand/Reinhold, New York (1991).

Schaechter M., Maaloe O., Kjeldgaard N. O. Dependency on medium and temperature of cell size and chemical composition during balanced growth of *Salmonella typhimurium*, *J. Gen. Microbiol.*, **19**, 592–606 (1958).

Schleiden M. Beiträge zur Phytogenesis, in *Müllers Archiv für Anatomie und Physiologie*, (1838).

Schneider T. D. Theory of molecular machines. II. Energy dissipation from molecular machines, *J. Theoret. Biol.*, **148**, 125–137 (1991a).

Schneider T. D. Theory of molecular machines. I. Channel capacity of molecular machines, *J. Theoret. Biol.*, **148**, 83–123 (1991b).

Schrodinger E. *What is Life?* Cambridge University Press, London (1945).

Schuster S., Heinrich R. Minimization of intermediate concentrations as a suggested optimality principle for biochemical networks. I, Theoretical analysis, *J. Math. Biol.*, **29**, 425–442 (1991).

Schwann T. *Die Übereinstimmung in der Struktur und dem Wachstum der Tiere und der Pflanzen*, Sandersche Buchhandlung (Reiner), Berlin (1839).

Segel I. H. *Enzyme Kinetics*, John Wiley & Sons, New York (1975).

Segel I. H., Martin R. L. The general modifier ('allosteric') unireactant enzyme mechanism: redundant conditions for reduction of the steady state velocity equation to one that is first degree in substrate and effector, *J. Theoret. Biol.*, **135**, 445–453 (1988).

Segel L. A. *Mathematical Models in Molecular and Cellular Biology*, Cambridge University Press, Cambridge (1980).

Segel L. A. On the validity of the steady state assumption of enzyme kinetics, *Bull. Math. Biol.*, **50**, 6, 579–593 (1988).

Segel L. A., Perelson A. S. Plasmid copy number control: a case study of quasi-steady-state assumption, *J. Theoret. Biol.*, **158**, 481–494 (1992).

Sellers P. H. The theory and computation of evolutionary distances: pattern recognition, *J. Algorithms*, **1**, 359–373 (1980).

Sen A. K. On the time course of the reversible Michaelis–Menten reaction, *J. Theoret Biol.*, **135**, 483–493 (1988).

Shabalina S. A., Yurieva O. V., Kondrashov A. S. On the frequencies of nucleotides and nucleotide substitutions in conservative regulatory DNA sequences, *J. Theoret. Biol.*, **149**, 43–54 (1991).

Shannon C. E., Weaver W. *The Mathematical Theory of Communication*, University of Illinois Press, Urbana, Ill (1972).

Shcherbak V. I. Rumer's rule and transformation in the context of the co-operative symmetry of the genetic code, *J. Theoret. Biol.*, **139**, 271–276 (1989a).

Shcherbak V. I. Ways of wobble pairing are formalized with the co-operative symmetry of the genetic code, *J. Theoret. Biol.*, **139**, 277–281 (1989b).

Shcherbak V. I. The 'START' and 'STOP' of the genetic code: why exactly ATG and TAG, TAA?, *J. Theoret. Biol.*, **139**, 283–286 (1989c).

Sibbald P. R., Banerjee S., Maze J. Calculating higher order DNA sequence information measures, *J. Theoret. Biol.*, **136**, 475–483 (1989).

Siemion I. Z., Stefanowicz P. Periodical changes of amino acid reactivity within the genetic code, *Biosystems* **27**, 77–84 (1992).

Small J. R., Fell D. A. The matrix method of metabolic control analysis: its validity for complex pathway structures, *J. Theoret. Biol.*, **136**, 181–197 (1989).

Smillie F., Bains W. Repetition structure of mammalian nuclear DNA, *J. Theoret. Biol.*, **142**, 463–471 (1990).

Smolle J., Stettner H. Computer simulation of tumour cell invasion by a stochastic growth model, *J. Theoret. Biol.*, **160**, 63–72 (1993).

Sorribas A., Savageau M. A. Strategies for representing metabolic pathways within biochemical systems theory: reversible pathways, *Math. Biosciences*, **94**, 239–269 (1989a).

Sorribas A., Savageau M. A. A comparison of variant theories of intact biochemical systems. I. Enzyme–enzyme interactions and biochemical systems theory, *Math. Biosciences*, **94**, 161–193 (1989b).

Sorribas A., Savageau M. A. A comparison of variant theories of intact biochemical systems. II. Flux-oriented and metabolic control theories, *Math. Biosciences*, **94**, 195–238 (1989c).

Spivey H. O. Evidence of NADH channeling between dehydrogenases, *J. Theoret. Biol.*, **152**, 103–107 (1991).

Srere P. A. Channeling: The pathway that cannot be beaten, *J. Theoret. Biol.*, **152**, 23 (1991).

Stadler P. F. Dynamics of autocatalytic reaction networks IV: Inhomogeneous replicator networks, *Biosystems* **26**, 1–19 (1991).

Stadler P. F., Schuster P. Mutation in autocatalytic reaction networks. An analysis based on perturbation theory, *J. Math. Biol.*, **30**, 597–632 (1992).

Steel G. G., Hanes S. The technique of labelled mitoses analysis by automatic curve-fitting, *Cell Tissue Kinet.*, **4**, 93–105 (1971).

Stettler U. H., Weber H., Koller T., Weissmann C. Preparation and characterisation of form V DNA, the duplex DNA resulting from association of complementary circular single-stranded DNA, *J. Mol. Biol.*, **131**, 21–40 (1979).

Takahashi M. Theoretical basis for cell cycle analysis, *J. Theoret. Biol.*, **18**, 195–209 (1968).

Takahashi M. A fractal model of chromosomes and chromosomal DNA replication, *J. Theoret. Biol.*, **141**, 117–136 (1989).

Tarumi K., Mueller E. Wavelength selection mechanism in the Gierer–Meinhardt model, *Bull. Math. Biol.*, **51**, 207–216 (1989).

Tauro P., Halvorson H. O., Epstein R. L. Time of gene expression in relation to centromere distance during the cell cycle of *Saccharomyces cerevisae*, *Proc. Natl Acad. Sci. USA*, **59**, 277–284 (1968).

Thom R. *Stabilité structurelle et morphogenèse*, Ediscience, Paris (1972).

Thom R. *Modeles mathématique de la morphogenèse*, Collection 10–18, Union générale d'editions, Paris (1974).

Thoma J. U. Bond graphs for thermal energy transport and entropy flow, *J. Franklin Inst.*, **292**, 109–120 (1971).

Thomas D. Artificial enzyme membranes, transport, memory, and oscillatory phenomena, in *Analysis and Control of Immobilized Enzyme Systems*, Eds Thomas D., Kernevez J.-P., Springer-Verlag, Berlin (1975).

Thron C. D. The secant condition for instability in biochemical feedback control. I, The role of cooperativity and saturability, *Bull. Math. Biol.*, **53**, 383–401 (1991).

Topham C. M. Ill-conditioning associated with the 'end-point' method for the determination of kinetic parameters describing irreversible enzyme inactivation by an unstable inhibitor, *J. Theoret. Biol.* **135**, 169–173 (1988).

Treinin M., Feitelson D. G. Unequal cell division as a driving force during differentiation, *J. Theoret. Biol.*, **160**, 85–95 (1993).

Trucco E. Mathematical models for cellular systems: the Von Foerster equation, *Bull. Math. Biophys.*, **27**, 285–304 (1965).

Turing A. M. The chemical basis of morphogenesis, *Phi. Trans. Roy. Soc. (London)*, **B237**, 37–72 (1952).

Turner J. S. Finite fluctuations and multiple steady-states far from equilibrium, *Bull. Math. Biol.*, **36**, 205–213 (1974).

Tyrcha J. Asymptotic stability in a generalized probabilistic/deterministic model of the cell cycle, *J. Math. Biol.*, **26**, 465–475 (1988).

Tyson J. *The Belousov–Zhabotinskii Reaction, Lecture Notes in Biomathematics*, Vol. 10, Springer-Verlag, Berlin (1976).

Tyson J. J. (Book review): *Cell Cycle Control in Eukaryotes* by Beach D., Basilico C., Newport J., *Bull. Math. Biol.*, **51**, 649–651 (1989).

Tyson J. J., Chen C., Lederman M., Bates R. C. Analysis of the kinetic hairpin transfer model for parvoviral DNA replication, *J. Theoret. Biol.*, **144**, 155–169 (1990).

Valleron A.-J., Frindel E. Computer simulation of growing cell populations, *Cell Tissue Kinet.*, **1**, 69–79 (1973).

Valleron A.-J., MacDonald P. D. *Biomathematics and Cell Kinetics*, Elsevier North-Holland, Amsterdam (1978).

Varon R., Garcia Canovas F., Garcia Carmona F., Tudela J., Roman A., Vazquez A. Kinetics of a model for zymogen activation: the case of high activating enzyme concentrations, *J. Theoret. Biol.*, **132**, 51–59 (1988).

Vidybida A. K. Selectivity and sensitivity improvement in co-operative systens with a threshold in the presence of noise, *J. Theoret. Biol.*, **152**, 159–164 (1991).

Viniegra-Gonzalez G. Stability properties of metabolic pathways with feed-back interactions, in *Biological and Biochemical Oscillators*, Ed. Change B., Academic Press, New York (1973).

Vinograd J., Lebowitz J., Radloff R., Watson R., Laipis P., *Proc. Natl Acad. Sci. USA*, **53**, 1104 (1965).

Von Foerster H. Some remarks on changing populations, in *The Kinetics of Cellular Poliferation*, Ed. Stohlman F., Grüne and Stratton, New York (1959).

Von Foerster H. On self-organizing systems and their environments, in *Self-organizing Systems*, Eds Yovitz M. C., Cameron S., Pergamon Press, London (1960).

Von Foerster H. Computation in neural nets, *Currents in Modern Biology*, **1**, 47–93 (1967).

Waddington C. H. *Strategy of the Genes*, Allen and Unwin, London (1957).

Waddington C. H. *New Patterns in Genetics and Development*, Columbia University Press, New York (1962).

Waddington C. H. Concept and theories of growth, development, differentiation and morphogenesis, in *Towards a Theoretical Biology 3–Drafts*, Edinburgh University Press, Edinburgh, New York (1970).

Wake R. G., Baldwin R. L. Physical studies on the replication of DNA *in vitro*, *J. Mol. Biol.*, **5**, 201–216 (1962).

Walter C. F. The absolute stability of certain types of controlled biological systems, *J. Theoret. Biol.*, **23**, 39–52 (1969).

Walter C. F. Stability of controlled biological systems, *J. Theoret. Biol.*, **23**, 23–38 (1969).

Walter C. F. Kinetic and thermodynamic aspects of biological and biochemical control mechanisms, in *Biochemical Regulatory Mechanisms in Neurological Cells*, Eds Kun E., Grisxia S., John Wiley, New York (1970).

Walter C. F. Some dynamic properties of linear hyperbolic and sigmoidal multi-enzyme systems with feedback control, *J. Theoret. Biol.*, **44**, 219–240 (1974).

Walter, G. G. Stability and structure of compartmental models of ecosystems, *Math. Biosci.*, **51**, 1–10 (1980).

Walter, G. G. Some equivalent compartmental models, *Math. Biosci.*, **64**, 273–293 (1983b).

Wang J. C. DNA topoisomerases: enzymes that catalyse the concerted breaking and rejoining of DNA back bone bonds, in *Molecular Genetics*, Ed. Taylor J. M., Academic Press, New York (1979a).

Wang J. C. Helical repeat of DNA in solution, *Proc. Natl Acad. Sci. USA*, **76**, 200–203 (1979b).

Wang J. C. DNA topoisomerases, *Sci. Am.*, **247** (1) 94–109 (1982).

Wang Z.-X Two theoretical problems concerning the irreversible modification kinetics of enzyme activity, *J. Theoret. Biol.*, **142**, 551–563 (1990a).

Wang Z.-X. Some applications of statistical mechanics in enzymology. 1. Elementary principle, *J. Theoret. Biol.*, **143**, 445–453 (1990b).

Wang Z.-X. Some applications of statistical mechanics in enzymology 3. Effects of non-specific inhibitors and activators on enzyme reaction, *J. Theoret. Biol.*, **143**, 465–472 (1990c).

Wang Z.-X. Theoretical considerations of the Tsou plot, *J. Theoret. Biol.*, **150**, 437–450 (1991).

Wang Z.-X., Kihara H. Some applications of statistical mechanics in enzymology 2. Statistical mechanical explanation on allosteric enzyme models, *J. Theoret. Biol.*, **143**, 455–464 (1990).

Wang Z.-X., Tsou C. An alternative method for determining inhibition rate constants by following the substrate reaction, *J. Theoret. Biol.*, **142**, 531–549 (1990).

Watson J. *Biologie moléculaire du gène*, Interéditions, Paris (1978).

Watson J. D., Hopkins N. H., Roberts J. W., Steitz J. A., Weiner A. M. *Molecular Biology of the Gene*, Benjamin Cummings, Menlo Park, CA (1987).

Weiner N. *Cybernetics, or Control and Communication in the Animal and the Machine*, John Wiley, New York (1948).

Welch G. R. On the role of organized multienzyme systems in cellular metabolism: a general synthesis, *Prog. Biophys. Molec. Biol.*, **32**, 103–191 (1977).

Welch G. R., Keleti T., Vertessy B. The control of cell metabolism for homogeneous vs. heterogeneous enzyme systems, *J. Theoret. Biol.*, **130**, 407–422 (1988).

Westerhoff H. V., Welch G. R. Enzyme organization and the direction of metabolic flow: physico-chemical considerations, *Curr. Top. Cell. Regul.*, **33**, 361–390 (1992).

White J. H. Self-linking and the Gauss integral in higher dimensions, *Am. J. Math.*, **91**, 693–728 (1969).

White R. A. The use of population projection matrices in cell kinetics, *J. Theoret. Biol.*, **74**, 49–67 (1978).

White R. A. Computing multiple cell kinetic properties from a single time point, *J. Theoret. Biol.*, **141**, 429–446 (1989).

White R. A. (Book review): Cell kinetic modelling and the chemotherapy of cancer, *Bull. Math. Biol.*, **52**, 322 (1990).

White R. A. A theory for analysis of cell populations with non-cycling S phase cells, *J. Theoret. Biol.*, **150**, 201–214 (1991).

Wilbur W. J., Lipman D. J. Rapid similarity searches of nucleic acid and protein data banks, *Proc. Natl Acad. Sci. USA*, **80**, 726–730 (1983).

Wilson E. B., Decius J. C., Cross P. C. *Molecular Vibrations*, Dover Publications, New York (1980).

Wolpert L. The French Flag problem: a contribution to the discussion on pattern development and regulation, in *Towards a Theoretical Biology, 1. Prolegomena*, Ed. Waddington C. H., Edinburgh University Press, Edinburgh (1968).

Wolpert L. Positional information and the spatial pattern of cellular differentiation, *J. Theoret. Biol.*, **25**, 1–47 (1969).

Wolpert L. Positional information and pattern formation, *Phil. Trans. Roy. Soc. (Lond.)*, **B295**, 441–450 (1980).

Wong A. J. Development of a spin-glass model of prebiotic evolution: environmental effects on ensembles of genetic polymers, *J. Theoret. Biol.*, **146**, 523–543 (1990).

Wong A. K. C., Chan S. C., Chiu D. K. Y. A multiple sequence comparison method, *Bull. Math. Biol.*, **55**, 465–486 (1993).

Wyman J. On allosteric models, *Curr. Top. Cell Reg.*, **6**, 209–226 (1972).

Yan J. F., Yan A. K., Yan, B. C. Prime numbers and the amino acid code: Analogy in coding properties, *J. Theoret. Biol.*, **151**, 333–341 (1991).

Yockey H. P. Some introductory ideas concerning the application of information theory to biology, in *Symposium on Information Theory in Biology*, Eds Yockey H., Platzman R. L., Quastler H., Pergamon Press, New York (1958).

Yon J. *La conformation des protéines*, Hermann, Paris (1969).

Zeeman E. C. *Catastrophe Theory*, Addison-Wesley, New York (1977).

Zimmerman J. M., Simha R. The kinetics of cooperative unwinding and template replication of biological macromolecules, *J. Theoret. Biol.*, **13**, 106–130 (1966).

Zimmerman J. M., Simha R. The kinetics of multicenter macromolecule growth along a template, *J. Theoret. Biol.*, **91**, 156–185 (1965).

Zooneveld C., Kooijman S. A. L. M. Comparative kinetics of embryo development, *Bull. Math. Biol.*, **55**, 609–635 (1993).

Zuker M., Somorjai, R. L. The alignment of protein structures in three dimensions, *Bull. Math. Biol.*, **51**, 55–78 (1989).

Zuckerkandl E. Programs of gene action and progressive evolution, in *Molecular Anthropology*, Eds Goodman M., Tashian R. D., Plenum Press, New York (1976).

Zuckerkandl E. Multilocus enzymes, gene regulation, and genetic sufficiency, *J. Mol. Evol.*, **12**, 57–89 (1978).

Zuckerkandl E. Molecular evolution as a pathway to man, *Z. Morph. Anthrop.*, **69**, 117–142 (1978b).

Index

acrasine 318, 319, 320
activated complex 85, 99
activation 39, 49, 295, 340
activator 15, 291
activator–inhibitor model 340
active site 38, 52, 149, 153, 156, 221, 295
active transport 337, 341
adaptation 249, 250
adenine 6, 7, 8, 115, 431
adiabatic approximation 122, 123, 124
age 430
  classes 380
  -distribution 385
  of a cell 372
aggregations 320
alkaline phosphatase 51
allosteric control 162
allosteric effects 16, 20, 21, 25, 27, 35, 36,
  38, 52, 54, 115, 149, 162, 163, 215,
  222, 301
alphabet 131, 133, 194, 226
ambiguity 129, 132, 133, 134, 221
ambiguous coding 218, 219
amino acids 8, 9, 10, 11, 178, 185, 194,
  219, 220, 221
  essential 221
  pool 402
  sequence 133
anisotropic 157
anisotropic medium 91, 155, 157
anticodon 186
anti-cooperative 34
antigen–antibody interaction 53, 54, 55
aporepressor 304, 305
association 39, 167, 168, 169, 269, 270,
  280, 359, 360, 361
  processes 269
  –dissociation constant 23, 32
asymmetry 390
atlas 449
atoms 121
attractors 106, 355, 356, 414, 443
autocatalysis 65, 66, 86, 228, 229, 257, 345
automaton theory 136
autonomy 123
auto-organisation 116

bacteria 197, 303
bacteriophage 288
basin 105, 356
bifurcation 95, 172, 174, 262, 406
  set 107, 108, 109, 111
  point 128
binding matrix 24, 25, 26
  polynomial 25, 26, 30, 32, 34
binomial distribution 32, 386
biochemical kinetics 355
  oscillations 67, 369, 370, 429
  pathway 224
  systems theory 293
biofeedback 264
biological clock 339
biopolymerisation 198, 201
biosynthesis 177
  protein 130, 132, 133, 183, 187, 194,
    200, 217, 222, 240, 305, 309, 358
bit 127
blastomeres 323
blastocyst 323, 327
blastula 324
blood-clotting 43
Boltzmann constant 18, 92, 425
bond graphs 101, 103
Briggs–Haldane equations 41, 116
Britten and Davidson's model 311
Brownian motion 50
Brusselator 66, 68, 96, 116, 295, 350
butterfly catastrophe 109, 111

cancer 378
capacitance 98
  chemical 100
catalytic 9
  cycle 241, 242, 243
  network 240
  protein cycles 242
catastrophes 104, 108, 109, 110, 111, 113,
  116
  elementary 116, 356, 357
  theory 71, 135, 354
catastrophic set 104, 108, 109, 116, 356
Ca^{++}-transport 153
causal loop model 368, 429

cell 5
  clock 368, 369, 370
  cycle 364, 368, 371
  death 378
  differentiation 251, 334, 336
  division 363, 366, 368, 386, 392, 397
  –ECM interactions 404
  embryonic 402
  growth 297, 390, 394, 397, 429
  kinetics 350
  membranes 155
  mesenchymal 401
  organisation 287
  population 372, 376
  density 403
  proliferation 372
  stem 321
  theory 283
  -to-cell contacts 331
cellular sociology 396
centre 446, 447
chance 248, 250
channel 129
  transmission 130, 135
  gating 36
channelling 155, 156, 157, 161, 165, 166, 176
chemical kinetics 51
  network 98
  potential 100
  reactions 20, 99, 101
  coupled 102
  reactor 66
  resistances 98
  species 238, 239
chemotaxis 318, 319, 320, 429
chreod 355, 356
chromatids 368
chromatin 179, 367
chromosome 179, 180, 214, 367, 368
chronobiology 222
chymotrypsin 21
circular chain 210
cistrons 302, 432
clone 398
closed system 88
closed trajectory index 445
coat patterns 351
co-dimension 109, 110, 112
codons 178, 186, 191, 260
co-factors 52
coiling 211, 367
collective behaviour 143, 146, 222, 241
communication 126
  channels 195
compartment 103, 135, 153, 156, 157, 165, 176, 373, 433
competition 240, 356
complementary instruction 236, 238
complex system 101
complexation 39

complexions 33
complexity 71, 72, 73, 134, 197, 198, 254, 257, 262, 361, 408
condensation 406
  cell 406
configuration integrals 18, 115
conformation 9, 16
  stable 17
  change 19, 21, 53, 115, 295
  classes 17
  states 13, 20, 25, 26, 34, 52
conservation of energy 89
  matter 40
  mass 59
conservative 441
constitutive 98, 302, 303
contractibility 218
control 72, 120, 263
  space 108
convection 403
cooperative effects 21, 27, 28, 34, 257
cooperativity 16, 30, 33, 115
  coefficient of 33
copying error 131, 236, 261
co-rank 110, 112
crease catastrophe 109
critical point 442
cross-regulation 308
Curie's principle 155
critical points 106, 108, 116, 444
  degenerate 108
culture medium 396
current 96, 97
cusp catastrophe 107
cycle 242, 243, 365
cyclic AMP 153, 320, 332
cytochrome-c 51
cytological age 376
cytosine 6, 7, 8, 115, 431
cytosol 157
cytotoxic drug 376

Darwinism 249, 250, 251, 255, 259, 412
death 380
degenerate code 195
deletions 189, 432
deleton 191
denaturation 198
deoxyribonucleic acid (DNA) 6, 7, 8, 11, 12, 17, 115
determinism 254
developmental biology 251, 283, 297, 298, 407, 429
Dictyostelium 317, 331, 334, 358
diffeomorphisms 448, 449
differential form 435
differentiation 283, 297, 298, 315, 316, 320, 331, 333, 334, 350, 363, 364, 365, 386, 388, 429
  asymmetrical 388

diffusion 60, 91, 102, 104, 122, 154, 156, 157, 160, 161, 168, 169, 174, 203, 207, 272, 318, 320, 330, 403, 429
  barrier 160
  coefficients 338
  constant 62
  equation 61, 447
diffusive coupling 372
digestion 119
dilution factor 273
directed assembly 288
dispersion relations 404
dissipation function 98
dissipative instabilities 254
  structures 88, 95, 155, 253, 348, 350
divergence 436, 438
diversity 250
DNA 6, 7, 8, 11, 12, 17, 115, 130, 178, 179, 183, 187, 191, 196, 197, 199, 202, 213, 216, 236, 279, 313, 363, 345, 392, 397, 431
  circular 209, 212, 213, 216
  condensation 180
  conformational 313
  replication 177, 182, 198, 202, 205, 206, 207, 209, 298, 309, 392, 395, 396
  sequencing properties of 189
  supercoiled 213
  synthesis 366, 367, 369, 391
  structure 211
  uncoiling 200
double helix 6, 115, 188, 199, 201, 204, 210, 213, 214, 215
dual representations 147, 148
dynamic instability 368
dynamic system 104, 121, 123, 125, 168, 174, 291, 356, 400, 440, 442, 449, 450
dynamics 121, 164

ecological systems 263
ectoderm 324
Eigen 83, 87, 227, 241, 245, 249, 253, 254, 263, 266, 273, 277
Eigen–Goodwin system 147
eigenvalues 125, 233
electric field 162
electrostatic attraction 13
elliptic umbilic catastrophe 109, 112
elongation 188
  factor 430
embryoblast 323
embryogenesis 315, 322, 355, 363, 401
embryology 283
endocrine system 297
endoderm 324
endotropic graph 81
energy 88, 93, 441
  equipartition 426
  pathway 101
enlacements 210, 211, 214, 215
enthalpies 178

entropy 88, 89, 95, 178, 252, 253, 425, 426, 427
  production 89
  production of 94
environmental constraint 231
enzymes 37, 42, 44, 52, 56, 149, 220, 221, 225, 240, 249
  activating 219
  activity 367
  catalysis 28, 38, 149, 165, 174
  chain of 292
  essential 219
  glycolytic 153, 154
  kinetics 50, 153, 295
  organisation 152, 157, 158, 173
  quenching 153
  regulated 16, 222
  sites 295
  synthesis 301
  –enzyme interactions 158
  –substrate complex 38
  –substrate system 37, 42
  –substrate association 53
epigenetic system 222, 223, 267, 326, 355
epimorphosis 329, 330, 337
epitopes 54
equilibrium 94, 106, 108, 111
  distribution 33
  point 442
  state 40, 105, 301
  surface 107, 109, 111
  thermodynamics 88
equilibrium constant 17, 19, 22, 99
  effective 29
  macroscopic 23
  microscopic 31
equipotence principle 407, 426
erythrocytes 320
erythropoietic system 321
eucaryotes 285
Euclidean dimension 50
evolution 72, 73, 78, 120, 121, 122, 137, 139, 189, 231, 236, 242, 245, 247, 250, 251, 253, 265, 358
  of populations 273, 276
evolutionary potential 258
exchange constants 162
exons 177, 183
expectation 126, 452
extinction 249
extracellular matrix (ECM) 403, 404
extremum hypothesis 413, 414, 422, 426, 427

facilitated transport 336
feed-back 170
fertilisation 322, 324
Fick's law 60, 61, 318
field theory 73, 148, 277
Fischer's hypothesis 38
flagella, 288, 289

fluctuation 253, 257, 262, 433
focus 447
fold catastrophe 109, 111
fractal 50, 68, 182
    dimension 50
fraction labelled mitosis (FLM) 378, 379,
    383, 385, 386
free energy variation 19, 20, 115
French flag problem 329, 330
fuga 312, 314, 322
function 137, 139, 141
    elementary 139
function of state 28
functional association 141, 146, 168, 176,
    272
    biology 262, 265
    complexity 73, 127, 198, 259, 261, 359
    interaction 69, 122, 136, 137, 140, 141,
        143, 146, 147, 149, 150, 152, 165,
        167, 279, 357, 358, 417
    order 421, 422, 426, 433
    organisation 136, 143, 144, 257, 357, 409
    self-association hypothesis 360
    units 179, 431
    units of gene action (fuga) 312, 314

gastrulation 324
Gatlin 197
gene 178, 302, 303, 397
    activation 388
    activator 346
    code 244
    control 313
    expression 434
    inductor 302
    operator 303
    productor 310
    receptor 310, 311, 313
    regulation 197, 304, 306, 311
    regulator 304, 313, 358
    regulatory 183, 313
    sequences 181, 269
    sensor 310, 311
    splicing 183
    split 181
    structural 183, 222, 302, 305, 309, 311,
        313
    structure 221, 358
genealogical tree 189
generation time 375, 390, 391, 392, 393, 395
genetic code 186, 194, 221, 245, 247, 253,
    254, 262
genetic recombination 303
genetic transcription 177
genome 182, 261
Glansdorff–Prigogine functional 93
global structure 354
glucose-6-phosphate pathway 289
gluconeogenesis 290
glycine 10

glycolysis 68, 291, 154, 293
glycolytic oscillations 370
Golgi apparatus 287
Goodwin model 222, 266, 271, 277
gradient 435
graph theory 136
Green's theorem 59, 90, 438
guanine 6, 7, 8, 115, 431
Guchenkeimer's theorem 356
gyrase 216

haemoglobin 5, 15, 16, 34, 35, 189, 322
Hamiltonian 441, 443
haptotaxis 403
harmonic balancing technique 164
    oscillator 443
Hebbian learning algorithm 254
helicase 214
helicoidal chain 210
helix 14, 17, 179, 182, 184,
    double 6, 115, 188, 201, 204, 213
Helmholtz free energy 19
    function 89
Henri–Michaelis–Menten kinetics 30, 31,
    39, 42, 43, 45, 56, 62, 116, 124, 158,
    159
heredity 6, 7, 431
Hermitian matrix 451
Hessian 30, 33, 444, 445
heterogeneity 341, 373
heterotropic effects 21
hierarchical 145, 165
    organisation 221, 358
    set 267
    system 72, 136, 141, 144, 146, 150, 222,
        224, 269, 277, 359, 361, 408, 409,
        421
hierarchisation 419
    operator 419, 420
hierarchy 123, 124, 126, 140, 142, 148,
    149, 341, 357, 428, 433
    functional 142, 223
    structural 142
Hill graph 30
    coefficient 163
histogenesis 298, 315
histone 179, 180, 368
homotropic effects 21, 22, 30
hormones 295, 363
hyaloplasm 287
hydrogen bonds 8, 12, 181
hyperbolic umbilic catastrophe 109, 112,
    285
hypercycles 240, 241, 243, 244, 245, 253,
    254, 256, 257, 260, 261, 262, 266,
    280, 292, 295

idempotence 366
immune system 321
immunology 53

implantation 323
induction 328
inductors 363
information 127, 128, 129, 130, 132, 134,
    196, 197, 255, 258, 265, 316, 425,
    429, 433
  carriers 226, 240
  theory 122, 126, 131, 177, 191, 194, 253,
    257, 259, 452, 453
  transfer 130, 225
inhibition 44, 49, 295, 340
  allosteric feedback 223
  competitive 45, 46
  contact 372
  lateral 357
  non-competitive 47, 48
  uncompetitive 48
inhibitor 15, 20, 45, 291
  –activator system 348
insertions 189, 432
instability 348, 388, 389
instruction loop 310
insulin 43, 290
intercellular contacts 332
interphase 367
introns 177, 183
inverse regulation 296, 297
ions 5
  heavy metal 5
irreversible processes 71, 88, 95, 103, 116,
    135, 309
islets of Langerhans 290
isotropic medium 155

Jacob and Monod 304
  model 305, 306, 308, 309, 336, 337, 341
Jacobian 444, 445, 446
  matrix 125, 297

kinetic theory of fluids 156, 157
Kirchhoff's laws 100, 101
KNF model 27, 28, 33
knot 446, 447
Koshland–Nemethy–Filmer models 116
Krebs cycle 154, 290
Kronecker delta 231

Lamarckism 249, 250
Laplace function 438, 439
law of mass action 16, 39, 80
Leslie matrices 84, 379, 382, 384, 430
level of organisation 143, 266, 420, 421
ligand 15
ligases 182
limit cycle 66, 68, 87, 369, 371, 429, 444,
    445
local alinments 189
Lyapunov function 92, 93, 94, 95, 413,
    414, 426, 427, 442, 443, 444
lysosomes 286

macromolecules 120
macromutation 251
malonic acid 45
Markov chains 196
mass factor 394, 395, 430
  transport 58
master equations 124
matrix algebra 451
maturity 373, 375, 430
Maxwell's convention 107, 108
mechano-chemical approach 400
membrane 60, 102, 103, 286
Mendelian population 248
message 177, 225, 255, 453
  length encoding 190
metabolic chain 152, 160
  channelling 154
  control 35, 293, 294
  fields 356
  flux 160
  networks 163, 174, 267, 271, 293
  organisation 154
  pathway 158, 160, 162, 164, 165, 166,
    168, 173, 174, 176, 289, 304, 358, 397
  system 75, 76, 78, 169, 223
  units 167, 222
metabolism 78, 248, 252, 255
metaphase 366
Michaelis constants 29, 41, 42, 45, 116, 160
  curve 32
  equation 34
Michaelis–Menten (see Henri–Michaelis–
    Menten) 50
microcompartmentation 152
microenvironment 149, 152, 156, 165
microfilaments 286, 287
micromutagenesis 277
micromutations 249, 263, 270, 280
microtubules 286
mitochondria 154, 286
mitogen 369
mitosis 180, 363, 364, 369, 370, 371, 372,
    374, 378, 385, 387, 388, 398
  cycle 364
mitotic division 377
  oscillator 369, 371
mobility 62
Möbius strip 210
molecular biology 263
Monod 304
Monod–Wyman–Changeux 116
  model 295
monomers 15, 21, 32
Monte Carlo method 378
morphallaxis 329, 330
morphogen 329, 336, 337, 341, 342, 343,
    344, 348, 351, 353, 357, 405, 432
morphogenesis 113, 135, 251, 297, 299,
    338, 340, 348, 350, 357, 429
  mesenchymal 400

morphogenetic fields 327, 328, 334, 336, 339, 355, 405, 429
  movement 331
  rules 405
morphology 108, 339, 356
morula 324
M-R system 76, 78, 217, 219, 279
Murray's model 400
mutagenesis 248, 265
mutations 12, 251, 252, 255, 262, 407
muton 432
MWC model 27, 28, 33, 34
myoglobin 13, 15, 35

natural selection 247, 250
negentropy 426
Neo-Darwinism 251, 268
nervous system 148
network 98, 149, 206, 207, 372
  theory 71, 240
  thermodynamics 96, 98, 103, 116, 187, 291
neural network 136, 254
neural plate 325
neurulation 325
noise 128, 129, 130, 132, 134, 195, 258, 279
non-equilibrium 66
  thermodynamics 88
non-instantaneity 136, 137
non-linear dynamics 136
non-linear system 68, 440
non-linear thermodynamics 92, 103
non-locality 136, 137, 149, 279
non-symmetry 136, 137, 409, 417, 425
normal modes 446
novagenesis 264, 265, 277, 280
nuclear envelope 287
nucleic acids 6, 7, 11, 115, 225, 242, 249
nucleic acid sequences 192
nucleolus 286
nucleosomes 177, 179, 367
nucleotides 6, 198, 431
  bases 194
nucleus 286, 397
NWS algorithm 189, 190

observed functional biological system (O-FBS) 136, 407, 408, 409, 417, 422, 424, 426
oligomer 15
oncogenic proteins 191
Onsager coefficients 91
open system 308, 309
operator 309
operon 309, 316, 397, 432
optimality 295
optimisation 231, 234, 236, 239, 242, 245, 247, 426, 427, 433
organisation 122, 127, 226, 411

  degree of 410
  state of 409
organising centre 356
organising principle 119
organogenesis 325, 326, 356, 359
organs 148
orgatropy 414, 416, 425, 426
orthogenesis 251
oscillations 84, 85, 86, 87, 96, 191, 233, 239, 275, 295, 372, 445
Oster–Murray mechano-chemical 338, 357
  theory 404

parabolic umbilic catastrophe 109, 112
parallel computer 147
parameters 104, 105, 440
  of action 124
  of order 124, 125, 126
partial derivatives 435, 447
parvoviruses 182
pattern formation 135, 405
Paynter graph 187
Pearson's law 386
peptide 43
  bond 11, 12
periodic solutions 172, 173
phages 197
phase space 123, 440, 448
3′-5′ phosphodiester bonds 6
phosphofructokinase 158
physiological control 148
  function 9, 11, 12, 141, 143, 144, 145, 165, 174, 280
  time 373
placenta 323
plasmids 214, 309
pleat catastrophe 111
polarity 329, 337
polaron 432
polymer 6, 32
polymerisation 6, 32, 80, 201, 202, 203, 207
polynucleotide 6, 7
polypeptide 12, 15, 185
polysomes 221
population 401
  biology 379
  development 374
  genetics 248
  models 427
  theory 378, 383
porphyrins 5
positional information 328, 329, 332, 337, 339, 343, 348, 429, 432
post-transcriptional control 335, 370
post-translational control 335
potential 411, 436, 442
  energy 18, 51
potential of organisation 73, 359, 410, 411, 412, 414, 421, 422, 423, 427
prebiotic 248, 254, 256, 259

preformation theory 326
primary structure 12, 115
primitive streak 325, 326
procaryotes 285
productivity 235, 236
proliferative cell cycle 365
prophase 366
protein 6, 7, 9, 11, 115, 133, 178, 179
  cycles 240
  –ligand interactions 21, 115
  –ligand associations 22
  synthesis 206, 253, 398
  –nucleic acid interaction 254
  –protein networks 155
proteolytic enzymes 42
protomer 15, 16
pseudo-steady state hypothesis 43, 44
purine 6, 115
pyrimidine 6, 115

quantum dynamics 72
quasi-species 234, 236, 249, 259, 260, 277
quaterary structure 15, 52, 115

Raman spectroscopy 19
rate constant 148
rate of reaction 44, 46
reaction–diffusion 389
  coupling 155, 157
  equation 54, 58, 60, 350, 405
  mechanism 353
  processes 186
  system 62, 67, 104, 351, 357
reciprocal copying 236
recognition 54, 55, 156
recombinations 270, 271
recon 432
recursion 219
red blood cells 320
redundancy 130, 134, 195, 196, 258, 261,
  279, 310, 312
reference class 138, 139
regeneration 315
regionalisation 299
regressive evolution 251
regulation 37, 72, 115, 120, 123, 166, 291,
  301, 302, 305, 308, 310, 327, 429
regulator 309, 312
regulatory gene systems 266
relational structure 139, 142
relational theory 71, 116, 217
relaxation process 242
reliability 134, 135
replicases 182
replications 181, 199, 200, 216, 237, 270,
  431, 432
  function 220
replication–translation 177
  apparatus 221, 224, 256, 271
  mechanism 255

repressor 304, 307, 336, 346
reproduction 379, 381, 384
resistance 98
respiration 101, 119
respiratory function 148, 150
restorability 219, 220
ribonucleic acids (RNA) 6, 7
ribosome 8, 185, 186, 207, 286, 287, 288
Riemann–Hugoniot surface 108
RNA 6, 7, 115, 178, 182, 183, 194, 197,
  236, 256, 313, 334
  messanger (mRNA) 8, 130, 133, 181,
    183, 186, 223, 305
  replication 236
  ribisomal (rRNA) 8, 183, 184
  synthesis 198, 370
  transfer (tRNA) 8, 183, 185, 186, 206,
    218, 220, 255
  viral 7

saddle 105, 446, 447
saturation 41
  curve 29, 35
  function 28
β-sheet 13
Schrödinger's equation 18
secondary structure 12, 13, 115
segmentation 323
segregation 158
selection 226, 228, 234, 236, 248, 249
  natural 247, 250
selective advantage 225
selective value 230, 231
self-assembly 287
self-association 167, 171, 173, 268, 358,
  361, 362, 407, 418
self-association hypothesis 165, 166
self-inductance 98
  chemical 100
self-instruction 242
self-organisation 67, 72, 119, 121, 122,
  123, 128, 129, 135, 225, 245, 247,
  248, 249, 252, 254, 256, 257, 258,
  259, 262, 264, 265, 274, 275, 422
  associative 140
self-replicative units 147, 243
self-reproduction 78, 240, 248, 252, 255
S-graph 86
shock wave 355
sinks 410, 411, 426
singular perturbation 43, 124
singular point 91, 106, 108, 442
singular solution 44
singularity 106, 107, 297, 354, 356, 357,
  446, 450
sink 73, 106, 145
slave variables 126
slime moulds 317
Smoluchwski constant 157
source 73, 106, 145, 410, 411, 426, 455

spatial compartmentation 176, 260
spatial organisation 103, 149, 353
spatiotemporal structure 254, 320, 355, 403
specialisation 417, 418, 419, 420, 422
speciation 251
specificity 5
  enzymic 9
  functional 6
spermatozoon 283
SPICE 103
spin-glass 254
S-systems 68
stable 171
stability 62, 68, 92, 125, 143, 158, 161, 163, 166, 221, 275, 291, 337, 361, 362, 364, 407, 408, 442
  asymptotically 161
  domain of, 164, 171, 172, 174
  Lyapunov 63
  marginal 95
  orbital 63, 64
  structural 64, 108, 309, 354, 355, 356
  topological 143
stationary state 40, 67, 84, 88, 92, 93, 99, 104, 105, 123, 125, 158, 161, 164, 230, 233, 239, 252, 307, 349, 350, 351, 372, 442
  multiple stable 390
statistical mechanics 16, 136, 427
  of gas kinetics 52
stem cells 321
stoichiometry 18, 101, 102
structural association 141, 146
  complexity 127
  discontinuity 136, 149, 152
  discontinuities 137, 154, 166
  disorder 257
  equivalence 148
  hierarchy 142
  levels 146
  organisation 128
  transformations 407
  unit 73, 140, 143, 158, 165, 170, 174, 267, 431
structure 137, 138, 140
β-structure 14
substrate 37, 52
substrate–enzyme complex 44
succinic acid 45
supercoiling 180, 209, 211, 212, 214, 216
superhelix 180
swallowtail catastrophe 109, 111
symbols 126, 131, 133, 196, 225, 226, 236
symmetry 155
  destruction 389
synaptic efficacy 148
synchronous population 386
synergetics 96, 122, 248

Taylor's approximation 91
telophase 366
template 200, 206
termination 187, 188
tertiary structure 13, 15, 115
theorem of connectivity 294
theorem of summation 294
thermodynamic branch 87, 95, 104, 319
  energy function 19
  theory 87
Thom's theory (see catastrophe theory)
thymine 6, 7, 8, 115, 431
thyroxine 137
Tikhonov's theorem 147
tissues 148
  differentiation 270
  specialisation 359, 360
topoisomerases 216, 279
topological dimension 50
  reality 202
  structure 215
topology 136, 209, 217, 354, 408
torsion 211, 212
tortility 211, 214
total structure 139
trajectories 442
transcription 131, 132, 181, 182, 183, 198, 216, 279, 312
transcriptional control 334
transformable elements 138
transformation systems 71, 78, 79, 81, 82, 84, 87, 96, 116, 135, 147, 148, 226, 227, 228, 229, 296, 378, 380, 381, 383
transitional state theory 39
translation 131, 132, 177, 181, 184, 185, 186, 243, 279
transmission channel 130
transpiration 119
tree-graph 85
true graphs 116
Turing's theory 338, 348, 349, 350

uncoiling 200, 203, 210
unitary matrix 451
uracil 7, 115
Ussing's equation 160

Van der Waals' forces 13, 18
variability 266
variation 249, 431
variational principles 407
vector 435, 436, 437
  analysis 435
  field 435
  gradient 436
ventilatory function 150
vertebrates 196
vertex 74
virus 119, 196, 287
viscosity 13, 154

vital coherence 166, 167, 262, 263, 264, 267, 268, 270, 271, 272, 274, 277, 280, 358, 407
vital field 354
Volterra's dynamic system 292
volume 430

wild type 236
writhing number 211, 212, 214, 215

X-ray diffraction 6, 8, 53

Yates–Pardee metabolic pathway 158, 162, 163
Yockey–Atlan theory 129

zebra's stripes 352
zymogens 43